T0027115

# Gray's Anatomy

With original illustrations
by Henry Carter

Introduction by George Davidson

**About George Davidson**
George Davidson studied languages and linguistics at the universities of Glasgow, Edinburgh and Strasbourg, and is a graduate of Edinburgh University. A former senior editor with Chambers Harrap, he is now a freelance compiler and editor of dictionaries and other reference books. He is an elder of the Church of Scotland, and lives in Edinburgh.

This edition published in 2023 by Arcturus Publishing Limited
26/27 Bickels Yard, 151–153 Bermondsey Street,
London SE1 3HA



AD007177UK

Printed in the US

## Introduction

The book that has come to be known simply as *Gray's Anatomy* is now over 150 years old. *Henry Gray's Anatomy Descriptive and Surgical* was first published in the United Kingdom in 1858, and in the United States the following year. An immediate success in both countries, and quickly becoming a medical bestseller (indeed, it is surely the medical bestseller of all time), it has never been out of print in the past century and a half. There is no other textbook that has been so widely used by medical students and medical professionals, and, frequently revised and updated over the years, it is now in its 40th edition.

Although other books on anatomy existed at the time of its publication and many others have been written since then, no other book has ever approached *Gray's Anatomy* in terms of either popularity or authority. It became, and remains, a classic, the standard work on the subject, the 'Bible' of anatomy. Moreover, its fame has spread well beyond the medical profession, and *Gray's Anatomy* is a book found on the bookshelves of many people whose only connection with the world of medicine is that of patient or client but who find in its clear text and detailed illustrations much that helps them understand the structure and functioning of their own bodies. Artists, too, find the illustrations of bones and muscles of immense help when drawing, moulding or sculpting the human body.

Since other anatomy textbooks were available in the mid-19th century, why did Henry Gray feel the need to produce yet another one? At the time, most of the anatomical textbooks available to students and doctors were small pocket-sized books, with small and cramped illustrations. Gray felt that a more detailed anatomy book, with a larger page-size allowing larger and clearer illustrations, would better meet the needs of the anatomy students of his day. Moreover, the purpose of the book was, according to Gray, not only to provide an accurate description of the anatomy of the human body, but to relate this information to actual surgical practice, to show how the knowledge of anatomy acquired from the book should be applied. With the clarity and detail of its text, its superb illustrations, and its emphasis on the surgical aspects of anatomy, the book simply outshone all its competitors. It has been suggested that the skilful way in which text and illustrations are interwoven could hardly have been improved on, and, more than that, that none of the subsequent editions have ever surpassed the beauty, simplicity and lucidity of the original work. It was a remarkable achievement by two remarkable young men, Henry Gray and Henry Vandyke Carter.

Who were these two young men, Gray and Carter? About Henry Gray's early life, little is known. He was born in London, probably in 1827. His father, Thomas, was a King's Messenger in the service of George IV and then William IV. In 1845, at the age of 18, Gray became a medical student at St George's Hospital in London. A dedicated student, painstaking and methodical in his work, he showed a particular interest in anatomy, which he learned not just from textbooks but from dissections of his own.

Gray soon produced some significant research papers in his chosen field. In 1848 he became a Member of the Royal College of Surgeons (he was later to become a Fellow), and in 1849 he was awarded the College's triennial prize for an essay on 'The Origin, Connexions and Distribution of nerves to the human eye and its appendages, illustrated by comparative dissections of the eye in other vertebrate animals'. In 1850, he presented a paper to the Royal Society on the development of the optic and auditory nerves, for which he was awarded a Fellowship of the Society in 1852, a particular distinction for someone who was only 25 and still at the start of his career. And in 1853, he won the Astley Cooper Prize for 'A Dissertation on the Structure and Use of the Spleen', based on his own medical and anatomical observations; this was followed in 1854 by the publication of his book on the same subject.

In 1850 Gray was appointed house surgeon at St George's Hospital, and was subsequently demonstrator, and then lecturer, in anatomy at its medical school. He was still in his twenties when he began work on his *Anatomy*, and only just into his thirties when it was published. Tragically, while looking after a nephew who was suffering from smallpox, Gray himself contracted a severe form of the disease, and died in 1861 at the age of 34. Fortunately he had by that time prepared a revised, corrected and enlarged 2nd edition of the book, which was published in 1860.

Although the book has long been known as *Gray's Anatomy*, it was very much a work of collaboration between Gray and his talented illustrator Henry Vandyke Carter. While the conception for the book may have been Gray's and the text Gray's, there is no doubt that the book owed much of its success to the high quality of Carter's illustrations. Carter produced all the drawings from which the engravings were made, and in many cases these drawings were made on the basis of dissections made by Gray and Carter themselves, although some were taken or adapted (with any necessary corrections) from other anatomical works of the day, such as *Quain's Anatomical Plates*.

In asking his colleague Carter to illustrate his book, Gray made a happy choice. Henry Vandyke Carter was not only an excellent anatomist but also a gifted artist. His father, Henry Barlow Carter, was a well-known artist (one may assume that the younger Henry was not given the name Vandyke by mere chance!), and his younger brother Joseph was also an artist. However, although Vandyke Carter, the older son, had also shown himself to be a talented painter, it was to medicine that he turned as a career. Born in 1831 and educated at Hull Grammar School, he went to London in 1847 to study medicine at St George's Hospital. In 1852 he became a Member of the Royal College of Surgeons and a Licentiate of the Society of Apothecaries, in 1853 he obtained a studentship in human and comparative anatomy at the Royal College of Surgeons, and in 1856 he became a Doctor of Medicine of the University of London. At the time of his collaboration with Gray, he was Demonstrator of Anatomy at St George's Hospital.

While Gray died tragically young, Carter went on to have a distinguished medical career – but not in England. In 1858, just before the publication of the book to which he had made such a major contribution, Carter joined the Indian Medical Service and sailed for Bombay (Mumbai), where he became Professor of Anatomy and Physiology (and subsequently Dean) at the Grant Medical College. He made significant contributions in the field of tropical medicine, especially with regard to leprosy and relapsing fever, and was instrumental in the establishment of leper colonies. On his retirement in 1888, he returned to England. His contributions to medicine were rewarded by appointments as Honorary Deputy Surgeon-General in the Indian Medical Service and as Honorary Surgeon to Queen Victoria. He died in Scarborough in 1897. (It is perhaps also worth mentioning that while Gray received a royalty of £150 for every 1000 copies of the *Anatomy* sold, Carter was paid only a one-off fee of £150 for the 363 illustrations he produced. He used the money to buy a microscope.)

The text of *Gray's Anatomy* that is published here is that of the 2nd edition, revised, corrected and enlarged by Gray before his untimely death and published in 1860. It contains the first extensive sections on Osteology, The Articulations, The Muscles and Fasciæ, The Arteries, The Veins and The Lymphatics.

*Gray's Anatomy* is undoubtedly the most famous medical textbook ever published. While one might speculate about what more Henry Gray might have achieved had he not died so young, it is quite possible that even had he lived for another 50 years, he would never have made any further achievement comparable to his *Anatomy Descriptive and Surgical.* We can only be thankful that he had the vision and talent to produce such a great work, that he had the benefit of the gifted artist/doctor H. Vandyke Carter as his co-worker, and that he lived long enough to see his great work into print. There is no doubt that for the 21st century as much as for the 19th and 20th centuries, the anatomy textbook is still *Gray's Anatomy*.

George Davidson

## Osteology.



### THE SPINE.



### THE SKULL.



### THE THORAX.



### THE UPPER EXTREMITY.



### THE LOWER EXTREMITY.



## The Articulations.



### ARTICULATIONS OF THE TRUNK.

ARTICULATIONS OF THE UPPER EXTREMITY.



ARTICULATIONS OF THE LOWER EXTREMITY.



**Muscles and Fasciæ.**



***Muscles and Fasciæ of the Head and Face.***



EPICRANIAL REGION.



AURICULAR REGION.



PALPEBRAL REGION.



ORBITAL REGION.



NASAL REGION.



SUPERIOR MAXILLARY REGION.



INFERIOR MAXILLARY REGION.



INTERMAXILLARY REGION.



TEMPORO-MAXILLARY REGION.



PTERYGO-MAXILLARY REGION.



***Muscles and Fasciæ of the Neck.***



SUPERFICIAL REGION.



INFRA-HYOID REGION.



SUPRA-HYOID REGION.



LINGUAL REGION.



PHARYNGEAL REGION.



PALATAL REGION.



VERTEBRAL REGION (ANTERIOR).



VERTEBRAL REGION (LATERAL).



***Muscles and Fasciæ of the Trunk.***



MUSCLES OF THE BACK.



FIRST LAYER.

## SECOND LAYER.



## THIRD LAYER.



## FOURTH LAYER.



## FIFTH LAYER.



## MUSCLES OF THE ABDOMEN.



***Muscles and Fasciæ of the Thorax.***



## DIAPHRAGMATIC REGION.



***Muscles and Fasciæ of the Upper Extremity.***



## ANTERIOR THORACIC REGION.



## LATERAL THORACIC REGION.



## ACROMIAL REGION.



## ANTERIOR SCAPULAR REGION.



## POSTERIOR SCAPULAR REGION.



## ANTERIOR HUMERAL REGION.



## POSTERIOR HUMERAL REGION.



## MUSCLES OF FORE-ARM.



## ANTERIOR BRACHIAL REGION, SUPERFICIAL LAYER.



## ANTERIOR BRACHIAL REGION, DEEP LAYER.



## RADIAL REGION.



## POSTERIOR BRACHIAL REGION, SUPERFICIAL LAYER.



## POSTERIOR BRACHIAL REGION, DEEP LAYER.



## MUSCLES AND FASCIÆ OF THE HAND.



## MUSCLES OF THE HAND.



## SURGICAL ANATOMY OF THE MUSCLES OF THE UPPER EXTREMITY.

## The Veins.

GENERAL ANATOMY.

# ANATOMY, DESCRIPTIVE AND SURGICAL.

## Osteology.

In the construction of the human body, it would appear essential, in the first place, to provide some dense and solid texture capable of forming a framework for the support and attachment of the softer parts of the frame, and of forming cavities for the protection of the more important vital organs; and such a structure we find provided in the various bones, which form what is called the Skeleton (σκέλλω, *to dry up*).

*Structure and Physical Properties of Bone.* Bone is one of the hardest structures of the animal body; it possesses also a certain degree of toughness and elasticity. Its colour, in a fresh state, is of a pinkish white externally, and deep red within. On examining a section of any bone, it is seen to be composed of two kinds of tissue, one of which is dense and *compact* in texture, like ivory; the other consisting of slender fibres and lamellæ, which join, to form a reticular structure; this, from its resemblance to lattice work, is called *cancellated.* The compact tissue is always placed on the exterior of a bone; the cancellous tissue is always internal. The relative quantity of these two kinds of tissue varies in different bones, and in different parts of the same bone, as strength or lightness is requisite. Close examination of the compact tissue shows it to be extremely porous, so that the difference in structure between it and the cancellous tissue depends merely upon the different amount of solid matter, and the size and number of the spaces in each; in the compact tissue the cavities being small, and the solid matter between them abundant; whilst in the cancellous tissue the spaces are large, and the solid matter diminished in quantity.

*Chemical Analysis.* Bone consists of an *organic*, or animal, and an *inorganic*, or earthy material, intimately combined together; the animal matter giving to bone its elasticity and toughness, the earthy part its hardness and solidity. The animal constituent may be separated from the earthy, by steeping bone in a dilute solution of nitric or muriatic acid: by this process, the earthy constituents are gradually dissolved out, leaving a tough semi-transparent substance, which retains, in every respect, the original form of the bone. This is often called cartilage, but differs from it in being softer, more flexible, and, when boiled under a high pressure, it is almost entirely resolved into gelatine. Cartilage does, however, form the animal basis of bone in certain parts of the skeleton. Thus, according to Tomes and De Morgan, it occurs in the petrous part of the temporal bone; and, according to Dr. Sharpey, on the articular ends of adult bones, lying underneath the natural cartilage of the joint. The earthy constituent may be obtained by subjecting a bone to strong heat in an open fire with free access of air. By these means, the animal matter is entirely consumed, the earthy part remaining as a white brittle substance still preserving the original shape of the bone. Both constituents present the singular property of remaining unaltered in chemical composition after a lapse of centuries.

The organic constituent of bone, forms about *one-third*, or 33.3 per cent.; the inorganic matter, *two-thirds*, or 66.7 per cent. as is seen in the subjoined analysis by Berzelius:—

| | | |
|---|---|---|
| *Organic Matter,* | Gelatine and Blood-vessels | 33.30 |
| *Inorganic,* or *Earthy Matter,* | Phosphate of Lime. | 51.04 |
| | Carbonate of Lime | 11.30 |
| | Fluoride of Calcium | 2.00 |
| | Phosphate of Magnesia | 1.16 |
| | Soda and Chloride of Sodium | 1.20 |
| | | 100.00 |

Some chemists add to this about one per cent. of fat.

The relative proportions of the two constituents of bone are found to differ in *different bones of the skeleton*, as shown by Dr. Owen Rees. Thus, the bones of the head, and the long bones of the extremities, contain more earthy matter than those of the trunk; and those of the upper extremity somewhat more than the corresponding bones of the lower extremity. The humerus contains more earthy matter than the bones of the fore-arm; and the femur more than the tibia and fibula. The vertebræ, ribs and clavicle, contain nearly the same proportion of earthy matter. The metacarpal and metatarsal bones contain about the same proportion as those of the trunk.

Much difference exists in the analyses given by chemists as to the proportion between the two constituents of bone at *different periods of life*. According to Schreger, and others, there is a considerable increase in the earthy constituents of the bones with advancing years. Dr. Rees states, that this is especially marked in the long bones, and the bones of the head, which, in the fœtus, do not contain the excess of earthy matter found in those of the adult. But the bones of the trunk in the fœtus, according to this analyst, contain as much earthy matter as those of the adult. On the other hand, the analyses of Stark and Von Bibra show, that the proportions of animal and earthy matter are almost precisely the same at different periods of life. According to the analyses of Von Bibra, Valentin, and Dr. Rees, the compact substance contains more earthy matter than the cancellous. The comparative analysis of the same bones in both sexes shows no essential difference between them.

There are facts of some practical interest, bearing upon the difference which seems to exist in the amount of the two constituents of bone, at different periods of life. Thus, in the child, where the animal matter predominates, it is not uncommon to find, after an injury to the bones, that they become bent, or only partially broken, from the large amount of flexible animal matter which they contain. Again, also in aged people, where the bones contain a large proportion of earthy matter, the animal matter at the same time being deficient in quantity and quality, the bones are more brittle, their elasticity is destroyed; and, hence, fracture takes place more readily. Some of the diseases, also, to which bones are liable, mainly depend on the disproportion between the two constituents of bone. Thus, in the disease called rickets, so common in the children of scrofulous parents, the bones become bent and curved, either from the superincumbent weight of the body, or under the action of certain muscles. This depends upon some deficiency of the nutritive system, by which bone becomes minus its normal proportion of earthy matter, whilst the animal matter is of unhealthy quality. In the vertebra of a rickety subject, Dr. Bostock found in 100 parts 79.75 animal, and 20.25 earthy matter.

*Form of Bones.* The various mechanical purposes for which bones are employed in the animal economy, require them to be of very different forms. All the scientific principles of Architecture and Dynamics are more or less exemplified in the construction of this part of the human body. The power of the arch in resisting superincumbent pressure is well exhibited in various parts of the skeleton, such as the human foot, and more especially in the vaulted roof of the cranium.

Bones are divisible into four classes: *Long*, *Short*, *Flat*, and *Irregular*.

The *Long Bones* are found chiefly in the limbs, where they form a system of levers, which have to sustain the weight of the trunk, and to confer extensive powers of locomotion. A long bone consists of a lengthened cylinder or shaft, and two extremities. The *shaft* is a hollow cylinder, the walls consisting of dense compact tissue of great thickness in the middle, and becoming thinner towards the extremities; the spongy tissue is scanty, and the bone is hollowed out in its interior to form the *medullary canal*. The *extremities* are generally somewhat expanded for greater convenience of mutual connexion, for the purposes of articulation, and to afford a broad surface for muscular attachment. Here the bone is made up of spongy tissue with only a thin coating of compact substance. The long bones are, the *clavicle*, *humerus*, *radius*, *ulna*, *femur*, *tibia*, *fibula*, *metacarpal*, and *metatarsal* bones, and the *phalanges*.

*Short Bones.* Where a part is intended for strength and compactness, and the motion at the same time slight and limited, it is divided into a number of small pieces united together by ligaments, and the separate bones are short and compressed, such as the bones of the *carpus* and *tarsus*. These bones, in their structure, are spongy throughout, excepting at their surface, where there is a thin crust of compact substance.

*Flat Bones.* Where the principal requirement is either extensive protection, or the provision of broad surfaces for muscular attachment, we find the osseous structure remarkable for its slight thickness, becoming expanded into broad flat plates, as is seen in the bones of the skull and shoulder-blade. These bones are composed of two thin layers of compact tissue, enclosing between them a variable quantity of cancellous tissue. In the cranial bones, these layers of compact tissue are familiarly known as the *tables* of the skull; the outer one is thick and tough, the inner one thinner, denser, and more brittle, and hence termed the *vitreous table*. The intervening cancellous tissue is called the *diploë*. The flat bones are, the *occipital*, *parietal*, *frontal*, *nasal*, *lachrymal*, *vomer*, *scapulæ*, *ossa innominata*, *sternum*, and *ribs*.

The *Irregular* or *Mixed* bones are such as, from their peculiar form, cannot be grouped under either of the preceding heads. Their structure is similar to that of other bones, consisting of an external layer of compact, and of spongy cancellous tissue within. The irregular bones are, the *vertebræ*, *sacrum*, *coccyx*, *temporal*, *sphenoid*, *ethmoid*, *superior maxillary*, *inferior maxillary*, *palate*, *inferior turbinated*, and *hyoid*.

*Surfaces of Bones.* If the surface of any bone is examined, certain eminences and depressions are seen, to which descriptive anatomists have given the following names.

A prominent process projecting from the surface of a bone, which it has never been separate from, or moveable upon, is termed an *apophysis* (from ἀπόφυσις, *an excrescence*); but if such process is developed as a separate piece from the rest of the bone to which it is afterwards joined, it is termed an *epiphysis* (from ἐπίφυσις, *an accretion*).

These eminences and depressions are of two kinds: *articular*, and *non-articular*. Well-marked examples of articular eminences are found in the heads of the humerus and femur; and of articular depressions, in the glenoid cavity of the scapula, and the acetabulum. Non-articular eminences are designated according to their form. Thus, a broad, rough, uneven elevation is called a *tuberosity*; a small rough prominence, a *tubercle*; a sharp, slender, pointed eminence, a *spine*; a narrow rough elevation, running some way along the surface, a *ridge*, or *line*.

The non-articular depressions are also of very variable form, and are described as fossæ, grooves, furrows, fissures, notches, etc. These non-articular eminences and depressions serve to increase the extent of surface for the attachment of ligaments and muscles, and are usually well marked in proportion to the muscularity of the subject.

*Microscopic Structure.* If a thin transverse section from the shaft of a long bone be examined with a power of about 20 diameters, a number of apertures, surrounded by a series of concentric rings, are observed, with small dark spots grouped around them, also in a concentric manner. The apertures are sections of the *Haversian canals* (so called after their discoverer, Clopton Havers); the concentric rings are sections of the *lamellæ*, which are developed around the Haversian canals; the dark spots are small cavities in the substance of the bone, called *lacunæ*.

The Haversian canals are channelled out of the compact substance for the purpose of conveying blood-vessels for its nutrition. They vary in size from the $\frac{1}{200}$ to the $\frac{1}{2000}$ of an inch in diameter, the average size being about $\frac{1}{500}$. They are generally round or oval, sometimes angular. Those nearest to the outer surface, where the bone is most compact, are very small; but towards the medullary canal, they gradually acquire a larger size, and open into it, or into the cells of the cancellous tissue. The Haversian canals are lined by a delicate membrane continuous with the periosteum; the smallest canals contain a single capillary vessel; those larger in size contain a network of vessels; whilst the largest contain blood-vessels and marrow. If a thin longitudinal section of the shaft of a long bone be examined, the Haversian canals will be found to run in the long axis of the bone, and parallel with each other, communicating freely by transverse or oblique canals, so as to form, for the most part, rectangular meshes. Some of these canals open on the outer surface, to admit blood-vessels from the periosteum; others communicate with the medullary canal, receiving blood-vessels from the interior of this part. By this means, the Haversian canals establish a free communication between the blood-vessels of the periosteum, and those of the medullary membrane.

If a higher power is now applied to the same transverse section, each Haversian canal appears surrounded by a series of concentric rings, varying in number from eight to fifteen; these rings are termed the *lamellæ*, and their appearance is produced by transverse sections of concentric layers of bone that have been developed around the Haversian canal, the last formed layer being deposited on that surface next to the blood-vessel. This concentric arrangement is not complete around all the canals; for here and there one set of lamellæ may be seen ending between two adjacent ones. Besides the lamellæ surrounding the Haversian canals, some are disposed parallel with the outer and inner surfaces of the bone; these are termed *circumferential lamellæ*, and may be considered as concentric with the medullary canal. Others, again, penetrate between the Haversian systems; these are termed *interstitial lamellæ*. Each Haversian canal, together with its concentric lamellæ of bone, lacunæ, etc., is called an *Haversian system*, the blood-vessel contained in the central canal being the source of nutrition to the lamellæ which surround it. Nearly the whole of the compact tissue is made up of these *Haversian systems*, each one being, to a certain extent, independent of the rest. In a longitudinal section, the lamellæ are seen running in lines parallel with the Haversian canal which they surround, except when the section passes transversely or obliquely across a canal, in which case an appearance is seen, somewhat similar to that observed in a transverse section. This lamellated structure may be easily demonstrated on a piece of bone softened in dilute acid, when the lamellæ may be peeled from the surface of the bone in a longitudinal direction. According to Dr. Sharpey, the lamellæ, in structure, consist of fine transparent fibres decussating each other, so as to form a delicate network, the fibres apparently coalescing at their point of junction. The lamellæ are perforated, in certain situations, by bundles of fibres which penetrate them in a more or less oblique direction, serving to securely approximate the several plates. The lamellæ are also perforated by numerous minute apertures placed at regular distances apart, which are, probably, transverse sections of the canaliculi. In this fibrous basis of the lamellæ, the inorganic elements of bone are intimately united.

A transverse section of compact bone sometimes exhibits certain vacuities or spaces, termed, by Messrs. Tomes and De Morgan, *Haversian spaces*. These spaces are found at all periods of life, but especially in young and growing bones. They are characterised by an irregular or jagged outline, and are apparently produced by the absorption of parts of several Haversian systems, which have been, to a greater or less extent, removed in order to form them. These spaces may exist in various conditions: in some, the process of absorption is evidently going on; in others, the spaces are lined by newly-formed lamellæ which fill up the peripheral portion of the space; in others, the lamellæ fill in the whole of the space, leaving a Haversian canal in the centre. It would, thus, appear, that portions of the Haversian systems are, from time to time, removed by absorption, and a new system of lamellæ re-formed in place of those previously existing. Sometimes, these spaces may be seen filled in, at one part, by the deposition of lamellæ; while, at another part, they are extending themselves by absorption.

We have already said, that the dark spots seen in and between the lamellæ, arranged in concentric circles around the Haversian canals, are the *lacunæ*. They are minute cavities existing in the osseous substance, having numerous fine tubes called *canaliculi* issuing from all parts of their circumference. In fresh bones, each lacuna contains a delicate cell, with pellucid contents, and a single nucleus; and from the cell numerous fine processes are given off, which fill the canaliculi. These are the *bone cells*, discovered by Virchow. The lacunæ are oval flattened spaces, lying parallel to the direction of the lamellæ. The canaliculi issuing from them are extremely minute, their diameter ranging from $\frac{1}{14000}$ to $\frac{1}{20000}$ of an inch. They communicate freely with the canaliculi of adjoining lacunæ, some opening into the Haversian canals, or in the cancelli of the spongy substance, and some upon the free surface of the bone. By this communication between the lacunæ and canaliculi traversing the entire substance of the bone, the plasma of the blood is carried into every part.

*Vessels of Bone.* The blood-vessels of bone are very numerous. Those of the compact tissue are derived from a close and dense network of vessels, which ramify in a fibrous membrane termed the *periosteum*, which covers the surface of the bone in nearly every part. From this membrane, vessels pass through the minute orifices in the compact tissue, running through the canals which traverse its substance. The cancellous tissue is supplied in a similar way, but by a less numerous set of larger vessels, which, perforating the outer compact tissue, are distributed to the cavities of the spongy portion of the bone. In the long bones, numerous apertures may be seen at the ends near the articular surfaces, some of which give passage to the arteries referred to; but the greater number, and these are the largest of them, are for the veins of the cancellous tissue which run separately from the arteries. The medullary canal in the shafts of the long bones is supplied by one large artery (or sometimes more), which enters the bone at the nutritious foramen (situated, in most cases, near the centre of the shaft), and perforates obliquely the compact substance. This vessel, usually accompanied by one or two veins, sends branches upwards and downwards, to supply the medullary membrane, which lines the central cavity and the adjoining canals. The ramifications of this vessel anastomose with the arteries both of the cancellous and compact tissues. In most of the flat, and in many of the short spongy bones, one or more large apertures are observed, which transmit, to the centre of the bone, vessels which correspond to the medullary arteries and veins.

The veins emerge from the long bones in three places (Kölliker). 1. By a large vein which accompanies the nutrient artery; 2. by numerous large and small veins at the articular extremities; 3. by many small veins which arise in the compact substance. In the flat cranial bones, the veins are large, very numerous, and run in tortuous canals in the diploic tissue, the sides of which are constructed of a thin lamella of bone, perforated here and there for the passage of branches from the adjacent cancelli. The veins thus enclosed and supported by the osseous structure, have exceedingly thin coats; and when the bony structure is divided, they remain patulous, and do not contract in the canals in which they are contained. Hence the constant occurrence of purulent absorption after amputation, in those cases where the stump becomes inflamed, and the cancellous tissue is infiltrated and bathed in pus.

*Lymphatic* vessels have been traced, by Cruikshank, into the substance of bone, but Kölliker doubts their existence. *Nerves* are distributed freely to the periosteum, and accompany the nutritious arteries into the interior of the bone. They are said, by Kölliker, to be most numerous in the articular extremities of the long bones, in the vertebræ, and the larger flat bones.

*Periosteum.* The bones are covered by a tough fibrous membrane, the periosteum, which adheres to their surface in nearly every part, excepting at their cartilaginous extremities, and where strong tendons are attached. It is highly vascular; and from it, numerous vessels pass into minute orifices which cover the entire surface of the bone. It consists of two layers closely united together; the outer one formed chiefly of connective tissue, and occasionally a few fat-cells; the inner one, of elastic fibres of the finer kind, which form dense elastic membranous networks, superimposed in several layers (Kölliker). In young bones, this membrane is thick, very vascular, intimately connected at either end of the bone with the epiphysal cartilage; but less closely connected with the shaft, from which it is separated by a layer of soft blastema, in which ossification proceeds on the exterior of the young bone. Later in life, the periosteum is thinner, less vascular, and more closely connected with the adjacent bone, this adhesion growing stronger as age advances. The periosteum serves as a nidus for the ramification of the vessels previous to their distribution in the bone; hence the liability of bone to exfoliation or necrosis, when, from injury, it is denuded of this membrane.

*Marrow.* The medullary canal of adult long bones, the cavities of the cancellous tissue, and the larger Haversian canals, are filled with a substance called *marrow;* and lined by a highly vascular areolar tissue, the medullary membrane, or internal periosteum. It is by means of the vessels which ramify through this membrane, that the nourishment of the medulla and contiguous osseous tissue is effected.

The marrow differs in composition at different periods of life, and in different bones. In young bones, it is a transparent reddish fluid, of tenacious consistence, free from fat; and contains numerous minute roundish poly-nucleated cells. In the shafts of adult long bones, the marrow is of a *yellow* colour, and contains, in 100 parts, 96.0 fat, 1.0 areolar tissue and vessels, and 3.0 of fluid with extractive matters; whilst, in the flat and short bones, in the articular ends of the long bones, in the bodies of the vertebræ, the base of the cranium, and in the sternum and ribs, it is of a *red* colour, and contains, in 100 parts, 75.0 water, and 25.0 solid matter, consisting of albumen, fibrin, extractive matter, salts, and a mere trace of fat. It consists of fat-cells with a large quantity of fluid, containing numerous poly-nucleated cells, similar to those found in fœtal marrow.

*Development of Bone.* From the peculiar uses to which bone is applied in forming a hard skeleton or framework for the softer materials of the body, and in enclosing and protecting some of the more important vital organs, we find its development takes place at a very early period. Hence, the parts that appear soonest in the embryo, are the vertebral column and the skull, the great central column, to which the other parts of the skeleton are appended. At an early period of embryonic life, the parts destined to become bone consist of a congeries of cells, connected together by an amorphous blastema which constitutes the simplest form of cartilage. This *temporary cartilage*, as it is termed, is an exact miniature of the bone which, in due course, is to take its place; and as the process of ossification is slow, and not completed until adult life, it increases in bulk by an interstitial development of new cells. The next step in this process is the ossification of the intercellular substance, and of the cells composing the cartilage. Ossification commences in the interior of the cartilage at certain points, called *points* or *centres of ossification*, from which it extends into the surrounding substance. This mode of ossification is called *intra-cartilaginous*, to distinguish it from that which takes place in a membranous tissue, quite different in its nature from cartilage. The latter mode of ossification is called *intra-membranous.* Examples of it are seen, according to Kölliker, in—the upper half of the expanded portion of the occipital bone; the parietal and frontal bones; the squamous portion and tympanic ring of the temporal bone; the internal lamella of the pterygoid process of the sphenoid; the cornua sphenoidalia; in all the bones of the face, excepting the inferior turbinated; and according to Bruck, in the clavicle.

The period of ossification is different in different bones. The order of succession may be thus arranged (Kölliker):—

In the second month, first, in the clavicle, and lower jaw (fifth to seventh week); then, in the vertebræ, humerus, femur, the ribs, and the cartilaginous portion of the occipital bone.

At the end of the second, and commencement of the third month, the frontal bone, the scapula, the bones of the fore-arm and leg, and upper jaw, make their appearance,

In the third month, the remaining cranial bones, with few exceptions, begin to ossify, the metatarsus, the metacarpus, and the phalanges.

In the fourth month, the iliac bones, and the ossicula auditus.

In the fourth or fifth month, the ethmoid, sternum, pubis, and ischium.

From the sixth to the seventh month, the calcaneum, and astragalus.

In the eighth month, the hyoid bone.

At birth, the epiphyses of all cylindrical bones, occasionally with the exception of those of the femur and tibia; all the bones of the carpus; the five smaller ones of the tarsus; the patella; sesamoid bones; and the last pieces of the coccyx, are still unossified.

From the time of birth to the fourth year, osseous nuclei make their appearance also in these parts.

At twelve years, in the pisiform bone.

The number of ossific centres is different in different bones. In most of the short bones, ossification commences by a single point in the centre, and proceeds towards the circumference. In the long bones, there is a central point of ossification for the shaft or diaphysis; and one or more for each extremity, the epiphyses. That for the shaft is the first to appear; those for the extremities appear later. For a long period after birth, a thin layer of unossified cartilage remains between the diaphysis and epiphyses, until their growth is finally completed, their junction taking place either at the period of puberty, or towards the end of the period of growth. The union of the epiphyses with the shaft takes place in the inverse order to that in which their ossification began; for, although ossification commences latest in those epiphyses towards which the nutritious artery in the several bones is directed, they become joined to the diaphyses sooner than the epiphyses at the opposite extremity, with the exception of the fibula, the lower end of which commences to ossify at an earlier period than the upper end, but, nevertheless, is joined to the shaft earliest.

The order in which the epiphyses become united to the shaft, appears to be regulated by the direction of the nutritious artery of the bone. Thus the arteries of the bones of the arm and forearm are directed towards the elbow, and the epiphyses of the bones forming this joint become united to the shaft before those at the opposite extremity. In the lower extremities, on the contrary, the nutritious arteries pass in a direction from the knee; that is upwards in the femur, downwards in the tibia and fibula; and in them it is observed, that the upper epiphysis of the femur, and the lower epiphyses of the tibia and fibula, become first united to the shaft.

Where there is only one epiphysis, the medullary artery is directed towards that end of the bone where there is no additional centre: as, towards the acromial end in the clavicle; towards the distal end of the metacarpal bone of the thumb and great toe; and towards the proximal end of the other metacarpal and metatarsal bones.

A knowledge of the exact periods when the epiphyses become joined to the shaft, aids the surgeon in the diagnosis of many of the injuries to which the joints are liable; for it not unfrequently happens, that on the application of severe force to a joint, the epiphyses become separated from the shaft, and such injuries may be mistaken for fracture.

*Growth of Bone.* Increase in the length of a bone is provided for by the development of new bone in the cartilage at either end of the shaft (diaphysis); and in the thickness, by the deposition of soft ossifying blastema in successive layers upon the inner surface of the periosteum.

The entire skeleton in an adult consists of 294 distinct bones. These are –

| | |
|---|---|
| Vertebral column (sacrum and coccyx included) | 26 |
| Cranium | 8 |
| Ossicula auditûs | 6 |
| Face | 14 |
| Os hyoides, sternum, and ribs | 26 |
| Upper extremities | 64 |
| Lower extremities | 60 |
| | 204 |

In this enumeration, the patellæ and other sesamoid bones, as well as the Wormian bones are excluded, as also are the teeth, which differ from bone both in structure, development, and mode of growth.

## THE SPINE.

The Spine is a flexuous column, formed of a series of bones called *Vertebræ*.

The Vertebræ are thirty-three in number, exclusive of those which form the skull, and have received the names *cervical, dorsal, lumbar, sacral*, and *coccygeal* according to the position which they occupy; seven being found in the cervical region, twelve in the dorsal, five in the lumbar, five in the sacral, and four in the coccygeal.

This number is sometimes found increased by an additional segment in one region, or the number may be diminished in one region, the deficiency being supplied by an additional segment in another. These observations do not apply to the cervical portion of the spine, the number of segments forming which is seldom increased or diminished.

The Vertebræ in the three uppermost regions of the spine are separate segments throughout the whole of life; but those found in the sacral and coccygeal regions, are, in the adult, firmly united, so as to form two bones—five entering into the formation of the upper bone or *sacrum*, and four into the terminal bone of the spine or *coccyx*.

## General Characters of a Vertebra.

Each vertebra consists of two essential parts, an anterior solid segment or body, and a posterior segment, the arch. The arch is formed of two pedicles and two laminæ, supporting seven processes; viz, four articular, two transverse, and one spinous process.

The Bodies of the vertabræ are piled one upon the other, forming a strong pillar, for the support of the cranium and trunk; the arches forming behind these a hollow cylinder for the protection of the spinal cord. The different segments are connected together by means of the articular processes, and the transverse and spinous processes serve as levers for the attachment of muscles which move the different parts of the spine. Lastly between each pair of vertebræ apertures exist through which the spinal nerves pass from the cord. Each of these constituent parts must now be separately examined.

*The Body* is the largest and most solid part of a vertebra. Above and below, it is slightly concave, presenting a rim around its circumference; and its surfaces are rough, for the attachment of the intervertebral fibro-cartilages. In front it is convex from side to side, concave from above downwards. Behind, flat from above downwards and slightly concave from side to side. Its anterior surface is perforated by a few small apertures, for the passage of nutrient vessels; whilst on the posterior surface is a single irregular-shaped, or occasionally several large apertures, for the exit of veins from the body of the vertebra, the *venæ basis vertebræ*.

The *Pedicles* project backwards, one on each side, from the upper part of the body of the vertebra, at the line of junction of its posterior and lateral surfaces. The concavities above and below the pedicles are the *intervertebral notches;* they are four in number, two on each side, the inferior ones being generally the deeper. When the vertebræ are articulated, the notches of each contiguous pair of bones form the intervertebral foramina which communicate with the spinal canal and transmit the spinal nerves.

The *Laminæ* are two broad plates of bone, which complete the vertebral arch behind, enclosing a foramen which serves for the protection of the spinal cord; they are connected to the body by means of the pedicles. Their upper and lower borders are rough, for the attachment of the *ligamenta subflava*.

The *Articular Processes*, four in number, two on each side, spring from the junction of the pedicles with the laminæ. The two superior project upwards, their articular surfaces being directed more or less backwards, the two inferior project downwards, their articular surfaces looking more or less forwards.

The *Spinous Process* projects backwards from the junction of the two laminæ, and serves for the attachment of muscles.

The *Transverse Processes*, two in number, project one at each side from the point where the articular processes join the pedicle. They also serve for the attachment of muscles.

## Characters of the Cervical Vertebræ (Fig. 1).

*The Body* is smaller than in any other region of the spine, and broader from side to side than from before backwards. The anterior and posterior surfaces are flattened and of equal depth; the former is placed on a lower level than the latter, and its inferior border is prolonged downwards so as to overlap the upper and fore part of the vertebra below. Its upper surface is concave transversely, and presents a projecting lip on each side; its lower surface being convex from side to side; concave from before backwards, and presenting laterally a shallow concavity, which receives the corresponding projecting lip of the adjacent vertebra. The *pedicles* are directed obliquely outwards, and the superior intervertebral notches are deeper, but narrower, than the inferior. The *laminæ* are narrow, long, thinner above than below, and overlap each other; enclosing the spinal foramen, which is very large, and of a triangular form. The *spinous processes* are short, bifid at the extremity, to afford greater extent of surface for the attachment of muscles, the two divisions being often of unequal size. They increase in length from the fourth to the seventh. The *transverse processes* are short, directed downwards, outwards, and forwards, are bifid at their extremity, and marked by a groove along their upper surface, which runs downwards and outwards from the superior intervertebral notch, and serves for the transmission of one of the cervical nerves. The transverse processes are pierced at their base by a foramen, for the transmission of the vertebral artery, vein, and plexus of nerves. Each process is formed by two roots: the anterior root arises from the side of the body, and corresponds to the ribs: the posterior root springs from the junction of the pedicle with the lamina, and corresponds with the transverse processes in the dorsal region. It is by the junction of these two processes, that the vertebral foramen is formed. The extremities of each of these roots form the *anterior* and *posterior tubercles* of the transverse processes. The *articular processes* are oblique: the superior are of an oval form, flattened, and directed upwards and backwards; the inferior downwards and forwards.

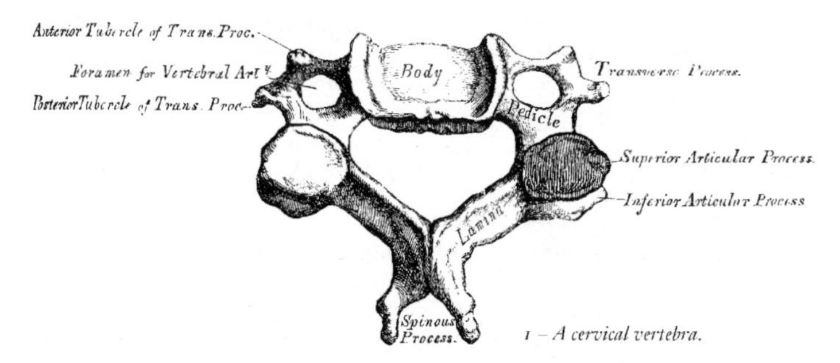


1 – *A cervical vertebra.*

The peculiar vertebræ in the cervical region are the first or *Atlas;* the second or *Axis;* and the seventh or *Vertebra prominens*. The great modifications in the form of the atlas and axis are to admit of the nodding and rotatory movements of the head.

The *Atlas* (fig. 2) (so named from supporting the globe of the head). The chief pecularities of this bone are, that it has neither body, nor spinous process. The body is detached from the rest of the bone, and forms the odontoid process of the second vertebra, the parts corresponding to the pedicles pass in front and join to form the anterior arch. The atlas consists of an anterior arch, a posterior arch, and two lateral masses. The *anterior* arch, forms about one-fifth of the bone; its anterior surface is convex, and presents about its centre a tubercle, for the attachment of the Longus colli muscle; posteriorly it is concave, and marked by a smooth oval or circular facet, for articulation with the odontoid process of the axis. The *posterior* arch forms about two-fifths of the circumference of the bone; it terminates behind in a tubercle, which is the rudiment of a spinous process, and gives origin to the Rectus capitis posticus minor. The diminutive size of this process prevents any interference in the movements between it and the cranium. The posterior part of the arch presents above a rounded edge; whilst in front, immediately behind each superior articular process, is a groove, sometimes converted into a foramen by a delicate bony spicula which arches backwards from the posterior extremity of the superior articular process. These grooves represent the superior inter-

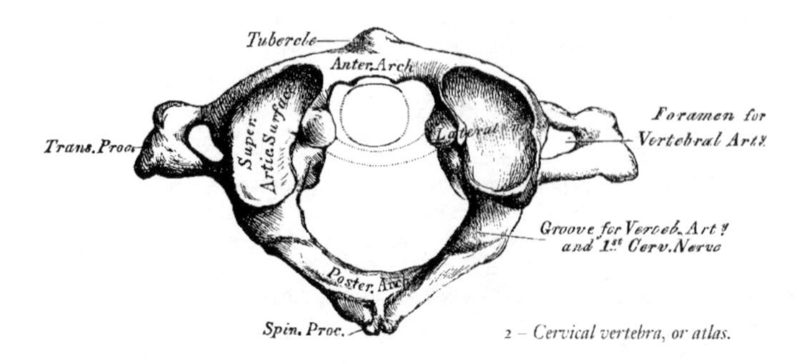


2 – *Cervical vertebra, or atlas.*

vertebral notches, and are peculiar from being situated behind the articular processes, instead of before them, as in the other vertebræ. They serve for the transmission of the vertebral artery, which, ascending through the foramen in the transverse process, winds round the lateral mass in a direction backwards and inwards. They also transmit the sub-occipital nerves. On the under surface of the posterior arch, in the same situation, are two other grooves, placed behind the lateral masses, and representing the inferior intervertebral notches of other vertebræ. They are much less marked than the superior. The *lateral masses*, are the most bulky and solid parts of the atlas, in order to support the weight of the head; they present two articulating processes above, and two below. The two superior are of large size, oval, concave, and approach towards one another in front, but diverge behind; they are directed upwards, inwards, and a little backwards, forming a kind of cup for the condyles of the occipital bone, and are admirably adapted to the nodding movements of the head. Not unfrequently they are partially subdivided by a more or less deep indentation which encroaches upon each lateral margin; the inferior articular processes are circular in form, flattened, or slightly concave, and directed downwards, inwards, and a little backwards, articulating with the axis, and permitting the rotatory movements. Just below the inner margin of each superior articular surface, is a small tubercle, for the attachment of a ligament which, stretching across the ring of the atlas, divides it into two unequal parts; the anterior or smaller segment receiving the odontoid process of the axis, the posterior allowing the transmission of the spinal cord and its membranes. This part of the spinal canal is of considerable size, to afford space for the spinal cord; and hence lateral displacement of the atlas may occur without compression of the spinal cord. This ligament and the odontoid process are marked in figure 2 in dotted outline. The transverse processes are of large size, for the attachment of special muscles which assist in rotating the head—long, not bifid, perforated at their base by a canal for the vertebral artery, which is directed from below, upwards and backwards.

The *Axis* (fig. 3) (so named from forming the pivot upon which the head rotates). The most distinctive character of this bone is the strong prominent process, tooth-like in form (hence the name odontoid), which rises perpendicularly from the upper part of the body. The body is of a triangular form; deeper in front than behind, and prolonged downwards anteriorly so as to overlap the upper and fore part of the adjacent vertebra. It presents in front a median longitudinal ridge, separating two lateral depressions for the attachment of the Longi colli muscles. The odontoid process presents two articulating surfaces: one in front of an oval form, for articulation with the atlas; another behind, for the transverse ligament; the latter frequently encroaches on the sides of the process; the apex is pointed. Below the apex this process is somewhat enlarged, and presents on either side a rough impression for the attachment of the odontoid or check ligaments, which connect it to the occipital bone; the base of the process, where attached to the body, is constricted, so as to prevent displacement from the transverse ligament, which binds it in this situation to the anterior arch of the atlas. Sometimes, however, this process does become displaced, especially in children, where the ligaments are more relaxed, instant death is the result. The pedicles are broad and strong, especially their anterior extremities which coalesce with the sides of the body and the root of the odontoid process. The laminæ are thick and strong, and the spinal foramen very large. The superior articular surfaces are round, slightly convex, directed upwards and outwards, and are peculiar in being supported on the body, pedicles, and transverse processes. The inferior articular surfaces, have the same direction as those of the other cervical vertebræ. The superior intervertebral notches are very shallow, and lie behind the articular processes; the inferior in front of them, as in the other cervical vertebræ. The transverse processes are very small, not bifid, and perforated by the vertebral foramen, which is directed obliquely upwards, and outwards. The spinous process is of large size, very strong, deeply chanelled on its under surface, and presents a bifid tubercular extremity for the attachment of muscles, which serve to rotate the head upon the spine.

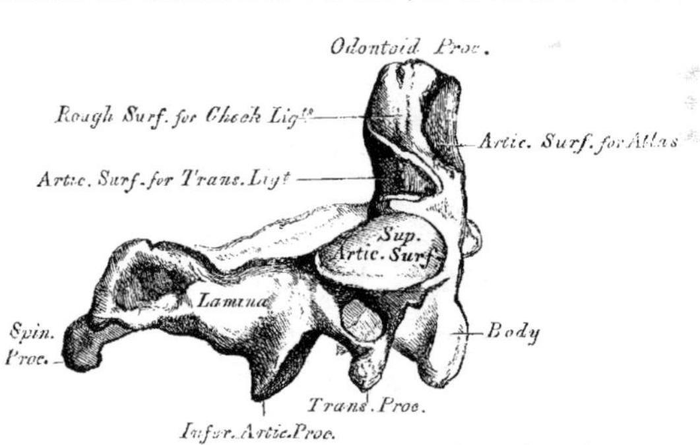

*3 – 2nd cervical vertebra, or axis.*

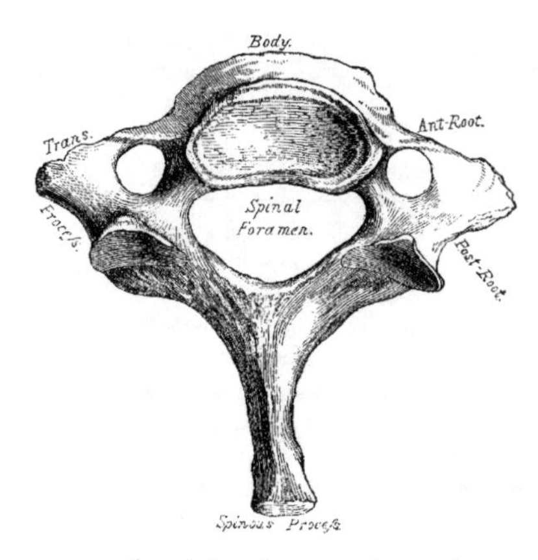


4 – *7th cervical vertebra, or vertebra prominens.*

*Seventh Cervical* (fig. 4). The most distinctive character of this vertebra is the existence of a very long, and prominent spinous process; hence the name 'Vertebra prominens.' This process is thick, nearly horizontal in direction, not bifurcated, and has attached to it the ligamentum nuchæ. The transverse process is usually of large size, especially its posterior root, its upper surface has usually a shallow groove, and seldom presents more than a trace of bifurcation at its extremity. The vertebral foramen is sometimes as large as in the other cervical vertebræ, usually smaller, on one or both sides and sometimes wanting. On the left side, it occasionally gives passage to the vertebral artery; more frequently the vertebral vein traverses it on both sides; but the usual arrangement is for both artery and vein to pass through the foramen in the transverse process of the sixth cervical.

## Characters of the Dorsal Vertebræ.

The bodies of the dorsal vertebræ resemble those in the cervical and lumbar regions at the respective ends of this portion of the spine; but in the middle of the dorsal region their form is very characteristic, being heart-shaped, and broader in the antero-posterior than in the lateral direction. They are thicker behind than in front, flat above and below, convex and prominent in front, deeply concave behind, slightly constricted in front and at the sides, and marked on each side, near the root of the pedicle, by two demi-facets, one above, the other below. These are covered with cartilage in the recent state; and, when articulated with the adjoining vertebræ, form oval surfaces for the reception of the heads of the corresponding ribs. The pedicles are directed backwards, and the inferior intervertebral notches are of large size, and deeper than in any other region of the spine. The laminæ are broad and thick, and the spinal foramen small, and of a circular form. The articular processes are flat, nearly vertical in direction, and project from the upper and lower part of the pedicles, the superior being directed backwards and a little outwards and upwards, the inferior forwards and a little inwards and downwards. The transverse processes arise from the same parts of the arch as the posterior roots of the transverse processes in the neck; they are thick, strong, and of great length, directed obliquely backwards and outwards, presenting a clubbed extremity, lipped on its anterior part by a small concave surface, for articulation with the tubercle of a rib. Besides the articular facet for the rib, two indistinct tubercles may be seen rising from the extremity of the transverse processes, one near the upper, the other near the lower border. In man they are comparatively of small size, and serve only for the attachment of muscles. But in some animals, they attain considerable magnitude either for the purpose of more closely connecting the segments of this portion of the spine, or for muscular and ligamentous attachment. The spinous processes are long, triangular in form, directed obliquely downwards, and terminate by a tubercular margin. They overlap one another from the fifth to the eighth, but are less oblique in direction above and below.

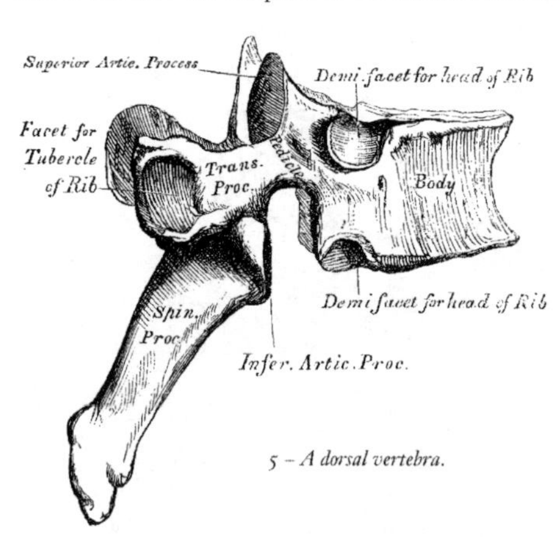


5 – *A dorsal vertebra.*

The peculiar dorsal vertebræ are the *first, ninth, tenth, eleventh, and twelfth* (fig. 6.)

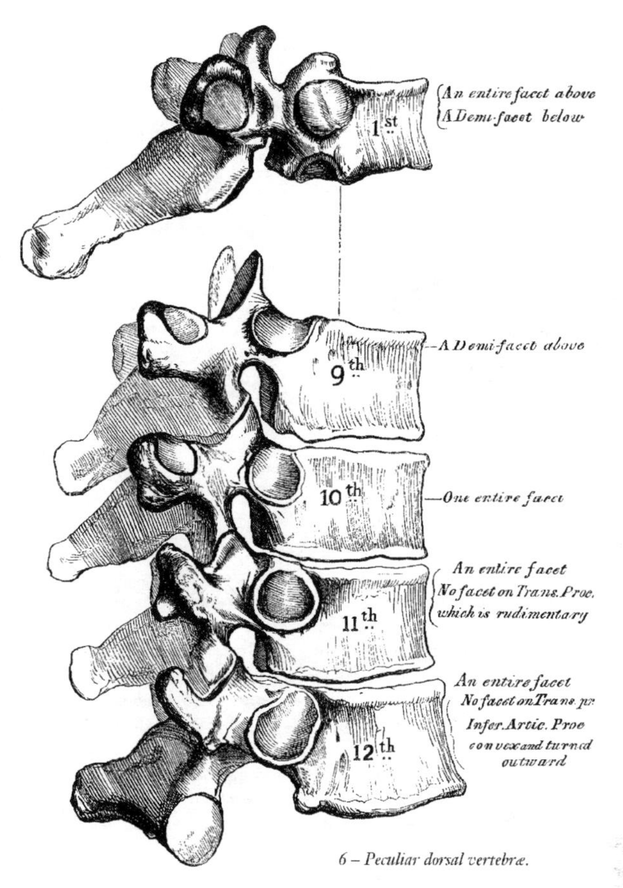


*6 – Peculiar dorsal vertebræ.*

The *First Dorsal Vertebra* presents on each side of the body, a single entire articular facet for the head of the first rib, and a half facet for the upper half of the second. The upper surface of the body is like that of a cervical vertebra, being broad transversely, concave, and lipped on each side. The *articular surfaces* are oblique, and the *spinous process* thick, long, and almost horizontal.

The *Ninth Dorsal* has no demi-facet below. In some subjects, the ninth has two demi-facets on each side, then the tenth has a demi-facet at the upper part; none below.

The *Tenth Dorsal* has an entire articular facet on each side above; no demi-facet below.

In the *Eleventh Dorsal*, the body approaches in its form and size to the lumbar. The articular facets for the heads of the ribs, one on each side, are of large size, and placed chiefly on the pedicles, which are thicker and stronger in this and the next vertebra, than in any other part of the dorsal region. The *transverse* processes are very short, tubercular at their extremities, and have no articular facets for the tubercles of the ribs. The spinous process is short, nearly horizontal in direction, and presents a slight tendency to bifurcation at its extremity.

The *Twelfth Dorsal* has the same general characters as the eleventh; but may be distinguished from it by the inferior articular processes being convex and turned outwards, like those of the lumbar vertebræ; by the general form of the body, laminæ, and spinous process, approaching to that of the lumbar vertebræ; and by the transverse processes being shorter, and the tubercles at their extremities more marked.

## Characters of the Lumbar Vertebræ.

The Lumbar Vertebræ (fig. 7) are the largest segments of the vertebral column. The body is large, broader from side to side than from before backwards, and about equal in depth in front and behind, flattened or slightly concave above and below, concave behind, and deeply constricted in front and at the sides, presenting prominent margins which afford a broad basis for the support of the superincumbent weight. The pedicles are very strong, directed backwards from the upper part of the bodies; consequently the inferior intervertebral notches are of large size. The laminæ are short, but broad and strong; and the foramen triangular, larger than in the dorsal, smaller than in the cervical region. The superior articular processes are concave, and look almost directly inwards; the inferior, convex, look outwards and a little forwards; the former are separated by a much wider interval than the latter, embracing the lower articulating processes of the vertebra above. The transverse processes are long, slender, directed transversely outwards in the upper three lumbar vertebræ, slanting a little upwards in the lower two. By some anatomists they are considered homologous with the ribs. Of the two tubercles noticed in connection with the transverse processes in the dorsal region, the superior ones become connected in this region with the back part of the superior articular processes, the inferior ones with the posterior part of the base of the transverse processes. Although in man they are comparatively small, in some animals they attain considerable size, and serve to lock the vertebræ more closely together. The spinous processes are thick and broad, somewhat quadrilateral, horizontal in direction, thicker below than above, and terminate by a rough uneven border.

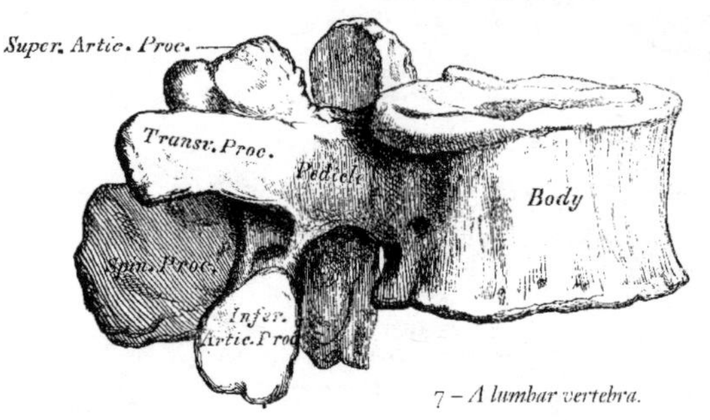

7 – A lumbar vertebra.

The *Fifth Lumbar* vertebra is characterised by having the body much thicker in front than behind, which accounts for the prominence of the sacro-vertebral articulation, by the smaller size of its spinous process, by the wide interval between the inferior articulating processes, and by the greater size and thickness of its transverse processes.

*Structure of the Vertebræ.* The structure of a vertebra differs in different parts. The body is composed of light spongy cancellous tissue, having a thin coating of compact tissue on its external surface perforated by numerous orifices, some of large size, for the passage of vessels, its interior being traversed by one or two large canals for the reception of veins, which converge towards a single large irregular or several small apertures at the posterior part of the body of each bone. The arch and processes projecting from it have, on the contrary, an exceedingly thick covering of compact tissue.

*Development.* Each vertebra is formed of three primary cartilaginous portions (fig. 8); one for each lamina and its processes, and one for the body. Ossification commences in the laminæ about the sixth week of fœtal life, in the situation where the transverse processes afterwards project, the ossific granules shooting backwards to the spine, forwards to the body, and outwards into the transverse and articular processes. Ossification in the body makes its appearance in the middle of the cartilage about the eighth week. At birth, these three pieces are perfectly separate. During the first year, the laminæ become united behind, by a portion of cartilage in which the spinous process is ultimately formed, and thus the arch is completed. About the third year, the body is joined to the arch on each side, in such a manner that the body is formed from the three original centres of ossification, the amount contributed by the pedicles increasing in extent from below upwards. Thus the bodies of the sacral vertebræ are formed almost entirely from the central nuclei, the bodies of the lumbar segments are formed laterally and behind by the pedicles. In the dorsal region the pedicles advance as far forwards as the articular depressions for the heads of the ribs, forming these cavities of reception; and in the neck the whole of the lateral portions of the bodies are formed by the advance of the pedicles. Before puberty, no other changes occur, excepting a gradual increase in the growth of these primary centres, the upper and under surface of the bodies, and the ends of the transverse and spinous processes, being tipped with cartilage, in which ossific granules are not as yet deposited. At sixteen years (fig. 9), four secondary centres appear, one for the tip of each transverse process, and two (sometimes united into one) for the end of the spinous process. At twenty-one years (fig. 10), a thin circular plate of bone is formed in the thin layer of cartilage situated on the upper and under surface of the body, the former being the thicker of the two. All these become joined; and the bone is completely formed about the thirtieth year of life.

Exceptions to this mode of development occur in the first, second, and seventh cervical, and in the vertebræ of the lumbar region.

The *Atlas* (fig. 11) is developed by *two* primary centres, and by *one* or more epiphyses. The two primary centres consist of the two lateral or neural masses, ossification of which commences before birth, near the articular processes, and extending backwards, they are separated from one another behind, at birth, by a narrow interval filled in with cartilage. Between the second and third years, they unite either directly or through the medium of an epiphysal centre, developed in the cartilage near their point of junction. The anterior arch, at birth, is altogether cartilaginous, and this portion of the atlas is completed by the gradual extension forwards and ultimate junction of the two neural processes. Occasionally a separate nucleus is developed in the anterior arch, which, extending laterally, joins the neural processes in front of the pedicles; or, there are two nuclei developed in the anterior arch, one on either side of the median line, they join to form a single mass, which is afterwards united to the lateral portions in front of the articulating processes.

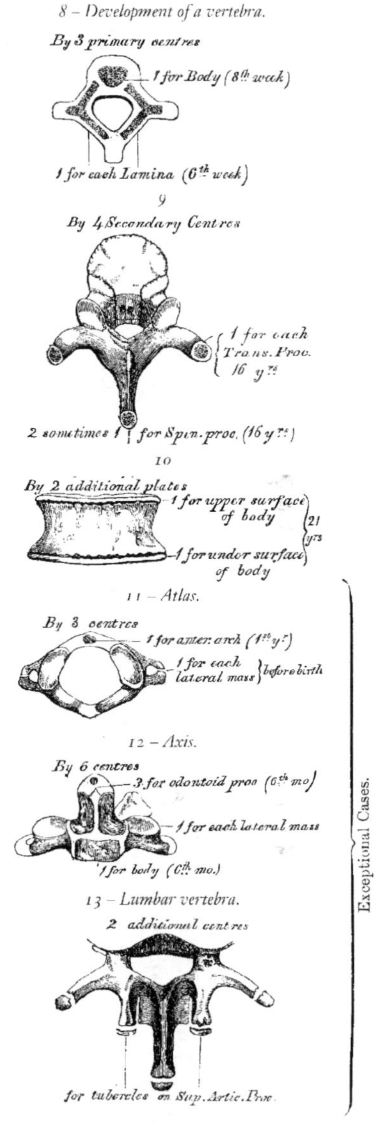


The *Axis* (fig. 12) is developed by *six* centres. The body and arch of this bone are formed in the same manner as the corresponding parts in the other vertebræ: one centre for the lower part of the body, and one for each lamina. The odontoid process, which is really the centrum or body of the axis, consists originally of an extension upwards of the cartilaginous mass, in which the lower part of the body is formed. At about the sixth month of fœtal life, two osseous nuclei make their appearance in the base of this process: they are placed laterally, and join before birth to form a conical-shaped bi-lobed mass, deeply cleft above; the interval between the cleft and the summit of the process, is formed by a wedge-shaped piece of cartilage; the base of the process being separated from the body by a cartilaginous interval, which gradually becomes ossified, sometimes by a separate epiphysal nucleus. Finally, as Mr. Humphry has lately demonstrated, the apex of the odontoid process has a separate nucleus.

*The Seventh Cervical.* The anterior or costal part of the transverse process of the seventh cervical, is developed from a separate osseous centre at about the sixth month of fœtal life, and joins the body and posterior division of the transverse process between the fifth and sixth years. Sometimes this process continues as a separate piece, and becoming lengthened outwards, constitutes what is known as a cervical rib.

*The Lumbar Vertebræ* (fig. 13) have *two additional centres* (besides those peculiar to the vertebræ generally), for the tubercles, which project from the back part of the superior articular processes. The transverse process of the first lumbar is sometimes developed as a separate piece, which may remain permanently unconnected with the remaining portion of the bone; thus forming a lumbar rib, a peculiarity which is sometimes, though rarely, met with.

Progress of Ossification in the Spine generally. Ossification of the laminæ of the vertebræ commences at the upper part of the spine, and proceeds gradually downwards; hence the frequent occurrence of spina bifida in the lower part of the spinal column. Ossification of the bodies, on the other hand, commences a little below the centre of the spinal column (about the ninth or tenth dorsal vertebræ), and extends both upwards and downwards. Although, however, the ossific nuclei make their first appearance in the lower dorsal vertebræ, the lumbar and first sacral are those in which these nuclei are largest at birth.

*Attachment of Muscles.* To the *Atlas* are attached the Longus colli, Rectus anticus minor, Rectus lateralis, Rectus posticus minor, Obliquus superior and inferior, Splenius colli, Levator anguli scapulæ, Interspinous, and Intertransverse.

To the *Axis* are attached the Longus colli, Obliquus inferior, Rectus posticus major, Semi-spinalis colli, Multifidus spinæ, Levator anguli scapulæ, Splenius colli, Transversalis colli, Scalenus posticus, Intertransversales, Interspinales.

To the remaining Vertebræ generally are attached, *anteriorly*, the Rectus anticus major, Longus colli, Scalenus anticus and posticus, Psoas magnus, Psoas parvus, Quadratus lumborum, Diaphragm, Obliquus internus and Transversalis,—*posteriorly*, the Trapezius, Latissimus dorsi, Levator anguli scapulæ, Rhomboideus major and minor, Serratus posticus superior and inferior, Splenius, Sacro-lumbalis, Longissimus dorsi, Spinalis dorsi, Cervicalis ascendens, Transversalis colli, Trachelo-mastoid, Complexus, Semi-spinalis dorsi and colli, Multifidus spinæ, Interspinales, Supraspinales, Intertransversales, Levatores costarum.

## Sacral and Coccygeal Vertebræ.

The Sacral and Coccygeal Vertebræ consist, at an early period of life, of nine separate pieces, which are united in the adult, so as to form two bones, five entering into the formation of the sacrum, four the coccyx.

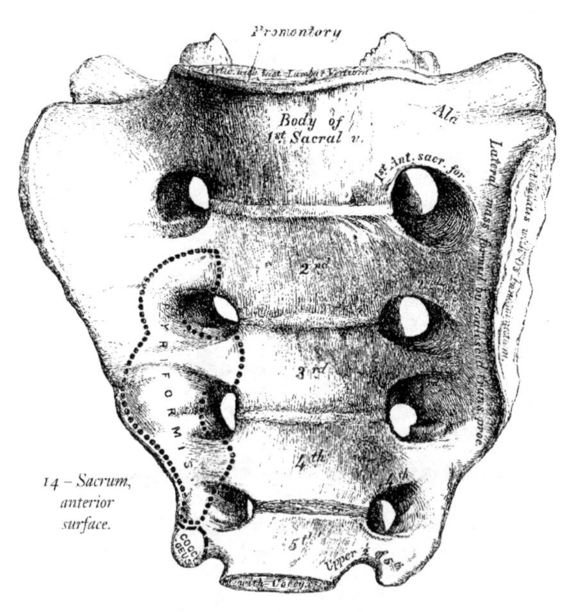


14 – *Sacrum, anterior surface.*

The Sacrum (so called from its having been offered in sacrifice, and hence considered *sacred*), (fig. 14), is a large triangular bone, situated at the lower part of the vertebral column, and at the upper and back part of the pelvic cavity, where it is inserted like a wedge between the two ossa innominata; its upper part, or base, articulating with the last lumbar vertebra, its apex with the coccyx. The sacrum is curved upon itself, and placed very obliquely, its upper extremity projecting forwards, forming, with the last lumbar vertebra, a very prominent angle, called the *promontory* or *sacro-vertebral angle*, whilst its central part is directed backwards, so as to give increased capacity to the pelvic cavity. It presents for examination an anterior and posterior surface, two lateral surfaces, a base, an apex, and a central canal.

*The Anterior Surface* is concave from above downwards, and slightly so from side to side. In the middle are seen four transverse ridges, indicating the original division of the bone into five separate pieces. The portions of bone intervening between the ridges correspond to the bodies of the vertebræ. The body of the first segment is of large size, and in form resembles that of a lumbar vertebra; the succeeding ones diminish in size from above downwards, are flattened from before backwards, and curved so as to accommodate themselves to the form of the sacrum, being concave in front, convex behind. At each end of the ridges above-mentioned, are seen the *anterior sacral foramina*, analogous to the intervertebral foramina, four in number on each side, somewhat rounded in form, diminishing in size from above downwards, and directed outwards and forwards; they transmit the anterior branches of the sacral nerves. External to these foramina, is the *lateral mass*, consisting, at an early period of life, of separate segments, which correspond to the anterior transverse processes, these become blended, in the adult, with the bodies, with each other, and with the posterior transverse processes. Each lateral mass is traversed by four broad shallow grooves, which lodge the anterior sacral nerves as they pass outwards, the grooves being separated by prominent ridges of bone, which give attachment to the slips of the Pyriformis muscle.

15 – *Vertical section of the sacrum.*

If a vertical section is made through the centre of the bone (fig. 15), the bodies are seen to be united at their circumference by bone, a wide interval being left centrally, which, in the recent state, is filled by intervertebral substance. In some bones, this union is more complete between the lower segments, than between the upper ones.

The *Posterior Surface* (fig. 16) is convex, and much narrower than the anterior. In the middle line, are three or four tubercles, which represent the rudimentary spinous processes of the sacral vertebræ. Of these tubercles, the first is usually prominent, and perfectly distinct from the rest; the second and third, are either separate, or united into a tubercular ridge, which diminishes in size from above downwards; the fourth usually, and the fifth always, remaining undeveloped. External to the spinous processes on each side, are the *laminæ*, broad and well marked in the three first pieces; sometimes the fourth, and generally the fifth, being undeveloped; in this situation the lower end of the sacral canal is exposed. External to the laminæ are a linear series of indistinct tubercles representing the *articular processes*; the upper pair are large, well developed, and correspond in shape and direction to the superior articulating processes of a lumbar vertebra; the second and third are small; the fourth and fifth (usually blended together) are situated on each side of the sacral canal: they are called the *sacral cornua*, and articulate with the cornua of the coccyx. External to the articular processes are the four *posterior sacral foramina*; they are smaller in size, and less regular in form than the anterior, and transmit the posterior branches of the sacral nerves. On the outer side of the posterior sacral foramina are a series of

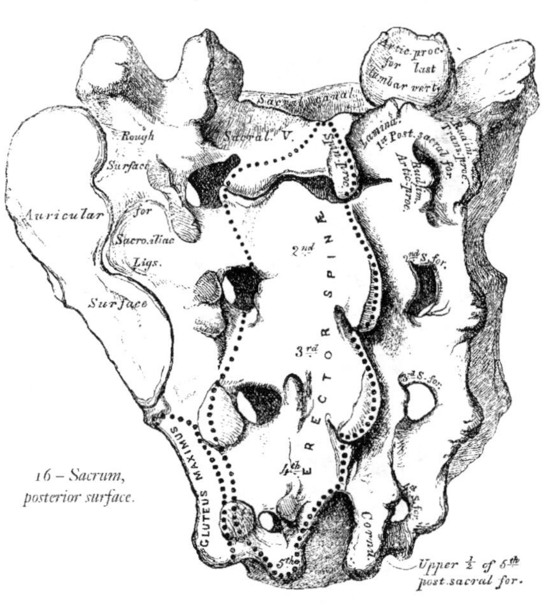


16 – *Sacrum, posterior surface.*

tubercles, the rudimentary posterior *transverse processes* of the sacral vertebræ. The first pair of transverse tubercles are of large size, very distinct, and correspond with each superior angle of the bone; the second, small in size, enter into the formation of the sacro-iliac articulation; the third give attachment to the oblique sacro-iliac ligaments; and the fourth and fifth to the great sacro-ischiatic ligaments. The interspace between the spinous and transverse processes on the back of the sacrum, presents a wide shallow concavity, called the *sacral groove*; it is continuous above with the vertebral groove, and lodges the origin of the Erector spinæ.

The *Lateral surface*, broad above, becomes narrowed into a thin edge below. Its upper half presents in front a broad ear-shaped surface for articulation with the ilium. This is called the *auricular* or *ear-shaped* surface, and in the fresh state is coated with cartilage. It is bounded posteriorly by deep and uneven impressions, for the attachment of the posterior sacro-iliac ligaments. The lower half is thin and sharp, and gives attachment to the greater and lesser sacro-ischiatic ligaments, and to some fibres of the Glutæus maximus; below, it presents a deep notch, which is converted into a foramen by articulation with the transverse process of the upper piece of the coccyx, and transmits the anterior branch of the fifth sacral nerve.

17 – *Development of sacrum.*

*Formed by union of 5 Vertebræ.*
*2 characteristic points.*
*2 Additional centres for the first 3 pieces.**

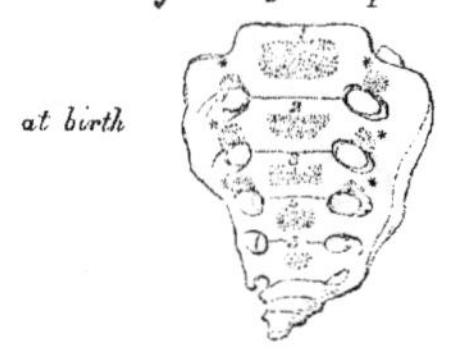


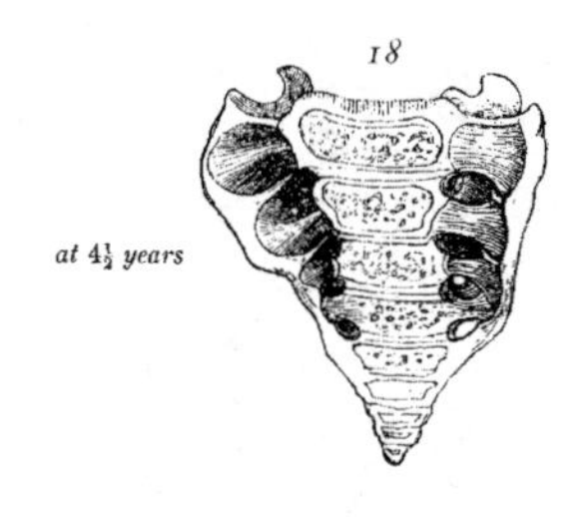


19
*2 Epiphyseal laminæ for each lateral surface.* *

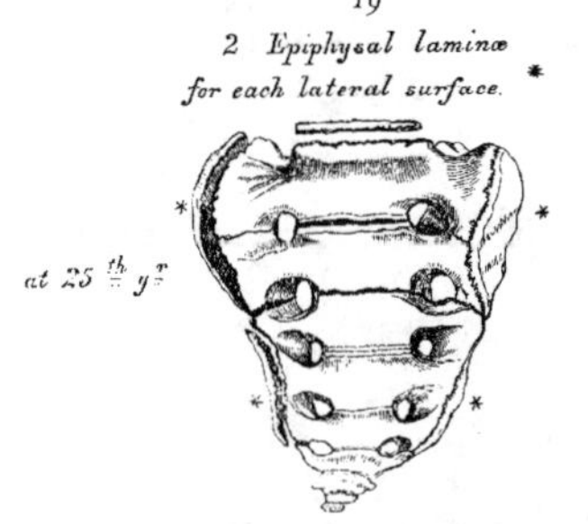


The *Base* of the sacrum, which is broad and expanded, is directed upwards and forwards. In the middle is seen an oval articular surface, which corresponds with the under surface of the body of the last lumbar vertebra, bounded behind by the large triangular orifice of the sacral canal. This orifice is formed behind by the spinous process and laminæ of the first sacral vertebra, whilst projecting from it on each side are the superior articular processes; they are oval, concave, directed backwards and inwards, like the superior articular processes of a lumbar vertebra; in front of each articular process is an intervertebral notch, which forms the lower half of the last intervertebral foramen. Lastly, on each side of the articular surface is a broad and flat triangular surface of bone, which extends outwards, and is continuous on each side with the iliac fossa.

The *Apex*, directed downwards and forwards, presents a small oval concave surface for articulation with the coccyx.

The *Sacral Canal* runs throughout the greater part of the bone; it is large and triangular in form above, small and flattened from before backwards below. In this situation, its posterior wall is incomplete, from the non-development of the laminæ and spinous processes. It lodges the sacral nerves, and is perforated by the anterior and posterior sacral foramina, through which these pass out.

*Structure.* It consists of much loose spongy tissue within, invested externally by a thin layer of compact tissue.

Differences in the Sacrum of the Male and Female. The sacrum in the female is usually wider than in the male; and it is much less curved, the upper half of the bone being nearly straight, the lower half presenting the greatest amount of curvature. The bone is also directed more obliquely backwards; which increases the size of the pelvic cavity, and forms a more prominent sacro-vertebral angle. In the male, the curvature is more evenly distributed over the whole length of the bone, and is altogether greater than in the female.

Peculiarities of the Sacrum. This bone, in some cases, consists of six pieces; occasionally the number is reduced to four. Sometimes the bodies of the first and second segments are not joined, or the laminæ and spinous processes have not coalesced. Occasionally, the upper pair of transverse tubercles are not joined to the rest of the bone on one or both sides; and, lastly, the sacral canal may be open for nearly the lower half of the bone, in consequence of the imperfect development of the laminæ and spinous processes. The sacrum, also, varies considerably with respect to its degree of curvature. From the examination of a large number of skeletons, it would appear, that, in one set of cases, the anterior surface of this bone was nearly straight, the curvature, which was very slight, affecting only its lower end. In another set of cases, the bone was curved throughout its whole length, but especially towards its middle. In a third set, the degree of curvature was less marked, and affected especially the lower third of the bone.

*Development* (fig. 17). The sacrum, formed by the union of five vertebræ, has *thirty-five* centres of ossification.

The *bodies* of the sacral vertebræ have each three ossific centres; one for the central part, and one for the epiphysal plates on its upper and under surface.

The *laminæ* of the sacral vertebræ are each developed by two centres; these meet behind to form the arch, and subsequently join the body.

The *lateral masses* have six additional centres, two for each of the first three vertebræ. These centres make their appearance above and to the outer side of the anterior sacral foramina (fig. 17), and are developed into separate segments, which correspond with the anterior transverse processes (fig. 18); they are subsequently blended with each other, and with the bodies and the posterior transverse processes, to form the lateral mass.

Lastly, each *lateral surface* of the sacrum is developed by two epiphysal plates (fig. 19); one for the auricular surface, and one for the remaining part of the thin lateral edge of the bone.

*Period of Development.* At about the eighth or ninth week of fœtal life, ossification of the central part of the bodies of the first three vertebræ commences; and, at a somewhat later period, that of the last two. Between the sixth and eighth months, ossification of the laminæ takes place; and, at about the same period, the characteristic osseous tubercles for the three first sacral vertebræ make their appearance. The laminæ join to form the arch, and are united to the bodies, first, in the lowest vertebræ. This occurs about the second year, the uppermost segment appearing as a single piece about the fifth or sixth year. About the sixteenth year, the epiphyses for the upper and under surfaces of the bodies are formed; and, between the eighteenth and twentieth years, those for each lateral surface of the sacrum make their appearance. At about this period, the last two segments are joined to one another; and this process gradually extending upwards, all the pieces become united, and the bone completely formed from the twenty-fifth to the thirtieth year of life.

*Articulations.* With four bones: the last lumbar vertebra, coccyx, and the two ossa innominata.

*Attachment of Muscles.* The Pyriformis and Coccygeus on either side; behind, the Gluteus maximus, and Erector spinæ.

## The Coccyx.

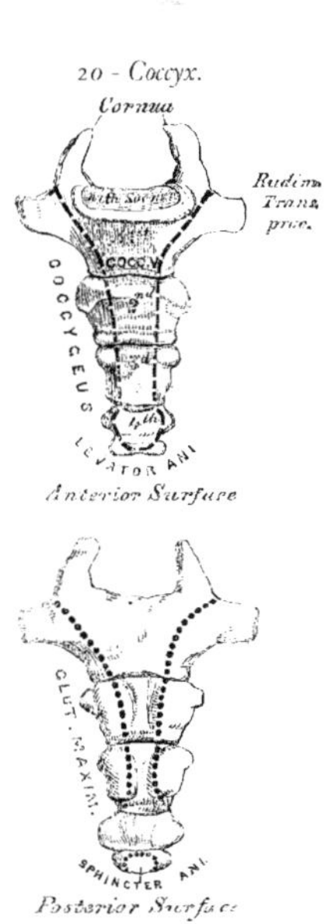

20 - *Coccyx.*

The Coccyx (κόκκυξ, *cuckoo*), so called from resembling a cuckoo's beak (fig. 20), is usually formed of four small segments of bone, the most rudimentary parts of the vertebral column. In each of the first three segments may be traced a rudimentary body, articular and transverse processes; the last piece (sometimes the third) being merely a rudimentary nodule of bone, without distinct processes. All the segments are destitute of laminæ and spinous processes; and, consequently, of spinal canal, and intervertebral foramina. The first segment is the largest, resembles the lowermost sacral vertebra, and often exists as a separate piece; the last three, diminishing in size from above downwards, are usually blended together so as to form a single bone. The gradual diminution in the size of the pieces gives this bone a triangular form, articulating by its base with the end of the sacrum. It presents for examination an anterior and posterior surface, two borders, a base, and an apex. The *anterior surface* is slightly concave, and marked with three transverse grooves, indicating the points of junction of the different pieces. It has attached to it the anterior sacro-coccygeal ligament, the Levator ani muscle, and supports the lower end of the rectum. The *posterior surface* is convex, marked by transverse grooves similar to those on the anterior surface; and presents on each side a linear row of tubercles, the rudimentary articular processes of the coccygeal vertebræ. Of these, the superior pair are very large; and are called the *cornua of the coccyx*; they project upwards, and articulate with the cornua of the sacrum, the junction between these two bones completing the fifth sacral foramen for the transmission of the posterior branch of the fifth sacral nerve. The *lateral borders* are thin, and present a series of small eminences, which represent the transverse processes of the coccygeal vertebræ. Of these, the first on each side is of large size, flattened from before backwards; and often ascends to join the lower part of the thin lateral edge of the sacrum, thus completing the fifth sacral foramen: the others diminish in size from above downwards, and are often wanting. The borders of the coccyx are narrow, and give attachment on each side to the sacro-sciatic ligaments and Coccygeus muscle. The *base* presents an oval surface for articulation with the sacrum. The *apex* is rounded, and has attached to it the tendon of the external Sphincter muscle. It is occasionally bifid, and sometimes deflected to one or other side.

*Development.* The coccyx is developed by *four* centres, one for each piece. Occasionally, one of the first three pieces of this bone is developed by two centres, placed side by side. The ossific nuclei make their appearance in the following order: in the first segment, at birth; in the second piece, at from five to ten years; in the third, from ten to fifteen years; in the fourth, from fifteen to twenty years. As age advances, these various segments become united in the following order: the first two pieces join; then the third and fourth; and, lastly, the bone is completed by the union of the second and third. At a late period of life, especially in females, the coccyx becomes joined to the end of the sacrum.

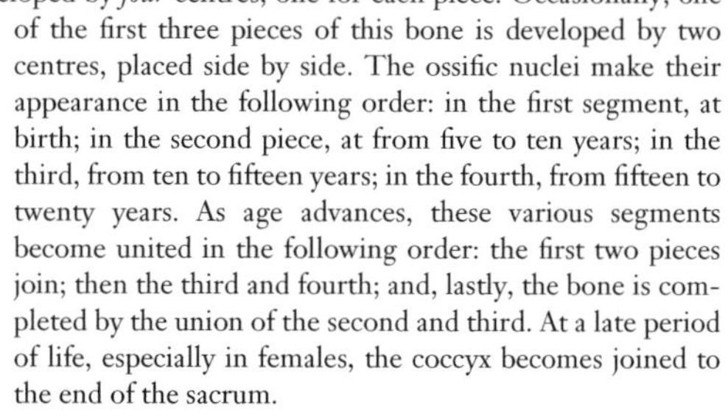

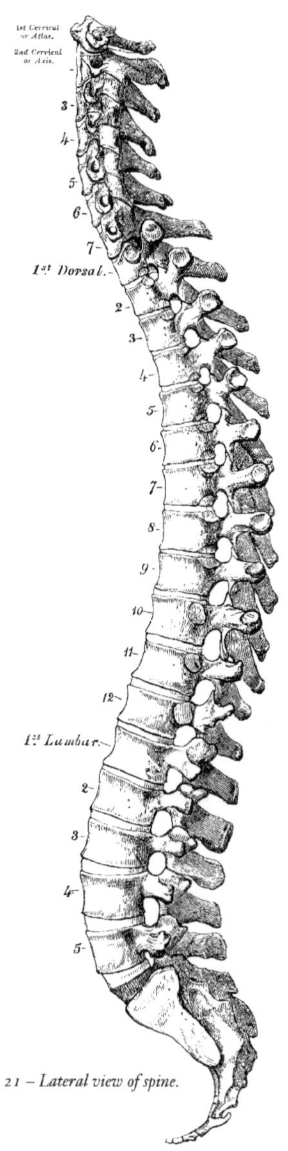


21 – *Lateral view of spine.*

*Articulation.* With the sacrum.

*Attachment of Muscles.* On either side, the Coccygeus; behind, the Gluteus maximus; at its apex, the Sphincter ani; and in front, the Levator ani.

Of the Spine in general.—The spinal column, formed by the junction of the vertebræ, is situated in the median line, at the posterior part of the trunk: its average length is about two feet two or three inches; the lumbar region contributing seven parts, the dorsal eleven, and the cervical five.

Viewed in front, it presents two pyramids joined together at their bases, the upper one being formed by all the vertebræ from the second cervical to the last lumbar; the lower one by the sacrum, and coccyx. Viewed somewhat more closely, the uppermost pyramid is seen to be formed of three smaller pyramids. Of these, the most superior one consists of the six lower cervical vertebræ; its apex being formed by the axis or second cervical; its base, by the first dorsal. The second pyramid, which is inverted, is formed by the four upper dorsal vertebræ, the base being at the first dorsal, the smaller end at the fourth. The third pyramid commences at the fourth dorsal, and gradually increases in size to the fifth lumbar.

Viewed laterally (fig. 21), the spinal column presents several curves, which correspond to the different regions of the column, and are called *cervical, dorsal, lumbar*, and *pelvic*. The *cervical* curve commences at the apex of the odontoid process, and terminates at the middle of the second dorsal vertebra; it is convex in front, but the least marked of all the curves. The *dorsal* curvature, which is concave forwards, commences at the middle of the second, and terminates at the middle of the twelfth dorsal. Its most prominent point behind corresponds to the body of the seventh or eighth vertebra. The *lumbar* curve commences at the middle of the last dorsal, and terminates at the sacro-vertebral angle. It is convex anteriorly; the convexity of the lower three vertebræ being much greater than that of the upper ones. The *pelvic* curve commences at the sacro-vertebral articulation, and terminates at the point of the coccyx. It is concave posteriorly. These curves are partly due to the shape of the bodies of the vertebræ, and partly to the intervertebral substances, as will be explained in the *Articulations of the Spine*.

The spine has also a slight lateral curvature, the convexity of which is directed toward the right side. This is most probably produced, as Bichat first explained, from the effect of muscular action; most persons using the right arm in preference to the left, especially in making long-continued efforts, when the body is curved to the right side. In support of this explanation, it has been found, by Beclard, that in one or two individuals who were left-handed, the lateral curvature was directed to the left side.

The spinal column presents for examination an anterior, a posterior, and two lateral surfaces; a base, summit, and vertebral canal.

The *anterior surface* presents the bodies of the vertebræ separated in the recent state by the intervertebral discs. The bodies are broad in the cervical region, narrow in the upper part of the dorsal, and broadest in the lumbar region. The whole of this surface is convex transversely, concave from above downwards in the dorsal region, and convex in the same direction in the cervical and lumbar regions.

The *posterior surface* presents in the median line the spinous processes. These are short, horizontal, with bifid extremities in the cervical region. In the dorsal region, they are directed obliquely above, assume almost a vertical direction in the middle, and are horizontal, like the spines of the lumbar vertebræ, below. They are separated by considerable intervals in the loins, by narrower intervals in the neck, and are closely approximated in the middle of the dorsal region. Occasionally one of these processes deviates a little from the median line, a fact to be remembered, as irregularities of this sort are attendant on fractures or displacements of the spine. On either side of the spinous processes, extending the whole length of the column, is the vertebral groove, formed by the laminæ in the cervical and lumbar regions, where it is shallow, and by the laminæ and transverse processes in the dorsal region, where it is deep and broad. In the recent state, these grooves lodge the deep muscles of the back. External to the vertebral grooves are the articular processes, and still more externally the transverse processes. In the dorsal region, the latter processes stand backwards, on a place considerably posterior to the same processes in the cervical and lumbar regions. In the cervical region, the transverse processes are placed in front of the articular processes, and between the intervertebral foramina. In the lumbar, they are placed also in front of the articular process, but behind the intervertebral foramina. In the dorsal region, they are posterior both to the articular processes and foramina.

The *lateral surfaces* are separated from the posterior by the articular processes in the cervical and lumbar regions, and by the transverse processes in the dorsal. These surfaces present in front the sides of the bodies of the vertebræ, marked in the dorsal region by the facets for articulation with the heads of the ribs. More posteriorly are the intervertebral foramina, formed by the juxtaposition of the intervertebral notches, oval in shape, smallest in the cervical and upper part of the dorsal regions, and gradually increasing in size to the last lumbar. They are situated between the transverse processes in the neck, and in front of them in the back and loins, and transmit the spinal nerves. The *base* of the vertebral column is formed by the under surface of the body of the fifth lumbar vertebra; and the *summit* by the upper surface of the atlas. The *vertebral canal* follows the different curves of the spine; it is largest in those regions in which the spine enjoys the greatest freedom of movement, as in the neck and loins, where it is wide and triangular; and narrow and rounded in the back, where motion is more limited.

## THE SKULL.

The Skull, or superior expansion of the vertebral column, is composed of four vertebræ, the elementary parts of which are specially modified in form and size, and almost immoveably connected, for the reception of the brain, and special organs of the senses. These vertebræ are the occipital, parietal, frontal, and nasal. Descriptive anatomists, however, divide the skull into two parts, the Cranium and the Face. The Cranium (*κράνος, a helmet*), is composed of *eight* bones: viz., the *occipital, two parietal, frontal, two temporal, sphenoid, and ethmoid.* The face is composed of *fourteen* bones; viz., the *two nasal, two superior maxillary, two lachrymal, two malar, two palate, two inferior turbinated, vomer, inferior maxillary.* The *ossicula auditûs*, the *teeth*, and *Wormian bones*, are not included in this enumeration.

*Skull*, 22 *bones*.
- *Cranium*, 8 *bones*.
  - Occipital.
  - Two Parietal.
  - Frontal.
  - Two Temporal.
  - Sphenoid.
  - Ethmoid.
- *Face*, 14 *bones*.
  - Two Nasal.
  - Two Superior Maxillary.
  - Two Lachrymal.
  - Two Malar.
  - Two Palate.
  - Two Inferior Turbinated.
  - Vomer.
  - Inferior Maxillary.

## The Occipital Bone.

The *Occipital Bone* (fig. 22) is situated at the back part and base of the cranium, is trapezoid in form, curved upon itself, and presents for examination two surfaces, four borders, and four angles.

The *External Surface* is convex. Midway between the summit of the bone and the posterior margin of the foramen magnum is a prominent tubercle, the external occipital protuberance, for the attachment of the Ligamentum nuchæ; and descending from it, as far as the foramen, a vertical ridge, the external occipital crest. This tubercle and crest, vary in prominence in different skulls. Passing outwards from the occipital protuberance on each side are two semicircular ridges, the superior curved lines; and running parallel with these from the middle of the crest, are the two inferior curved lines. The surface of the bone above the superior curved lines is smooth on each side, and in the recent state, is covered by the Occipito-frontalis muscle, whilst the ridges, as well as the surface of the bone between them, serve for the attachment of numerous muscles. The superior curved line gives attachment internally to the Trapezius, externally to the Occipito-frontalis, and Sterno-cleido mastoid; to the extent shewn in the figure. The depressions between the curved lines to the Complexus internally, the Splenius capitis and Obliquus superior externally. The inferior curved line, and the depressions below it, afford insertion to the Rectus capitis posticus, major and minor.

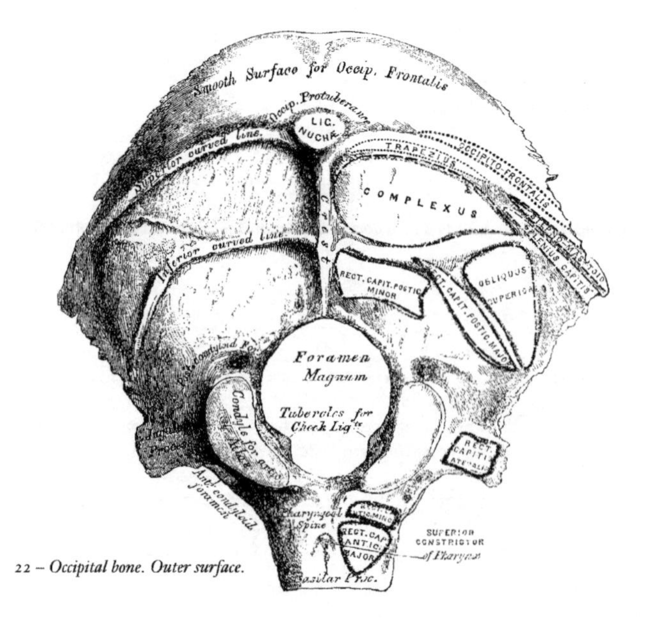


22 – *Occipital bone. Outer surface.*

The *foramen magnum* is a large oval aperture, its long diameter extending from before backwards. It transmits the spinal cord and its membranes, the spinal accessory nerves, and the vertebral arteries. Its back part is wide for the transmission of the cord, and the corresponding margin rough for the attachment of the dura mater enclosing it; the fore part is narrower, being encroached upon by the condyles; it has projecting towards it from below the odontoid process, and its margins are smooth and bevelled internally to support the medulla oblongata. On each side of the foramen magnum are the condyles, for articulation with the atlas; they are convex, oblong or reniform in shape, and directed downwards and outwards; they converge in front, and encroach slightly upon the anterior segment of the foramen. On the inner border of each condyle is a rough tubercle for the attachment of the ligaments (check) which connect this bone with the odontoid process of the axis; whilst external to them is a rough tubercular prominence, the transverse or jugular process, (the representative of the transverse process of a vertebra) channelled in front by a deep notch, which forms part of the jugular foramen. The under surface of this process affords attachment to the Rectus capitis lateralis; its upper or cerebral surface presents a deep groove, which lodges part of the lateral sinus, whilst its prominent extremity is marked by a quadrilateral rough surface, covered with cartilage in the fresh state, and articulating with a similar surface on the petrous portion of the temporal bone. On the outer side of each condyle, near its fore part, is a foramen, the anterior condyloid; it is directed downwards, outwards, and forwards, and transmits the hypoglossal nerve. This foramen is sometimes double. Behind each condyle is a fossa,* perforated at the bottom by a foramen, the posterior condyloid, for the transmission of a vein to the lateral sinus. In front of the foramen magnum is a strong quadrilateral plate of bone, the basilar process, wider behind than in front; its under surface, which is rough, presenting in the median line a tubercular ridge, the pharyngeal spine, for the attachment of the tendinous raphe and Superior constrictor of the pharynx; and on each side of it, rough depressions for the attachment of the Recti capitis antici, major and minor.

The *Internal or Cerebral Surface* (fig. 23) is deeply concave. The posterior or occipital part is divided by a crucial ridge into four fossæ. The two superior, the smaller, receive the posterior lobes

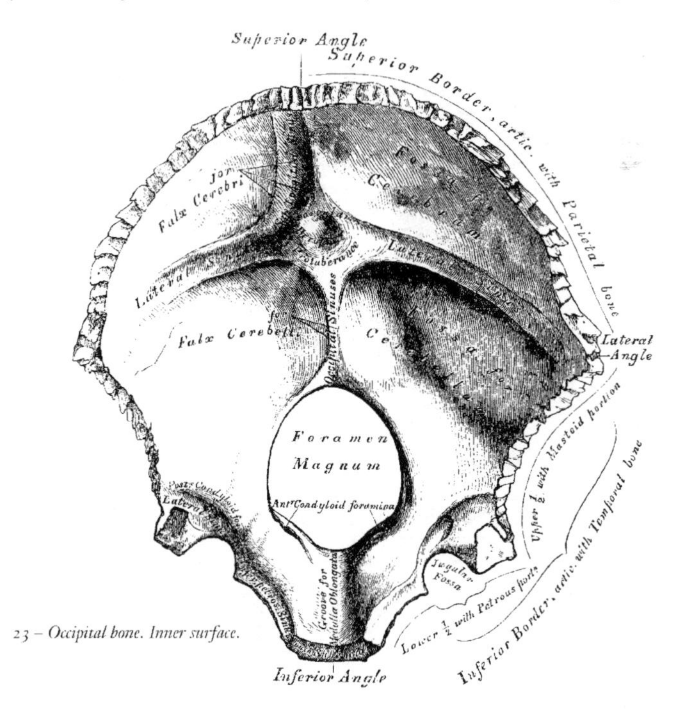


23 – *Occipital bone. Inner surface.*

* This fossa presents many variations in size. It is usually shallow; and the foramen small; occasionally wanting, on one, or both sides. Sometimes both fossa and foramen are large, but confined to one side only; more rarely, the fossa and foramen are very large on both sides.

of the cerebrum, and present slight eminences and depressions corresponding to their convolutions. The two inferior, which receive the lateral lobes of the cerebellum, are larger than the former, and comparatively smooth; both are marked by slight grooves for the lodgment of arteries. At the point of meeting of the four divisions of the crucial ridge is an eminence, the internal occipital protuberance. It nearly corresponds to that on the outer surface, and is perforated by one or more large vascular foramina. From this eminence, the superior division of the crucial ridge, runs upwards to the superior angle of the bone; it presents occasionally a deep groove for the superior longitudinal sinus, the margins of which give attachment to the falx cerebri. The inferior division, the internal occipital crest, runs to the posterior margin of the foramen magnum, on the edge of which it becomes gradually lost: this ridge, which is bifurcated below, serves for the attachment of the falx cerebelli. The transverse grooves pass outwards to the lateral angles; they are deeply channelled, for the lodgment of the lateral sinuses, their prominent margins affording attachment to the tentorium cerebelli.† At the point of meeting of these grooves is a depression, the 'Torcular Herophili',‡ placed a little to one or the other side of the internal occipital protuberance. More anteriorly is the foramen magnum, and on each side of it, but nearer its anterior than its posterior part, the internal openings of the anterior condyloid foramina; the internal openings of the posterior condyloid foramina being a little external and posterior to them, protected by a small arch of bone. In front of the foramen magnum is the basilar process, presenting a shallow depression, the basilar groove, which slopes from behind, upwards and forwards, and supports the medulla oblongata; and on each side of the basilar process is a narrow channel, which, when united with a similar channel on the petrous portion of the temporal bone, forms a groove, which lodges the inferior petrosal sinus.

*Angles.* The *superior* angle is received into the interval between the posterior superior angles of the two parietal bones: it corresponds with that part of the skull in the fœtus which is called the *posterior fontanelle*. The *inferior* angle is represented by the square-shaped surface of the basilar process. At an early period of life, a layer of cartilage separates this part of the bone from the sphenoid; but in the adult, the union between them is osseous. The *lateral angles* correspond to the outer ends of the transverse grooves, and are received into the interval between the posterior inferior angles of the parietal and the mastoid portion of the temporal.

*Borders.* The *superior* extends on each side from the superior to the lateral angle, is deeply serarated for articulation with the parietal bone, and forms by this union the lambdoid suture. The *inferior* border extends from the lateral to the inferior angle; its upper half is rough, and articulates with the mastoid portion of the temporal, forming the masto-occipital suture: the inferior half articulates with the petrous portion of the temporal, forming the petro-occipital suture: these two portions are separated from one another by the jugular process. In front of this process is a deep notch, which, with a similar one on the petrous portion of the temporal, forms the foramen lacerum posterius. This notch is occasionally subdivided into two parts by a small process of bone, and presents an aperture at its upper part, the internal opening of the posterior condyloid foramen.

*Structure.* The occipital bone consists of two compact laminæ, called the *outer* and *inner tables*, having between them the diploic tissue; this bone is especially thick, at the ridges, protuberances, condyles, and anterior part of the basilar process; whilst at the bottom of the fossæ, especially the inferior, it is thin, semitransparent, and destitute of diploë.

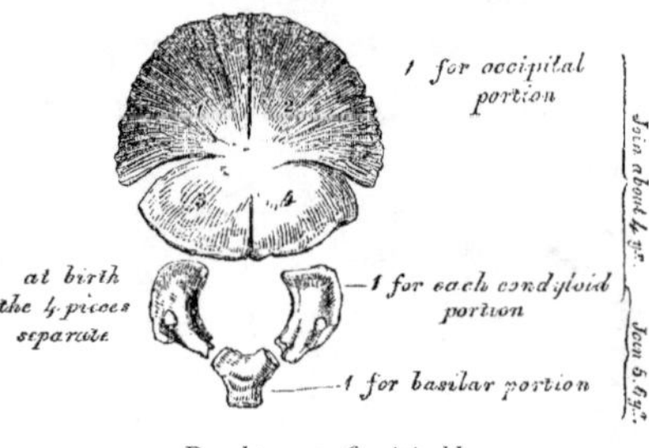

24 – *Development of occipital bone.*

*Development.* (fig. 24). The occipital bone has *four* centres of development; one for the posterior or occipital part, which is formed in membrane, one for the basilar portion; and one for each condyloid portion, which are formed in cartilage.

The centre for the occipital portion appears about the tenth week of fœtal life; and consists, according to Blandin and Cruvelhier,

† Usually one of the transverse grooves is deeper and broader than the other; this seems in nearly equal proportion on the two sides, occasionally both grooves are of equal depth and breadth, or both equally indistinct. The broader of the two transverse grooves is nearly always continuous with the vertical groove for the superior longitudinal sinus, and occupies the corresponding side of the median line.

‡ The columns of blood coming in different directions were supposed to be *pressed* together at this point.

of a small oblong plate which appears in the situation of the occipital protuberance.* The condyloid portions then ossify, and lastly the basilar portion. At birth, the bone consists of four parts, separate from one another, the occipital portion being fissured in the direction above indicated. At about the fourth year, the occipital and the two condyloid pieces join; and about the sixth year the bone consists of a single piece. At a later period, between the eighteenth and twenty-fifth years, the occipital and sphenoid become united, forming a single bone.

*Articulations.* With six bones; two parietal, two temporal, sphenoid, and atlas.

*Attachment of Muscles.* To the superior curved line are attached the Occipito-frontalis, Trapezius, and Sterno-cleido-mastoid. To the space between the curved lines, the Complexus, Splenius capitis, and Obliquus superior; to the inferior curved line, and the space between it and the foramen magnum, the Rectus posticus major and minor; to the transverse process, the Rectus lateralis; and to the basilar process, the Recti antici majores and minores, and Superior Constrictor of the pharynx.

## The Parietal Bones.

The *Parietal Bones* (*paries*, a wall), form by their union the sides and roof of the skull; each bone is of an irregular quadrilateral form, and presents for examination two surfaces, four borders, and four angles.

*Surfaces.* The *external surface* (fig. 25) is convex, smooth, and marked about its centre by an eminence, called the parietal eminence, which indicates the point where ossification commenced. Crossing the centre of the bone in an arched direction is a curved ridge, the temporal ridge, for the attachment of the temporal fascia. Above this ridge, the surface of the bone is rough and porous, and covered by the aponeurosis of the Occipito-frontalis; below it the bone is smooth, forms part of the temporal fossa, and affords attachment to the Temporal muscle. At the back part of the superior border, close to the sagittal suture, is a small foramen, the parietal foramen, which transmits a vein to the superior longitudinal sinus. Its existence is not constant, and its size varies considerably.

The *internal surface* (fig. 26), concave, presents eminences and depressions for lodging the

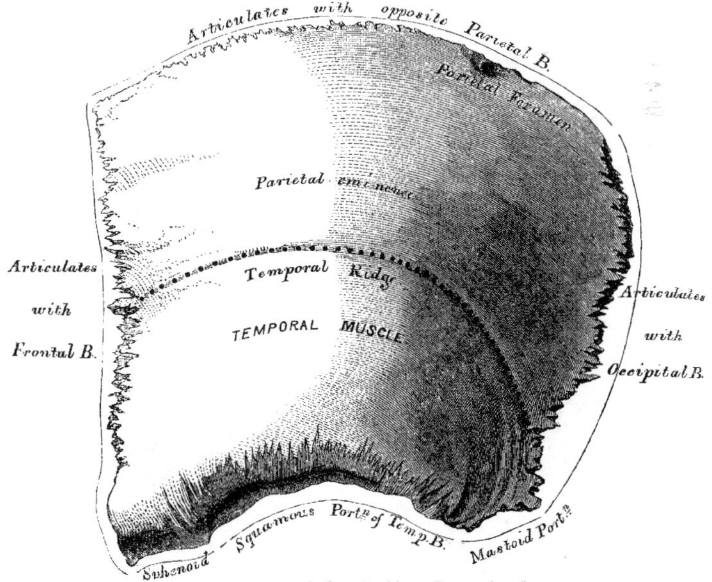


25 – *Left parietal bone. External surface.*

* Beclard considers this segment to have four centres of ossification, arranged in pairs, two above, and two below the curved lines, and Meckel describes eight, four of which correspond in situation with those above described, of the other four, two are placed in juxta-position, at the upper angle of the bone, and the remaining two, one at each side, in the lateral angles.

convolutions of the cerebrum, and numerous furrows for the ramifications of the meningeal arteries; the latter run upwards and backwards from the anterior inferior angle, and from the central and posterior part of the lower border of the bone. Along the upper margin is part of a shallow groove, which, when joined to the opposite parietal, forms a channel for the superior longitudinal sinus, the elevated edges of which afford attachment to the falx cerebri. Near the groove are seen several depressions; they lodge the Pacchionian bodies. The internal opening of the parietal foramen is also seen when that aperture exists.

*Borders.* The *superior*, the longest and thickest, is dentated to articulate with its fellow of the opposite side, forming the sagittal suture. The *inferior* is divided into three parts; of these, the anterior is thin and pointed, bevelled at the expense of the outer surface, and overlapped by the tip of the great wing of the sphenoid; the middle portion is arched, bevelled at the expense of the outer surface, and overlapped by the squamous portion of the temporal; the posterior portion being thick and serrated for articulation with the mastoid portion of the temporal. The *anterior border*, deeply serrated, is bevelled at the expense of the outer surface above, and of the inner below; it articulates with the frontal bone, forming the coronal suture. The *posterior* border, deeply denticulated, articulates with the occipital, forming the lambdoid suture.

*Angles.* The *anterior superior*, thin and pointed, corresponds with that portion of the skull which in the fœtus is membranous, and is called the *anterior fontanelle*. The *anterior inferior angle* is thin and lengthened, being received in the interval between the great wing of the sphenoid and the frontal. Its inner surface is marked by a deep groove, sometimes a canal, for the anterior branch of the middle meningeal artery. The *posterior superior angle* corresponds with the junction of the sagittal and lambdoid sutures. In the fœtus this part of the skull is membranous, and is called the *posterior fontanelle*. The *posterior inferior angle* articulates with the mastoid portion of the temporal bone, and generally presents on its inner surface a broad shallow groove for lodging part of the lateral sinus.

*Development.* The parietal bone is formed in membrane, being developed by *one* centre, which corresponds with the parietal eminence, and makes its first appearance about the fifth or sixth week of fœtal life. Ossification gradually extends from the centre to the circumference of the bone, the angles are consequently the parts last formed, and it is in their situation, that the fontanelles exist, previous to the completion of the growth of the bone.

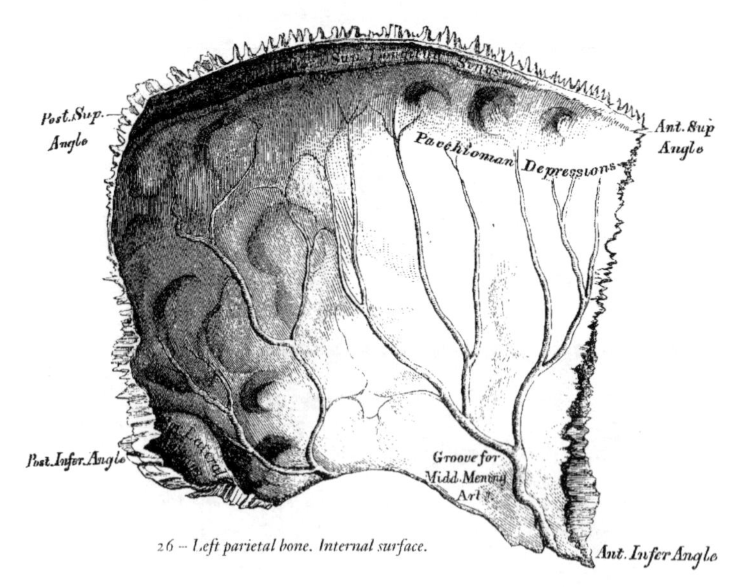


26 – *Left parietal bone. Internal surface.*

*Articulations.* With five bones; the opposite parietal, the occipital, frontal, temporal, and sphenoid.

*Attachment of Muscles.* To one only, the Temporal.

## The Frontal Bone.

This bone, which resembles a cockle-shell in form, consists of two portions—a *vertical* or *frontal* portion, situated at the anterior part of the cranium, forming the forehead; and a *horizontal* or *orbito-nasal* portion, which enters into the formation of the roof of the orbits and nose.

*Vertical Portion. External Surface* (fig. 27). In the median line, traversing the bone from the upper to its lower part, is occasionally seen a slightly elevated ridge, and in young subjects a suture, which represents the point of union of the two lateral halves of which the bone consists at an early period of life: in the adult, this suture usually disappears, excepting below. On either side of this ridge, a little below the centre of the bone, is a rounded eminence, the frontal eminence. These eminences vary in size in different individuals, and are occasionally unsymmetrical in the same

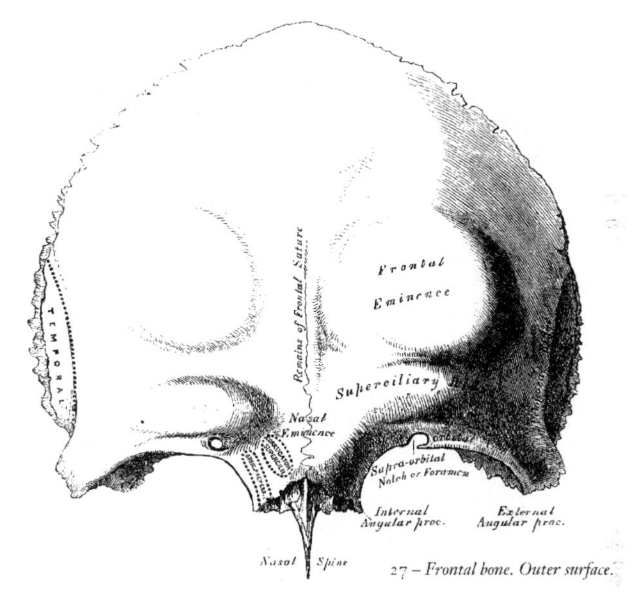


27 – *Frontal bone. Outer surface.*

subject. They are especially prominent in cases of well marked cerebral development. The whole surface of the bone above this part is smooth, and covered by the aponeurosis of the Occipito-frontalis muscle. Below the frontal eminence, and separated from it by a slight groove, is the superciliary ridge, broad internally where it is continuous with the nasal eminence, but less distinct as it arches outwards. These ridges are caused by the projection outwards of the frontal sinuses. Beneath the superciliary ridge is the supra-orbital arch, a curved and prominent margin, which forms the upper boundary of the orbit, and separates the vertical from the horizontal portion of the bone. The outer part of the arch is sharp and prominent, affording to the eye, in that situation, considerable protection from injury; the inner part is less prominent. At the inner third of this arch is a notch, sometimes converted into a foramen by a bony process or ligament, and called the *supra-orbital notch* or *foramen*. It transmits the supra-orbital artery, veins, and nerve. A small aperture is seen in the upper part of the notch, which transmits a vein from the diploë to join the ophthalmic vein. The supra-orbital arch terminates externally in the external angular process, and internally in the internal angular process. The external angular process is strong, prominent, and articulates with the malar bone: running upwards and backwards from it is a sharp curved crest, the temporal ridge, for the attachment of the temporal fascia; and beneath it a slight concavity, that forms the anterior part of the temporal fossa, and gives origin to the Temporal muscle. The internal angular processes are less marked than the external, and articulate with the lachrymal bones. Between the two is a rough, uneven interval, the *nasal notch*, which articulates in the middle line with the nasal, and on either side with the nasal process of the superior maxillary bone. The notch is continuous below, with a long pointed process, the *nasal spine*.

*Vertical Portion. Internal Surface* (fig. 28). Along the middle line is a vertical groove, the edges of which unite below to form a ridge, the frontal crest; the groove lodges the superior

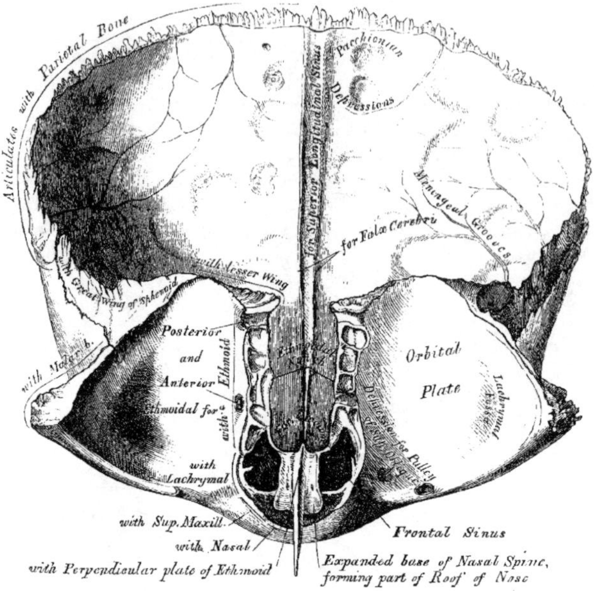

28 – *Frontal bone. Inner surface.*

longitudinal sinus, whilst its edges afford attachment to the falx cerebri. The crest terminates below at a small opening, the foramen cœcum, which is generally completed behind by the ethmoid. This foramen varies in size in different subjects, is usually partially, or completely impervious, lodges a process of the falx cerebri, and, when open, transmits a vein from the lining membrane of the nose to the superior longitudinal sinus. On either side of the groove, the bone is deeply concave, presenting eminences and depressions for the convolutions of the brain, and numerous small furrows for lodging the ramifications of the anterior meningeal arteries. Several small, irregular fossæ are also seen on either side of the groove, for the reception of the Pacchionian bodies.

*Horizontal Portion. External Surface.* This portion of the bone consists of two thin plates, which form the vault of the orbits, separated from one another by the ethmoidal notch. Each orbital vault consists of a smooth, concave, triangular plate of bone, marked at its anterior and external part (immediately beneath the external angular process) by a shallow depression, the lachrymal fossa, for lodging the lachrymal gland; and at its anterior and internal part, by a depression (sometimes a small tubercle), for the attachment of the fibrous pulley of the Superior oblique muscle. The ethmoidal notch separates the two orbital plates: it is quadrilateral; and filled up, when the bones are united, by the cribriform plate of the ethmoid. The margins of this notch present several half-cells, which, when united with corresponding half-cells on the upper surface of the ethmoid, complete the ethmoidal cells: two grooves are also seen crossing these edges transversely; they are converted into canals by articulation with the ethmoid, and are called the *anterior* and *posterior* ethmoidal canals; they open on the inner wall of the orbit. The anterior one transmits the nasal nerve and anterior ethmoidal vessels, the posterior one, the posterior ethmoidal vessels. In front of the ethmoidal notch is the nasal spine, a sharp-pointed eminence which projects downwards and forwards, and articulates in front with the crest of the nasal bones; behind it is marked by two grooves, separated by a vertical ridge: the ridge articulates with the perpendicular lamella of the ethmoid, the grooves form part of the roof of the nasal fossæ. On either side of the base of the nasal spine are the openings of the frontal sinuses. These are two irregular cavities, which extend upwards and outwards, a variable distance, between the two tables of the skull, and are separated from one another by a thin bony septum. They give rise to the prominences above the root of the nose, called the *nasal eminences* and *superciliary ridges*. In the child they are generally absent, and they become gradually developed as age advances. These cavities vary in size in different persons, are larger in men than in women, and are frequently of unequal size on the two sides, the left being commonly the larger. Occasionally they are subdivided by incomplete bony laminæ. They are lined by mucous membrane; communicate with the nose by the infundibulum, and occasionally with each other by apertures in the septum.

The *Internal Surface* of the *Horizontal Portion* presents the convex upper surfaces of the orbital plates, separated from each other in the middle line by the ethmoidal notch, and marked by eminences and depressions for the convolutions of the anterior lobes of the brain.

*Borders.* The border of the vertical portion is thick, strongly serrated, bevelled at the expense of the internal table above, where it rests upon the parietal, at the expense of the external table at each side, where it receives the lateral pressure of those bones: this border is continued below, into a triangular rough surface, which articulates with the great wing of the sphenoid. The border of the horizontal portion is thin, serrated, and articulates with the lesser wing of the sphenoid.

*Structure.* The vertical portion, and external angular processes, are very thick, consisting of diploic tissue contained between two compact laminæ. The horizontal portion is thin, translucent, and composed entirely of compact tissue; hence the facility with which instruments can penetrate the cranium through this part of the orbit.

*Development* (fig. 29). The frontal bone is formed in membrane, being developed by *two* centres, one for each lateral half, which make their appearance, at an early period of fœtal life, in the situation of the orbital arches. From this point, ossification extends, in a radiating manner, upwards into the forehead, and backwards over the orbit. At birth, it consists of two pieces, which afterwards become united along the middle line, by a suture which runs from the vertex to the root of the nose. This suture becomes obliterated within a few years after birth; but it occasionally remains throughout life.

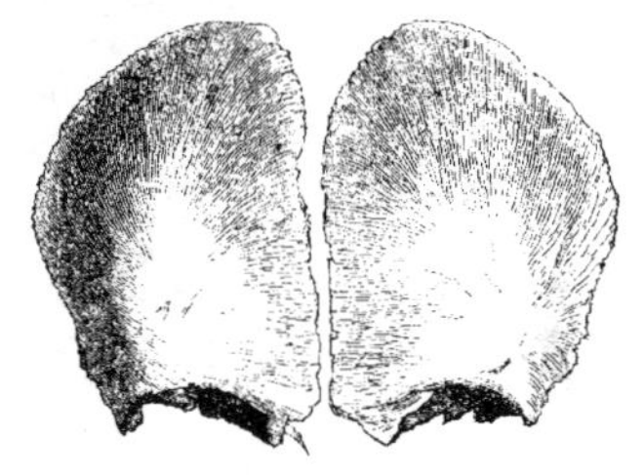

*29 – Frontal bone at birth. Developed by two lateral halves.*

*Articulations.* With twelve bones: two parietal, sphenoid, ethmoid; two nasal, two superior maxillary, two lachrymal, and two malar.

*Attachment of Muscles.* To three pairs: the Corrugator supercilii, Orbicularis palpebrarum, and Temporal.

## The Temporal Bones.

The Temporal Bones are so called because they occupy that part of the head on which the hair first begins to turn gray, thus indicating the age. They are situated at the side and base of the skull, and present for examination a *squamous*, *mastoid*, and *petrous* portion.

The *Squamous Portion* (*squama*, a scale), (fig. 30), the most anterior and superior part of the bone, is scale-like in form, thin and translucent in texture. Its outer surface is smooth, convex, and grooved at its back part for the deep temporal arteries; it affords attachment to the Temporal muscle, and forms part of the temporal fossa. At its back part may be seen a curved ridge—part of the temporal ridge; it serves for the attachment of the temporal fascia, limits the origin of the Temporal muscle, and marks the boundary between the squamous and mastoid portions of the bone. Projecting from the lower part of the squamous portion, is a long arched outgrowth of bone, the zygomatic process. It is at first directed outwards, its two surfaces looking upwards and downwards; it then appears as if twisted upon itself, and takes a direction forwards, its surfaces now looking inwards and outwards. The superior border of this process is long, thin, and sharp, and serves for the attachment of the temporal fascia. The inferior, short, thick, and arched, has attached to it some fibres of the Masseter muscle. Its outer surface is convex and subcutaneous. Its inner, concave; also affords attachment to the Masseter. The extremity, broad, and deeply serrated, articulates with the malar bone. This process is connected to the temporal bone by three divisions, called *the roots of the zygomatic process*, an anterior, middle, and posterior. The anterior, which is short, but broad and strong, runs transversely inwards into a rounded eminence, the eminentia articularis. This eminence forms the front boundary of the glenoid fossa, and in the recent state is covered with cartilage. The middle root forms the outer margin of the glenoid cavity; running obliquely inwards, it terminates at the commencement of a well-marked fissure, the Glaserian fissure; whilst the posterior root, which is strongly marked, runs from the upper border of the zygoma, in an arched direction, upwards and backwards, forming the posterior part of the temporal ridge. At the junction of the anterior root with the zygoma, is a projection, called the *tubercle*, for the attachment of the external lateral ligament of the lower jaw; and between the anterior and middle roots is an oval depression,

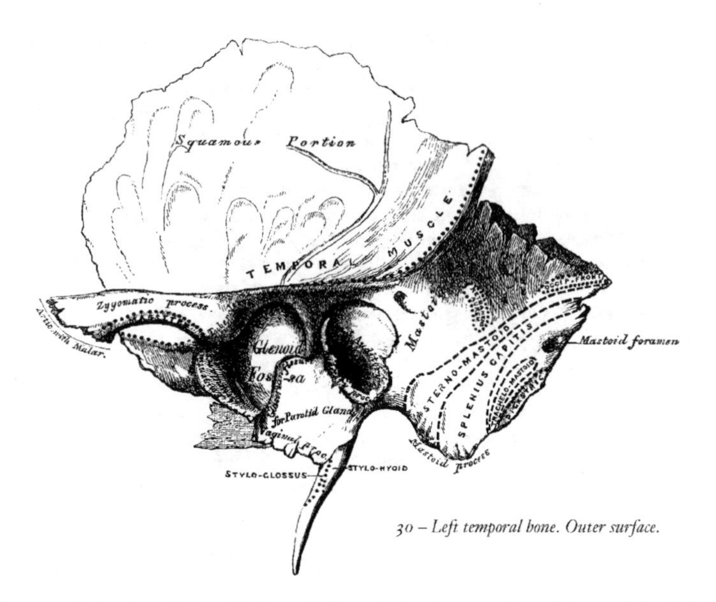


30 – *Left temporal bone. Outer surface.*

forming part of the glenoid fossa (*γλήνη*, *a shallow pit*), for the reception of the condyle of the lower jaw. This fossa is bounded, in front, by the eminentia articularis; behind, by the vaginal process; and, externally, by the auditory process, and middle root of the zygoma; and is divided into two parts by a narrow slit, the Glaserian fissure: the anterior part, formed by the squamous portion of the bone, is smooth, covered in the recent state with cartilage, and articulates with the condyle of the lower jaw. This part of the glenoid fossa is separated from the auditory process, by a small tubercle, the *post glenoid process*, the representative of a prominent tubercle which, in some of the mammalia, descends behind the condyle of the jaw, and prevents it being displaced backwards during mastication (Humphry). The posterior part of the glenoid fossa is formed chiefly by the vaginal process of the petrous portion, and lodges part of the parotid gland. The Glaserian fissure, which leads into the tympanum, lodges the processus gracilis of the malleus, and transmits the Laxator tympani muscle and the anterior tympanic artery. The chorda tympani nerve passes through a separate canal parallel to the Glaserian fissure (canal of Huguier), on the outer side of the Eustachian tube, in the retiring angle between the squamous and petrous portions of the temporal bone.

The *internal surface* of the squamous portion (fig. 31) is concave, presents numerous eminences and depressions for the convolutions of the cerebrum, and two well-marked grooves for branches of the middle meningeal artery.

*Borders.* The superior border is thin, bevelled at the expense of the internal surface, so as to overlap the lower border of the parietal bone, forming the squamous suture. The anterior inferior border is thick, serrated, and bevelled alternately at the expense of the inner and outer surfaces, for articulation with the great wing of the sphenoid.

The *Mastoid Portion* (*μαστος*, *a nipple* or *teat*) is situated at the posterior part of the bone; its outer surface is rough, and perforated by numerous foramina: one of these, of large size, situated at the posterior border of the bone, is termed the *mastoid foramen;* it transmits a vein to the lateral sinus and a small artery. The position and size of this foramen are very variable, being sometimes situated in the occipital bone, or in the suture between the temporal and the occipital. The mastoid portion is continued below into a conical projection, the mastoid process, the size and form of which varies somewhat in different individuals. This process serves for the attachment of the Sterno-mastoid, Splenius capitis, and Trachelo-mastoid muscles. On the inner side of the mastoid process is a deep groove, the digastric fossa, for the attachment of the Digastric muscle; and running parallel with it, but more internal, the occipital groove, which lodges the occipital artery. The internal surface of the mastoid portion presents a deeply curved groove, which lodges part of the lateral sinus; and into it may be seen opening the mastoid foramen. A section of the mastoid process shows it to be hollowed out into a number of cellular spaces, communicating with each other, called the *mastoid cells;* they open by a single or double orifice into the back of the tympanum; are lined by a prolongation of its lining membrane; and, probably, form some secondary part of the organ of hearing. The mastoid

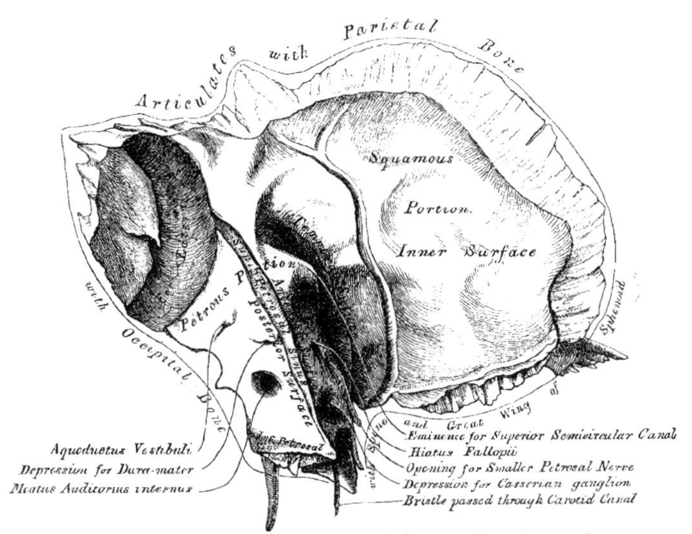


31 – Left temporal bone. Inner surface.

cells, like the other sinuses of the cranium, are not developed until after puberty; hence the prominence of this process in the adult.

*Borders.* The superior border of the mastoid portion is broad and rough, its serrated edge sloping outwards, for articulation with the posterior inferior angle of the parietal bone. The posterior border, also uneven and serrated, articulates with the inferior border of the occipital bone between its lateral angle and jugular process.

The *Petrous Portion* (*πέτρος, a rock*), so named from its extreme density and hardness, is a pyramidal process of bone, wedged in at the base of the skull between the sphenoid and occipital bones. Its direction from without is forwards, inwards, and a little downwards. It presents for examination a base, an apex, three surfaces, and three borders; and contains, in its interior, the essential parts of the organ of hearing. The *base* is applied against the internal surface of the squamous and mastoid portions, its upper half being concealed; but its lower half is exposed by their divergence, which brings into view the oval expanded orifice of a canal leading into the tympanum, the meatus auditorius externus. This canal is situated between the mastoid process and the posterior and middle roots of the zygoma; its upper margin is smooth and rounded, but the greater part of its circumference is surrounded by a curved plate of bone, the auditory process, the free margin of which is thick and rough for the attachment of the cartilage of the external ear.

The *apex* of the petrous portion, rough and uneven, is received into the angular interval between the spinous process of the sphenoid, and the basilar process of the occipital; it presents the anterior orifice of the carotid canal, and forms the posterior and external boundary of the foramen lacerum medium.

The *anterior surface* of the petrous portion (fig. 31), forms the posterior boundary of the middle fossa of the skull. This surface is continuous with the squamous portion, to which it is united by a suture, the temporal suture, the remains of which are distinct at a late period of life. This surface presents six points for examination. 1. An eminence near the centre which indicates the situation of the superior semicircular canal. 2. On the outer side of this eminence is a depression, indicating the position of the tympanum, the layer of bone which separates the tympanum from the cranial cavity being extremely thin. 3. A shallow groove, sometimes double, leading backwards to an oblique opening, the hiatus Fallopii, for the passage of the petrosal branch of the Vidian nerve. 4. A smaller opening, occasionally seen external to the latter for the passage of the smaller petrosal nerve. 5. Near the apex of the bone is seen the termination of the carotid canal, the wall of which in this situation is deficient in front. 6. Above this canal is a shallow depression for the reception of the Casserian ganglion.

The *posterior surface* forms the front boundary of the posterior fossa of the skull, and is continuous with the inner surface of the mastoid portion of the bone. It presents three points for examination. 1. About its centre is a large orifice, the meatus auditorius internus. This aperture varies

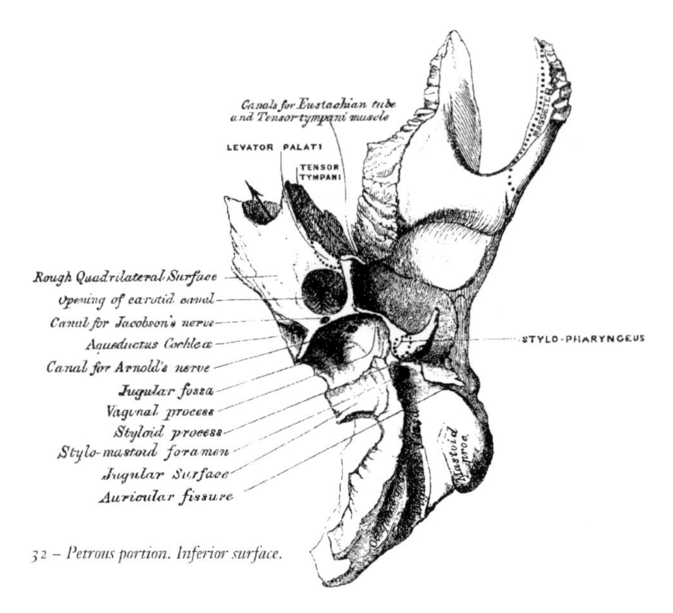


32 – *Petrous portion. Inferior surface.*

considerably in size; its margins are smooth and rounded; and it leads into a short canal, about four lines in length, which runs directly outwards. The end of the canal is closed by a vertical plate, divided by a horizontal crest into two unequal portions. It transmits the auditory and facial nerves, and auditory artery. 2. Behind the meatus auditorius is a small slit, almost hidden by a thin plate of bone, leading to a canal, the aquæductus vestibuli; it transmits a small artery and vein, and lodges a process of the dura mater. 3. In the interval between these two openings, but above them, is an angular depression which lodges a process of the dura mater, and transmits a small vein into the cancellous tissue of the bone.

The *inferior* or *basilar surface* (fig. 32) is rough and irregular, and forms part of the base of the skull. Passing from the apex to the base, this surface presents eleven points for examination. 1. A rough surface, quadrilateral in form, which serves partly for the attachment of the Levator palati, and Tensor tympani muscles. 2. The opening of the carotid canal, a large circular aperture, which ascends at first vertically upwards, and then making a bend, runs horizontally forwards and inwards. It transmits the internal carotid artery, and the carotid plexus. 3. The aquæductus cochleæ, a small triangular opening, lying on the inner side of the latter, close to the posterior border of the petrous portion; it transmits a vein from the cochlea, which joins the internal jugular. 4. Behind these openings is a deep depression, the jugular fossa, which varies in depth and size in different skulls; it lodges the internal jugular vein, and with a similar depression on the margin of the occipital bone, forms the foramen lacerum posterius. 5. A small foramen for the passage of Jacobson's nerve (the tympanic branch of the glosso-pharyngeal). This is seen in front of the bony ridge dividing the carotid canal from the jugular fossa. 6. A small foramen seen on the inner wall of the jugular fossa, for the *entrance* of the auricular branch of the pneumogastric (Arnold's) nerve. 7. Behind the jugular fossa is a smooth square-shaped facet, the jugular surface; it is covered with cartilage in the recent state, and articulates with the jugular process of the occipital bone. 8. The vaginal process, a very broad sheath-like plate of bone, which extends from the carotid canal to the mastoid process; it divides behind into two laminæ, receiving between them the 9th point for examination, the styloid process; a long sharp spine, about an inch in length, continuous with the vaginal process, between the laminæ of which it is received, and directed downwards, forwards, and inwards. It varies in size and shape; and sometimes consists of several pieces united by cartilage. It affords attachment to three muscles, the Stylo-pharyngeus, Stylo-glossus, and Stylo-hyoideus; and two ligaments, the stylo-hyoid, and stylo-maxillary. 10. The stylo-mastoid foramen, a rather large orifice, placed between the styloid and mastoid processes; it is the termination of the aquæductus Fallopii, and transmits the facial nerve, and stylo-mastoid artery. 11. The auricular fissure, situated between the vaginal and mastoid processes, for the *exit* of the auricular branch of the pneumogastric nerve.

*Borders.* The *superior*, the longest, is grooved for the superior petrosal sinus, and has attached to it the tentorium cerebelli: at its inner extremity is a semilunar notch, upon which reclines the fifth nerve. The *posterior* border is intermediate in length between the superior and the anterior. Its inner half is marked by a groove, which, when completed by its articulation with the occipital, forms the channel for the inferior petrosal sinus. Its outer half presents a deep excavation the jugular fossa, which, with a similar notch on the occipital, forms the foramen lacerum posterius. A projecting eminence of bone occasionally stands out from the centre of the notch, and divides the foramen into two parts. The *anterior* border is divided into two parts, an outer, joined to the squamous portion by a suture, the remains of which are distinct; an inner, free, articulating with the spinous process of the sphenoid. At the angle of junction of these two parts, are seen two canals, separated from one another by a thin plate of bone, the processus cochleariformis; they both lead into the tympanum, the upper one transmitting the Tensor tympani muscle, the lower one the Eustachian tube.

*Structure.* The squamous portion is like that of the other cranial bones, the mastoid portion cellular, and the petrous portion dense and hard.

*Development* (fig. 33). The temporal bone is developed by *four* centres, exclusive of those for the internal ear and the ossicula, viz.:—one for the squamous portion including the zygoma, one for the petrous and mastoid parts, one for the styloid, and one for the auditory process (tympanic bone). The first traces of the development of this bone appear in the squamous portion, about the time when osseous matter is deposited in the vertebræ; the auditory process succeeds next; it consists of an elliptical portion of bone, forming about three-fourths of a circle, the deficiency being above; it is grooved along its concave surface for the attachment of the membrana tympani, and becomes united by its extremities to the sqamous portion during the last months of intra-uterine life. The petrous and mastoid portions then become ossified, and lastly the styloid process, which remains separate a considerable period, and is occasionally never united to the rest of the bone. At birth, the temporal bone, excluding the styloid process, is formed of three pieces, the squamous and zygomatic, the petrous and mastoid, and the auditory. The auditory process joins with the squamous at about the ninth month. The petrous and mastoid join with the squamous during the first year, and the styloid process becomes united between the second and third years. The subsequent changes in this bone are the extension outwards of the auditory process, so as to form the meatus auditorius; the glenoid fossa becomes deeper; and the mastoid part, which at an early period of life is quite flat, enlarges from the development of numerous cellular cavities in its interior.

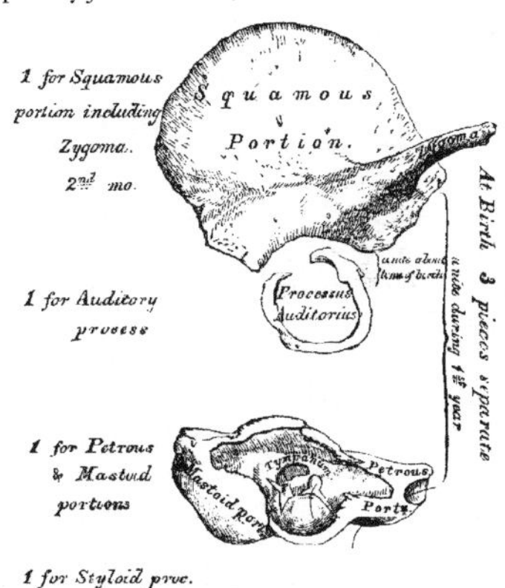


33 – *Development of temporal bone. By four centres.*

*Articulations.* With five bones, occipital, parietal, sphenoid, inferior maxillary and malar.

*Attachment of Muscles.* To the squamous portion, the Temporal; to the zygoma, the Masseter; to the mastoid portion, the Occipito-frontalis, Sterno-mastoid, Splenius capitis, Trachelo-mastoid, Digastricus and Retrahens aurem; to the styloid process, the Stylo-pharyngeus, Stylo-hyoideus and Stylo-glossus; and to the petrous portion, the Levator palati, Tensor tympani, and Stapedius.

## The Sphenoid Bone.

The Sphenoid bone (σφὴν, a *wedge;* εἶδος, *likeness*) is situated at the anterior part of the base of the skull, articulating with all the other cranial bones, which it binds firmly and solidly together. In its form it somewhat resembles a bat, with its wings extended; and is divided into a central portion or body, two greater and two lesser wings extending outwards on each side of the body; and two processes, the pterygoid processes, which project from it below.

The *Body* is of large size, quadrilateral in form, and hollowed out in its interior so as to form a mere shell of bone. It presents for examination *four* surfaces—a superior, an inferior, an anterior, and a posterior.

The *superior surface* (fig. 34). From before, backwards, is seen a prominent spine, the ethmoidal spine, for articulation with the ethmoid; behind this a smooth surface presenting, in the median line, a slight longitudinal eminence, with a depression on each side, for lodging the olfactory nerves. A

34 – *Sphenoid bone, superior surface.*

narrow transverse groove, the optic groove, bounds the above-mentioned surface behind; it lodges the optic commissure, and terminates on either side in the optic foramen, for the passage of the optic nerve and ophthalmic artery. Behind the optic groove is a small eminence, olive-like in shape, the olivary process; and still more posteriorly, a deep depression, the pituitary fossa, or 'sella Turcica,' which lodges the pituitary body. This fossa is perforated by numerous foramina, for the transmission of nutrient vessels to the substance of the bone. It is bounded in front by two small eminences, one on either side, called the middle clinoid processes (κλίνη, 'a bed'), and behind by a square-shaped plate of bone, terminating at each superior angle in a tubercle, the posterior clinoid processes, the size and form of which vary considerably in different individuals. These processes deepen the pituitary fossa, and serve for the attachment of prolongations from the tentorium. The sides of the plate of bone supporting the posterior clinoid processes are notched, for the passage of the sixth pair of nerves; and behind, it presents a shallow depression, which slopes obliquely backwards, and is continuous with the basilar groove of the occipital bone; it supports the medulla oblongata. On either side of the body is a broad groove, curved somewhat like the italic letter *f*; it lodges the internal carotid artery and the cavernous sinus, and is called the *cavernous groove*. The *posterior surface*, quadrilateral in form, articulates with the basilar process of the occipital bone. During childhood, a separation between these bones exists by means of a layer of cartilage; but in after-life this becomes ossified, ossification commencing above, and extending downward, and the two bones are then immoveably connected together. The *anterior surface* (fig. 35) presents, in the middle line, a vertical lamella of bone, which articulates in front with the perpendicular plate of the ethmoid, forming part of the septum of the nose. On either side of it are the irregular openings leading into the sphenoidal sinuses. These are two large irregular cavities, hollowed out of the interior of the body of the sphenoid bone, and separated from one another by a more or less complete perpendicular bony septum. Their form and size vary considerably, they are seldom

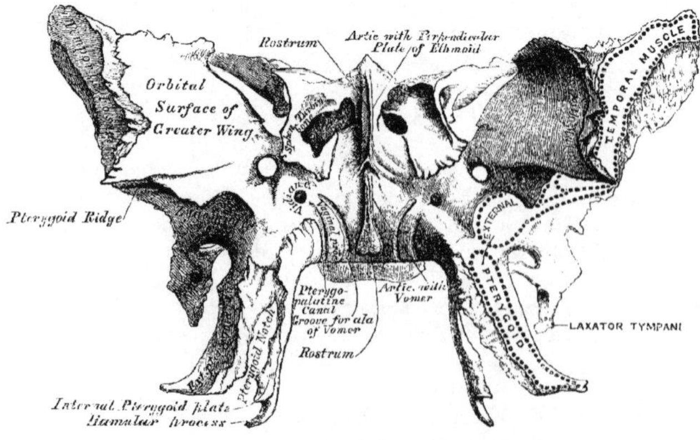


35 – *Sphenoid bone, anterior surface.*

symmetrical, and are often partially subdivided by irregular osseous laminæ. Occasionally they extend into the basilar process of the occipital nearly as far as the foramen magnum. The septum is seldom quite vertical, commonly being bent to one or the other side. These sinuses do not exist in children; but they increase in size as age advances. They are partially closed, in front and below, by two thin curved plates of bone; the sphenoidal turbinated bones, leaving a round opening at their upper parts, by which they communicate with the upper and back part of the nose, and occasionally with the posterior ethmoidal cells. The lateral margins of this surface present a serrated edge, which articulates with the os planum of the ethmoid, completing the posterior ethmoidal cells; the lower margin, also rough and serrated, articulates with the orbital process of the palate bone; and the upper margin with the orbital plate of the frontal bone. The *inferior surface* presents, in the middle line, a triangular spine, the rostrum, which is continuous with the vertical plate on the anterior surface, and is received into a deep fissure between the alæ of the vomer. On each side may be seen a projecting lamina of bone, which runs horizontally inwards from near the base of the pterygoid process: these plates, termed the vaginal processes, articulate with the edges of the vomer. Close to the root of the pterygoid process is a groove, formed into a complete canal when articulated with the sphenoidal process of the palate bone; it is called the pterygo-palatine canal, and transmits the pterygo-palatine vessels and pharyngeal nerve.

The *Greater Wings* are two strong processes of bone, which arise at the sides of the body, and are curved in a direction upwards, outwards, and backwards; being prolonged behind into a sharp-pointed extremity, the *spinous process of the sphenoid.* Each wing presents three surfaces and a circumference. The *superior* or *cerebral* surface forms part of the middle fossa of the skull; it is deeply concave, and presents eminences and depressions for the convolutions of the brain. At its anterior and internal part is seen a circular aperture, the foramen rotundum, for the transmission of the second division of the fifth nerve. Behind and external to this, a large oval foramen, the foramen ovale, for the transmission of the third division of the fifth, the small meningeal artery, and the small petrosal nerve. At the inner side of the foramen ovale, a small aperture may occasionally be seen opposite the root of the pterygoid process; it is the foramen Vesalii, transmitting a small vein. Lastly, in the apex of the spine of the sphenoid is a short canal, sometimes double, the foramen spinosum; it transmits the middle meningeal artery. The *external* surface is convex, and divided by a transverse ridge, the pterygoid ridge, into two portions. The superior or larger, convex from above downwards, concave from before backwards, enters into the formation of the temporal fossa, and attaches part of the Temporal muscle. The inferior portion, smaller in size and concave, enters into the formation of the zygomatic fossa, and affords attachment to the External pterygoid muscle. It presents, at its posterior part, a sharp-pointed eminence of bone, the spinous process, to which is connected the internal lateral ligament of the lower jaw, and the Laxator tympani muscle. The pterygoid ridge, dividing the temporal and zygomatic portions, gives attachment to part of the External pterygoid muscle. At its inner extremity is a triangular spine of bone, which serves to increase the extent of origin of this muscle. The *anterior* or *orbital* surface, smooth and quadrilateral in form, assists in forming the outer wall of the orbit. It is bounded above by a serrated edge, for articulation with the frontal bone; below, by a rounded border, which enters into the formation of the spheno-maxillary fissure; internally, it enters into the formation of the sphenoidal fissure; whilst externally it presents a serrated margin, for articulation with the malar bone. At the upper part of the inner border is a notch, for the transmission of a branch of the ophthalmic artery; and at its lower part a small pointed spine of bone, which serves for the attachment of part of the lower head of the External rectus. One or two small foramina may occasionally be seen, for the passage of arteries; they are called the *external orbitar foramina. Circumference:* from the body of the sphenoid to the spine (commencing from behind), the outer half of this margin is serrated, for articulation with the petrous portion of the temporal bone; whilst the inner half forms the anterior boundary of the foramen lacerum medium, and presents the posterior aperture of the Vidian canal. In front of the spine, the circumference of the great wing presents a serrated edge, bevelled at the expense of the inner table below, and of the external above, which articulates with the squamous portion of the temporal bone. At the tip of the great wing a triangular portion is seen, bevelled at the expense of the internal surface, for articulation with the anterior inferior angle of the parietal bone. Internal to this is a broad serrated surface, for articulation with the frontal bone: this surface is continuous internally with the sharp inner edge of the orbital plate, which assists in the formation of the sphenoidal fissure.

The *Lesser Wings* (processes of Ingrassias) are two thin triangular plates of bone, which arise from the upper and lateral parts of the body of the sphenoid; and, projecting transversely outwards, terminate in a more or less acute point. The superior surface of each is smooth, flat, broader internally than externally, and supports the anterior lobe of the brain. The inferior surface forms the back part of the roof of the orbit, and the upper boundary of the sphenoidal fissure or foramen lacerum anterius. This fissure is of a triangular form, and leads from the cavity of the cranium into the orbit; it is bounded internally by the body of the sphenoid; above, by the lesser wing; below, by the orbital surface of the great wing; and is converted into a foramen by the articulation of this bone with the frontal. It transmits the third, fourth, ophthalmic division of the fifth and sixth nerves, and the ophthalmic vein. The anterior border of the lesser wing is serrated, for articulation with the frontal bone; the posterior, smooth and rounded, is received into the fissure of Sylvius of the brain. The inner extremity of this border forms the anterior clinoid process. The lesser wing is connected to the side of the body by two roots, the upper thin and flat, the lower thicker, obliquely directed, and presenting on its outer side near its junction with the body a small tubercle, for the attachment of the common tendon of the muscles of the eye. Between the two roots is the optic foramen, for the transmission of the optic nerve and ophthalmic artery.

The *Pterygoid* processes (*πτέρυξ, a wing, εἶδος, likeness*), (fig. 36), one on each side, descend perpendicularly from the point where the body and great wing unite. Each process consists of an external and an internal plate, separated behind by an intervening notch the pterygoid fossa; but joined partially in front. The *external pterygoid plate* is broad and thin, turned a little outwards, and forms part of the inner wall of the zygomatic fossa. It gives attachment, by its outer surface, to the External pterygoid; its inner surface forms part of the pterygoid fossa, and gives attachment to the Internal pterygoid. The *internal pterygoid plate* is much narrower and longer, curving outwards, at its extremity, into a hook-like process of bone, the hamular process, around which turns the tendon of the Tensor-palati muscle. At the base of this plate is a small, oval, shallow depression, the scaphoid fossa, from which arises the Tensor-palati, and above which is seen the posterior orifice of the Vidian canal. The outer surface of this plate forms part of the pterygoid fossa, the inner surface forming the outer boundary of the posterior aperture of the nares. The two pterygoid plates are separated below by an angular interval, in which the pterygoid process; or tuberosity, of the palate bone is received. The anterior surface of the pterygoid process is very broad at its base, and forms the posterior wall of the spheno-maxillary fossa. It supports Meckel's ganglion. It presents, above, the anterior orifice of the Vidian canal; and below, a rough margin, which articulates with the perpendicular plate of the palate bone.

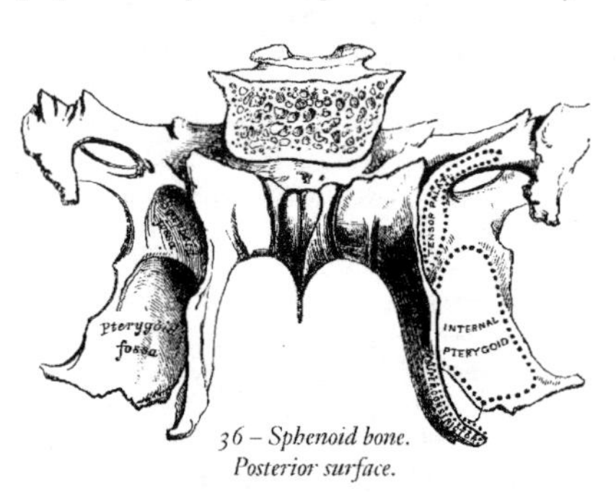

*36 – Sphenoid bone. Posterior surface.*

*Development.* The sphenoid bone is developed by *ten* centres, six for the posterior sphenoidal division, and four for the anterior sphenoid. The six centres for the post-sphenoid are, one for each greater wing and external pterygoid plate; one for each internal pterygoid plate; two for posterior part of the body. The four for the anterior sphenoid are, one for each lesser wing and anterior part of the body; and one for each sphenoidal turbinated bone. Ossification takes place in these pieces in the following order: the greater wing and external pterygoid plate are first formed, ossific granules being deposited close to the foramen rotundum on each side, at about the second month of fœtal life; ossification spreading outwards into the great wing, and downwards into the external pterygoid plate. Each internal pterygoid plate is then formed, and becomes united to the external about the middle of fœtal life. The two centres for the posterior part of the body appear as separate nuclei, side by side, beneath the sella Turcica; they join about the middle of fœtal life into a single piece, which remains ununited to the rest of the bone until after birth. Each lesser wing is formed by a separate centre, which appears on the outer side of the optic foramen, at about the third month; they become united and join with the body at about the eighth month of fœtal life. At about the end of the third year, ossification has made its appearance in the sphenoidal spongy bones.

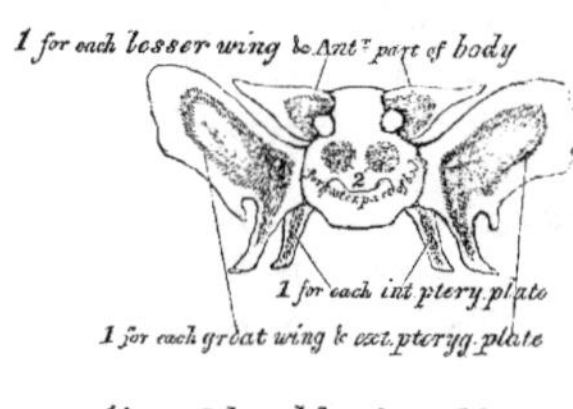

*37 – Plan of the development of sphenoid. By ten centres.*

At birth, the sphenoid consists of three pieces; viz. the greater wing and pterygoid processes on each side; the lesser wings and body united. At the first year after birth, the greater wings and body are united. From the tenth to the twelfth year, the spongy bones are partially united to the sphenoid, their junction being complete by the twentieth year. Lastly, the sphenoid joins the occipital.

*Articulations.* The sphenoid articulates with *all* the bones of the cranium, and five of the face; the two malar, two palate, and vomer: the exact extent of articulation with each bone is shewn in the accompanying figures.

*Attachment of Muscles.* The Temporal, External pterygoid, Internal pterygoid, Superior constrictor, Tensor-palati, Laxator-tympani, Levator-palpebræ, Obliquus superior, Superior rectus, Internal rectus, Inferior rectus, External rectus.

## The Sphenoidal Spongy Bones.

The *Sphenoidal Spongy Bones* are two thin, curved plates of bone, which exist as separate pieces until puberty, and occasionally are not joined to the sphenoid in the adult. They are situated at the anterior and inferior part of the body of the sphenoid, an aperture of variable size being left in their anterior wall, through which the sphenoidal sinuses open into the nasal fossæ. They are irregular in form, and taper to a point behind, being broader and thinner in front. Their inner surface, which looks towards the cavity of the sinus, is concave; their outer surface convex. Each bone articulates in front with the ethmoid, externally with the palate; behind, its point is placed above the vomer, and is received between the root of the pterygoid process on the outer side, and the rostrum of the sphenoid on the inner.

## The Ethmoid Bone.

The *Ethmoid* (ἠθμὸς, a sieve), is an exceedingly light spongy bone, of a cubical form, situated at the anterior part of the base of the cranium, between the two orbits, at the root of the nose, and contributing to form each of these cavities. It consists of three parts: a horizontal plate, which forms part of the base of the cranium; a perpendicular plate, which forms part of the septum nasi; and two lateral masses of cells.

The *Horizontal* or *Cribriform Plate* (fig. 38) forms part of the anterior fossa of the base of the skull, and is received into the ethmoid notch of the frontal bone between the two orbital plates. Projecting upwards from the middle line of this plate, is a thick smooth triangular process of bone, the crista galli, so called from its resemblance to a cock's comb. Its base joins the cribriform plate. Its posterior border, long, thin, and slightly curved, serves for the attachment of the falx cerebri. Its anterior border, short and thick, articulates with the frontal bone, and presents two small projecting alæ, which are received into corresponding depressions in the frontal, completing the foramen cœcum behind. Its sides are smooth, and sometimes bulging, when it is found to enclose a small sinus. On each side of the crista galli, the cribriform plate is narrow, and deeply grooved, to support the bulb of the olfactory nerves, and perforated by foramina for the passage of its filaments. These foramina are arranged in three rows; the innermost, which are the largest and least numerous, are lost in grooves on the upper part of the septum; the foramina of the outer row are continued on to the surface of the upper spongy bone. The foramina of the middle row are the smallest; they perforate the bone, and transmit nerves to the roof of the nose. At the front part of the cribriform plate, on each side of the crista galli, is a small fissure, which transmits the nasal branch of the ophthalmic nerve; and at its posterior part a triangular notch, which receives the ethmoidal spine of the sphenoid.

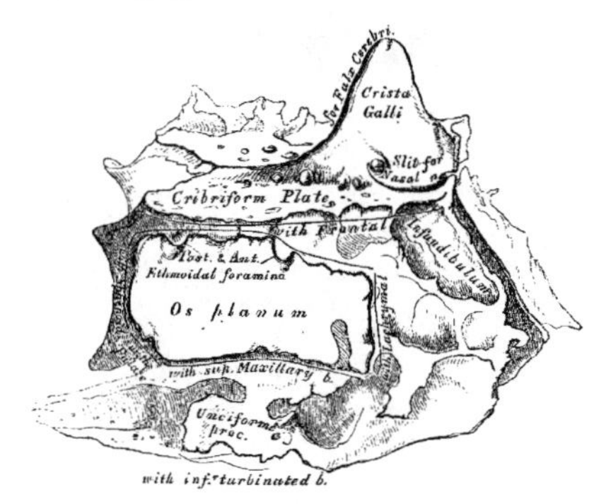


*38 – Ethmoid bone. Outer surface of right lateral mass (enlarged).*

The *Perpendicular Plate* (fig. 39) is a thin flattened lamella of bone, which descends from the under surface of the cribriform plate, and assists in forming the septum of the nose. It is much thinner in the middle, than at the circumference, and is generally deflected a little to one side. Its

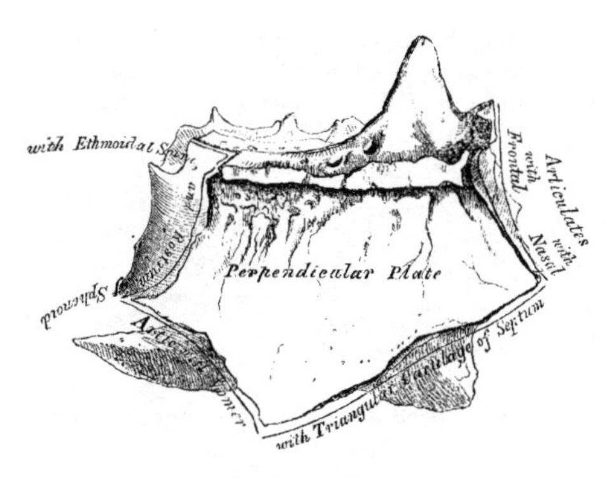


*39 – Perpendicular plate of ethmoid (enlarged). Shewn by removing the right lateral mass.*

anterior border articulates with the frontal spine and crest of the nasal bones. Its posterior, divided into two parts, is connected by its upper half with the rostrum of the sphenoid; by its lower half with the vomer. The inferior border serves for the attachment of the triangular cartilage of the nose. On each side of the perpendicular plate numerous grooves and canals are seen, leading from foramina on the cribriform plate; they lodge filaments of the olfactory nerves.

The *Lateral Masses* of the ethmoid consist of a number of thin walled cellular cavities, the *ethmoidal cells*, interposed between two vertical plates of bone, the outer one of which forms part of the orbit, and the inner one part of the nasal fossa of the corresponding side. In the disarticulated bone, many of these cells appear to be broken; but when the bones are articulated, they are closed-in in every part. The upper surface of each lateral mass presents a number of apparently half-broken cellular spaces; these, however, are completely closed-in when articulated with the edges of the ethmoidal fissure of the frontal bone. Crossing this surface are two grooves on each side, converted into canals by articulation with the frontal; they are the anterior and posterior ethmoidal foramina; they open on the inner wall of the orbit. The posterior surface also presents large irregular cellular cavities, which are closed in by articulation with the sphenoidal turbinated bones, and orbital process of the palate. The cells at the anterior surface are completed by the lachrymal bone and nasal process of the superior maxillary, and those below also by the superior maxillary. The outer surface of each lateral mass is formed of a thin smooth square plate of bone, called the *os planum*; it forms part of the inner wall of the orbit, and articulates above with the orbital plate of the frontal; below, with the superior maxillary and orbital process of the palate; in front, with the lachrymal; and behind, with the sphenoid.

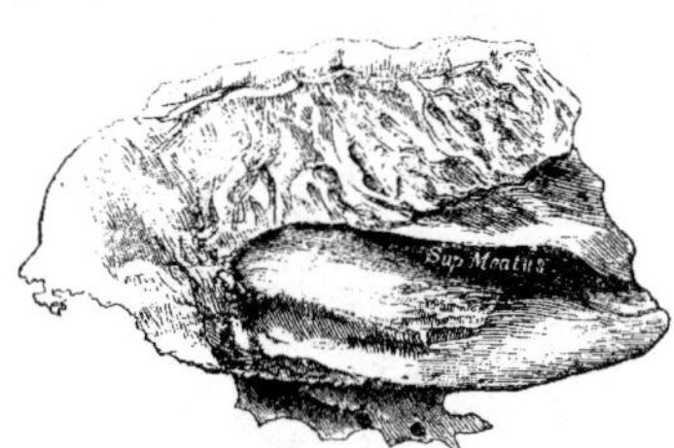


*40 – Ethmoid bone. Inner surface of right lateral mass (enlarged).*

From the inferior part of each lateral mass, immediately beneath the os planum, there projects downwards and backwards an irregular lamina of bone, called the *unciform process*, from its hook-like form: it serves to close in the upper part of the orifice of the antrum, and articulates with the ethmoidal process of the inferior turbinated bone.

The inner surface of each lateral mass forms part of the outer wall of the nasal fossa of the corresponding side. It is formed of a thin lamella of bone, which descends from the under surface of the cribriform plate, and terminates below in a free convoluted margin, the middle turbinated bone. The whole of this surface is rough, and marked above by numerous grooves which run nearly vertically downwards from the cribriform plate; they lodge branches of the olfactory nerve, which are distributed on the mucous membrane, covering the bone. The back part of this surface is subdivided by a narrow oblique fissure, the superior meatus of the nose, bounded above by a thin curved plate of bone—the superior turbinated bone. By means of an orifice at the upper part of this fissure, the posterior ethmoidal cells open into the nose. Below and in front of the superior meatus is seen the convex surface of another thin convoluted plate of bone—the middle turbinated bone. It extends along the whole length of the inner surface of each lateral mass; its lower margin is free and thick, and its concavity, directed outwards, assists in forming the middle meatus. It is by means of a large orifice at the upper and front part of the middle meatus, that the anterior ethmoid cells, and through them the frontal sinuses, by means of a funnel-shaped canal, the infundibulum, communicate with the nose. The cellular cavities of each lateral mass, thus walled in by the os planum on the outer side, and by its articulation with the other bones already mentioned, are divided by a thin transverse bony partition into two sets, which do not communicate with each other; they are termed the *anterior* and *posterior ethmoidal cells*; the former the smallest but the most numerous, communicate with the frontal sinuses above, and the middle meatus below, by means of a long flexuous cellular canal, the *infundibulum*; the posterior, the largest and least numerous, open into the superior meatus, and communicate (occasionally) with the sphenoidal sinuses.

*Development.* By *three* centres; one for the perpendicular lamella, and one for each lateral mass.

The lateral masses are first developed, ossific granules making their first appearance in the os planum between the fourth and fifth months of fœtal life, and afterwards in the spongy bones. At birth, the bone consists of the two lateral masses, which are small and ill-developed; but when the perpendicular and horizontal plates begin to ossify, as they do about the first year after birth, the lateral masses become joined to the cribriform plate. The formation and increase in the ethmoidal cells, which complete the formation of the bone, take place about the fifth or sixth year.

*Articulations.* With fifteen bones; the sphenoid, two sphenoidal turbinated, the frontal, and eleven of the face—two nasal, two superior maxillary, two lachrymal, two palate, two inferior turbinated, and vomer.

## Development of the Cranium

The development of the cranium takes place at a very early period, on account of the importance of the organ it is intended to protect. In its most rudimentary state, it consists of a thin membranous capsule; enclosing the cerebrum, and accurately moulded upon its surface. This capsule is placed external to the dura mater, and in close contact with it; its walls are continuous with the canal for the spinal cord, and the chorda dorsalis, or primitive part of the vertebral column, is continued forwards, from the spine, along the base, to its fore part, where it terminates in a tapering point. The next step in the process of development is the formation of cartilage. This is deposited in the base of the skull, in two symmetrical segments, one on either side of the median line; these subsequently coalesce, so as to enclose the chorda dorsalis: the chief part of the cerebral capsule still retaining its membranous form. Ossification first takes place in the roof, and is preceded by the deposition of a membranous blastema upon the surface of the cerebral capsule, in which the ossifying process extends; the primitive membranous capsule becoming the internal periosteum, and being ultimately blended with the dura mater. Although the bones of the vertex of the skull appear before those at the base, and make considerable progress in their growth: at birth, ossification is more advanced in the base, this portion of the skull forming a solid immoveable groundwork.

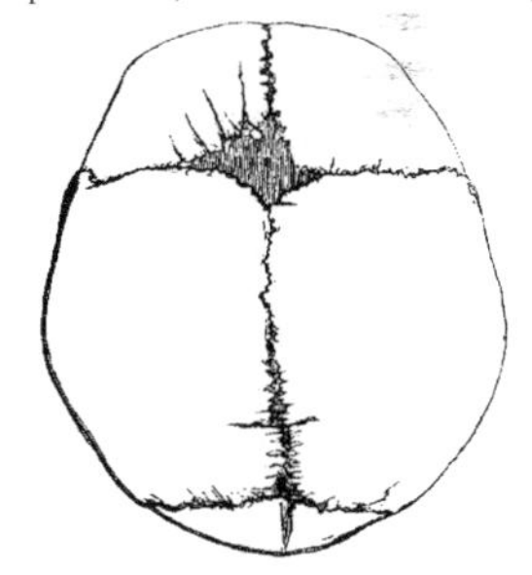

*41 – Skull at birth, shewing the anterior and posterior fontanelles.*

## The Fontanelles (Figs. 41, 42)

Before birth, the bones at the vertex and side of the skull are separated from each other by membranous intervals, in which bone is deficient. These intervals, at certain parts, are of considerable size, and are termed the *fontanelles*, so called from the pulsations of the brain, which resemble the rising of water at a fountain head. The fontanelles are four in number, and correspond to the junction of the four angles of the parietal with the contiguous bones. The anterior fontanelle is the largest, and corresponds to the junction of the sagittal and coronal sutures; the posterior fontanelle, of smaller size, is situated at the junction of the sagittal and lambdoid sutures; the two remaining ones are situated at the inferior angles of the parietal bone. The latter are closed soon after birth; the two at the superior angles remain open longer: the posterior one being closed in a few months after birth; the anterior one remaining open until the first or second year. These spaces are gradually filled in by an extension of the ossifying process, or by the development of a Wormian bone. Fine specimens of large Wormian bones closing in the anterior and posterior fontanelles, and replacing the anterior inferior angle of the parietal bones, exist in the St. George's Hospital Museum. Sometimes, the anterior fontanelle remains open beyond two years, and is occasionally persistent throughout life.

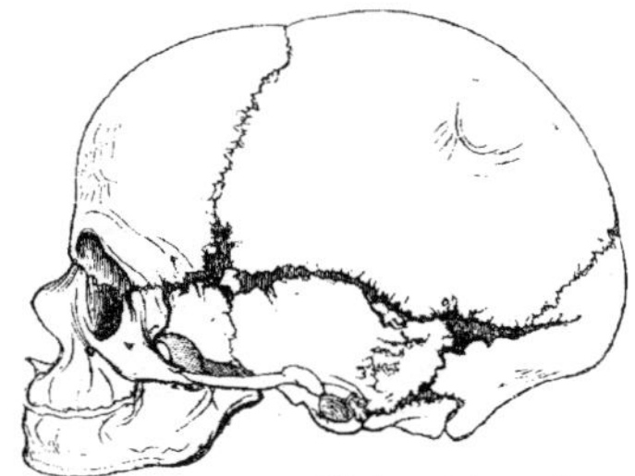

*42 – The lateral fontanelles.*

### Supernumerary or Wormian* Bones

When ossification of any of the tabular bones of the skull proves abortive, the membranous interval left unclosed, is usually filled in by a supernumerary piece of bone, which is developed from a separate centre, and gradually extends until it fills in the vacant space. These supernumerary pieces are called Wormian bones; they are called also, from their form, *ossa triquetra*, presenting much variation in situation, number, and size.

They occasionally occupy the situation of the fontanelles. Bertin, Cruvelhier, and Cuvier have each noticed the presence of one in the anterior fontanelle. There are two specimens in the Museum of St. George's Hospital, which present Wormian bones in this situation. In one, the skull of a child, the supernumerary piece is of considerable size, and of a quadrangular form.

They are occasionally found in the posterior fontanelle, appearing to replace the superior angle of the occipital bone. Not unfrequently, there is one replacing the extremity of the great wing of the sphenoid, or the anterior inferior angle of the parietal bone, in the fontanelle there situated.

They have been found in the different sutures on the vertex and side of the skull, and in some of those at the base. They are most frequent in the lambdoid. Ward mentions an instance 'in which one half of the lambdoid suture was formed by large Wormian bones disposed in a double row, and jutting deeply into each other'; and refers to similar specimens described by Dumontier and Bourgery.

A deficiency in the ossification of the flat bones would appear in some cases to be *symmetrical* on the two sides of the skull; for it is not uncommon to find these supernumerary bones corresponding in form, size, and situation on each side. Thus, in several instances, I have seen a pair of large Wormian bones symmetrically placed in the lambdoid suture; in another specimen, a pair in the coronal suture, with a supernumerary bone in the spheno-parietal suture of both sides.

The size of these supernumerary pieces varies, in some cases not being larger than a pin's head, and confined to the outer table; in other cases so large, that one pair of these bones formed the whole of that portion of the occipital bone above the superior curved lines, as described by Beclard and Ward. Their number is generally limited to two or three; but more than a hundred have been found in the skull of an adult hydrocephalic skeleton. In their development, structure, and mode of articulation, they resemble the other cranial bones.

### Congenital Fissures and Gaps

Mr. Humphry has called attention to the existence of *congenital fissures*, not unfrequently being found in the cranial bones, the result of incomplete ossification. These fissures have been noticed in the frontal, parietal, and squamous portion of the temporal bones; they extend from the margin towards the middle of the bone; and are of great interest in a medico-legal point of view, as they are liable to be mistaken for fractures. An arrest of the ossifying process may also give rise to the *deficiencies* or *gaps* occasionally found in the cranial bones. Such deficiencies are said to occur most frequently when ossification is imperfect, and to be situated near the natural apertures for vessels. Mr. Humphry describes such deficiencies to exist in a calvarium, in the Cambridge Museum, where a gap sufficiently large to admit the end of the finger, is seen on either side of the sagittal suture, in the place of the parietal foramen. There is a specimen precisely similar to this in the Museum of St. George's Hospital; and another, in which a small circular gap exists in the parietal bone of a young child, just above the parietal eminence. Similar deficiencies are not unfrequently met with in hydrocephalic skulls; being most frequent, according to Mr. Humphry, in the frontal bones; and in the parietal bones, on either side of the sagittal suture.

## Bones of the Face.

The Facial Bones are fourteen in number, viz., the

| | |
|---|---|
| Two Nasal, | Two Palate, |
| Two Superior Maxillary, | Two Inferior Turbinated, |
| Two Lachrymal, | Vomer, |
| Two Malar, | Inferior Maxillary. |

## Nasal Bones.

The Nasal are two small oblong bones, varying in size and form in different individuals; they are placed side by side at the middle and upper part of the face, forming, by their junction, the 'bridge' of the nose. Each bone presents for examination two surfaces, and four borders. The *outer* surface is concave from above downwards, convex from side to side; it is covered by the Compressor nasi muscle, marked by numerous small arterial furrows, and perforated about its centre by a foramen,

* Wormius, a physician in Copenhagen, is said to have given the first detailed description of these bones.

sometimes double, for the transmission of a small vein. Sometimes this foramen is absent on one or both sides, and occasionally the foramen cœcum opens on this surface. The *inner* surface is concave from side to side, convex from above downwards; in which direction it is traversed by a longitudinal groove (sometimes a canal), for the passage of a branch of the nasal nerve. The superior border is narrow, thick, and serrated for articulation with the nasal notch of the frontal bone. The inferior border is broad, thin, sharp, directed obliquely downwards, outwards, and backwards, and serves for the attachment of the lateral cartilage of the nose. This border presents about its centre a notch, which transmits the branch of the nasal nerve above referred to; and is prolonged at its inner extremity into a sharp spine, which, when articulated with the opposite bone, forms the nasal angle. The external border is serrated, bevelled at the expense of the internal surface above, and of the external below, to articulate with the nasal process of the superior maxillary. The internal border, thicker above than below, articulates with its fellow of the opposite side, and is prolonged behind into a vertical crest, which forms part of the septum of the nose; this crest articulates with the nasal spine of the frontal above, and the perpendicular plate of the ethmoid below.

*Development.* By *one* centre for each bone, which appears about the same period as in the vertebræ.

*Articulations.* With four bones: two of the cranium, the frontal and ethmoid; and two of the face, the opposite nasal and the superior maxillary.

No muscles are directly attached to this bone.

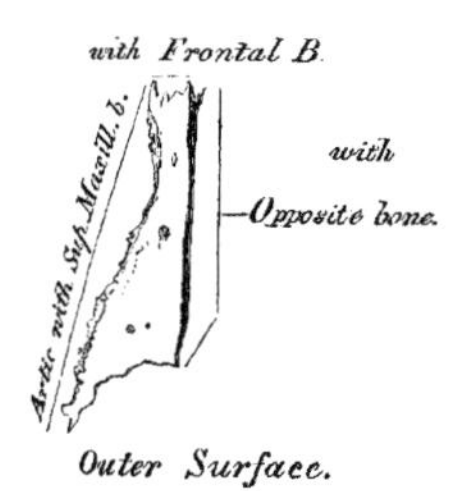


43 – *Right nasal bone.*

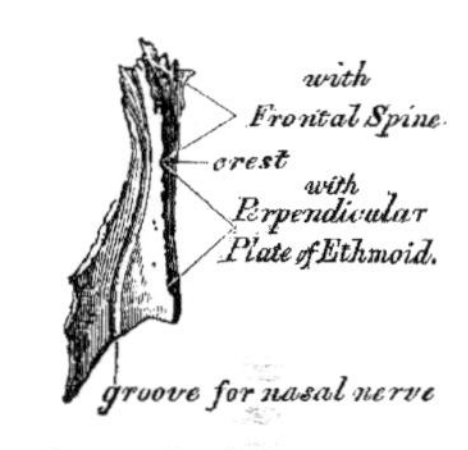


44 – *Right nasal bone.*

## Superior Maxillary Bone.

The Superior Maxillary is one of the most important bones of the face in a surgical point of view, on account of the number of diseases to which some of its parts are liable. Its minute examination becomes, therefore, a matter of considerable interest. It is the largest bone of the face, excepting the lower jaw; and forms, by its union with its fellow of the opposite side, the whole of the upper jaw. Each bone assists in the formation of three cavities, the roof of the mouth, the floor and outer wall of the nose, and the floor of the orbit; enters into the formation of two fossæ, the zygomatic, and spheno-maxillary; and two fissures, the spheno-maxillary, and pterygo-maxillary. Each bone presents for examination a body, and four processes, malar, nasal, alveolar, and palatine.

The body is somewhat quadrilateral, and is hollowed out in its interior to form a large cavity, the antrum of Highmore. It presents for examination four surfaces, an external or facial, a posterior or zygomatic, a superior or orbital, and an internal.

The *external* or *facial surface* (fig. 45) is directed forwards and outwards. In the median line of the bone, just above the incisor teeth, is a depression, the incisive or myrtiform fossa, which gives origin to the Depressor alæ nasi. Above and a little external to it, the Compressor naris arises. More external, is another depression, the canine fossa, larger and deeper than the incisive fossa, from which it is separated by a vertical ridge, the canine eminence, corresponding to the socket of the canine tooth. The canine fossa gives origin to the Levator anguli oris. Above the canine fossa is the infra-orbital foramen, the termination of the infra-orbital canal; it transmits the infra-orbital nerve and artery. Above the infra-orbital foramen is the margin of the orbit, which affords partial attachment to the Levator labii superioris proprius.

The *posterior* or *zygomatic surface* is convex, directed backwards and outwards, and forms part of the zygomatic fossa. It presents about its centre several apertures leading to canals in the substance of the bone; they are termed the *posterior dental canals*, and transmit the posterior dental vessels and nerves. At the lower part of this surface is a rounded eminence, the maxillary tuberosity, especially prominent after the growth of the wisdom-tooth, rough on its inner side for articulation with the tuberosity of the palate bone. Immediately above the rough surface is a groove, which, running obliquely down on the inner surface of the bone, is converted into a canal by articulation with the palate bone, forming the posterior palatine canal.

The *superior* or *orbital surface* is thin, smooth, triangular, and forms part of the floor of the orbit.

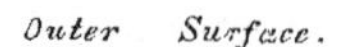


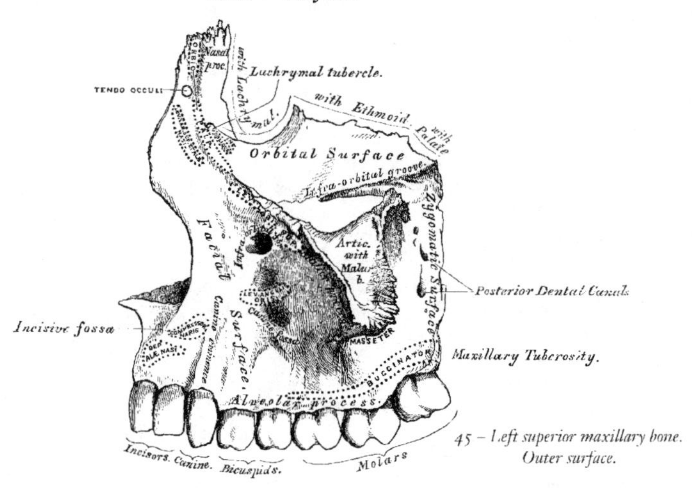


45 – *Left superior maxillary bone. Outer surface.*

It is bounded internally by an irregular margin which articulates, in front, with the lachrymal; in the middle, with the os planum of the ethmoid; behind, with the orbital process of the palate bone; bounded externally by a smooth rounded edge which enters into the formation of the spheno-maxillary fissure, and which sometimes articulates at its anterior extremity with the orbital plate of the sphenoid; bounded, in front, by part of the circumference of the orbit, which is continuous, on the inner side, with the nasal, on the outer side, with the malar process. Along the middle line of the orbital surface is a deep groove, the infra-orbital, for the passage of the infra-orbital nerve and artery. This groove commences at the middle of the outer border of this surface, and, passing forwards, terminates in a canal which subdivides into two branches; one of the canals, the infra-orbital, opens just below the margin of the orbit; the other, which is smaller, runs in the substance of the anterior wall of the antrum; it is called the anterior dental canal, transmitting the anterior dental vessels and nerves to the front teeth of the upper jaw. At the inner and fore part of the orbital surface, just external to the lachrymal canal, is a minute depression, which gives origin to the Inferior oblique muscle of the eye.

The *internal surface* (fig. 46) is unequally divided into two parts by a horizontal projection of bone, the palate process; that portion above the palate process forms part of the outer wall of the nose; the portion below it forms part of the cavity of the mouth. The superior division of this surface

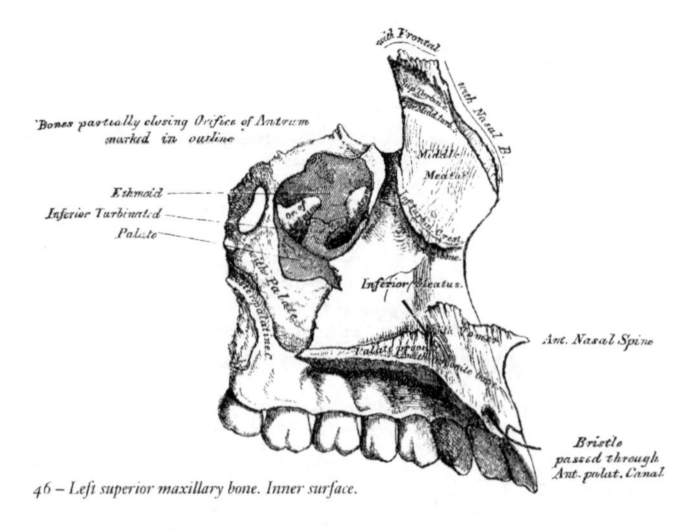


46 – *Left superior maxillary bone. Inner surface.*

presents a large irregular shaped opening leading into the antrum of Highmore. At the upper border of this aperture are a number of broken cellular cavities, which, in the articulated skull, are closed in by the ethmoid and lachrymal bones. Below the aperture is a smooth concavity which forms part of the inferior meatus of the nose, traversed by a fissure, the maxillary fissure, which runs from the lower part of the orifice of the antrum obliquely downwards and forwards, and receives the maxillary process of the palate bone. Behind it is a rough surface which articulates with the perpendicular plate of the palate bone, traversed by a groove which, commencing near the middle of the posterior border, runs obliquely downwards and forwards, and forms, when completed by its articulation with the palate bone, the posterior palatine canal. In front of the opening in the antrum is a deep groove, converted into a canal by the lachrymal and inferior turbinated bones, and lodging the nasal duct. More anteriorly is a well-marked rough ridge, the inferior turbinated crest, for articulation with the inferior turbinated bone. The concavity above this ridge forms part of the middle meatus of the nose; whilst that below it, forms part of the inferior meatus. The inferior division of this surface is concave, rough and uneven, and perforated by numerous small foramina for the passage of nutrient vessels.

The *Antrum of Highmore*, or Maxillary Sinus, is a large triangular-shaped cavity, hollowed out of the body of the maxillary bone; its apex, directed outwards, is formed by the malar process; its base, by the outer wall of the nose. Its walls are everywhere exceedingly thin, its roof being formed by the orbital plate; its floor by the alveolar process, bounded in front by the facial surface, and behind by the zygomatic. Its inner wall, or base, presents, in the disarticulated bone, a large irregular aperture, which communicates with the nasal fossæ. The margins of this aperture are thin and ragged, and the aperture itself is much contracted by its articulation with the ethmoid above, the inferior turbinated below, and the palate bone behind. In the articulated skull, this cavity communicates with the middle meatus of the nose generally by two small apertures left between the above-mentioned bones. In the recent state, usually only one small opening exists, near the upper part of the cavity, sufficiently large to admit the end of a probe, the other being closed by the lining membrane of the sinus.

Crossing the cavity of the antrum are often seen several projecting laminæ of bone, similar to those seen in the sinuses of the cranium; and on its posterior wall are the posterior dental canals, transmitting the posterior dental vessels and nerves to the teeth. Projecting into the floor are several conical processes, corresponding to the roots of the first and second molar teeth; in some cases, the floor is perforated by the teeth in this situation. It is from the extreme thinness of the walls of this cavity, that we are enabled to explain how a tumour, growing from the antrum, encroaches upon the adjacent parts, pushing up the floor of the orbit, and displacing the eyeball, projecting inward into the nose, protruding forwards on to the cheek, and making its way backwards into the zygomatic fossa, and downwards into the mouth.

The *Malar Process* is a rough triangular eminence, situated at the angle of separation of the facial from the zygomatic surface. In front, it is concave, forming part of the facial surface; behind, it is also concave, and forms part of the temporal fossa; above, it is rough and serrated for articulation with the malar bone; whilst below, a prominent ridge marks the division between the facial and zygomatic surfaces.

The *Nasal Process* is a thick triangular plate of bone, which projects upwards, inwards, and backwards, by the side of the nose, forming part of its lateral boundary. Its external surface is concave, smooth, perforated by numerous foramina, and gives attachment to the Levator labii superioris alæque nasi, the Orbicularis palpebrarum, and Tendo oculi. Its internal surface forms part of the inner wall of the nose; it articulates above with the frontal, and presents a rough uneven surface, which articulates with the ethmoid bone, closing in the anterior ethmoid cells; below this is a transverse ridge, the superior turbinated crest, for articulation with the middle turbinated bone of the ethmoid, bounded below by a smooth concavity, which forms part of the middle meatus; below this is the inferior turbinated crest (already described), for articulation with the inferior turbinated bone; and still more inferiorly, the concavity which forms part of the inferior meatus. The anterior border of the nasal process is thin, directed obliquely downwards and forwards, and presents a serrated edge for articulation with the nasal bone: its posterior border is thick, and hollowed into a groove for the nasal duct; of the two margins of this groove, the inner one articulates with the lachrymal bone, the outer one forms part of the circumference of the orbit. Just where the latter joins the orbital surface is a small tubercle, the lachrymal tubercle; this serves as a guide to the surgeon in the performance of the operation for fistula lachrymalis. The lachrymal groove in the articulated skull is converted into a canal by the lachrymal bone, and lachrymal process of the inferior turbinated; it is directed downwards, and a little backwards and outwards, is about the diameter of a goose-quill, slightly narrower in the middle than at either extremity, and lodges the nasal duct.

The *Alveolar Process* is the thickest and most spongy part of the bone, broader behind than in front, and excavated into deep cavities for the reception of the teeth. These cavities are eight in number, and vary in size and depth according to the teeth they contain: those for the canine teeth being the deepest; those for the molars being widest, and subdivided into minor cavities; those for the incisors being single, but deep and narrow.

The *Palate Process*, thick and strong, projects horizontally inwards from the inner surface of the bone. It is much thicker in front than behind, and forms a considerable part of the floor of the nares, and the roof of the mouth. Its upper surface is concave from side to side, smooth, and forms part of the floor of the nose. In front is seen the upper orifice of the anterior palatine (incisor) canal, which leads into a fossa formed by the junction of the two superior maxillary bones, and situated immediately behind the incisor teeth. It transmits the anterior palatine vessels, the naso-palatine nerves passing through the inter-maxillary suture. The inferior surface, also concave, is rough and uneven, and forms part of the roof of the mouth. This surface is perforated by numerous foramina for the passage of nutritious vessels, channelled at the back part of its aveolar border by a longitudinal groove, sometimes a canal, for the transmission of the posterior palatine vessels, and a large nerve, and presents little depressions for the lodgment of the palatine glands. This surface presents anteriorly the lower orifice of the anterior palatine fossa. In some bones, a delicate linear suture may be seen extending from the anterior palatine fossa, to the interval between the lateral incisor and the canine teeth. This marks out the intermaxillary bone, which in some animals exists permanently as a separate piece. It includes the whole thickness of the alveolus, the corresponding part of the floor of the nose, and the anterior nasal spine, and contains the sockets of the incisor teeth. The outer border of the palate process is firmly united with the rest of the bone. The inner border is thicker in front than behind, raised above into a ridge, which, with the corresponding ridge in the opposite bone, forms a groove for the reception of the vomer. The anterior margin is bounded by the thin concave border of the opening of the nose, prolonged forwards internally into a sharp process, forming, with a similar process of the opposite bone, the anterior nasal spine. The posterior border is serrated for articulation with the horizontal plate of the palate bone.

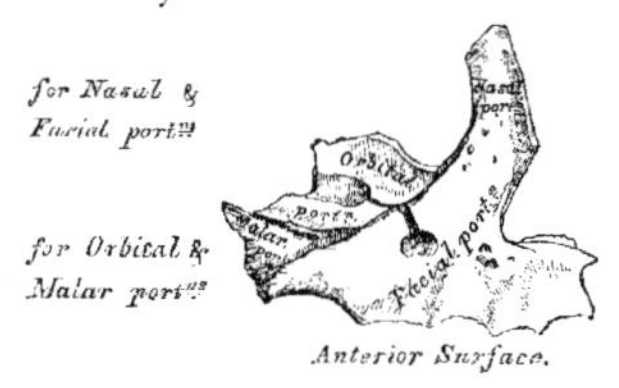


47 – *Development of superior maxillary bone. By four centres.*

*Development.* This bone is formed at such an early period, and ossification proceeds in it with such rapidity, that it has been found impracticable hitherto to determine with accuracy its number of centres. It appears, however, probable that it has *four* centres of development, viz., one for the nasal and facial portions, one for the orbital and malar, one for the incisive, and one for the palatal portion, including the entire palate except the incisive segment. The incisive portion is indicated in young bones by a fissure, which marks off a small segment of the palate, including the two incisor teeth. In some animals, this remains permanently as a separate piece, constituting the intermaxillary bone; and in the human subject, where the jaw is malformed, as in cleft palate, this segment may be separated from the maxillary bone by a deep fissure extending backwards between the two into the palate. If the fissure be on both sides, both segments are quite isolated from the maxillary bones, and hang from the end of the vomer, not unfrequently being much displaced, and often accompanied by congenital fissure of the upper lip, either on one or both sides of the median line. The maxillary sinus appears at an earlier period than any of the other sinuses, its development commencing about the fourth month of fœtal life.

*Articulations.* With *nine* bones; two of the cranium—the frontal and ethmoid, and seven of the face, viz., the nasal, malar, lachrymal, inferior turbinated, palate, vomer, and its fellow of the opposite side. Sometimes it articulates with the orbital plate of the sphenoid.

*Attachment of Muscles.* Orbicularis palpebrarum, Obliquus inferior oculi, Levator labii superioris alæque nasi, Levator labii superioris proprius, Levator anguli oris, Compressor nasi, Depressor alæ nasi, Masseter, Buccinator.

## The Lachrymal Bones.

The *Lachrymal* are the smallest and most fragile bones of the face, situated at the front part of the inner wall of the orbit, and resemble somewhat in form, thinness, and size, a finger-nail; hence they are termed the *ossa unguis*. Each bone presents for examination, two surfaces and four borders. The external (fig. 48) or orbital surface is divided by a vertical ridge into two parts. The portion of bone in front of this ridge presents a smooth, concave, longitudinal groove, the free margin of which unites with the nasal process of the superior maxillary bone, completing the lachrymal groove. The upper part of this groove lodges the lachrymal sac; the lower part assists in the formation of the lachrymal canal, and lodges the nasal duct. The portion of bone behind the ridge is smooth, slightly concave, and forms part of the inner wall of the orbit. The ridge, and part of the orbital surface immediately behind it, affords attachment to the Tensor tarsi: the ridge terminates below in a small hook-like process, which articulates with the lachrymal tubercle of the superior maxillary bone, and completes the upper orifice of the lachrymal canal. It sometimes exists as a separate piece, which is then called the *lesser lachrymal bone*. The internal or nasal surface presents a depressed furrow, corresponding to the ridge on its outer surface. The surface of bone in front of this forms part of the middle meatus; and that behind it articulates with the ethmoid bone, filling in the anterior ethmoidal cells. Of the *four borders*, the anterior is the longest, and articulates with the nasal process of the superior maxillary bone. The posterior, thin and uneven, articulates with the os planum of the ethmoid. The superior, the shortest and thickest, articulates with the internal angular process of the frontal bone. The inferior is divided by the lower edge of the vertical crest into two parts, the posterior part articulating with the orbital plate of the superior maxillary bone; the anterior portion being prolonged downwards into a pointed process, which articulates with the lachrymal process of the inferior turbinated bone, assisting in the formation of the lachrymal canal.

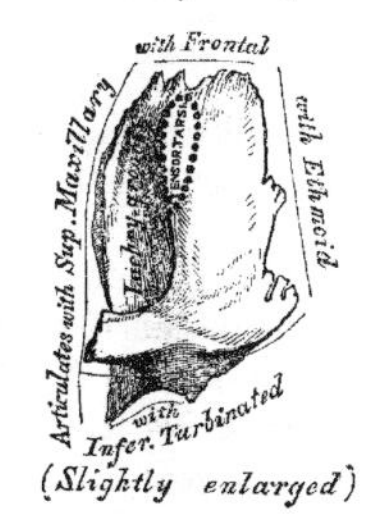


*48 – Left lachrymal bone. External surface.*

*Development.* By a single centre, which makes its appearance soon after ossification of the vertebræ has commenced.

*Articulations.* With four bones; two of the cranium, the frontal and ethmoid, and two of the face, the superior maxillary and the inferior turbinated.

*Attachment of Muscles.* The Tensor tarsi.

## The Malar Bones.

The *Malar* are two small quadrangular bones, situated at the upper and outer part of the face, forming the prominence of the cheek, part of the outer wall and floor of the orbit, and part of the temporal and zygomatic fossæ. Each bone presents for examination an external and an internal surface; four processes, the frontal, orbital, maxillary, and zygomatic; and four borders. The external surface (fig. 49) is smooth, convex, perforated near its centre by one or two small apertures, the malar foramina, for the passage of nerves and vessels, covered by the Orbicularis palpebrarum muscle, and affords attachment to the Zygomaticus major and minor muscles.

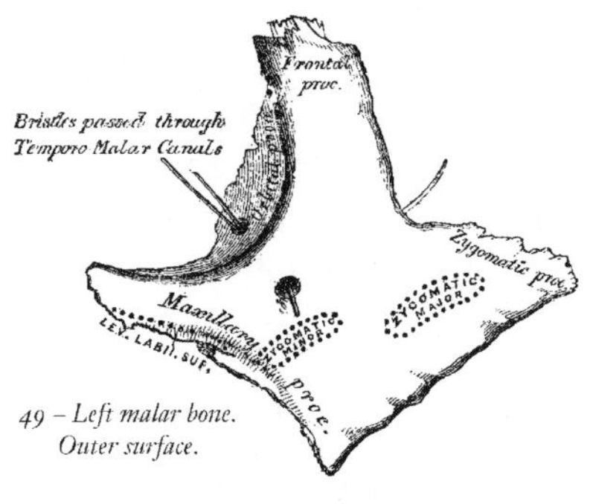


*49 – Left malar bone. Outer surface.*

The internal surface (fig. 50), directed backwards and inwards, is concave, presenting internally a rough triangular surface, for articulation with the superior maxillary bone; and externally, a smooth concave surface, which forms the anterior boundary of the temporal fossa above, wider below, where it forms part of the zygomatic fossa. This surface presents, a little above its centre, the aperture of one or two malar canals, and affords attachment to part of two muscles, the Temporal above, and the Masseter below. Of the four processes, the *frontal* is thick and serrated, and articulates with the external angular process of the frontal bone. The *orbital* process is a thick and strong plate, which projects backwards from the orbital margin of the bone. Its upper surface, smooth and concave, forms, by its junction with the great ala of the sphenoid, the outer wall of the

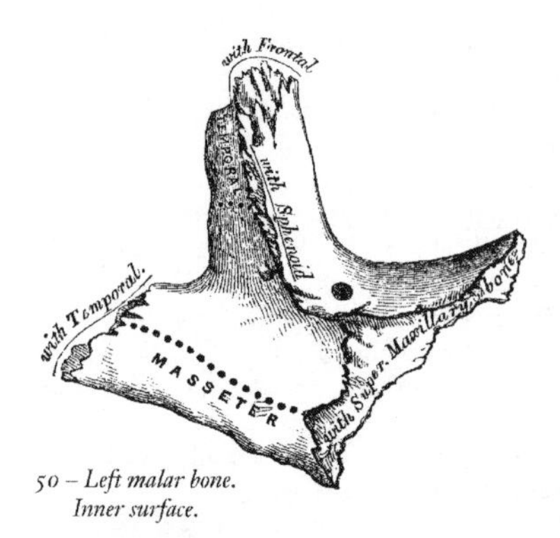


50 – *Left malar bone. Inner surface.*

orbit. Its under surface, smooth and convex, forms part of the temporal fossa. Its anterior margin is smooth and rounded, forming part of the circumference of the orbit. Its superior margin, rough, and directed horizontally, articulates with the frontal bone behind the external angular process. Its posterior margin is rough and serrated, for articulation with the sphenoid; internally it is also serrated for articulation with the orbital surface of the superior maxillary. At the angle of junction of the sphenoidal and maxillary portions, a short rounded non-articular margin is sometimes seen; this forms the anterior boundary of the spheno-maxillary fissure: occasionally, no such non-articular margin exists, the fissure being completed by the direct junction of the maxillary and sphenoid bones, or by the interposition of a small Wormian bone in the angular interval between them. On the upper surface of the orbital process are seen the orifices of one or two temporo-malar canals; one of these usually opens on the posterior surface, the other (occasionally two), on the facial surface: they transmit filaments (temporo-malar) of the orbital branch of the superior maxillary nerve. The *maxillary* process is a rough triangular surface, which articulates with the superior maxillary bone. The *zygomatic* process, long, narrow, and serrated, articulates with the zygomatic process of the temporal bone. *Of the four borders*, the superior or orbital, is smooth, arched, and forms a considerable part of the circumference of the orbit. The inferior, or zygomatic, is continuous with the lower border of the zygomatic arch, affording attachment by its rough edge to the Masseter muscle. The anterior or maxillary border is rough, and bevelled at the expense of its inner table, to articulate with the superior maxillary bone; affording attachment by its outer margin to the Levator labii superioris proprius, just at its point of junction with the superior maxillary. The posterior or temporal border, curved like an italic *f*, is continuous above with the commencement of the temporal ridge; below, with the upper border of the zygomatic arch; it affords attachment to the temporal fascia.

*Development.* By a single centre of ossification, which appears at about the same period when ossification of the vertebræ commences.

*Articulations.* With four bones: three of the cranium, frontal, sphenoid, and temporal; and one of the face, the superior maxillary.

*Attachment of Muscles.* Levator labii superioris proprius, Zygomaticus major and minor, Masseter, and Temporal.

## The Palate Bones.

The Palate Bones are situated at the back part of the nasal fossæ; they are two in number, one on each side, wedged in between the superior maxillary and the pterygoid process of the sphenoid. Each bone assists in the formation of three cavities; the floor and outer wall of the nose, the roof of the mouth, and the floor of the orbit; and enters into the formation of three fossæ: the zygomatic, spheno-maxillary, and pterygoid. In form, the palate bone somewhat resembles the letter L, and may be divided into an inferior or horizontal plate, and a superior or vertical plate.

The *Horizontal Plate* is thick, of a quadrilateral form, and presents two surfaces and four borders. The superior surface, concave from side to side, forms the back part of the floor of the nares. The inferior surface, slightly concave and rough, forms the back part of the hard palate. At its posterior part may be seen a transverse ridge, more or less marked, for the attachment of the aponeurosis of the Tensor palati muscle. At the outer extremity of this ridge is a deep groove, converted into a canal by its articulation with the tuberosity of the superior maxillary bone, and forming the posterior palatine canal. Near this groove, the orifices of one or two small canals, accessory posterior palatine, may frequently be seen. The anterior border is serrated, bevelled at the expense of its inferior surface, and articulates with the palate process of the superior maxillary bone. The posterior border is concave, free, and serves for the attachment of the soft palate. Its inner extremity is sharp and pointed, and when united with the opposite bone, forms a projecting process, the posterior nasal spine, for the attachment of the Azygos uvulæ. The external border is united with the lower part of the perpendicular plate almost at right angles. The internal border, the thickest, is serrated for articulation with its fellow of the opposite side; its superior edge is raised into a ridge, which, united with the opposite bone, forms a crest in which the vomer is received.

The *Vertical Plate* (fig. 51) is thin, of an oblong form, and directed upwards and a little inwards. It presents two surfaces, an external and an internal, and four borders.

The *internal surface* presents at its lower part a broad shallow depression, which forms part of the inferior meatus of the nose. Immediately above this is a well marked horizontal ridge, the inferior turbinated crest, for articulation with the inferior turbinated bone; above this, a second broad shallow depression, which forms part of the middle meatus, surmounted above by a horizontal ridge, less prominent than the inferior, the superior turbinated crest, for articulation with the middle turbinated bone. Above the superior turbinated crest is a narrow horizontal groove, which forms part of the superior meatus.

51 – *Left palate bone. Internal view (enlarged).*

The *external surface* is rough and irregular throughout the greater part of its extent, for articulation with the inner surface of the superior maxillary bone, its upper and back part being smooth where it enters into the formation of the zygomatic fossa; it is also smooth in front, where it covers the orifice of the antrum. Towards the back part of this surface is a deep groove, converted into a canal, the posterior palatine, by its articulation with the superior maxillary bone. It transmits the posterior palatine vessels and a large nerve. The anterior border is thin, irregular, and presents opposite the inferior turbinated crest, a pointed projecting lamina, the maxillary process, which is directed forwards, and closes in the lower and back part of the opening of the antrum, being received into a fissure that exists at the inferior part of this aperture. The posterior border (fig. 52) presents a deep groove, the edges of which are serrated for articulation with the pterygoid process of the sphenoid. At the lower part of this border is seen a pyramidal process of bone, the *pterygoid process* or tuberosity of the palate, which is received into the angular interval between the two pterygoid plates of the sphenoid at their inferior extremity. This process presents at its back part three grooves, a median and two lateral ones. The former is smooth, and forms part of the pterygoid fossa, affording attachment to the Internal pterygoid muscle; whilst the lateral grooves are rough and uneven, for articulation with the anterior border of each pterygoid plate. The base of this process, continuous with the horizontal portion of the bone, presents the apertures of the accessory descending palatine canals; whilst its outer surface is rough, for articulation with the inner surface of the body of the superior maxillary bone. The superior border of the vertical plate presents two well marked processes, separated by an intervening notch or foramen. The anterior, or larger, is called the *orbital process*; the posterior, the *sphenoidal*.

The *Orbital Process*, directed upwards and outwards, is placed on a higher level than the sphenoidal. It presents five surfaces, which enclose a hollow cellular cavity, and is connected to the perpendicular plate by a narrow constricted neck. Of these five surfaces, three are articular, two non-articular, or free surfaces. The three articular are the anterior or maxillary surface, which is directed forwards, outwards, and downwards, is of an oblong form, and rough for articulation with the superior maxillary bone. The posterior or sphenoidal surface, is directed backwards, upwards, and inwards. It ordinarily presents a small half-cellular cavity which communicates with the sphenoidal sinus, and the margins of which are serrated for articulation with the vertical part of the sphenoidal turbinated bone. The internal or ethmoidal surface is directed inwards, upwards and

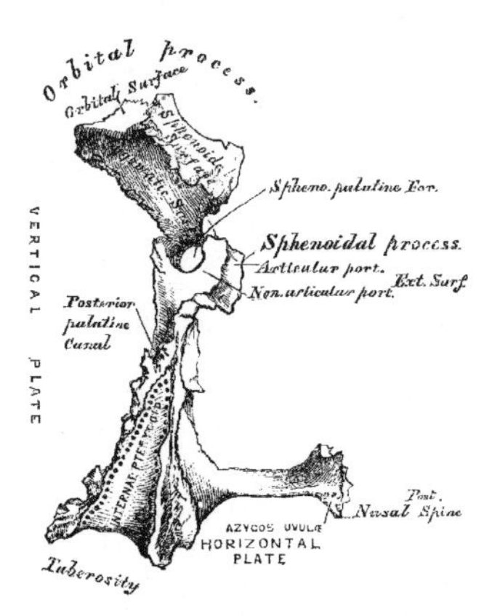


52 – *Left palate bone. Posterior view (enlarged).*

forwards, and articulates with the lateral mass of the ethmoid bone. In some cases, the cellular cavity above-mentioned opens on this surface of the bone; it then communicates with the posterior ethmoidal cells. More rarely it opens on both surfaces, and then communicates with the posterior ethmoidal cells, and the sphenoidal sinus. The non-articular or free surfaces are the superior or orbital, directed upwards and outwards, of triangular form, concave, smooth, articulating with the superior maxillary bone, and forming the back part of the floor of the orbit. The external or zygomatic surface, directed outwards, backwards and downwards, is of an oblong form, smooth, and forms part of the zygomatic fossa. This surface is separated from the orbital by a smooth rounded border, which enters into the formation of the spheno-maxillary fissure.

The *Sphenoidal Process* of the palate bone is a thin compressed plate, much smaller than the orbital, and directed upwards and inwards. It presents three surfaces and two borders. The superior surface, the smallest of the three, articulates with the horizontal part of the sphenoidal turbinated bone; it presents a groove which contributes to the formation of the pterygo-palatine canal. The internal surface is concave, and forms part of the outer wall of the nasal fossa. The external surface is divided into an articular, and a non-articular portion; the former is rough for articulation with the inner surface of the pterygoid process of the sphenoid; the latter is smooth, and forms part of the zygomatic fossa. The anterior border forms the posterior boundary of the spheno-palatine foramen. The posterior border, serrated at the expense of the outer table, articulates with the inner surface of the pterygoid process.

The orbital and sphenoidal processes are separated from one another by a deep notch, which is converted into a foramen, the spheno-palatine, by articulation with the sphenoidal turbinated bone. Sometimes the two processes are united above, and form between them a complete foramen, or the notch is crossed by one or more spiculæ of bone, so as to form two or more foramina. In the articulated skull, this foramen opens into the back part of the outer wall of the superior meatus, and transmits the spheno-palatine vessels and nerves.

*Development.* From a single centre, which makes its appearance at the angle of junction of the two plates of the bone. From this point ossification spreads; inwards, to the horizontal plate; downwards, into the tuberosity; and upwards, into the vertical plate. In the fœtus, the horizontal plate is much longer than the vertical; and even after it is fully ossified, the whole bone is remarkable for its shortness.

*Articulations.* With seven bones; the sphenoid, ethmoid, superior maxillary, inferior turbinated, vomer, opposite palate, and sphenoidal turbinated.

*Attachment of Muscles.* The Tensor palati, Azygos uvulæ, Internal and External pterygoid.

## The Inferior Turbinated Bones.

The *Inferior Turbinated* bones are situated one on each side of the outer wall of the nasal fossæ. Each bone consists of a layer of thin spongy bone, curled upon itself like a scroll, hence its name 'turbinated'; and extends horizontally across the outer wall of the nasal fossa, immediately below the orifice of the antrum. Each bone presents two surfaces, two borders, and two extremities.

The *internal surface* (fig. 53) is convex, perforated by numerous apertures, and traversed by longitudinal grooves and canals for the lodgment of arteries and veins. In the recent state it is covered by the lining membrane of the nose. The *external surface* is concave (fig. 54), and forms part of the inferior meatus. Its upper border is thin, irregular, and connected to various bones along the outer wall of the nose. It may be divided into three portions; of these, the anterior articulates with the inferior turbinated crest of the superior maxillary bone; the posterior with the inferior turbinated crest of the palate bone; the middle portion of the superior border presents three well marked processes, which vary much in their size and form. Of these the anterior and smallest, is situated at the junction of the anterior fourth with the posterior three-fourths of the bone; it is small and pointed, and is called the *lachrymal process*, for it articulates with the

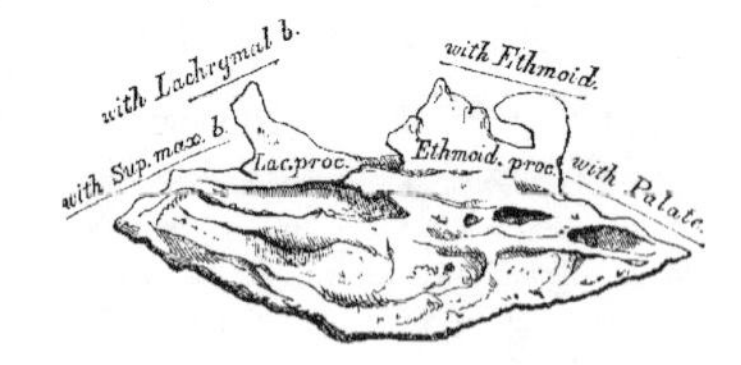

53 – *Right inferior turbinated bone. Inner surface.*

54 – *Right inferior turbinated bone. Outer surface.*

anterior inferior angle of the lachrymal bone, and by its margins, with the groove on the back of the nasal process of the superior maxillary, and thus assists in forming the lachrymal canal. At the junction of the two middle fourths of the bone, but encroaching on the latter, a broad thin plate, the *ethmoidal process*, ascends to join the unciform process of the ethmoid; from the lower border of this process, a thin lamina of bone curves downwards and outwards, hooking over the lower edge of the orifice of the antrum, which it narrows below; it is called the *maxillary process*, and fixes the bone firmly on to the outer wall of the nasal fossa. The inferior border is free, thick and cellular in structure, more especially in the centre of the bone. Both extremities are more or less narrow and pointed. If the bone is held so that its outer concave surface is directed backwards (i.e., towards the holder), and its superior border, from which the lachrymal and ethmoidal processes project, upwards, the lachrymal process will be directed to the side to which the bone belongs.

*Development.* By a single centre which makes its appearance about the middle of fœtal life.

*Articulations.* With four bones; one of the cranium, the ethmoid, and three of the face, the superior maxillary, lachrymal and palate.

No muscles are attached to this bone.

## The Vomer.

The *Vomer* is a single bone, situated vertically at the back part of the nasal fossæ, forming part of the septum of the nose. It is thin, somewhat like a ploughshare in form; but it varies in different individuals, being frequently bent to one or the other side; it presents for examination two surfaces and four borders. The lateral surfaces are smooth, marked with small furrows for the lodgment of blood-vessels, and by a groove on each side, sometimes a canal, the naso-palatine, which runs obliquely downwards and forwards to the intermaxillary suture between the two anterior palatine canals; it transmits the naso-palatine nerve. The superior border, the thickest, presents a deep groove, bounded on each side by a horizontal projecting ala of bone; the groove receives the rostrum of the sphenoid, whilst the alæ are overlapped and retained by laminæ (the vaginal processes) which project from the under surface of the body of the sphenoid at the base of the pterygoid processes. At the front of the groove a fissure is left for the transmission of blood-vessels to the substance of the bone. The inferior border, the longest, is broad and uneven in front, where it articulates with the two superior maxillary bones; thin and sharp behind where it joins with the palate bones. The upper half of the anterior border usually consists of two laminæ of bone, between which is received the perpendicular plate of the ethmoid, the lower half consisting of a single rough edge, also occasionally channelled, which is united to the triangular cartilage of the nose. The posterior border is free, concave, and separates the nasal fossæ behind. It is thick and bifid above, thin below.

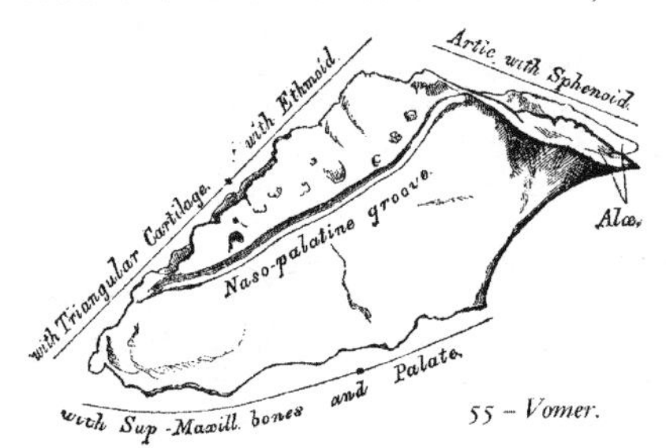

55 – Vomer.

*Development.* The vomer at an early period consists of two laminæ separated by a very considerable interval, and enclosing between them a plate of cartilage which is prolonged forwards to form the remainder of the septum. Ossification commences in it at about the same period as in the vertebræ, the coalescence of the laminæ taking place from behind forwards, but is not complete until after puberty.

*Articulations.* With six bones; two of the cranium, the sphenoid and ethmoid; and four of the face, the two superior maxillary, the two palate bones, and with the cartilage of the septum.

The vomer has no muscles attached to it.

## The Inferior Maxillary Bone.

The *Inferior Maxillary* Bone, the largest and strongest bone of the face, serves for the reception of the inferior teeth. It consists of a curved horizontal portion, the body, and of two perpendicular portions, the rami, which join the former nearly at right angles behind.

The *Horizontal* portion, or body (fig. 56), is convex in its general outline, and curved somewhat

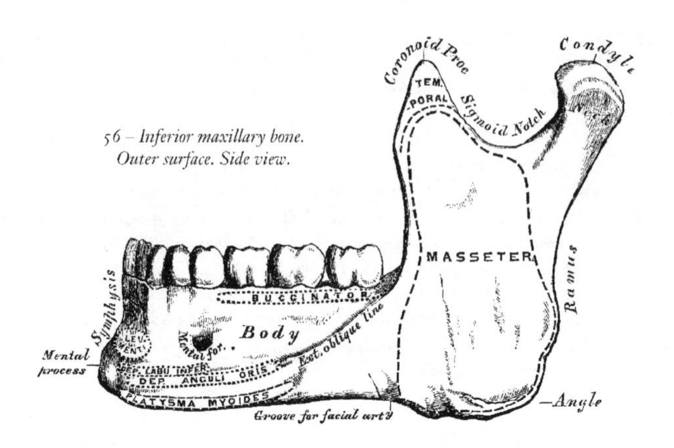


56 – *Inferior maxillary bone. Outer surface. Side view.*

like a horse-shoe. It presents for examination two surfaces and two borders. The *external surface* is convex from side to side, concave from above downwards. In the median line is a vertical ridge, the symphysis; it extends from the upper to the lower border of the bone, and indicates the point of junction of the two pieces of which the bone is composed at an early period of life. The lower part of the ridge terminates in a prominent triangular eminence, the mental process. On either side of the symphysis, just below the roots of the incisor teeth, is a depression, the incisive fossa, for the attachment of the Levator menti; and still more externally, a foramen, the mental foramen, for the passage of the mental nerve and artery. This foramen is placed just below the root of the second bicuspid tooth. Running outwards from the base of the mental process on each side, is a well marked ridge, the external oblique line. This ridge is at first nearly horizontal, but afterwards inclines upwards and backwards, and is continuous with the anterior border of the ramus; it affords attachment to the Depressor labii inferioris and Depressor anguli oris, below these the Platysma myoides is inserted. The external oblique line, and the internal or mylo-hyoidean line (to be afterwards described), divide the body of the bone into a superior or alveolar, and an inferior or basilar portion.

The *internal surface* (fig. 57) is concave from side to side, convex from above downwards. In the middle line is an indistinct linear depression, corresponding to the symphysis externally; on either side of this depression, just below its centre, are four prominent tubercles, placed in pairs, two above and two below; they are called the *genial tubercles*, and afford attachment, the upper pair to the Genio-hyoglossi muscles, the lower pair to the Genio-hyoidei muscles. Sometimes the tubercles on each side are blended into one, or they all unite into an irregular eminence of bone, or nothing but an irregularity may be seen on the surface of the bone at this part. On either side of the genial tubercles is an oval depression, the sublingual fossa, for lodging the sublingual gland; and beneath the fossa a rough depression on each side, which gives attachment to the anterior belly of the Digastric muscle. At the back part of the sublingual fossa, the internal oblique line (mylo-hyoidean) commences; it is at first faintly marked, but becomes more distinct as it passes upwards and outwards, and is especially prominent opposite the last two molar teeth; it divides the lateral surface of the bone into two portions, and affords attachment throughout its whole extent to the Mylo-hyoid muscle, the Superior constrictor being attached above its posterior extremity, nearer the alveolar margin. The portion of bone above this ridge is smooth, and covered by the mucous membrane of the mouth; whilst that below it presents an oblong depression, the submaxillary fossa, wider behind than in front, for the lodgment of the submaxillary gland. The *superior* or *alveolar border* is wider, and its margins thicker behind than in front. It is hollowed into numerous cavities, for the reception of the teeth; these are sixteen in number, and vary in depth and size according to the teeth which they contain. The *inferior border* is rounded, longer than the superior, and thicker in front than behind; it presents a shallow groove, just where the body joins the ramus, over which the facial artery turns.

The *Perpendicular Portions*, or *Rami*, are of a quadrilateral form. Each presents for examination two surfaces, four borders, and two processes. The *external surface* is flat, marked with ridges, and gives attachment throughout nearly the whole of its extent to the Masseter muscle. The *internal surface* presents about its centre the oblique aperture of the inferior dental canal, for the passage of the inferior dental vessels and nerve. The margin of this opening is irregular; it presents in front a prominent ridge, surmounted by a sharp spine, which gives attachment to the internal lateral ligament of the lower jaw; and at its lower and back part a notch leading to a groove, the

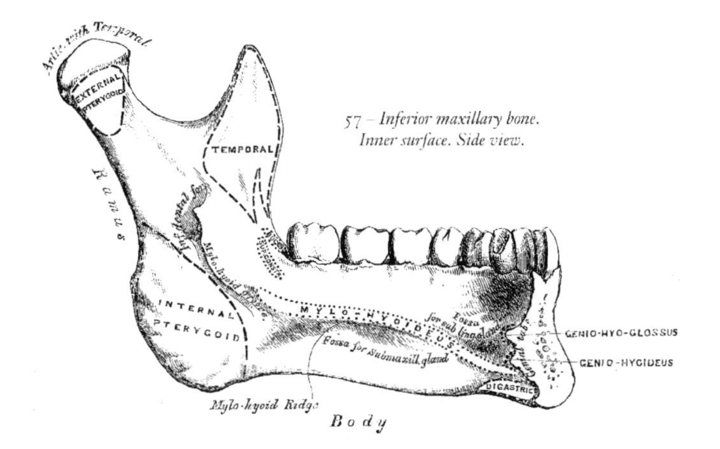


57 – *Inferior maxillary bone. Inner surface. Side view.*

mylo-hyoidean, which runs obliquely downwards to the back part of the submaxillary fossa; and lodges the mylo-hyoid vessels and nerve; behind the groove is a rough surface, for the insertion of the Internal pterygoid muscle. The inferior dental canal descends obliquely downwards and forwards in the substance of the ramus, and then horizontally forwards in the body; it is here placed under the alveoli, with which it communicates by small openings. On arriving at the incisor teeth, it turns back to communicate with the mental foramen, giving off two small canals, which run forward, to be lost in the cancellous tissue of the bone beneath the incisor teeth. This canal, in the posterior two thirds of the bone, runs nearest the internal surface of the jaw; and in the anterior third, nearer its external surface. Its walls are composed of compact tissue at either extremity, cancellous in the centre. It contains the inferior dental vessels and nerve, from which branches are distributed to the teeth through small apertures at the bases of the alveoli. The *upper border* of the ramus is thin, and presents two processes, separated by a deep concavity, the sigmoid notch. Of these processes, the anterior is the coronoid, the posterior the condyloid.

The *Coronoid Process* is a thin, flattened, triangular eminence of bone, which varies in shape and size in different subjects, and serves essentially for the attachment of the Temporal muscle. Its *external surface* is smooth, and affords attachment to the Masseter and Temporal muscles. Its *internal surface* gives attachment to the Temporal muscle, and presents the commencement of a longitudinal ridge, which is continued to the posterior part of the alveolar process. On the outer side of this ridge is a deep groove, continued below on the outer side of the alveolar process; this ridge and part of the groove afford attachment, above, to the Temporal; below, to the Buccinator muscle.

The *Condyloid Process*, shorter but thicker than the coronoid, consists of two portions: the condyle, and the constricted portion which supports the condyle, the neck. The condyle is of an oblong form, its long axis being transverse, and set obliquely on the neck in such a manner that its outer end is a little more forward and a little higher than its inner. It is convex from before backwards, and from side to side, the articular surface extending further on the posterior than on the anterior surface. The neck of the condyle is flattened from before backwards, and strengthened by ridges which descend from the fore part and sides of the condyle. Its lateral margins are narrow, and present externally a tubercle for the external lateral ligament. Its posterior surface is convex; its anterior is hollowed out on its inner side by a depression (the pterygoid fossa) for the attachment of the External pterygoid. The *lower border* of the ramus is thick, straight, and continuous with the body of the bone. At its junction with the posterior border is the angle of the jaw, which is either inverted or everted, and marked by rough oblique ridges on each side for the attachment of the Masseter externally, and the Internal pterygoid internally; and, between them, serving for the attachment of the stylo-maxillary ligament. The *anterior border* is thin above, thicker below, and continuous with the external oblique line. The *posterior border* is thick, smooth, rounded, and covered by the parotid gland.

The *Sigmoid Notch*, separating the two processes, is a deep semilunar depression, crossed by the masseteric artery and nerve.

*Development.* This bone is formed at such an early period of life, before, indeed, any other bone excepting the clavicle, that it has been found impossible at present to determine its earliest condition. It appears probable, however, that it is developed by *two* centres, one for each lateral half, the two segments meeting at the symphysis, where they become united. Additional centres have also been described for the coronoid process, the condyle, the angle, and the thin plate of bone which forms the inner side of the alveolus.

## Side-view of the Lower Jaw at different Periods of Life.

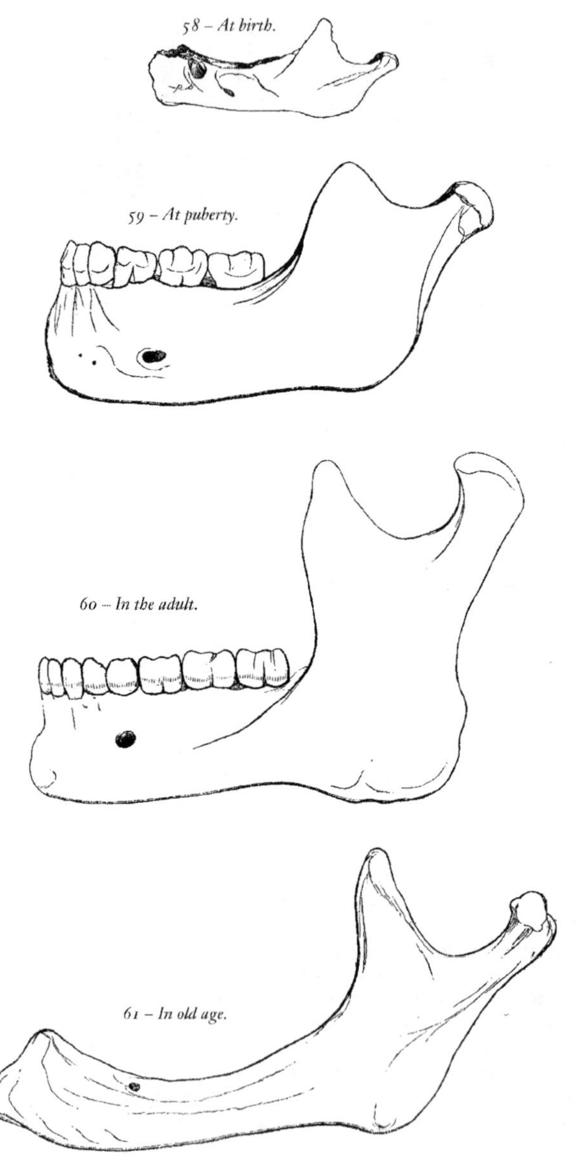


### Changes produced in the Lower Jaw by Age

The changes which the Lower Jaw undergoes after birth, relate—1. To the alterations effected in the body of the bone by the first and second dentitions, the loss of the teeth in the aged, and the subsequent absorption of the alveoli. 2. To the size and situation of the dental canal; and, 3. To the angle at which the ramus joins with the body.

*At birth* (fig. 58), the bone consists of two lateral halves, united by fibro-cartilaginous tissue, in which one or two osseous nuclei are generally found. The body is a mere shell of bone, containing the sockets of the two incisor, the canine, and the first molar teeth, imperfectly partitioned from one another. The dental canal is of large size, and runs near the lower border of the bone, the mental foramen opening beneath the socket of the first molar. The angle is obtuse, from the jaws not being as yet separated by the eruption of the teeth.

*After birth* (fig. 59), the two segments of the bone become joined at the symphysis, from below upwards, in the first year; but a trace of separation may be visible in the beginning of the second year, near the alveolar margin. The body becomes elongated in its whole length, but more especially behind the mental foramen, to provide space for the three additional teeth developed in this part. The depth of the body becomes greater, owing to increased growth of the alveolar part, to afford room for the fangs of the teeth, and by thickening of the subdental portion which enables the jaw to withstand the powerful action of the masticatory muscles; but the alveolar portion is the deeper of the two, and, consequently, the chief part of the body lies above the oblique line. The dental canal, after the second dentition, is situated just above the level of the mylo-hyoid ridge; and the mental foramen occupies the position usual to it in the adult. The angle becomes less obtuse, owing to the separation of the jaws by the teeth.

*In the adult* (fig. 60), the alveolar and basilar portions of the body are usually of equal depth. The mental foramen opens midway between the upper and lower border of the bone, and the dental canal runs nearly parallel with the mylo-hyoid line. The ramus is almost vertical in direction, and joins the body nearly at right angles.

*In old age* (fig. 61), the bone becomes greatly reduced in size; for, with the loss of the teeth, the alveolar process is absorbed, and the basilar part of the bone alone remains; consequently, the chief part of the bone is *below* the oblique line. The dental canal, with the mental foramen opening from it, is close to the alveolar border. The rami are oblique in direction, and the angle obtuse.

*Articulations.* With the glenoid fossæ of the two temporal bones.

*Attachment of Muscles.* By its external surface, commencing at the symphysis, and proceeding backwards: Levator menti, Depressor labii inferioris, Depressor anguli oris, Platysma myoides, Buccinator, Masseter. By its internal surface, commencing at the same point: Genio-hyo-glossus, Genio-hyoideus, Mylo-hyoideus, Digastric, Superior constrictor, Temporal, Internal pterygoid, External pterygoid.

## The Sutures.

The bones of the cranium and face are connected to each other by means of sutures. The dentations by which they are joined, are confined to the external table, the edges of the internal table lying merely in apposition with the contiguous bone. The *Cranial Sutures* may be divided into three sets: 1. Those at the vertex of the skull. 2. Those at the side of the skull. 3. Those at the base.

The sutures at the vertex of the skull are three, the sagittal, coronal, and lambdoid.

The *Sagittal Suture* (*interparietal*) is formed by the junction of the two parietal bones, and extends from the middle of the frontal bone, backwards to the superior angle of the occipital. In childhood, and occasionally in the adult, when the two halves of the frontal bone are not united, it is continued forwards to the root of the nose. This suture sometimes presents, near its posterior extremity, the parietal foramen on each side; and in front, where it joins the coronal suture, a space is occasionally left, which encloses a large Wormian bone.

The *Coronal Suture* (*fronto-parietal*) extends transversely across the vertex of the skull, and connects the frontal with the parietal bones. It commences at the extremity of the great wing of the sphenoid on one side, and terminates at the same point on the opposite side. The dentations of this suture are more marked at the sides than at the summit, and are so constructed that the frontal rests on the parietal above, whilst laterally the frontal supports the parietal.

The *Lambdoid Suture* (*occipito-parietal*), so called from its resemblance to the Greek letter *Λ*, connects the occipital with the parietal bones. It commences on each side at the mastoid portion of the temporal bone, and inclines upwards to the end of the sagittal suture. The dentations of this suture are very deep and distinct, and are often interrupted by several small Wormian bones.

The sutures at the side of the skull are also three in number: the spheno-parietal, squamo-parietal, and masto-parietal. They are subdivisions of a single suture, formed between the lower border of the parietal, and the temporal and sphenoid bones, and which extends from the lower end of the lambdoid suture behind, to the lower end of the coronal suture in front.

The *Spheno-parietal* is very short; it is formed by the tip of the great wing of the sphenoid, which overlaps the anterior inferior angle of the parietal bone.

The *Squamo-parietal*, or squamous suture, is arched. It is formed by the squamous portion of the temporal bone overlapping the middle division of the lower border of the parietal.

The *Masto-parietal* is a short suture, deeply dentated, formed by the posterior inferior angle of the parietal, and the superior border of the mastoid portion of the temporal.

The sutures at the base of the skull are, the basilar in the centre, and on each side, the petro-occipital, the masto-occipital, the petro-sphenoidal, and the squamo-sphenoidal.

The *Basilar Suture* is formed by the junction of the basilar surface of the occipital bone with the posterior surface of the body of the sphenoid. At an early period of life, a thin plate of cartilage exists between these bones; but in the adult they become inseparably united. Between the outer extremity of the basilar suture, and the termination of the lambdoid, an irregular suture exists which is subdivided into two portions. The inner portion, formed by the union of the petrous part of the temporal, with the occipital bone, is termed the *petro-occipital.* The outer portion, formed by the junction of the mastoid part of the temporal with the occipital, is called the *masto-occipital.* Between the bones forming the petro-occipital suture, a thin plate of cartilage exists; in the masto-occipital is occasionally found the opening of the mastoid foramen. Between the outer extremity of the basilar suture and the spheno-parietal, an irregular suture may be seen formed by the union of the sphenoid with the temporal bone. The inner and smaller portion of this suture is termed the *petro-sphenoidal*; it is formed between the petrous portion of the temporal, and the great wing of the sphenoid; the outer portion, of greater length, and arched, is formed between the squamous portion of the temporal, and the great wing of the sphenoid: it is called the *squamo-sphenoidal.*

The cranial bones are connected with those of the face, and the facial bones with each other, by numerous sutures, which, though distinctly marked, have received no special names. The only remaining suture deserving especial consideration, is the *transverse.* This extends across the upper part of the face, and is formed by the junction of the frontal with the facial bones; it extends from the external angular process of one side, to the same point on the opposite side, and connects the frontal with the malar, the sphenoid, the ethmoid, the lachrymal, the superior maxillary, and the nasal bones on each side.

The sutures remain separate for a considerable period after the complete formation of the skull. It is probable, that they serve the purpose of permitting the growth of the bones at their margins; while their peculiar formation, and the interposition of the sutural ligament between the bones forming them, prevents the dispersion of blows or jars received upon the skull. Mr. Humphry remarks, 'that, as a general rule, the sutures are first obliterated at the parts in which the ossification of the skull was last completed, viz., in the neighbourhood of the fontanelles; and the cranial bones seem in this respect to observe a similar law to that which regulates the union of the epiphyses to the shafts of the long bones.'

## The Skull.

The Skull, formed by the union of the several cranial and facial bones already described, when considered as a whole, is divisible into five regions: a superior region or vertex, an inferior region or base, two lateral regions, and an anterior region, the face.

## Vertex of the Skull.

The *Superior Region*, or *vertex*, presents two surfaces, an external, and an internal.

The *External Surface* is bounded, in front, by the nasal eminences, and superciliary ridges; behind, by the occipital protuberance and superior curved lines of the occipital bone; laterally, by an imaginary line extending from the outer end of the superior curved line, along the temporal ridge, to the external angular process of the frontal. This surface includes the vertical portion of the frontal, the greater part of the parietal, and the superior third of the occipital bone; it is smooth, convex, of an elongated oval form, crossed transversely by the coronal suture, and from before backwards by the sagittal, which terminates behind in the lambdoid. From before backwards may be seen the frontal eminences and remains of the suture connecting the two lateral halves of the frontal bone; on each side of the sagittal suture is the parietal foramen and parietal eminence, and still more posteriorly the smooth convex surface of the occipital bone.

The *Internal Surface* is concave, presents eminences and depressions for the convolutions of the cerebrum, and numerous furrows for the lodgment of branches of the meningeal arteries. Along the middle line of this surface is a longitudinal groove, narrow in front, where it terminates in the frontal crest; broader behind; it lodges the superior longitudinal sinus, and its margins afford attachment to the falx cerebri. On either side of it are several depressions for the Pacchionian bodies, and at its back part, the internal openings of the parietal foramina. This surface is crossed, in front, by the coronal suture; from before backwards, by the sagittal; behind, by the lambdoid.

## Base of the Skull.

The *Inferior Region*, or *base* of the skull presents two surfaces, an internal or cerebral, and an external or basilar.

The *Internal*, or *Cerebral Surface* (fig. 62), presents three fossæ on each side, called the *anterior*, *middle*, and *posterior* fossæ of the cranium.

The *Anterior Fossa* is formed by the orbital plate of the frontal, the cribriform plate of the ethmoid, the ethmoidal process and lesser wing of the sphenoid. It is the most elevated of the three fossæ, convex externally where it corresponds to the roof of the orbit, concave in the median line in the situation of the cribriform plate of the ethmoid. It is traversed by three sutures, the ethmoido-frontal, ethmo-sphenoidal, and fronto-sphenoidal; and lodges the anterior lobe of the cerebrum. It presents, in the median line, from before backwards, the commencement of the groove for the

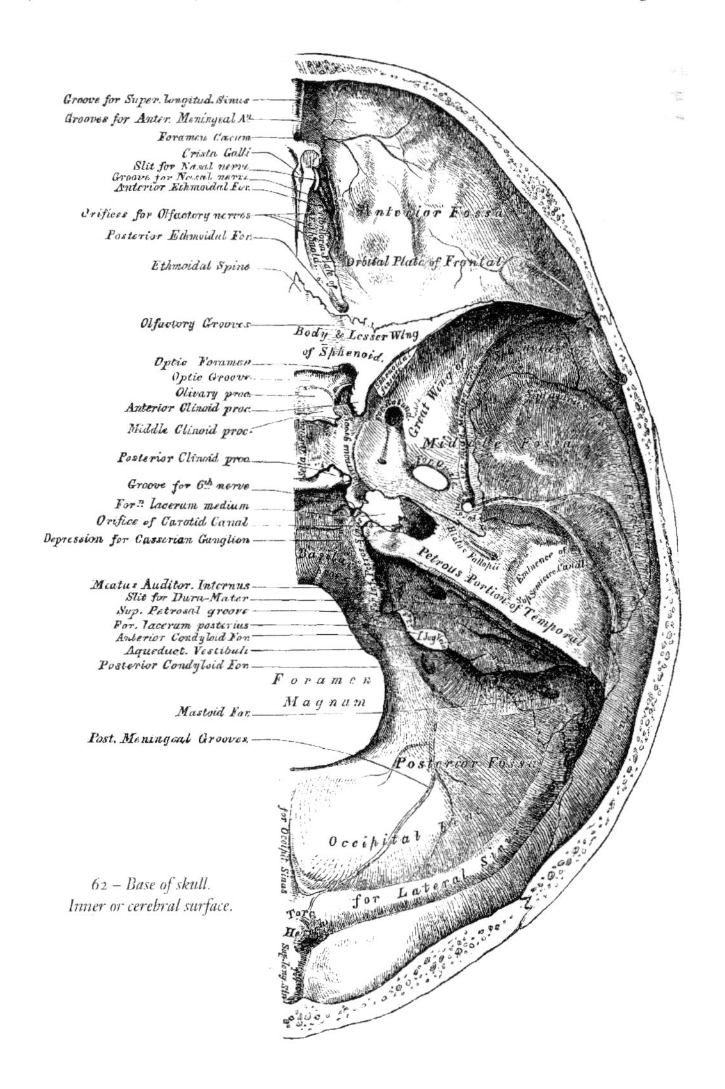


62 – *Base of skull. Inner or cerebral surface.*

superior longitudinal sinus, and crest for the attachment of the falx cerebri; the foramen cœcum, this aperture is formed by the frontal and crista galli of the ethmoid, and, if pervious, transmits a small vein from the nose to the superior longitudinal sinus. Behind the foramen cœcum is the crista galli, the posterior margin of which affords attachment to the falx cerebri. On either side of the crista galli is the olfactory groove, which supports the bulb of the olfactory nerve, perforated by three rows of orifices which give passage to its filaments; and in front by a slit-like opening, which transmits the nasal branch of the ophthalmic nerve. On the outer side of each olfactory groove are the internal openings of the anterior and posterior ethmoidal foramina; the former, situated about the middle of its outer margin, transmits the nasal nerve, which runs in a groove along its surface, to the slit-like opening above mentioned; whilst the latter, the posterior ethmoidal foramen, opens at the back part of this margin under cover of a projecting lamina of the sphenoid, it transmits the posterior ethmoidal artery and vein to the posterior ethmoidal cells. Further back in the middle line is the ethmoidal spine, bounded behind by an elevated ridge, separating a longitudinal groove on each side which supports the olfactory nerve. The anterior fossa presents laterally eminences and depressions for the convolutions of the brain, and grooves for the lodgment of the anterior meningeal arteries.

The *Middle Fossa*, somewhat deeper than the preceding, is narrow in the middle, and becomes wider as it expands laterally. It is bounded in front by the posterior margin of the lesser wing of the sphenoid, the anterior clinoid process, and the anterior margin of the optic groove; behind, by the petrous portion of the temporal, and basilar suture; externally, by the squamous portion of the temporal, and anterior inferior angle of the parietal bone, and is separated from its fellow by the sella Turcica. It is traversed by four sutures, the squamous, spheno-parietal, spheno-temporal, and petro-sphenoidal.

In the middle line, from before backwards, is the optic groove, which supports the optic commissure, terminating on each side in the optic foramen, for the passage of the optic nerve and ophthalmic artery; behind the optic groove is the olivary process, and laterally the anterior clinoid processes, which afford attachment to the folds of the dura mater, which form the cavernous sinuses. Separating the middle fossæ is the sella Turcica, a deep depression, which lodges the pituitary gland, bounded in front by a small eminence on either side, the middle clinoid process, and behind by a broad square plate of bone, surmounted at each superior angle by a tubercle, the posterior clinoid process; beneath the latter process is a groove, for the lodgment of the sixth nerve. On each side of the sella Turcica is the cavernous groove; it is broad, shallow, and curved somewhat like the italic letter *f*: it commences behind at the foramen lacerum medium, and terminates on the inner side of the anterior clinoid process. This groove lodges the cavernous sinus, the internal carotid artery, and the orbital nerves. The sides of the middle fossa are of considerable depth; they present eminences and depressions for the middle lobes of the brain, and grooves for lodging the branches of the middle meningeal artery; the latter commence on the outer side of the foramen spinosum, and consist of two large branches, an anterior and a posterior; the former passing upwards and forwards to the anterior inferior angle of the parietal bone, the latter passing upwards and backwards. The following foramina may also be seen from before backwards. Most anteriorly is the foramen lacerum anterius, or sphenoidal fissure, formed above by the lesser wing of the sphenoid; below, by the greater wing; internally, by the body of the sphenoid; and completed externally by the orbital plate of the frontal bone. It transmits the third, fourth, the three branches of the ophthalmic division of the fifth, the sixth nerve, and the ophthalmic vein. Behind the inner extremity of the sphenoidal fissure is the foramen rotundum, for the passage of the second division of the fifth or superior maxillary nerve; still more posteriorly is seen a small orifice, the foramen Vesalii; this opening is situated between the foramen rotundum and ovale, a little internal to both; it varies in size in different individuals, and transmits a small vein. It opens below in the pterygoid fossa, just at the outer side of the scaphoid depression. Behind and external to the latter opening is the foramen ovale, which transmits the third division of the fifth or inferior maxillary nerve, the small meningeal artery, and the small petrosal nerve. On the outer side of the foramen ovale is the foramen spinosum, for the passage of the middle meningeal artery; and on the inner side of the foramen ovale, the foramen lacerum medium. The lower part of this aperture is filled up with cartilage in the recent state. On the anterior surface of the petrous portion of the temporal bone is seen from without inwards, the eminence caused by the projection of the superior semicircular canal, the groove leading to the hiatus Fallopii, for the transmission of the petrosal branch of the Vidian nerve; beneath it, the smaller groove, for the passage of the smaller petrosal nerve; and near the apex of the bone, the depression for the Gasserian ganglion, and the orifice of the carotid canal, for the passage of the internal carotid artery and carotid plexus of nerves.

The *Posterior Fossa*, deeply concave, is the largest of the three, and situated on a lower level than either of the preceding. It is formed by the occipital, the petrous and mastoid portions of the temporal, and the posterior inferior angle of the parietal bone; is crossed by three sutures, the petro-occipital, masto-occipital, and masto-parietal; and lodges the cerebellum, pons Varolii, and medulla oblongata. It is separated from the middle fossa in the median line by the basilar suture, and on each side by the superior border of the petrous portion of the temporal bone. This serves for the attachment of the tentorium cerebelli, is grooved externally for the superior petrosal sinus, and at its inner extremity presents a notch, upon which rests the fifth nerve. Its circumference is bounded posteriorly by the grooves for the lateral sinuses. In the centre of this fossa is the foramen magnum, bounded on either side by a rough tubercle, which gives attachment to the odontoid ligaments; and a little above these are seen the internal openings of the anterior condyloid foramina. In front of the foramen magnum is the basilar process, grooved for the support of the medulla oblongata and pons Varolii, and articulating on each side with the petrous portion of the temporal bone, forming the petro-occipital suture, the anterior half of which is grooved for the inferior petrosal sinus, the posterior half being encroached upon by the foramen lacerum posterius, or jugular foramen. This foramen is partially subdivided into two parts; the posterior and larger division transmits the internal jugular vein, the anterior the eighth pair of nerves. Above the jugular foramen is the internal auditory foramen, for the auditory and facial nerves and auditory artery; behind and external to this is the slit-like opening leading into the aquæductus vestibuli; whilst between the two latter, and near the superior border of the petrous portion, is a small triangular depression, which lodges a process of the dura mater, and occasionally transmits a small vein into the substance of the bone. Behind the foramen magnum are the inferior occipital fossæ, which lodge the lateral lobes of the cerebellum, separated from one another by the internal occipital crest, which serves for the attachment of the falx cerebelli, and lodges the occipital sinuses. These fossæ are surmounted, above, by the deep transverse grooves for the lodgment of the lateral sinuses. These channels, in their passage outwards, groove the occipital bone, the posterior inferior angle of the parietal, the mastoid portion of the temporal, and the occipital just behind the jugular foramen, at the back part of which they terminate. Where this sinus grooves the mastoid part of the temporal bone, the orifice of the mastoid foramen may be seen; and, just previous to its termination, it has opening into it the posterior condyloid foramen.

The *External Surface* of the base of the Skull (fig. 63) is extremely irregular. It is bounded in front by the incisor teeth in the upper jaws; behind, by the superior curved lines of the occipital bone; and laterally, by the alveolar arch, the lower border of the malar bone, the zygoma, and an imaginary line, extending from the zygoma to the mastoid process and extremity of the superior curved line of the occiput. It is formed by the palate processes of the two superior maxillary and palate bones, the vomer, the pterygoid, under surface of the great wing, spinous process and part of the body of the sphenoid, the under surface of the squamous, mastoid, and petrous portions of the temporal, and occipital bones. The anterior part of the base of the skull is raised above the level of the rest of this surface (when the skull is turned over for the purpose of examination), surrounded by the alveolar process, which is thicker behind than in front, and excavated by sixteen depressions for lodging the teeth of the upper jaw; they vary in depth and size according to the teeth they contain. Immediately behind the incisor teeth is the anterior palatine fossa. At the bottom of this fossa may usually be seen four apertures, two placed laterally, which open above, one in the floor of each nostril, and transmit the anterior palatine vessels, and two in the median line of the intermaxillary suture, one in front of the other, the most anterior one transmitting the left, and the posterior one (the larger) the right naso-palatine nerve. These two latter canals are sometimes wanting, or they may join to form a single one, or one of them may open into one of the lateral canals above referred to. The palatine vault is concave, uneven, perforated by numerous foramina, marked by depressions for the palatal glands, and crossed by a crucial suture, which indicates the point of junction of the four bones of which it is composed. One or two small foramina, seen in the alveolar margin behind the incisor teeth, occasionally seen in the adult, almost constant in young subjects, are called the *incisive foramina*; they transmit nerves and vessels to the incisor teeth. At each posterior angle of the hard palate is the posterior palatine foramen, for the transmission of the posterior palatine vessels and anterior palatine nerve, and running forwards and inwards from it a groove, which lodges the same vessels and nerve. Behind the posterior palatine foramen is the tuberosity of the palate bone, perforated by one or more accessory posterior palatine canals, and marked by the commencement of a ridge, which runs transversely inwards, and serves for the attachment of the tendinous expansion of the Tensor palati muscle. Projecting backwards from the centre of the posterior border of the hard palate is the posterior nasal spine, for the attachment of the Azygos uvulæ. Behind and above the hard palate is the posterior aperture of the nares, divided into two parts by the vomer, bounded

63 – *Base of skull. External surface.*

above by the body of the sphenoid, below by the horizontal plate of the palate bone, and laterally by the pterygoid processes of the sphenoid. Each aperture measures about an inch in the vertical, and half an inch in the transverse direction. At the base of the vomer may be seen the expanded alæ of this bone, receiving between them the rostrum of the sphenoid. Near the lateral margins of the vomer, at the root of the pterygoid processes are the pterygo-palatine canals. The pterygoid process, which bounds the posterior nares on each side, presents near its base the pterygoid or Vidian canal, for the Vidian nerve and artery. Each process consists of two plates, which bifurcate at the extremity to receive the tuberosity of the palate bone, and are separated behind by the pterygoid fossa, which lodges the Internal pterygoid muscle. The internal plate is long and narrow, presenting on the outer side of its base the scaphoid fossa, for the origin of the Tensor palati muscle, and at its extremity the hamular process, around which the tendon of this muscle turns. The external pterygoid plate is broad, forms the inner boundary of the zygomatic fossa, and affords attachment to the External pterygoid muscle.

Behind the nasal fossæ in the middle line is the basilar surface of the occipital bone, presenting in its centre the pharyngeal spine for the attachment of the Superior constrictor muscle of the pharynx, with depressions on each side for the insertion of the Rectus anticus major and minor. At the base of the external pterygoid plate is the foramen ovale; behind this, the foramen spinosum, and the prominent spinous process of the sphenoid, which gives attachment to the internal lateral ligament of the lower jaw and the Laxator tympani muscle. External to the spinous process is the glenoid fossa, divided into two parts by the Glaserian fissure, the anterior portion being concave, smooth, bounded in front by the eminentia articularis, and serving for the articulation of the condyle of the lower jaw; the posterior portion rough, bounded behind by the vaginal process, and serving for the reception of part of the parotid gland. Emerging from between the laminæ of the vaginal process is the styloid process; and at the base of this process is the stylo-mastoid foramen, for the exit of the facial nerve, and entrance of the stylo-mastoid artery. External to the stylo-mastoid foramen is the auricular fissure for the auricular branch of the pneumogastric, bounded behind by the mastoid process. Upon the inner side of the mastoid process is a deep groove, the digastric fossa; and a little more internally, the occipital groove, for the occipital artery. At the base of the internal pterygoid plate is a large and somewhat triangular aperture, the foramen lacerum medium, bounded in front by the great wing of the sphenoid, behind by the apex of the petrous portion of the temporal bone, and internally by the body of the sphenoid and basilar process of the occipital bone; it presents in front the posterior orifice of the Vidian canal, behind the aperture of the carotid canal. The basilar surface of this opening is filled up in the recent state by a fibro-cartilaginous substance; across its upper or cerebral aspect passes the internal carotid artery and Vidian nerve. External to this aperture, the petrosphenoidal suture is observed, at the outer termination of which is seen the orifice of the canal for the Eustachian tube, and that for the Tensor tympani muscle. Behind this suture is seen the under surface of the petrous portion of the temporal bone, presenting, from within outwards, the quadrilateral rough surface, part of which affords attachment to the Levator palati and Tensor tympani muscles; external to this surface are the orifices of the carotid canal and the aquæductus cochleæ, the former transmitting the internal carotid artery and the ascending branches of the superior cervical ganglion of the sympathetic, the latter serving for the passage of a small artery and vein to the cochlea. Behind the carotid canal is a large aperture, the jugular fossa, formed in front by the petrous portion of the temporal, and behind by the occipital; it is generally larger on the right than on the left side; and towards its cerebral aspect is divided into two parts by a ridge of bone, which projects usually from the temporal, the anterior, or smaller portion, transmitting the three divisions of the eighth pair of nerves; the posterior, transmitting the internal jugular vein and the ascending meningeal vessels, from the occipital and ascending pharyngeal arteries. On the ridge of bone dividing the carotid canal from the jugular fossa, is the small foramen for the transmission of the tympanic nerve; and on the outer wall of the jugular foramen, near the root of the styloid process, is the small aperture for the transmission of Arnold's nerve. Behind the basilar surface of the occipital bone is the foramen magnum, bounded on each side by the condyles, rough internally for the attachment of the alar ligaments, and presenting externally a rough surface, the jugular process, which serves for the attachment of the Rectus lateralis. On either side of each condyle anteriorly is the anterior condyloid fossa, perforated by the anterior condyloid foramen, for the passage of the hypoglossal nerve. Behind each condyle are the posterior condyloid fossæ, perforated on one or both sides by the posterior condyloid foramina, for the transmission of a vein to the lateral sinus. Behind the foramen magnum is the external occipital crest, terminating above at the external occipital protuberance, whilst on each side are seen the superior and inferior curved lines; these, as well as the surfaces of the bone between them, being rough for the attachment of numerous muscles.

## Lateral Region of the Skull.

The *Lateral Region* of the Skull is somewhat of a triangular form, its base being formed by a line extending from the external angular process of the frontal bone along the temporal ridge backwards to the outer extremity of the superior curved line of the occiput: and the sides being formed by two lines, the one drawn downwards and backwards from the external angular process of the frontal bone to the angle of the lower jaw, the other from the angle of the jaw upwards and backwards to the extremity of the superior curved line. This region is divisible into three portions, temporal, mastoid, and zygomatic.

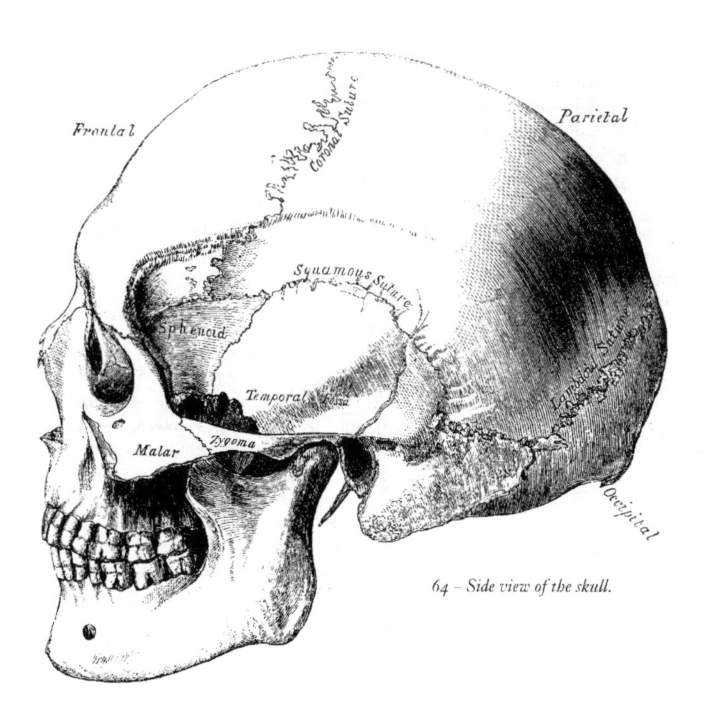


64 – *Side view of the skull.*

## The Temporal Fossæ.

The *Temporal* fossa, is bounded above and behind by the temporal ridge, which extends from the external angular process of the frontal upwards and backwards across the frontal and parietal bones, curving downwards behind to terminate at the root of the zygomatic process. In front, it is bounded by the frontal, malar, and great wing of the sphenoid: externally, by the zygomatic arch, formed conjointly by the malar and temporal bones; below, it is separated from the zygomatic fossa by the pterygoid ridge, seen on the outer surface of the great wing of the sphenoid. This fossa is formed by five bones, part of the frontal, great wing of the sphenoid, parietal, squamous portion of the temporal, and malar bones, and is traversed by five sutures, the transverse facial, coronal, spheno-parietal, squamo-parietal, and squamo-sphenoidal. It is deeply concave in front, convex behind, traversed by grooves for lodging branches of the deep temporal arteries, and filled by the Temporal muscle.

The *Mastoid Portion* is bounded in front by the anterior root of the zygoma; above, by a line which runs from the posterior root of the zygoma to the end of the masto-parietal suture; behind and below, by the masto-occipital suture. It is formed by the mastoid and part of the squamous portion of the temporal bone; its surface is convex and rough for the attachment of muscles, and presents, from behind forwards, the mastoid foramen, the mastoid process, the external auditory meatus, surrounded by the auditory process, and, most anteriorly, the glenoid fossa, bounded in front by the eminentia articularis; behind, by the vaginal process.

## The Zygomatic Fossæ.

The *Zygomatic* fossa is an irregular-shaped cavity, situated below, and on the inner side of the zygoma; bounded, in front, by the tuberosity of the superior maxillary bone and the ridge which descends from its malar process; behind, by the posterior border of the pterygoid process; above, by the pterygoid ridge on the outer surface of the great wing of the sphenoid and squamous portion of the temporal; below, by the alveolar border of the superior maxilla; internally, by the external pterygoid plate; and externally, by the zygomatic arch and ramus of the jaw. It contains the lower part of the Temporal, the External, and Internal pterygoid muscles, the internal maxillary artery, the inferior maxillary nerve, and their branches. At its upper and inner part may be observed two fissures, the spheno-maxillary and pterygo-maxillary.

The *Spheno-maxillary* fissure, horizontal in direction, opens into the outer and back part of the orbit. It is formed above by the lower border of the orbital surface of the great wing of the sphenoid; below, by the external border of the orbital surface of the superior maxilla and a small part of the palate bone; externally, by a small part of the malar bone; internally, it joins at right angles with the pterygo-maxillary fissure. This fissure opens a communication from the orbit into three fossæ, the temporal, zygomatic, and spheno-maxillary; it transmits the superior maxillary nerve, infra-orbital artery, and ascending branches from Meckel's ganglion.

The *Pterygo-maxillary* fissure is vertical, and descends at right angles from the inner extremity of the preceding; it is an elongated interval, formed by the divergence of the superior maxillary bone from the pterygoid process of the sphenoid. It serves to connect the spheno-maxillary fossa with the zygomatic, and transmits branches of the internal maxillary artery.

## The Spheno-maxillary Fossa.

The Spheno-maxillary fossa is a small triangular space situated at the angle of junction of the spheno-maxillary and pterygo-maxillary fissures, and placed beneath the apex of the orbit. It is formed above by the under surface of the body of the sphenoid; in front, by the superior maxillary bone; behind, by the pterygoid process of the sphenoid; internally by the vertical plate of the palate. This fossa has three fissures terminating in it, the sphenoidal, spheno-maxillary, and pterygo-maxillary; it communicates with three fossæ, the orbital, nasal, and zygomatic, and with the cavity of the cranium, and has opening into it five foramina. Of these there are three on the posterior wall, the foramen rotundum above, the Vidian below and internal, and still more inferior and internal, the pterygo-palatine. On the inner wall is the spheno-palatine foramen by which it communicates with the nasal fossa, and below, the superior orifice of the posterior palatine canal, besides occasionally the orifices of two or three accessory posterior palatine canals.

## Anterior Region of the Skull.

The Anterior Region of the Skull, which forms the face, is of an oval form, presents an irregular surface, and is excavated for the reception of the two principal organs of sense, the eye and the nose. It is bounded above by the nasal eminences and margins of the orbit; below, by the prominence of the chin; on each side, by the malar bone, and anterior margin of the ramus of the jaw. In the median line are seen from above downwards, the nasal eminences, which indicate the situation of the frontal sinuses; diverging outwards from the nasal eminences are the superciliary ridges which support the eyebrows. Beneath the nasal eminences is the arch of the nose, formed by the nasal bones, and the nasal processes of the superior maxillary. The nasal arch is convex from side to side, concave from above downwards, presenting in the median line the inter-nasal suture, formed between the nasal bones, laterally the naso-maxillary suture, formed between the nasal and the nasal process of the superior maxillary bones, both these sutures terminating above in that part of the transverse suture which connects the nasal bones and nasal processes of the superior maxillary with the frontal. Below the nose is seen the heart-shaped opening of the anterior nares, the narrow end upwards, and broad below; it presents laterally the thin sharp margins which serve for the attachment of the lateral cartilages of the nose, and in the middle line below, a prominent process, the anterior nasal spine, bounded by two deep notches. Below this is the intermaxillary suture, and on each side of it the incisive fossa. Beneath this fossa is the alveolar process of the upper and lower jaw, containing the incisor teeth, and at the lower part of the median line, the symphysis of the chin, the mental eminence, and the incisive fossa of the lower jaw.

Proceeding from above downwards, on each side, is the supra-orbital ridge, terminating externally in the external angular process at its junction with the malar, and internally in the internal angular process; towards the inner third of this ridge is the supra-orbital notch or foramen, for the passage of the supra-orbital vessels and nerve, and at its inner side a slight depression for the attachment of the cartilaginous pulley of the Superior oblique muscle. Beneath the supra-orbital ridge is the opening of the orbit, bounded externally by the orbital ridge of the malar bone; below, by the orbital ridge formed by the malar, superior maxillary, and lachrymal bones; internally, by the nasal process of the superior maxillary, and the internal angular process of the frontal bone. On the

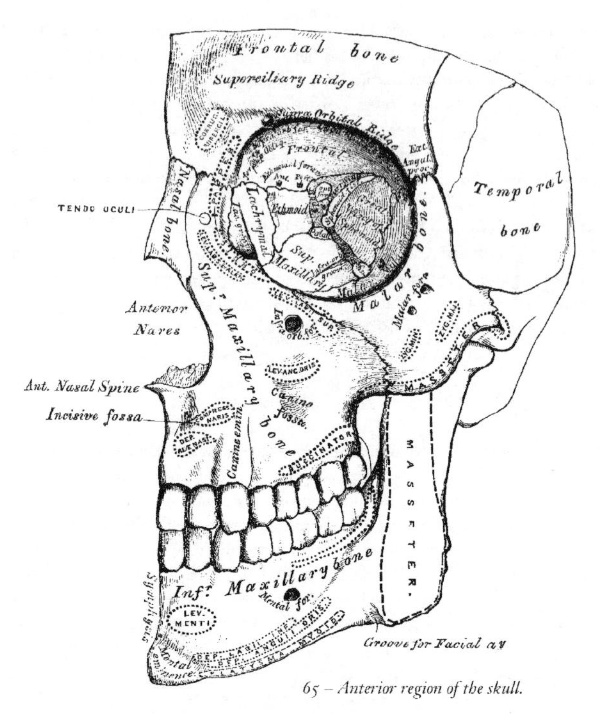

65 – *Anterior region of the skull.*

outer side of the orbit, is the quadrilateral anterior surface of the malar bone, perforated by one or two small malar foramina. Below the inferior margin of the orbit, is the infra-orbital foramen, the termination of the infra-orbital canal, and beneath this, the canine fossa, which gives attachment to the Levator anguli oris; bounded below by the alveolar processes, containing the teeth of the upper and lower jaw. Beneath the alveolar arch of the lower jaw is the mental foramen for the passage of the mental nerve and artery, the external oblique line, and at the lower border of the bone, at the point of junction of the body with the ramus, a shallow groove for the passage of the facial artery.

## The Orbits.

The Orbits (fig. 65) are two quadrilateral hollow cones, situated at the upper and anterior part of the face, their bases being directed forwards and outwards, and their apices backwards and inwards. Each orbit is formed of *seven* bones, the frontal, sphenoid, ethmoid, superior maxillary, malar, lachrymal and palate; but three of these, the frontal, ethmoid and sphenoid, enter into the formation of *both* orbits, so that the two cavities are formed of *eleven* bones only. Each cavity presents for examination, a roof, a floor, an inner and an outer wall, a circumference or base, and an apex. The *Roof* is concave, directed downwards and forwards, and formed in front by the orbital plate of the frontal; behind, by the lesser wing of the sphenoid. This surface presents internally the depression for the fibro-cartilaginous pulley of the Superior oblique muscle; externally, the depression for the lachrymal gland, and posteriorly, the suture connecting the frontal and lesser wing of the sphenoid.

The *Floor* is nearly flat, and of less extent than the roof; it is formed chiefly by the orbital process of the superior maxillary; in front, to a small extent, by the orbital process of the malar, and behind, by the orbital surface of the palate. This surface presents at its anterior and internal part, just external to the lachrymal canal, a depression for the attachment of the Inferior oblique muscle; externally, the suture between the malar and superior maxillary bones; near its middle, the infra-orbital groove; and posteriorly, the suture between the maxillary and palate bones.

The *Inner Wall* is flattened, and formed from before backwards by the nasal process of the superior maxillary, the lachrymal, os planum of the ethmoid, and a small part of the body of the sphenoid. This surface presents the lachrymal groove, and crest of the lachrymal bone, and the sutures connecting the ethmoid, in front, with the lachrymal, behind, with the sphenoid.

The *Outer Wall* is formed in front by the orbital process of the malar bone; behind, by the orbital plate of the sphenoid. On it are seen the orifices of one or two malar canals, and the suture connecting the sphenoid and malar bones.

*Angles.* The *superior external angle* is formed by the junction of the upper and outer walls; it presents, from before backwards, the suture connecting the frontal with the malar in front, and with the orbital plate of the sphenoid behind; quite posteriorly is the foramen lacerum anterius, or sphenoidal fissure, which transmits the third, fourth, ophthalmic division of the fifth and sixth nerves, and the ophthalmic vein. The *superior internal angle* is formed by the junction of the upper and inner wall, and presents the suture connecting the frontal with the lachrymal in front, and with the ethmoid behind. This suture is perforated by two foramina, the anterior and posterior ethmoidal, the former transmitting the anterior ethmoidal artery and nasal nerve, the latter the posterior ethmoidal artery and vein. The *inferior external angle*, formed by the junction of the outer wall and floor, presents the spheno-maxillary fissure, which transmits the infra-orbital vessels and nerve, and the ascending branches from the spheno-palatine ganglion. The *inferior internal angle* is formed by the union of the lachrymal and os planum of the ethmoid, with the superior maxillary and palate bones. The *circumference*, or base, of the orbit, quadrilateral in form, is bounded above by the supra-orbital arch; below, by the anterior border of the orbital plate of the malar, superior maxillary, and lachrymal bones; externally, by the external angular process of the frontal and the malar bone; internally, by the internal angular process of the frontal, and the nasal process of the superior maxillary. The circumference is marked by three sutures, the fronto-maxillary internally, the fronto-malar externally, and the malo-maxillary below; it contributes to the formation of the lachrymal groove, and presents above, the supra-orbital notch (or foramen), for the passage of the supra-orbital artery, veins, and nerve. The *apex*, situated at the back of the orbit, corresponds to the optic foramen, a short circular canal, which transmits the optic nerve and ophthalmic artery. It will thus be seen that there are *nine* openings communicating with each orbit, viz., the optic, foramen lacerum anterius; spheno-maxillary fissure, supra-orbital foramen, infra-orbital canal, anterior and posterior ethmoidal foramina, malar foramina, and lachrymal canal.

## The Nasal Fossæ.

The *Nasal Fossæ* are two large irregular cavities, situated in the middle line of the face, extending from the base of the cranium to the roof of the mouth, and separated from each other by a thin vertical septum. They communicate by two large apertures, the anterior nares, with the front of the face; and with the pharynx behind by the two posterior nares. These fossæ are much narrower above than below, and in the middle than at the anterior or posterior openings: their depth, which is considerable, is much greater in the middle than at either extremity. Each nasal fossa communicates with four sinuses, the frontal above, the sphenoidal behind, and the maxillary and ethmoidal on either side. Each fossa also communicates with four cavities: with the orbit by the lachrymal canal, with the mouth by the anterior palatine canal, with the cranium by the olfactory foramina, and with the spheno-maxillary fossa by the spheno-palatine foramen; and they occasionally communicate with each other by an aperture in the septum. The bones entering into their formation are fourteen in number: three of the cranium, the frontal, sphenoid, and ethmoid, and all the bones of the face excepting the malar and lower jaw. Each cavity is bounded by, a roof, a floor, an inner, and an outer wall.

The *upper wall*, or roof (fig. 66), is long, narrow, and concave from before backwards; it is formed in front by the nasal bones and nasal spine of the frontal, which are directed downwards and forwards; in the middle, by the cribriform lamella of the ethmoid, which is horizontal; and behind, by the under surface of the body of the sphenoid, and sphenoidal turbinated bones, which are directed downwards and backwards. This surface presents, from before backwards, the internal aspect of the nasal bones; on their outer side, the suture formed between the nasal, with the nasal process of the superior maxillary; on their inner side, the elevated crest which receives the nasal spine of the frontal, and the perpendicular plate of the ethmoid, and articulates with its fellow of the opposite side; whilst the surface of the bones is perforated by a few small vascular apertures, and presents the longitudinal groove for the nasal nerve: further back is the transverse suture, connecting the frontal with the nasal in front, and the ethmoid behind, the olfactory foramina on the under surface of the cribriform plate, and the suture between it and the sphenoid behind: quite posteriorly are seen the sphenoidal turbinated bones, the orifices of the sphenoidal sinuses, and the articulation of the alæ of the vomer with the under surface of the body of the sphenoid.

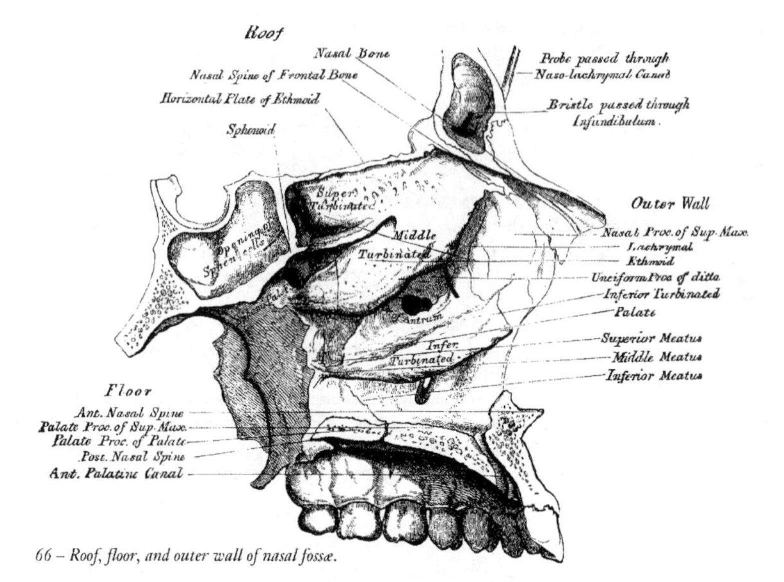


66 – *Roof, floor, and outer wall of nasal fossæ.*

The *floor* is flattened from before backwards, concave from side to side, and wider in the middle than at either extremity. It is formed in front by the palate process of the superior maxillary; behind, by the palate process of the palate bone. This surface presents, from before backwards, the anterior nasal spine; behind this, the upper orifice of the anterior palatine canal; internally, the elevated crest which articulates with the vomer; and behind, the suture between the palate and superior maxillary bones, and the posterior nasal spine.

The *inner wall*, or septum (fig. 67), is a thin vertical septum, which separates the nasal fossæ from one another; it is occasionally perforated so that the fossæ communicate, and it is frequently deflected considerably to one side. It is formed, in front, by the crest of the nasal bones and nasal spine of the frontal; in the middle, by the perpendicular lamella of the ethmoid; behind, by the vomer and rostrum of the sphenoid; below, by the crest of the superior maxillary and palate bones. It presents, in front, a large triangular notch, which receives the triangular cartilage of the nose; above, the lower orifices of the olfactory canals; and behind, the guttural edge of the vomer. Its surface is

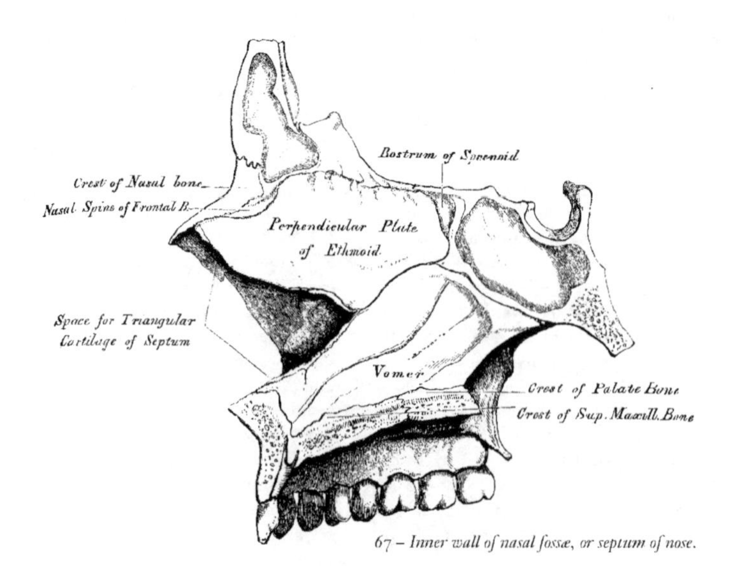


67 – *Inner wall of nasal fossæ, or septum of nose.*

marked by numerous vascular and nervous canals, and traversed by sutures connecting the bones of which it is formed.

The *outer wall* is formed, in front, by the nasal process of the superior maxillary and lachrymal bones; in the middle, by the ethmoid and inner surface of the superior maxillary and inferior turbinated bones; behind, by the vertical plate of the palate bone. This surface presents three irregular longitudinal passages, or meatuses, formed between three horizontal plates of bone that spring from it; they are termed the superior, middle, and inferior meatuses of the nose. The *superior meatus*, the smallest of the three, is situated at the upper and back part of each nasal fossa, occupying the posterior third of the outer wall. It is situated between the superior and middle turbinated bones, and has opening into it two formaina, the spheno-palatine at the back part of its outer wall, the posterior ethmoidal cells at the front part of the upper wall. The opening of the sphenoidal sinuses is usually at the upper and back part of the nasal fossæ, immediately behind the superior turbinated bone. The *middle meatus* is situated between the middle and inferior turbinated bones, and occupies the posterior two-thirds of the outer wall of the nasal fossa. It presents two apertures. In front is the orifice of the infundibulum, by which the middle meatus communicates with the anterior ethmoidal cells, and through these with the frontal sinuses. At the centre of the outer wall is the orifice of the antrum, which varies somewhat as to its exact position in different skulls. The *inferior meatus*, the largest of the three, is the space between the inferior turbinated bone and the floor of the nasal fossa. It extends along the entire length of the outer wall of the nose, is broader in front than behind, and presents anteriorly the lower orifice of the lachrymal canal.

## Os Hyoides.

The Hyoid bone is named from its resemblance to the Greek Upsilon; it is also called the *lingual bone*, from supporting the tongue, and giving attachment to its numerous muscles. It is a bony arch, shaped like a horse-shoe, and consisting of five segments, a central portion or body, two greater cornua, and two lesser cornua.

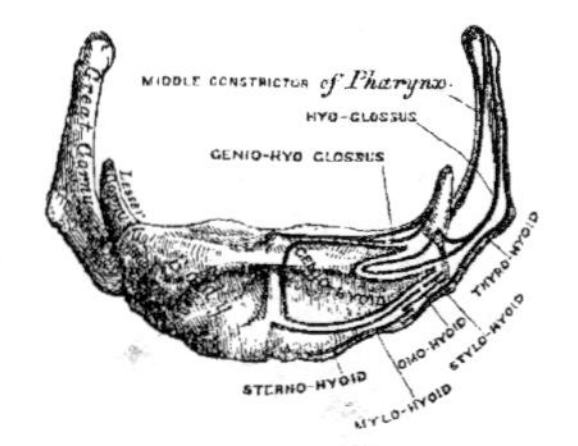


*68 – Hyoid bone. Anterior surface.*

The *Body* forms the central part of the bone, is of a quadrilateral form, its *anterior surface* (fig. 68) convex, directed forwards and upwards, is divided into two parts by a vertical ridge, which descends along the median line, and is crossed at right angles by a horizontal ridge, so that this surface is divided into four muscular depressions. At the point of meeting of these two lines is a prominent elevation, the tubercle. The portion above the horizontal ridge is directed upwards, and is sometimes described as the superior border. The anterior surface gives attachment to the Genio-hyoid in the greater part of its extent; above, to the Genio-hyo-glossus; below, to the Mylo-hyoid, Stylo-hyoid, and aponeurosis of the Digastric; and between these to part of the Hyo-glossus. The *posterior surface* is smooth, concave, directed backwards and downwards, and separated from the epiglottis by the thyro-hyoid membrane, and by a quantity of loose areolar tissue. The *superior border* is rounded, and gives attachment to the thyro-hyoid membrane, and part of the Genio-hyo-glossi muscles. The *inferior border* gives attachment, in front, to the Sterno-hyoid; behind, to part of the Thyro-hyoid, and to the Omo-hyoid at its junction with the great cornu. The *lateral surfaces* are small, oval, convex facets, covered with cartilage for articulation with the greater cornua.

The *Greater Cornua* project backwards from the lateral surfaces of the body; they are flattened from above downwards, diminish in size from before backwards, and terminate posteriorly in a tubercle for the attachment of the thyro-hyoid ligament. Their outer surface gives attachment to the Hyo-glossus; their upper border, to the Middle constrictor of the pharynx; their lower border, to part of the Thyro-hyoid muscle.

The *Lesser Cornua* are two small conical-shaped eminences, attached by their bases to the angles of junction between the body and greater cornua, and giving attachment by their apices to the stylo-hyoid ligaments. In youth, the cornua are connected to the body by cartilaginous surfaces, and held together by ligaments; in middle life, the body and greater cornua usually become joined; and in old age, all the segments are united together, forming a single bone.

*Development.* By *five* centres; one for the body, and one for each cornu. Ossification commences in the body and greater cornua towards the end of fœtal life, those of the cornua first appearing. Ossification of the lesser cornua commences some months after birth.

*Attachment of Muscles.* Sterno-hyoid, Thyro-hyoid, Omo-hyoid, aponeurosis of the Digastricus, Stylo-hyoid, Mylo-hyoid, Genio-hyoid, Genio-hyo-glossus, Hyo-glossus, Middle constrictor of the pharynx, and occasionally a few fibres of the Lingualis. It also gives attachment to the thyro-hyoidean membrane, and the stylo-hyoid, thyro-hyoid, and hyo-epiglottic ligaments.

## THE THORAX.

The Thorax, or chest, is an osseo-cartilaginous cage, intended to contain and protect the principal organs of respiration and circulation. It is the largest of the three cavities connected with the spine, and is formed by the sternum and costal cartilages in front, the twelve ribs on each side, and the bodies of the dorsal vertebræ behind.

### The Sternum.

The Sternum (figs. 69, 70) is a flat narrow bone, situated in the median line of the front of the chest, and consisting, in the adult, of three portions. Its form resembles an ancient sword: the upper piece, representing the handle, is termed the *manubrium*; the middle and largest piece, which represents the chief part of the blade, is termed the *gladiolus*; and the inferior piece, like the point of the sword, is termed the *ensiform or xiphoid appendix*. In its natural position, its direction is oblique from above, downwards, and forwards. It is flattened in front, concave behind, broad above, becoming narrowed at the point where the first and second pieces are connected; after which it again widens a little, and is pointed at its extremity. Its average length in the adult is six inches, being rather longer in the male than in the female.

The *First Piece* of the sternum, or *Manubrium*, is of a somewhat triangular form, broad and thick above, narrow below at its junction with the middle piece. Its *anterior surface*, convex from side to side, concave from above downwards, is smooth, and affords attachment on each side to the Pectoralis major and sternal origin of the Sterno-cleido-mastoid muscle. In well-marked bones, ridges limiting the attachment of these muscles are very distinct. Its *posterior surface*, concave and smooth, affords attachment on each side to the Sterno-hyoid and Sterno-thyroid muscles. The *superior border*, the thickest, presents at its centre the interclavicular notch; and, on each side, an oval articular surface, directed upwards, backwards, and outwards, for articulation with the sternal end of the clavicle. The *inferior border* presents an oval rough surface, covered in the recent state with a thin layer of cartilage, for articulation with the second portion of the bone. The *lateral borders* are marked above by an articular depression for the first costal cartilage, and below by a small facet, which, with a similar facet on the upper angle of the middle portion of the bone, forms a notch for the reception of the costal cartilage of the second rib. These articular surfaces are separated by a narrow curved edge which slopes from above downwards and inwards.

The *Second Piece* of the sternum, or *gladiolus*, considerably longer, narrower, and thinner than the superior, is broader below than above. Its *anterior surface* is nearly flat, directed upwards and forwards, and marked by three transverse lines which cross the bone opposite the third, fourth, and fifth articular depressions. These lines indicate the point of union of the four separate pieces of which this part of the bone consists at an early period of life. At the junction of the third and fourth pieces, is occasionally seen an orifice, the sternal foramen; it varies in size and form in different individuals, and pierces the bone from before backwards. This surface affords attachment on each side to the sternal origin of the Pectoralis major. The *posterior surface*, slightly concave, is also marked by three transverse lines; but they are less distinct than those in front: this surface affords attachment below, on each side, to the Triangularis sterni muscle, and occasionally presents the posterior opening of the sternal foramen. The *superior border* presents an oval surface for articulation with the manubrium. The *inferior border* is narrow and articulates with the ensiform appendix. Each *lateral border* presents at each superior angle a small facet, which, with a similar facet on the manubrium, forms a cavity for the cartilage of the second rib; the four succeeding angular depressions receive the cartilages of the third, fourth, fifth and sixth ribs, whilst each inferior angle presents a small facet, which, with a corresponding one on the ensiform appendix, forms a notch for the cartilage of the seventh rib. These articular depressions are separated by a series of curved inter-articular intervals, which diminish in length from above downwards, and correspond to the inter-costal spaces. The costal cartilage of each true rib, excepting the first, is thus seen to articulate with the sternum at the line of junction of two of its primitive component segments. This is well seen in

69 – *Sternum and costal cartilages. Anterior surface.*

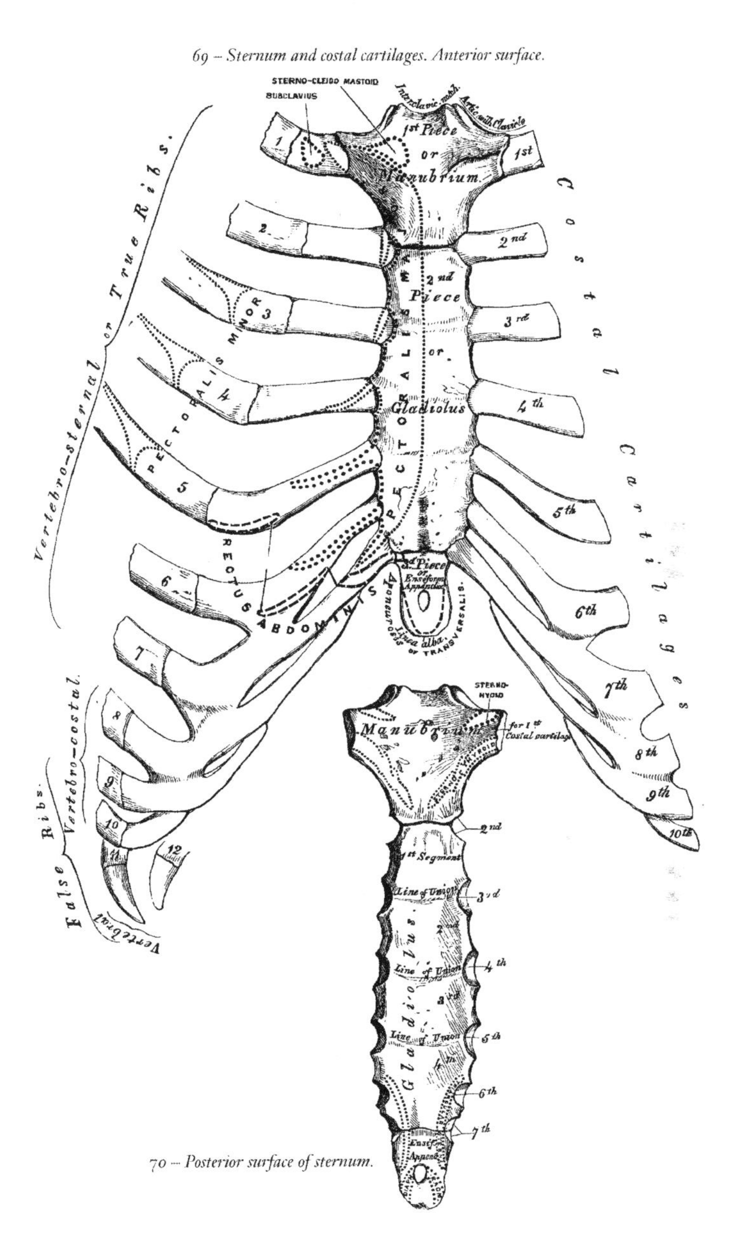


70 – *Posterior surface of sternum.*

many of the lower animals, where the separate parts of the bone remain ununited longer than in man. In this respect a striking analogy exists between the mode of connection of the ribs with the vertebral colum, and the connection of their cartilages with the sternal column.

The *Third Piece* of the sternum, the *ensiform* or *xiphoid appendix*, is the smallest of the three; it is thin and elongated in form, cartilaginous in structure in youth, but more or less ossified at its upper part in the adult. Its *anterior surface* affords attachment to the costo-xiphoid ligaments. Its *posterior surface*, to some of the fibres of the Diaphragm and Triangularis sterni muscles. Its *lateral borders*, to the aponeurosis of the abdominal muscles. Above, it is continuous with the lower end of the gladiolus; below, by its pointed extremity, it gives attachment to the linea alba, and at each superior angle

*71 – Development of sternum, by six centres.*

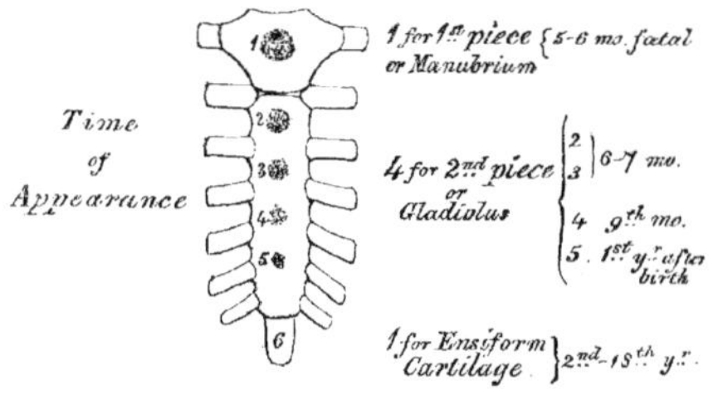

*72*

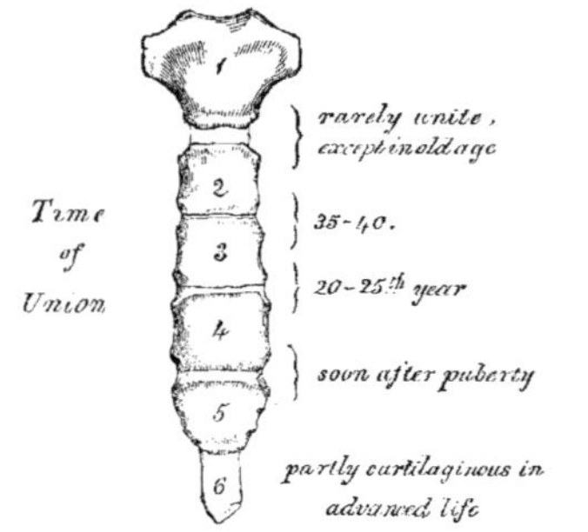

*73 – Peculiarities.*

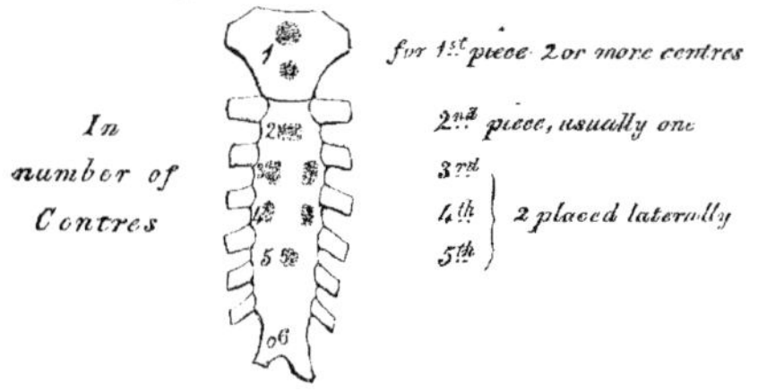

*74*

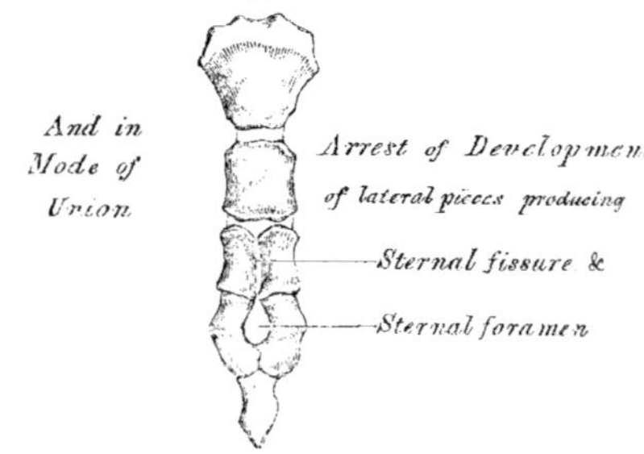

presents a facet for the lower half of the cartilage of the seventh rib. This portion of the sternum is very various in appearance, being sometimes pointed, broad and thin, sometimes bifid, or perforated by a round hole, occasionally curved, or deflected considerably to one or the other side.

*Structure.* This bone is composed of delicate cancellated texture, covered by a thin layer of compact tissue, which is thickest in the manubrium, between the articular facets for the clavicles.

*Development.* The sternum, including the ensiform appendix, is developed by six centres. One for the first piece or manubrium, four for the second piece or gladiolus, and one for the ensiform appendix. The sternum is entirely cartilaginous up to the middle of fœtal life, and when ossification takes place, the ossific granules are deposited in the middle of the intervals between the articular depressions for the costal cartilages, in the following order (fig. 71). In the first piece, between the fifth and sixth months; in the second and third, between the sixth and seventh months; in the fourth piece, at the ninth month; in the fifth, within the first year, or between the first and second years after birth; and in the ensiform appendix, between the second and the seventeenth or eighteenth years, by a single centre which makes its appearance at the upper part, and proceeds gradually downwards. To these may be added the occasional existence, as described by Breschet, of two small episternal centres, which make their appearance one on each side of the interclavicular notch. These are regarded by him as the anterior rudiments of a rib, of which the posterior rudiment is the anterior lamina of the transverse process of the seventh cervical vertebra. It occasionally happens that some of the segments are formed from more than one centre, the number and position of which vary (fig. 73). Thus the first piece may have two, three, or even six centres. When two are present, they are generally situated one above the other, the upper one being the larger; the second piece has seldom more than one; the third, fourth, and fifth pieces, are often formed from two centres placed laterally, the irregular union of which will serve to explain the occasional occurrence of the sternal foramen (fig. 74), or of the vertical fissure which occasionally intersects this part of the bone. Union of the various centres commences from below, and proceeds upwards, taking place in the following order (fig. 72). The fifth piece is joined to the fourth soon after puberty; the fourth to the third, between the twentieth and twenty-fifth years; the third to the second, between the thirty-fifth and fortieth years; the second is rarely joined to the first except in very advanced age.

*Articulations.* With the clavicles, and seven costal cartilages on each side.

*Attachment of Muscles.* The Pectoralis major, Sterno-cleido-mastoid, Sterno-hyoid, Sterno-thyroid, Triangularis sterni, aponeurosis of the Obliquus externus, Obliquus internus, and Transversalis muscles, Rectus and Diaphragm.

## The Ribs.

The Ribs are elastic arches of bone, which form the chief part of the thoracic walls. They are twelve in number on each side; but this number may be increased by the development of a cervical or lumbar rib, or may be diminished to eleven. The first seven are connected behind with the spine, and in front with the sternum, through the intervention of the costal cartilages; they are called *vertebro-sternal*, or true ribs. The remaining five are false ribs; of these the first three, being connected behind with the spine, and in front with the costal cartilages, are called the *vertebro-costal ribs:* the last two are connected with the vertebræ only, being free at their anterior extremities; they are termed *vertebral* or *floating ribs.* The ribs vary in their direction, the upper ones being placed nearly at right angles with the spine; the lower ones are placed obliquely, so that the anterior extremity is lower than the posterior. The extent of obliquity reaches its maximum at the ninth rib, gradually decreasing from that point towards the twelfth. The ribs are situated one beneath the other in such a manner that spaces are left between them; these are called *intercostal spaces.* Their length corresponds to the length of the ribs, their breadth is more considerable in front than behind, and between the upper than between the lower ribs. The ribs increase in length from the first to the seventh, when they again diminish to the twelfth. In breadth they decrease from above downwards; in each rib the greatest breadth is at the sternal extremity.

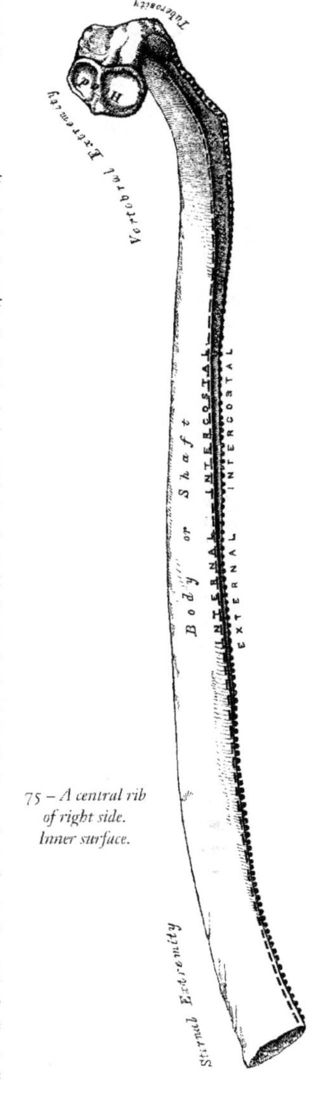


75 – *A central rib of right side. Inner surface.*

*Common characters of the Ribs* (fig. 75). Take a rib from the middle of the series in order to study its common characters. Each rib presents two extremities, a posterior or vertebral, an anterior or sternal, and an intervening portion, the body or shaft. The *posterior* or *vertebral extremity* presents for examination a head, neck, and tuberosity.

The *head* (fig. 76), is marked by a kidney-shaped articular surface, divided by a horizontal ridge into two facets for articulation with the costal cavity formed by the junction of the bodies of two contiguous dorsal vertebræ; the upper facet is small, the inferior one of large size; the ridge separating them serves for the attachment of the inter-articular ligament.

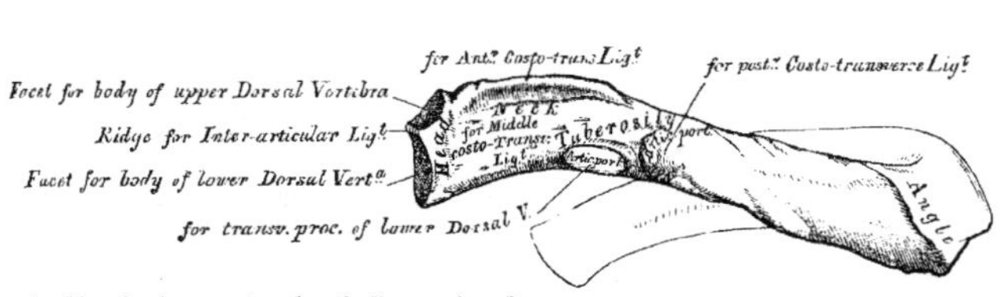


76 – *Vertebral extremity of a rib. External surface.*

The *neck* is that flattened portion of the rib which extends outwards from the head; it is about an inch long and rests upon the transverse process of the lower of the two vertebræ with which the head articulates. Its *anterior surface* is flat and smooth, its *posterior* rough, for the attachment of the middle costo-transverse ligament, and perforated by numerous foramina, the direction of which is less constant than those found on the inner surface of the shaft. Of its two borders, the *superior* presents a rough crest for the attachment of the anterior costo-transverse ligament; its *inferior border* is rounded. On the posterior surface of the neck, just where it joins the shaft, and nearer the lower than the upper border, is an eminence—the tuberosity; it consists of an articular and a non-articular portion. The *articular portion*, the most internal and inferior of the two, presents a small oval surface, for articulation with the extremity of the transverse process of the lower of the two vertebræ to which the head is connected. The *non-articular portion* is a rough elevation, which affords attachment to the posterior costo-transverse ligament. The tubercle is much more prominent in the upper than in the lower ribs.

The *shaft* is thin and flat, so as to present two surfaces, an external and an internal; and two borders, a superior and an inferior. The *external surface* is convex, smooth; and marked, at its back part, a little in front of the tuberosity, by a prominent line, directed obliquely from above, downwards and outwards; this gives attachment to a tendon of the Sacro-lumbalis muscle, and is called the *angle*. At this point, the rib is bent in two directions. If the rib is laid upon its lower border, it will be seen, that the anterior portion of the shaft, as far as the angle, rests upon this margin, while the vertebral end of the bone, beyond the angle, is bent inwards and at the same time tilted upwards. The interval between the angle and the tuberosity increases gradually from the second to the tenth rib. The portion of bone between these two parts is rounded, rough, and irregular, and serves for the attachment of the Longissimus dorsi. The portion of bone between the angle and sternal extremity is also slightly twisted upon its own axis, the external surface looking downwards behind the angle, a little upwards in front of it. This surface presents, towards its sternal extremity, an oblique line, the anterior angle. The *internal surface* is concave, smooth, directed a little upwards behind the angle; a little downwards in front of it. This surface is marked by a ridge which commences at the lower extremity of the head; it is strongly marked as far as the inner side of the angle, and gradually becomes lost at the junction of the anterior with the middle third of the bone. The interval between it and the inferior border is deeply grooved, to lodge the intercostal vessels and nerve. At the back part of the bone, this groove belongs to the inferior border, but just in front of the angle, where it is deepest and broadest, it corresponds to the internal surface. The superior edge of the groove is rounded; it serves for the attachment of the Internal intercostal muscle. The inferior edge corresponds to the lower margin of the rib, and gives attachment to the External intercostal. Within the groove are seen the orifices of numerous small foramina, which traverse the wall of the shaft obliquely from before backwards. The *superior border*, thick and rounded, is marked by an external and an internal lip, more distinct behind than in front; they serve for the attachment of the External and Internal intercostal muscles. The *inferior border*, thin and sharp, has attached the External intercostal muscle. The anterior or sternal extremity, is flattened, and presents a porous oval concave depression, into which the costal cartilage is received.

## Peculiar Ribs.

The ribs which require especial consideration, are five in number, viz., the first, second, tenth, eleventh and twelfth.

The *first rib* (fig. 77) is one of the shortest and the most curved of all the ribs; it is broad, flat, and placed horizontally at the upper part of the thorax, its surfaces looking upwards and downwards; and its borders inwards and outwards. The *head* is of small size, rounded, and presents only a single articular facet for articulation with the body of the first dorsal vertebra. The *neck* is narrow and rounded. The *tuberosity*, thick and prominent, rests on the outer border. There is no angle, and the shaft is not twisted on its axis. The upper surface of the shaft is marked by two shallow depressions, separated from one another by a ridge, which becomes more prominent towards the internal border, where it terminates in a Scalene tubercle; this tubercle and ridge serve for the attachment of the Scalenus anticus muscle, the groove in front of it transmitting the subclavian vein; that behind it, the subclavian artery. Between the groove for the subclavian artery and the tuberosity, is a depression for the attachment of the Scalenus medius muscle. The *under surface* is smooth, and destitute of the groove observed on the other ribs. The *outer border* is convex, thick, and rounded. The *inner*, concave, thin, and sharp, and marked about its centre by the tubercle before mentioned. The *anterior extremity* is larger and thicker than any of the other ribs.

The *second rib* (fig. 78) is much longer than the first, but bears a very considerable resemblance to it in the direction of its curvature. The non-articular portion of the tuberosity is occasionally only slightly marked. The *angle* is slight, and situated close to the tuberosity, and the shaft is not twisted, so that both ends touch any plane surface upon which it may be laid. The shaft is not horizontal, like that of the first rib; its *outer surface*, which is convex, looking upwards and a little outwards. It presents, near the middle, a rough eminence for the attachment of part of the first, and the second serration of the Serratus magnus. The *inner surface*, smooth and concave, is directed downwards and a little inwards; it presents a short groove towards its posterior part.

The *tenth rib* (fig. 79) has only a single articular facet on its head.

The *eleventh* and *twelfth ribs* (figs. 80 and 81) have each a single articular facet on the head, which is of rather large size; they have no neck or tuberosity, and are pointed at the extremity. The eleventh has a slight angle and a shallow groove on the lower border. The twelfth has neither, and is much shorter than the eleventh.

*Structure.* The ribs consist of cancellous tissue, enclosed in a thin compact layer.

*Development.* Each rib, with the exception of the last two, is developed by *three* centres, one for the shaft, one for the head, and one for the tubercle. The last two have only *two* centres, that for the tubercle being wanting. Ossification commences in the body of the ribs at a very early period, before its appearance in the vertebræ. The epiphysis of the head, which is of a slightly angular shape, and that for the tubercle, of a lenticular form, make their appearance between the sixteenth and twentieth years, and are not united to the rest of the bone until about the twenty-fifth year.

*Attachment of Muscles.* The Intercostals, Scalenus anticus, Scalenus medius, Scalenus posticus, Pectoralis minor, Serratus magnus, Obliquus externus, Transversalis, Quadratus lumborum, Diaphragm, Latissimus dorsi, Serratus posticus superior, Serratus posticus inferior, Sacro-lumbalis,

*Peculiar ribs.*

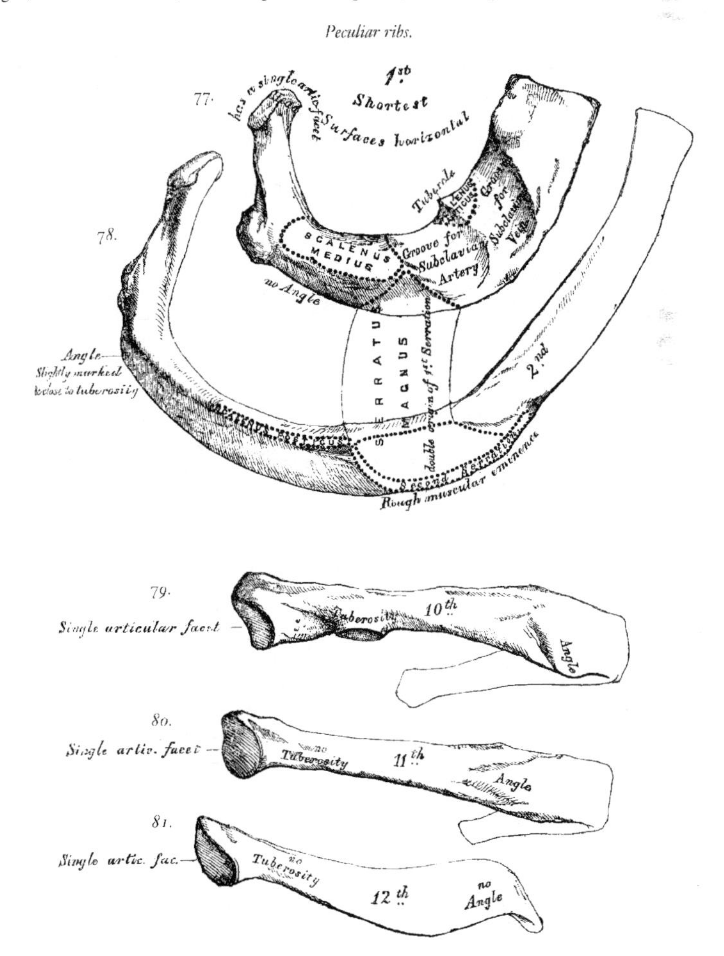

Musculus accessorius ad sacro-lumbalem, Longissimus dorsi, Cervicalis ascendens, Levatores costarum.

## The Costal Cartilages.

The *Costal Cartilages* (fig. 69) are white elastic structures, which serve to prolong the ribs forward to the front of the chest, and contribute very materially to the elasticity of this cavity. The first seven are connected with the sternum, the next three with the lower border of the cartilage of the preceding rib. The cartilages of the last two ribs, which have pointed extremities, float freely in the walls of the abdomen. Like the ribs, the costal cartilages vary in their length, breadth, and direction. They increase in length from the first to the seventh, then gradually diminish to the last. They diminish in breadth, as well as the intervals between them, from the first to the last. They are broad at their attachment to the ribs, and taper towards their sternal extremities, excepting the first two, which are of the same breadth throughout, and the sixth, seventh, and eighth, which are enlarged where their margins are in contact. In direction they also vary; the first descends a little, the second is horizontal, the third ascends slightly, whilst all the rest follow the course of the ribs for a short extent, and then ascend to the sternum or preceding cartilage. Each costal cartilage presents two surfaces, two borders, and two extremities. The *anterior surface* is convex, and looks forwards and upwards; that of the first gives attachment to the costo-clavicular ligament; that of the first, second, third, fourth, fifth, and sixth, at their sternal ends, to the Pectoralis major. The others are covered, and give partial attachment to some of the great flat muscles of the abdomen. The *posterior surface* is concave, and directed backwards and downwards, the six or seven inferior ones affording attachment to the Transversalis and Diaphragm muscles. Of the two borders, the superior is concave; the inferior, convex; they afford attachment to the Intercostal muscles, the upper border of the sixth giving attachment to the Pectoralis major muscle. The contiguous borders of the sixth, seventh, and eighth, and sometimes the ninth and tenth costal cartilages present smooth oblong surfaces at the points where they articulate. Of the two extremities, the outer one is continuous with the osseous tissue of the rib to which it belongs. The inner extremity of the first is continuous with the sternum; the six succeeding ones have rounded extremities, which are received into shallow concavities on the lateral margins of the sternum. The inner extremities of the eighth, ninth, and tenth costal cartilages are pointed, and lie in contact with the cartilage above. Those of the eleventh and twelfth are free, and pointed.

The costal cartilages are most elastic in youth, those of the false ribs being more so than the true. In old age, they become of a deep yellow colour. Under certain diseased conditions, they are prone to ossify. Mr. Humphry's observations on this subject have led him to regard the ossification of the costal cartilages as a sign of disease rather than of age. 'The ossification takes place in the first cartilage sooner than in the others; and in men more frequently, and at an earlier period of life, than in women.'

*Attachment of Muscles.* The Subclavius, Sterno-thyroid, Pectoralis major, Internal oblique, Transversalis, Rectus, Diaphragm, Triangularis sterni, Internal and External intercostals.

# OF THE EXTREMITIES.

The Extremities, or limbs, are those long-jointed appendages of the body, which are connected to the trunk by one end, being free in the rest of their extent. They are *four* in number: an *upper* or *thoracic pair*, connected with the thorax through the intervention of the shoulder, and subservient mainly to tact and prehension; and a *lower pair*, connected with the pelvis, intended for support and locomotion. Both pairs of limbs are constructed after one common type, so that they present numerous analogies; while, at the same time, certain differences are observed in each, dependent on the peculiar offices they severally perform.

## Of the Upper Extremity.

The upper extremity consists of the arm, the fore-arm, and the hand. Its continuity with the trunk is established by means of the shoulder, which is homologous with the innominate or haunch bone in the lower limb.

## Of the Shoulder.

The shoulder is placed upon the upper part and side of the chest, connecting the upper extremity to the trunk; it consists of two bones, the clavicle, and the scapula.

## The Clavicle.

The *Clavicle* (*clavis*, a 'key'), or collar-bone, forms the anterior portion of the shoulder. It is a long bone, curved somewhat like the italic letter *ſ*, and placed nearly horizontally at the upper and anterior part of the thorax, immediately above the first rib. It articulates internally with the upper border of the sternum, and with the acromion process of the scapula by its outer extremity; serving to sustain the upper extremity in the various positions which it assumes, whilst, at the same time, it allows it great latitude of motion. The horizontal plane of the clavicle is nearly straight; but in the vertical plane it presents a double curvature, the convexity being, in front, at the sternal end; and, behind, at the scapular end. Its outer third is flattened from above downwards, and extends, in the natural position of the bone, from the coracoid process to the acromion. Its inner two-thirds are of a cylindrical form, and extend from the sternum to the coracoid process of the scapula.

### *External or Flattened Portion*

The *outer third* is flattened from above downwards, so as to present two surfaces, an upper, and a lower; and two borders, an anterior, and a posterior.

The *upper surface* is flattened, rough, marked in front, for the attachment of the Deltoid; behind, for the Trapezius; between these two impressions, externally, a small portion of the bone is subcutaneous. The *under surface* is flattened. At its posterior border, at the junction of the prismatic with the flattened portion, is a rough eminence, the *conoid tubercle*; this, in the natural position of the bone, surmounts the coracoid process of the scapula, and gives attachment to the conoid ligament. From this tubercle, an oblique line, occasionally a depression, passes forwards and outwards to near the outer end of the anterior border; it is called the *oblique line*, and affords attachment to the trapezoid ligament. The *anterior border* is concave, thin, and rough; it limits the attachment of the Deltoid, and occasionally presents, near the centre, a tubercle, the *deltoid tubercle*, which is sometimes distinct in the living subject. The *posterior border* is convex, rough, broader than the anterior, and gives attachment to the Trapezius.

### *Internal or Cylindrical Portion*

The cylindrical portion forms the inner two-thirds of the bone. It is curved, so as to be convex in front, concave behind, and is marked by three borders separating three surfaces.

The *anterior border* is continuous with the anterior margin of the flat portion. At its commencement it is smooth, and corresponds to the unoccupied interval between the attachment of the Pectoralis major and Deltoid muscles; about the centre of the clavicle it divides to enclose an elliptical space for the attachment of the clavicular portion of the Pectoralis major. This space extends inwards as far as the anterior margin of the sternal extremity.

The *superior border* is continuous with the posterior margin of the flat portion, and separates the anterior from the posterior surface. At its commencement it is smooth and rounded, becomes rough towards the inner third for the attachment of the Sterno-mastoid muscle, and terminates at the upper angle of the sternal extremity.

The *posterior* or *subclavian border* separates the posterior from the inferior surface, and extends from the conoid tubercle to the rhomboid depression. It forms the posterior boundary of the groove for the Subclavius muscle, and gives attachment to the fascia which encloses it.

The *anterior surface* is included between the superior and anterior borders. It is directed forwards and a little upwards at the sternal end, outwards and still more upwards at the acromial extremity, where it becomes continuous with the upper surface of the flat portion. Externally, it is smooth, convex, nearly subcutaneous, being covered only by the Platysma; but, corresponding to the inner half of the bone, it is divided by a more or less prominent line into two parts: an anterior portion, elliptical in form, rough, and slightly convex, for the attachment of the Pectoralis major; and an upper part, which is rough behind, for the attachment of the Sterno-cleido mastoid. Between the two muscular impressions is a small subcutaneous interval.

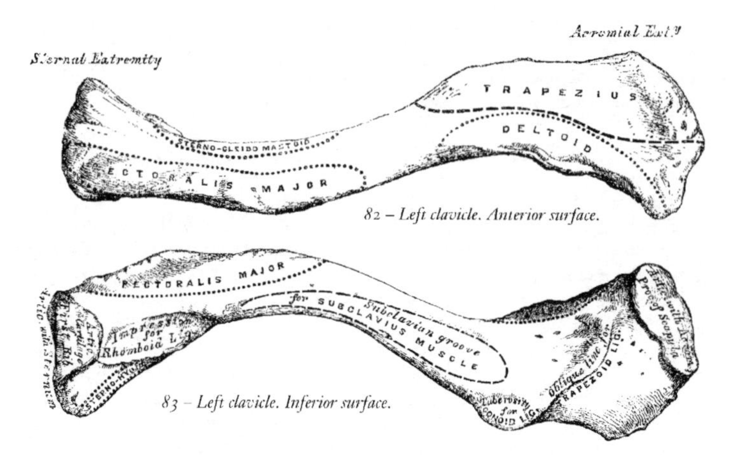


82 – Left clavicle. Anterior surface.

83 – Left clavicle. Inferior surface.

The *posterior* or *cervical surface* is smooth, flat, directed vertically, and looks backwards towards the root of the neck. It is limited, above, by the superior border; below, by the subclavian border; internally, by the margin of the sternal extremity; externally, it is continuous with the posterior border of the flat portion. It is concave from within outwards, and is in relation, by its lower part, with the supra-scapular vessels. It gives attachment, near the sternal extremity, to part of the Sterno-hyoid muscle; and presents, at or near the middle, a foramen, directed obliquely outwards, which transmits the chief nutrient artery of the bone. Sometimes, there are two foramina on the posterior surface, or one on the posterior; the other on the inferior surface.

The *inferior* or *subclavian surface* is bounded, in front, by the anterior border; behind, by the subclavian border. It is narrow internally, but gradually increases in width externally, and is continuous with the under surface of the flat portion. Commencing at the sternal extremity may be seen a small facet for articulation with the cartilage of the first rib. This is continuous with the articular surface at the sternal end of the bone. External to this is a broad rough impression, the rhomboid, rather more than an inch in length, for the attachment of the costo-clavicular (rhomboid) ligament. The remaining part of this surface is occupied by a longitudinal groove, the subclavian groove, broad and smooth externally; narrow and more uneven internally; it gives attachment to the Subclavius muscle, and, by its anterior margin, to the strong aponeurosis which encloses it. Not unfrequently this groove is subdivided into two parts, by a longitudinal line, which gives attachment to the intermuscular septum of the Subclavius muscle.

The *internal* or *sternal* end of the clavicle is triangular in form, directed inwards, and a little downwards and forwards; and presents an articular facet, concave from before backwards, convex from above downwards, which articulates with the sternum through the intervention of an inter-articular fibro-cartilage; the circumference of the articular surface is rough, for the attachment of numerous ligaments. This surface is continuous with the costal facet on the inner end of the inferior or subclavian surface, which articulates with the cartilage of the first rib.

The *outer* or *acromial extremity*, directed outwards and forwards, presents a small, flattened, oval facet, directed obliquely downwards and inwards, for articulation with the acromion process of the scapula. The direction of this surface serves to explain the greater frequency of dislocation of this bone upon, and not beneath, the acromion process. The circumference of the articular facet is rough, especially above, for the attachment of the acromio-clavicular ligaments.

*Peculiarities of the Bone in the Sexes and in Individuals.* In the female, the clavicle is generally less curved, smoother, and more slender than in the male. In those persons who perform considerable manual labour, which brings into constant action the muscles connected with this bone, it acquires considerable bulk, becomes shorter, more curved, its ridges for muscular attachment become prominently marked, and its sternal end of a prismatic or quadrangular form. The right clavicle is generally heavier, thicker, and rougher, and often shorter, than the left.

*Structure.* The shaft, as well as the extremities, consists of cancellous tissue, invested in a compact layer much thicker in the centre than at either end. The clavicle is highly elastic, by reason of its curves. From the experiments of Mr. Ward, it has been shown that it possesses sufficient longitudinal elastic force to raise its own weight nearly two feet on a level surface; and sufficient transverse elastic force, opposite the centre of its anterior convexity, to raise its own weight about a foot. This extent of elastic power must serve to moderate very considerably the effect of concussions received upon the point of the shoulder.

*Development.* By *two* centres: one for the shaft, and one for the sternal extremity. The centre for the shaft appears very early, before any other bone; the centre for the sternal end makes its appearance about the eighteenth or twentieth year, and unites with the rest of the bone a few years after.

*Articulations.* With the sternum, scapula, and cartilage of the first rib.

*Attachment of Muscles.* The Sterno-cleido mastoid, Trapezius, Pectoralis major, Deltoid, Subclavius, and Sterno-hyoid.

## The Scapula.

The *Scapula* forms the back part of the shoulder. It is a large flat bone, triangular in shape, situated at the posterior aspect and side of the thorax, between the first and eighth ribs, its posterior border or base, being about an inch from, and nearly parallel with, the spinous processes of the vertebræ. It presents for examination two surfaces, three borders, and three angles.

The *anterior surface*, or *venter* (fig. 84), presents a broad concavity, the subscapular fossa. It is marked, in the posterior two thirds, by several oblique ridges, which pass from behind obliquely forwards and upwards, the anterior third being smooth. The oblique ridges, above-mentioned, give attachment to the tendinous intersections; and the surfaces between them, to the fleshy fibres of the Subscapularis muscle. The anterior third of the fossa, which is smooth, is covered by, but does not afford attachment to, the fibres of this muscle. This surface is separated from the posterior border, by a smooth triangular margin at the superior and inferior angles, and in the interval between these, by a narrow edge which is often deficient. This marginal surface affords attachment throughout its entire extent to the Serratus magnus muscle. The subscapular fossa presents a transverse depression

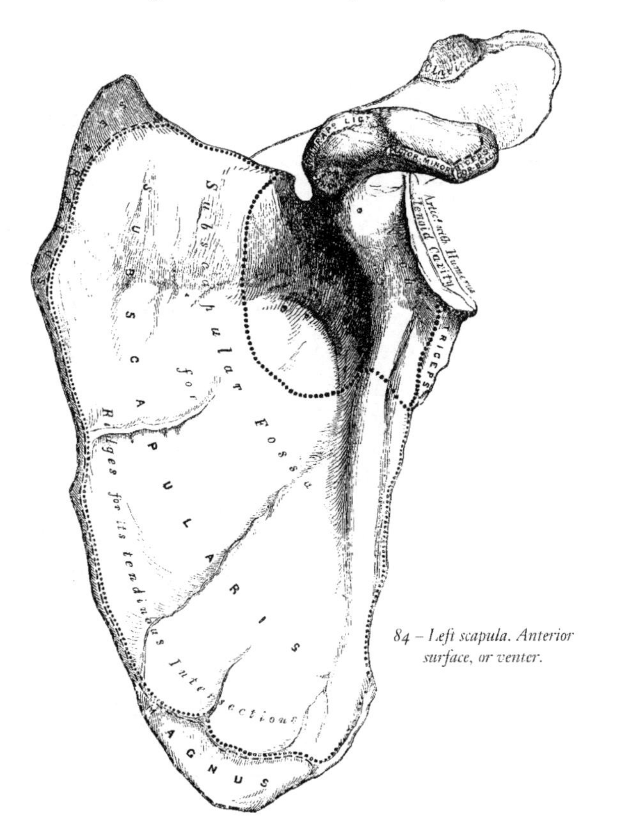


84 – Left scapula. Anterior surface, or venter.

at its upper part, called the *subscapular angle*; it is in this situation that the fossa is deepest; and consequently the thickest part of the Subscapularis muscle lies in a line parallel with the glenoid cavity, and must consequently operate most effectively on the humerus which is contained in it.

The *posterior surface*, or *dorsum* (fig. 85) is arched from above downwards, alternately convex and concave from side to side. It is subdivided unequally into two parts by the spine; that portion above the spine is called the supra-spinous fossa, and that below it, the infra-spinous fossa.

The *supra-spinous fossa*, the smaller of the two, is concave, smooth, and broader at the vertebral than at the humeral extremity. It affords attachment by its inner two-thirds to the Supra-spinatus muscle.

The *infra-spinous fossa* is much larger than the preceding; towards its vertebral margin a shallow concavity is seen at its upper part; its centre presents a prominent convexity, whilst towards the axillary border is a deep groove, which runs from the upper towards the lower part. The inner three-fourths of this surface afford attachment to the Infra-spinatus muscle; the outer fourth is only covered by it, without giving origin to its fibres. This surface is separated from the axillary border by an elevated ridge, which runs from the lower part of the glenoid cavity, downwards and backwards to the posterior border, about an inch above the inferior angle. This ridge serves for the attachment of a strong aponeurosis, which separates the Infra-spinatus from the two Teres muscles. The surface of bone between this line and the axillary border is narrow for the upper two-thirds of its extent, and traversed near its centre by a groove for the passage of the dorsalis scapulæ vessels; it affords attachment to the Teres minor. Its lower third presents a broader, somewhat triangular surface, which gives origin to the Teres major, and over which glides the Latissimus dorsi; sometimes the latter muscle takes origin by a few fibres from this part. The broad and narrow portions of bone above alluded to are separated by an oblique line, which runs from the axillary border,

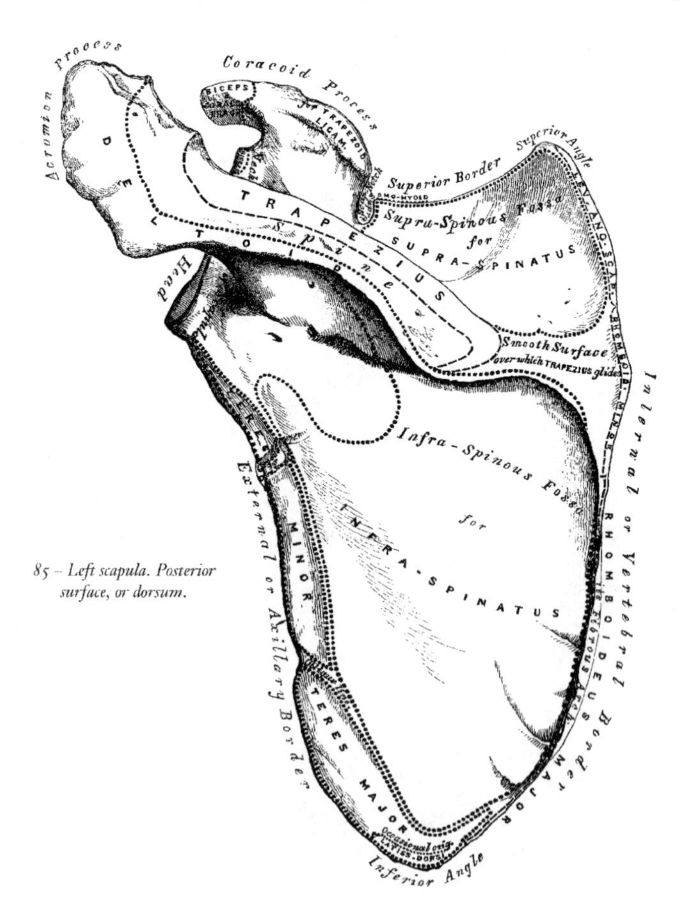


85 – *Left scapula. Posterior surface, or dorsum.*

downwards and backwards; to it is attached the aponeurosis separating the two Teres muscles from each other.

The *Spine* is a prominent plate of bone, which crosses obliquely the inner four-fifths of the dorsum of the scapula at its upper part, and separates the supra from the infra-spinous fossa: it commences at the vertebral border by a smooth triangular surface, over which the Trapezius glides, separated by a bursa; and, gradually becoming more elevated as it passes forwards, terminates in the acromion process which overhangs the shoulder joint. The spine is triangular and flattened from above downwards, its apex corresponding to the posterior border; its base, which is directed outwards, to the neck of the scapula. It presents two surfaces and three borders. Its *superior surface* is concave, assists in forming the supra-spinous fossa, and affords attachment to part of the Supra-spinatus muscle. Its *inferior surface* forms part of the infra-spinous fossa, gives origin to part of the Infra-spinatus muscle, and presents near its centre the orifice of a nutritious canal. Of the three borders, the *anterior* is attached to the dorsum of the bone; the *posterior*, or *crest* of the spine, is broad, and presents two lips, and an intervening rough interval. To the superior lip is attached the Trapezius, to the extent shown in the figure. A very rough prominence is generally seen occupying that portion of the spine which receives the insertion of the middle and inferior fibres of this muscle. To the inferior lip, throughout its whole length, is attached the Deltoid. The interval between the lips is also partly covered by the fibres of these muscles. The *external border*, the shortest of the three, is slightly concave, its edges thick and round, continuous above with the under surface of the acromion process; below, with the neck of the scapula. The narrow portion of bone external to this border, serves to connect the supra and infra-spinous fossæ.

The *Acromion process*, so called from forming the summit of the shoulder (ἄκρον, a summit; ὦμος, the shoulder), is a large, and somewhat triangular process flattened from behind forwards, directed at first a little outwards, and then curving forwards and upwards, so as to overhang the glenoid cavity. Its *upper surface* directed upwards, backwards, and outwards, is convex, rough, and gives attachment to some fibres of the Deltoid. Its *under surface* is smooth and concave. Its *outer border*, which is thick and irregular, affords attachment to the Deltoid muscle. Its *inner margin*, shorter than the outer, is concave, gives attachment to a portion of the Trapezius muscle, and presents about its centre a small oval surface, for articulation with the scapular end of the clavicle. Its *apex*, which corresponds to the point of meeting of these two borders in front, is thin, and has attached to it the coraco-acromion ligament.

Of the three borders or costæ of the scapula, the *superior* is the shortest, and thinnest; it is concave, terminating at its inner extremity at the superior angle, at its outer extremity at the coracoid process. At its outer part is a deep semicircular notch, the supra-scapular, formed partly by the base of the coracoid process. This notch is converted into a foramen by the transverse ligament, and serves for the passage of the supra-scapular nerve. The adjacent margin of the superior border affords attachment to the Omo-hyoid muscle. The *external*, or *axillary border*, is the thickest of the three. It commences above at the lower margin of the glenoid cavity, and inclines obliquely downwards and backwards to the inferior angle. Immediately below the glenoid cavity, is a rough depression about an inch in length, which affords attachment to the long head of the Triceps muscle; to this succeeds a longitudinal groove, which extends as far as its lower third, and affords origin to part of the Subscapularis muscle. The inferior third of this border which is thin and sharp, serves for the attachment of a few fibres of the Teres major behind, and of the Subscapularis in front. The *internal*, or *vertebral border*, also named the base, is the longest of the three, and extends from the superior to the inferior angle of the bone. It is arched, intermediate in thickness between the superior and the external borders, and that portion of it above the spine is bent considerably outwards, so as to form an obtuse angle with the lower part. The vertebral border presents an anterior lip, a posterior lip, and an intermediate space. The *anterior lip* affords attachment to the Serratus magnus; the *posterior lip*, to the Supra-spinatus above the spine, the Infra-spinatus below; the interval between the two lips, to the Levator anguli scapulæ above the triangular surface at the commencement of the spine; the Rhomboideus minor, to the edge of that surface; the Rhomboideus major being attached by means of a fibrous arch, connected above to the lower part of the triangular surface at the base of the spine, and below to the lower part of the posterior border.

Of the three angles, the *superior*, formed by the junction of the superior and internal borders, is thin, smooth, rounded, somewhat inclined outwards, and gives attachment to a few fibres of the Levator anguli scapulæ muscle. The *inferior* angle, thick and rough, is formed by the union of the vertebral and axillary borders, its outer surface affording attachment to the Teres major, and occasionally a few fibres of the Latissimus dorsi. The *anterior* angle is the thickest part of the bone, and forms what is called the *head* of the scapula. The head presents a shallow, pyriform, articular surface, the *glenoid cavity* (*γλήνη, a superficial cavity; εἶδος, like*); its longest diameter is from above downwards, and its direction outwards and forwards. It is broader below than above; at its apex is attached the long tendon of the Biceps muscle. It is covered with cartilage in the recent state; and its margins, slightly raised, give attachment to a fibro-cartilaginous structure, the glenoid ligament, by which its cavity is deepened. The neck of the scapula is the slightly depressed surface which surrounds the head; it is more distinct on the posterior than on the anterior surface, and below than above. In the latter situation, it has, arising from it, a thick prominence, the coracoid process.

The *Coracoid process,* so called from its fancied resemblance to a crow's beak (*κόραξ, a crow; εἶδος, like*), is a thick curved process of bone, which arises by a broad base from the upper part of the neck of the scapula; it ascends at first upwards and inwards; then, becoming smaller, it changes its direction and passes forwards and outwards. The ascending portion, flattened from before backwards, presents in front a smooth concave surface, over which passes the Sub-scapularis muscle. The horizontal portion is flattened from above downwards; its upper surface is convex and irregular; its under surface is smooth; its anterior border is rough, and gives attachment to the Pectoralis minor, its posterior border is also rough for the coraco-acromion ligament, while the apex is embraced by the conjoined tendon of origin of the short head of the Biceps and Coraco-brachialis muscles. At the inner side of the root of the coracoid process is a rough depression for the attachment of the conoid ligament, and, running from it obliquely forwards and outwards on the upper surface of the horizontal portion, an elevated ridge for the attachment of the trapezoid ligament.

*Structure.* In the head, processes, and all the thickened parts of the bone, it is cellular in structure, of a dense compact tissue in the rest of its extent. The centre and upper part of the dorsum, but especially the former, is usually so thin as to be semi-transparent; occasionally the bone is found wanting in this situation, and the adjacent muscles come into contact.

*Development* (fig. 86). By *seven* centres; one for the body, two for the coracoid process, two for the acromion, one for the posterior border, and one for the inferior angle.

Ossification of the body of the scapula commences about the second month of fœtal life, by the formation of an irregular quadrilateral plate of bone, immediately behind the glenoid cavity. This plate extends itself so as to form the chief part of the bone, the spine growing up from its posterior surface about the third month. At birth the chief part of the scapula is osseous, the coracoid and acromion processes, the posterior border, and inferior angle, being cartilaginous. About the first year after birth, ossification takes place in the middle of the coracoid process; which usually becomes joined with the rest of the bone at the time when the other centres make their appearance. Between the fifteenth and seventeenth years, ossification of the remaining centres takes place in quick succession, and in the following order: first, near the base of the acromion, and in the upper part of the coracoid process, the latter appearing in the form of a broad scale; secondly, in the inferior angle and contiguous part of the posterior border; thirdly, near the extremity of the acromion; fourthly, in the posterior border. The acromion process, besides being formed of two separate nuclei, has its base formed by an extension into it of the centre of ossification which belongs to the spine, the extent of which varies in different cases. The two separate nuclei unite, and then join with the extension carried in from the spine. These various epiphyses become joined to the bone between the ages of twenty-two and twenty-five years. Sometimes failure of union between the acromion process and spine occurs, the junction being effected by fibrous tissue, or by an imperfect articulation; in some cases of supposed fracture of the acromion with ligamentous union, it is probable the detached segment was never united to the rest of the bone.

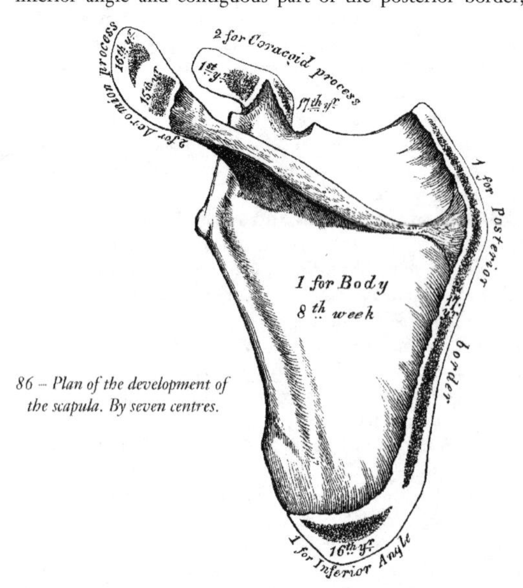


*86 – Plan of the development of the scapula. By seven centres.*

*Articulations.* With the humerus and clavicle.

*Attachment of Muscles.* To the anterior surface, the Subscapularis; posterior surface, Supra-spinatus, Infra-spinatus; spine, Trapezius, Deltoid; superior border, Omo-hyoid; vertebral border, Serratus magnus, Levator anguli scapulæ, Rhomboideus minor and major; axillary border, Triceps, Teres minor, Teres major; glenoid cavity, long head of the Biceps; coracoid process, short head of Biceps, Coraco-brachialis, Pectoralis minor; and to the inferior angle occasionally a few fibres of the Latissimus dorsi.

## The Humerus.

The *Humerus* is the longest and largest bone of the upper extremity; it presents for examination a shaft and two extremities.

The *Upper Extremity* is the largest part of the bone; it presents a rounded head, a constriction around the base of the head, the neck, and two other eminences, the greater and lesser tuberosities (fig. 87).

The *head*, nearly hemispherical in form, is directed upwards, inwards, and a little backwards; its surface is smooth, coated with cartilage in the recent state, and articulates with the glenoid cavity of the scapula. The circumference of its articular surface is slightly constricted, and is termed the *anatomical neck*, in contradistinction to the constriction which exists below the tuberosities, and is called the *surgical neck*, from its often being the seat of fracture. It should be remembered, however, that fracture of the *anatomical neck* does sometimes, though rarely, occur.

The *anatomical neck* is obliquely directed, forming an obtuse angle with the shaft. It is more distinctly marked in the lower half of its circumference than in the upper half, where it presents a narrow groove, separating the head from the tuberosities. Its circumference affords attachment to the capsular ligament, and is perforated by numerous vascular foramina.

The *greater tuberosity* is situated on the outer side of the head and lesser tuberosity. Its upper surface is rounded and marked by three flat facets, separated by two slight ridges, the most anterior facet gives attachment to the tendon of the Supra-spinatus; the middle one, to the Infra-spinatus; the posterior one, to the Teres minor. The outer surface of the great tuberosity is convex, rough, and continuous with the outer side of the shaft.

The *lesser tuberosity* is more prominent, although smaller than the greater; it is situated in front of the head, and is directed inwards and forwards. Its summit presents a prominent facet for the insertion of the tendon of the Subscapularis muscle. The tuberosities are separated from one another by a deep groove, the *bicipital groove*, so called from its lodging the long tendon of the Biceps muscle. It commences above between the two tuberosities, passes obliquely downwards and a little inwards, and terminates at the junction of the upper with the middle third of the bone. It is deep and narrow at its commencement, and becomes shallow and a little broader as it descends. In the recent state it is covered with a thin layer of cartilage, lined by a prolongation of the synovial membrane of the shoulder joint, and receives part of the tendon of insertion of the Latissimus dorsi about its centre.

The *Shaft* of the humerus is almost cylindrical in the upper half of its extent; prismatic and flattened below, it presents three borders and three surfaces for examination.

The *anterior border* runs from the front of the great tuberosity above, to the coronoid depression below, separating the internal from the external surface. Its upper part is very prominent and rough, forms the outer lip of the bicipital groove, and serves for the attachment of the tendon of the Pectoralis major. About its centre is seen the rough deltoid impression; below, it is smooth and rounded, affording attachment to the Brachialis anticus.

The *external border* runs from the back part of the greater tuberosity to the external condyle, and separates the external from the posterior surface. It is rounded and indistinctly marked in its upper half, serving for the attachment of the external head of the Triceps muscle; its centre is traversed by a broad but shallow oblique depression, the musculo-spiral groove; its lower part is marked by a prominent rough margin, a little curved from behind forwards, which presents an anterior lip for the attachment of the Supinator longus above, the Extensor carpi radialis longior below, a posterior lip for the Triceps, and an interstice for the attachment of the external intermuscular aponeurosis.

The *internal border* extends from the lesser tuberosity to the internal condyle. Its upper third is marked by a prominent ridge, forming the inner lip of the bicipital groove, and gives attachment from above downwards to the tendons of the Latissimus dorsi, Teres major, and part of the origin of

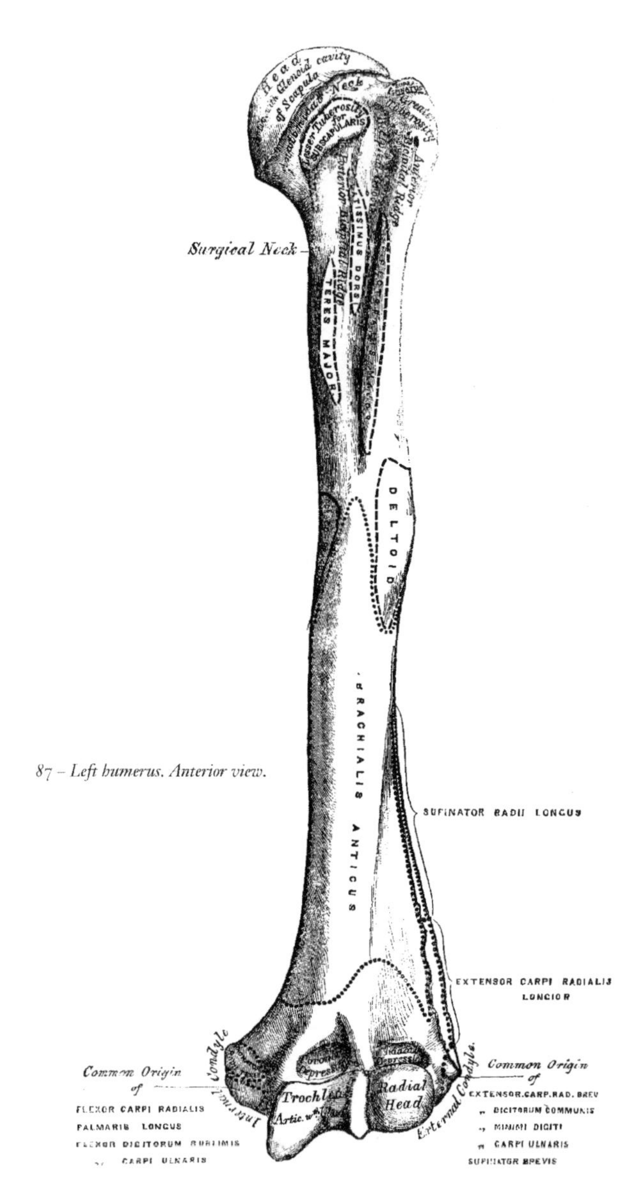


87 – *Left humerus. Anterior view.*

the inner head of the Triceps. About its centre is a rough ridge for the attachment of the Coraco-brachialis, and just below this is seen the entrance of the nutritious canal directed downwards. Sometimes there is a second canal higher up, which takes a similar direction. Its inferior third is raised into a slight ridge, which becomes very prominent below; it presents an anterior lip for the attachment of the Brachialis anticus, a posterior lip for the internal head of the Triceps, and an intermediate space for the internal intermuscular aponeurosis.

The *external surface* is directed outwards above, where it is smooth, rounded, and covered by the Deltoid muscle; forwards below, where it is slightly concave from above downwards, and gives origin to part of the Brachialis anticus muscle. About the middle of this surface, is seen a rough triangular impression for the insertion of the Deltoid muscle, and below it the musculo-spiral groove, directed obliquely from behind, forwards and downwards; it transmits the musculo-spiral nerve and superior profunda artery.

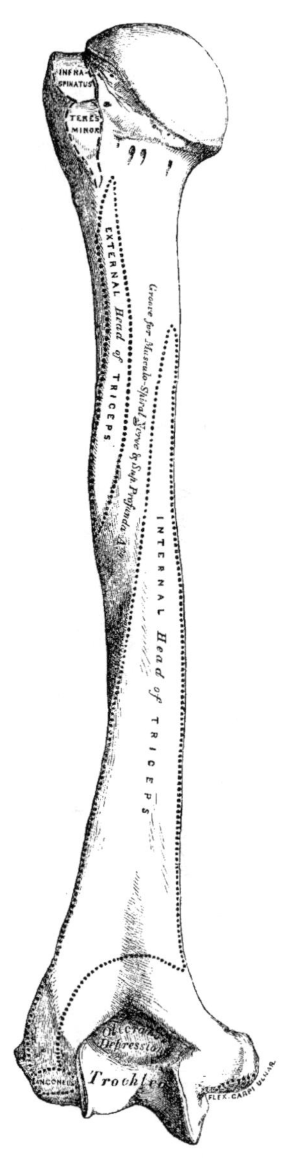


88 – *Left humerus. Posterior surface.*

The *internal surface*, less extensive than the external, is directed forwards above, forwards and inwards below: at its upper part it is narrow, and forms the bicipital groove. The middle part of this surface is slightly rough for the attachment of the Coraco-brachialis; its lower part is smooth, concave, and gives attachment to the Brachialis anticus muscle.*

The *posterior surface* (fig. 88) appears somewhat twisted, so that its upper part is directed a little inwards, its lower part backwards, and a little outwards. Nearly the whole of this surface is covered by the external and internal heads of the Triceps, the former being attached to its upper and outer part, the latter to its inner and back part, their origin being separated by the musculo-spiral groove.

The *Lower Extremity* is flattened from before backwards, and curved slightly forwards; it terminates below in a broad articular surface, which is divided into two parts by a slight ridge. On either side of the articular surface are the external and internal condyles. The articular surface extends a little lower than the condyles, and is curved slightly forwards, so as to occupy the more anterior part of the bone; its greatest breadth is in the transverse diameter, and it is obliquely directed, so that its inner extremity occupies a lower level than the outer. The outer portion of the articular surface presents a smooth rounded eminence, which has received the name of the *lesser* or *radial head* of the humerus; it articulates with the cup-shaped depression on the head of the radius, is limited to the front and lower part of the bone, not extending as far back as the other portion of the articular surface. On the inner side of this eminence is a shallow groove, in which is received the inner margin of the head of the radius. The inner or trochlear portion of the articular surface presents a deep depression between two well marked borders. This surface is convex from before backwards, concave from side to side, and occupies the anterior lower and posterior parts of the bone. The external border, less prominent than the internal, corresponds to the interval between the radius and ulna. The internal border is thicker, more prominent, and consequently, of greater length than the external. The grooved portion of the articular surface fits accurately within the greater sigmoid cavity of the ulna; it is broader and deeper on the posterior than on the anterior aspect of the bone, and is directed obliquely from behind forwards, and from without inwards. Above the back part of the trochlear surface, is a deep triangular depression, the olecranon depression, in which is received the summit of the olecranon process in extension of the fore-arm. Above the front part of the trochlear surface, is seen a small depression, the coronoid depression; it receives the coronoid process of the ulna during flexion of the fore-arm. These fossæ are separated from one another by a thin transparent lamina of bone, which is sometimes perforated; their

* A small hook-shaped process of bone, varying from $\frac{1}{10}$ to $\frac{3}{4}$ of an inch in length, is not unfrequently found projecting from the inner surface of the shaft of the humerus two inches above the internal condyle. It is curved downwards, forwards and inwards, and its pointed extremity is connected to the internal border just above the inner condyle, by a ligament or fibrous band; completing an arch, through which the median nerve and brachial artery pass, when these structures deviate from their usual course. Sometimes the nerve alone is transmitted through it, or the nerve may be accompanied by the ulnarinterosseous artery, in cases of high division of the brachial. A well marked groove is usually found behind the process, in which the nerve and artery are lodged. This space is analogous to the supra-condyloid foramen in many animals, and probably serves in them to protect the nerve and artery from compression during the contraction of the muscles in this region. A detailed account of this process is given by Dr. Struthers, in his 'Anatomical and Physiological Observations,' p. 202.

margins afford attachment to the anterior and posterior ligaments of the elbow joint, and they are lined in the recent state by the synovial membrane of this articulation. Above the front part of the radial tuberosity, is seen a slight depression which receives the anterior border of the head of the radius when the fore-arm is strongly flexed. The external condyle is a small tubercular eminence, less prominent than the internal, curved a little forwards, and giving attachment to the external lateral ligament of the elbow joint, and to a tendon common to the origin of some of the extensor and supinator muscles. The internal condyle, larger and more prominent than the external, is directed a little backwards, it gives attachment to the internal lateral ligament, and to a tendon common to the origin of some of the flexor muscles of the fore-arm. These eminences are directly continuous above with the external and internal borders. The greater prominence of the inner one renders it more liable to fracture.

*Structure.* The extremities consist of cancellous tissue, covered with a thin compact layer; the shaft is composed of a cylinder of compact tissue, thicker at the centre than at the extremities, and hollowed out by a large medullary canal, which extends along its whole length.

*Development.* By *seven* centres (fig. 89); one for the shaft, one for the head. one for the greater tuberosity, one for the radial, one for the trochlear portion of the articular surface, and one for each condyle. The centre for the shaft appears very early, soon after ossification has commenced in the clavicle, and soon extends towards the extremities. At birth, it is ossified nearly in its whole length, the extremities remaining cartilaginous. Between the first and second years, ossification commences in the head of the bone, and between the second and third years the centre for the tuberosities makes its appearance usually by a single ossific point, but sometimes, according to Beclard, by one for each tuberosity, that for the lesser being small, and not appearing until after the fourth year. By the fifth year, the centres for the head and tuberosities have enlarged and become joined, so as to form a single large epiphysis.

The lower end of the humerus is developed in the following manner: At the end of the second year, ossification commences in the radial portion of the articular surface, and from this point extends inwards, so as to form the chief part of the articular end of the bone, the centre for the inner part of the articular surface not appearing until about the age of twelve. Ossification commences in the internal condyle about the fifth year, and in the external one not until between the thirteenth or fourteenth year. About sixteen or seventeen years, the outer condyle and both portions of the articulating surface (having already joined) unite with the shaft; at eighteen years, the inner condyle becomes joined, whilst the upper epiphysis, although the first formed, is not united until about the twentieth year.

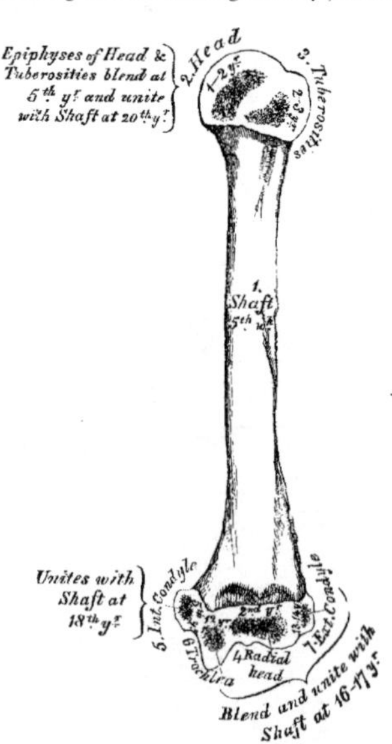

*89 – Plan of the development of the humerus. By 7 centres.*

*Articulations.* With the glenoid cavity of the scapula, and with the ulna and radius.

*Attachment of Muscles.* To the greater tuberosity, the Supra-spinatus, Infra-spinatus, and Teres minor; to the lesser tuberosity, the Subscapularis; to the anterior bicipital ridge, the Pectoralis major; to the posterior bicipital ridge and groove, the Latissimus dorsi and Teres major; to the shaft, the Deltoid, Coraco-brachialis, Brachialis anticus, external and internal heads of the Triceps; to the internal condyle, the Pronator radii teres, and common tendon of the Flexor carpi radialis, Palmaris longus, Flexor digitorum sublimis, and Flexor carpi ulnaris; to the external condyloid ridge, the Supinator longus, and Extensor carpi radialis longior; to the external condyle, the common tendon of the Extensor carpi radialis brevior, Extensor communis digitorum, Extensor minimi digiti, and Extensor carpi ulnaris, the Anconeus, and Supinator brevis.

The *Fore-arm* is that portion of the upper extremity, situated between the elbow and wrist. It is composed of two bones, the Ulna, and the Radius.

## The Ulna.

The *Ulna* (figs. 90, 91), so called from its forming the elbow (ὤλενη), is a long bone, prismatic in form, placed at the inner side of the fore-arm, parallel with the radius, being the largest and longest of the two. Its upper extremity, of great thickness and strength, forms a large part of the articulation of the elbow-joint; it diminishes in size from above downwards, its lower extremity being very small, and excluded from the wrist-joint by the interposition of an inter-articular fibro-cartilage. It is divisible into a shaft, and two extremities.

The *Upper Extremity*, the strongest part of the bone, presents for examination two large curved processes, the Olecranon process, and the Coronoid process; and two concave articular cavities, the greater and lesser Sigmoid cavities.

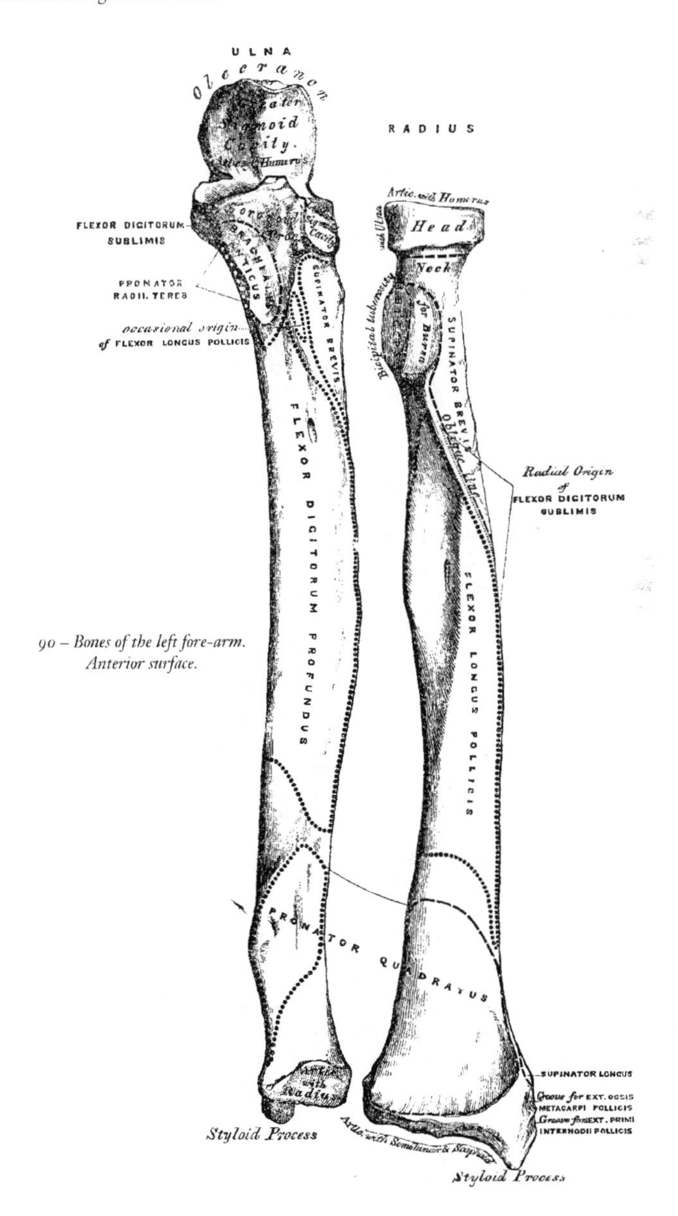


*90 – Bones of the left fore-arm. Anterior surface.*

The *Olecranon Process* (*ὤλενη, elbow; κράνον, head*) is a large, thick, curved eminence, situated at the upper and back part of the ulna. It rises somewhat higher than the coronoid, is curved forwards at the summit so as to present a prominent tip, its base being contracted where it joins the shaft. This is the narrowest part of the upper end of the ulna, and, consequently, the most usual seat of fracture. Its posterior surface, directed backwards, is of a triangular form, smooth, subcutaneous, and covered by a bursa. Its upper surface, directed upwards, is of a quadrilateral form, marked behind by a rough impression for the attachment of the Triceps muscle; and, in front, near the margin, by a slight transverse groove for the attachment of part of the posterior ligament of the elbow-joint. Its anterior surface is smooth, concave, covered with cartilage in the recent state, and forms the upper and back part of the great sigmoid cavity. The lateral borders present a continuation of the same groove that was seen on the margin of the superior surface; they serve for the attachment of ligaments, viz., the back part of the internal lateral ligament internally; the posterior ligament externally. The Olecranon process, in its structure as well as in its position and use, resembles the Patella in the lower limb; and, like it, sometimes exists as a separate piece, not united to the rest of the bone.*

The *Coronoid Process* (*κορώνη, a crow's beak; εἶδος*) is a rough triangular eminence of bone which projects horizontally forwards from the upper and front part of the ulna, forming the lower part of the great sigmoid cavity. Its base is continuous with the shaft, and of considerable strength, so much so, that fracture of it is an accident of rare occurrence. Its apex is pointed, slightly curved upwards, and received into the coronoid depression of the humerus in flexion of the fore-arm. Its upper surface is smooth, concave, and forms the lower part of the great sigmoid cavity. The under surface is concave and marked internally by a rough impression for the insertion of the Brachialis anticus. At the junction of this surface with the shaft, is a rough eminence, the tubercle of the ulna, for the attachment of the oblique ligament. Its outer surface presents a narrow, oblong, articular depression, the lesser sigmoid cavity. The inner surface by its prominent free margin, serves for the attachment of part of the internal lateral ligament. At the front part of this surface is a small rounded eminence for the attachment of one head of the Flexor digitorum sublimis. Behind the eminence, a depression for part of the origin of the Flexor profundus digitorum; and, descending from the eminence, a ridge, which gives attachment to one head of the Pronator radii teres.

The *Greater Sigmoid Cavity* (*σίγμα, εἶδος form*), so called from its resemblance to the Greek letter *Σ*, is a semilunar depression of large size, situated between the olecranon and coronoid processes, and serving for articulation with the trochlear surface of the humerus. About the middle of either lateral border of this cavity is a notch, which contracts it somewhat, and serves to indicate the junction of the two processes of which it is formed. The cavity is concave from above downwards, and divided into two lateral parts by a smooth elevated ridge, which runs from the summit of the olecranon to the tip of the coronoid process. Of these two portions, the internal is the largest; it is slightly concave transversely, the external portion being nearly plane from side to side.

The *Lesser Sigmoid Cavity* is a narrow, oblong, articular depression, placed on the outer side of the coronoid process, and serving for articulation with the head of the radius. It is concave from before backwards; and its extremities, which are prominent, serve for the attachment of the orbicular ligament.

The *Shaft* is prismatic in form at its upper part, and curved from behind forwards, and from within outwards, so as to be convex behind and externally; its central part is quite straight; its lower part rounded, smooth, and bent a little outwards; it tapers gradually from above downwards, and presents for examination three borders, and three surfaces.

The *anterior border* commences above at the prominent inner angle of the coronoid process, and terminates below in front of the styloid process. It is well marked above, smooth and rounded in the middle of its extent, and affords attachment to the Flexor profundus digitorum: sharp and prominent in its lower fourth for the attachment of the Pronator quadratus. It separates the anterior from the internal surface.

The *posterior border* commences above at the apex of the triangular surface at the back part of the olecranon, and terminates below at the back part of the styloid process; it is well marked in the upper three-fourths, and gives attachment to an aponeurosis common to the Flexor carpi ulnaris, the Extensor carpi ulnaris, and the Flexor profundus digitorum muscles; its lower fourth is smooth and rounded. This border separates the internal from the posterior surface.

* Professor Owen regards the olecranon to be homologous not with the patella, but with an extension of the upper end of the fibula above the knee-joint, which is met with in the Ornithorynchus, Echidna, and some other animals. (Owen, 'On the Nature of Limbs.')

The *external border* commences above by two lines, which converge one from each extremity of the lesser sigmoid cavity, enclosing between them a triangular space for the attachment of part of the Supinator brevis, and terminates below at the middle of the head of the ulna. Its two middle fourths are very prominent, and serve for the attachment of the interosseous membrane; its lower fourth is smooth and rounded. This border separates the anterior from the posterior surface.

The *anterior surface*, much broader above than below, is concave in the upper three-fourths of its extent, and affords attachment to the Flexor profundus digitorum; its lower fourth, also concave, to the Pronator quadratus. The lower fourth is separated from the remaining portion of the bone by a prominent ridge, directed obliquely from above downwards and inwards; this ridge marks the

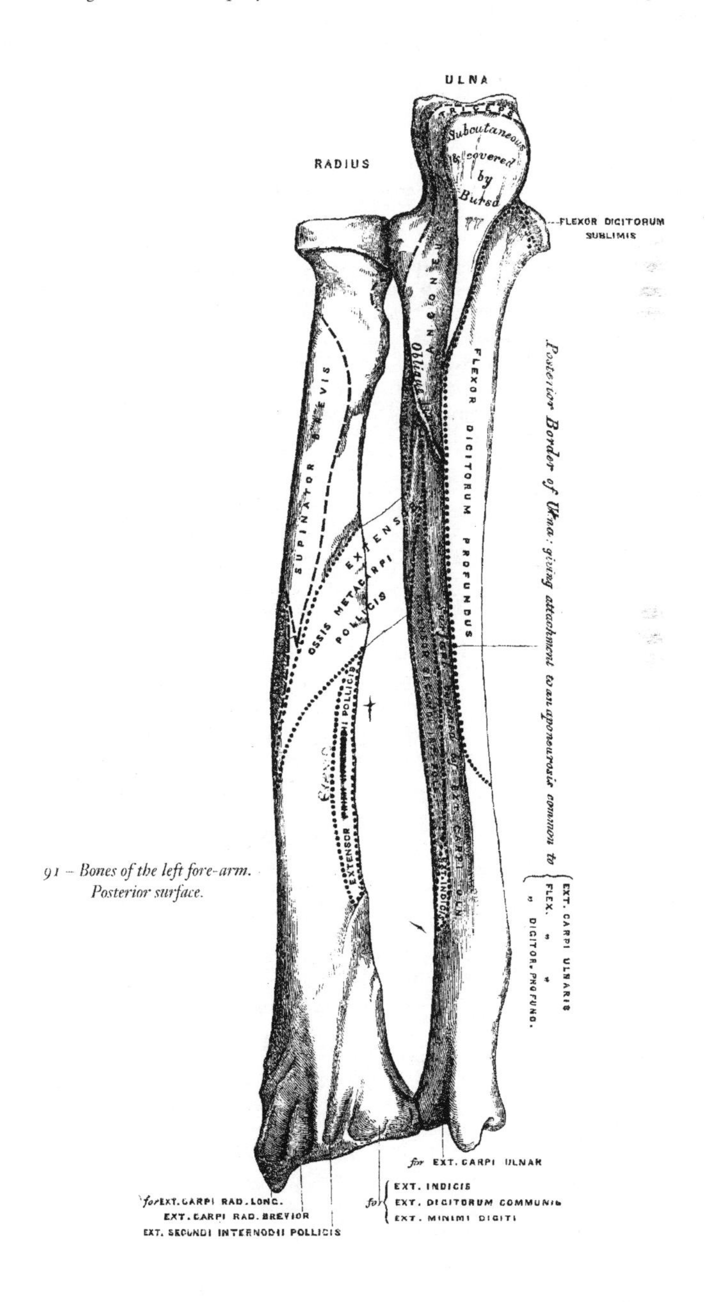


91 – *Bones of the left fore-arm. Posterior surface.*

extent of attachment of the Pronator above. At the junction of the upper with the middle third of the bone, is the nutritious canal, directed obliquely upwards and inwards.

The *posterior surface*, directed backwards and outwards, is broad and concave above, somewhat narrower and convex in the middle of its course, narrow, smooth, and rounded below. It presents above an oblique ridge, which runs from the posterior extremity of the lesser sigmoid cavity, downwards to the posterior border; the triangular surface above this ridge receives the insertion of the Anconeus muscle, whilst the ridge itself affords attachment to the Supinator brevis. The surface of bone below this is subdivided by a longitudinal ridge into two parts, the internal part is smooth, concave, and gives origin to (occasionally is merely covered by) the Extensor carpi ulnaris. The external portion, wider and rougher, gives attachment from above downwards to part of the Supinator brevis, the Extensor ossis metacarpi pollicis, the Extensor secundi internodii pollicis, and the Extensor indicis muscles.

The *internal surface* is broad and concave above, narrow and convex below. It gives attachment by its upper three-fourths to the Flexor profundus digitorum muscle; its lower fourth is subcutaneous.

The *Lower Extremity* of the ulna is of small size, and excluded from the articulation of the wrist joint. It presents for examination two eminences; the outer and larger is a rounded articular eminence, termed the head of the ulna. The inner narrower and more projecting, is a non-articular eminence, the styloid process. The *head* presents an articular facet, part of which, of an oval form, is directed downwards, and plays on the surface of the triangular fibro-cartilage, which separates this bone from the wrist joint; the remaining portion, directed outwards, is narrow, convex, and received into the sigmoid cavity of the radius. The *styloid process* projects from the inner and back part of the bone, and descends a little lower than the head, terminating in a rounded summit, which affords attachment to the internal lateral ligament of the wrist. The head is separated from the styloid process by a depression for the attachment of the triangular inter-articular fibro-cartilage; and behind, by a shallow groove for the passage of the tendon of the Extensor carpi ulnaris.

*Structure.* Similar to that of the other long bones.

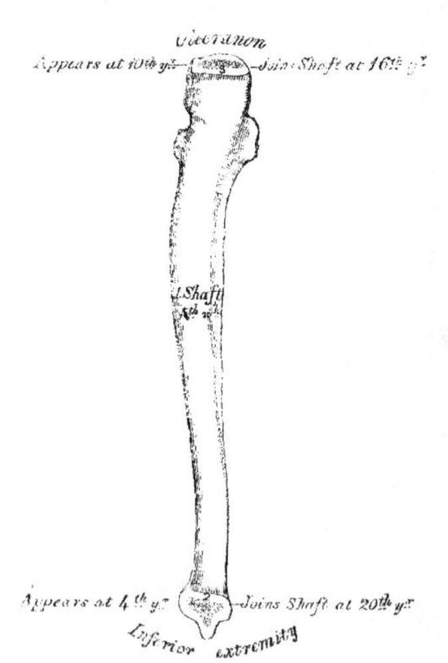

*92 – Plan of the development of the ulna. By 3 centres.*

*Development.* By *three* centres; one for the shaft, one for the inferior extremity, and one for the olecranon (fig. 92). Ossification commences near the middle of the shaft about the fifth week, and soon extends through the greater part of the bone. At birth, the ends are cartilaginous. About the fourth year, a separate osseous nucleus appears in the middle of the head, which soon extends into the styloid process. About the tenth year, ossific matter appears in the olecranon near its extremity, the chief part of this process being formed from an extension of the shaft of the bone into it. At about the sixteenth year, the upper epiphysis becomes joined, and at about the twentieth year the lower one.

*Articulations.* With the humerus and radius.

*Attachment of Muscles.* To the olecranon; the Triceps, Anconeus, and one head of the Flexor carpi ulnaris. To the coronoid process; the Brachialis anticus, Pronator radii teres, Flexor sublimis digitorum, and Flexor profundus digitorum. To the shaft; the Flexor profundus digitorum, Pronator quadratus, Flexor carpi ulnaris, Extensor carpi ulnaris, Anconeus, Supinator brevis, Extensor secundi ossis metacarpi pollicis, Extensor secundi internodii pollicis, and Extensor indicis.

## The Radius.

The *Radius* (so called from its fancied resemblance to the spoke of a wheel), is situated on the outer side of the fore-arm, lying parallel with the ulna, which exceeds it in length and size. Its upper end is small, and forms only a small part of the elbow-joint; but its lower end is large, and forms the chief part of the wrist. It is one of the long bones, having a prismatic form, slightly curved longitudinally, and presenting for examination a shaft and two extremities.

The *Upper Extremity* presents a head, neck, and tuberosity. The *head* is of a cylindrical form, depressed on its upper surface into a shallow cup, which articulates with the radial or lesser head of the humerus in flexion of the joint. Around the circumference of the head is a smooth articular surface, coated with cartilage in the recent state, broad internally where it rotates within the lessor sigmoid cavity of the ulna, narrow in the rest of its circumference, to play in the orbicular ligament. The head is supported on a round, smooth, and constricted portion of bone, called the *neck*, which presents, behind, a slight ridge, for the attachment of part of the Supinator brevis. Beneath the neck, at the inner and front aspect of the bone, is a rough eminence, the *tuberosity*. Its surface is divided into two parts by a vertical line—a posterior rough portion, for the insertion of the tendon of the Biceps muscle; and an anterior smooth portion, on which a bursa is interposed between the tendon and the bone.

The *Shaft* of the bone is prismoid in form, narrower above than below, and slightly curved, so as to be convex outwards. It presents three surfaces, separated by three borders.

The *anterior border* extends from the lower part of the tuberosity above, to the anterior part of the base of the styloid process below. It separates the anterior from the external surface. Its upper third is very prominent; and, from its oblique direction, downwards and outwards, has received the name of the *oblique line of the radius*. It gives attachment, externally, to the Supinator brevis; internally, to the Flexor longus pollicis, and between these to the Flexor digitorum sublimis. The middle third of the anterior border is indistinct and rounded. Its lower fourth is sharp, prominent, affords attachment to the Pronator quadratus, and terminates in a small tubercle, into which is inserted the tendon of the Supinator longus.

The *posterior border* commences above, at the back part of the neck of the radius, and terminates below, at the posterior part of the base of the styloid process; it separates the posterior from the external surface. It is indistinct above and below, but well marked in the middle third of the bone.

The *internal* or *interosseous border* commences above, at the back part of the tuberosity, where it is rounded and indistinct, becomes sharp and prominent as it descends, and at its lower part bifurcates into two ridges, which descend to the anterior and posterior margins of the sigmoid cavity. This border separates the anterior from the posterior surface, and has the interosseous membrane attached to it throughout the greater part of its extent.

The *anterior surface* is narrow and concave for its upper two-thirds, and gives attachment to the Flexor longus pollicis muscle; below, it is broad and flat, and gives attachment to the Pronator quadratus. At the junction of the upper and middle thirds of this surface is the nutritious foramen, which is directed obliquely upwards.

The *posterior surface* is rounded, convex, and smooth in the upper third of its extent, and covered by the Supinator brevis muscle. Its middle third is broad, slightly concave, and gives attachment to the Extensor ossis metacarpi pollicis above, the Extensor primi internodii pollicis below. Its lower third is broad, convex, and covered by the tendons of the muscles which subsequently run in the grooves on the lower end of the bone.

The *external surface* is rounded and convex throughout its entire extent. Its upper third gives attachment to the Supinator brevis muscle. About its centre is seen a rough ridge, for the insertion of the Pronator radii teres muscle. Its lower part is narrow, and covered by the tendons of the Extensor ossis metacarpi pollicis and Extensor primi internodii pollicis muscles.

The *Lower Extremity* of the radius is large, of quadrilateral form, and provided with two articular surfaces, one at the extremity for articulation with the carpus, and one at the inner side of the bone for articulation with the ulna. The carpal articular surface is of triangular form, concave, smooth, and divided by a slight antero-posterior ridge into two parts. Of these, the external is large, of a triangular form, and articulates with the scaphoid bone; the inner, smaller and quadrilateral, articulates with the semi-lunar. The articular surface for the ulna is called the *sigmoid cavity* of the radius; it is narrow, concave, smooth, and articulates with the head of the ulna. The circumference of this end of the bone presents three surfaces, an anterior, external, and posterior.

The *anterior surface*, rough and irregular, affords attachment to the anterior ligament of the wrist-joint. The *external surface* is prolonged obliquely downwards into a strong conical projection, the styloid process, which gives attachment by its base to the tendon of the Supinator longus, and by its apex to the external lateral ligament of the wrist-joint. The outer surface of this process is marked by two grooves, which run obliquely downwards and forwards, and are separated from one another by an elevated ridge. The most anterior one gives passage to the tendon of the Extensor ossis metacarpi pollicis, the posterior one to the tendon of the Extensor primi internodii pollicis. The *posterior surface* is convex, affords attachment to the posterior ligament of the wrist, and is marked by three grooves. The most external is broad, but shallow, and subdivided into two by a slightly elevated ridge. The external groove transmits the tendon of the Extensor carpi radialis longior, the inner one the tendon of the Extensor carpi radialis brevior. Near the centre of the bone is a deep, but narrow, groove, directed obliquely from above downwards and outwards; it transmits the tendon of the Extensor secundi internodii pollicis. Internally is a broad groove, for the passage of the tendons of the Extensor communis digitorum, and Extensor indicis; the tendon of the Extensor minimi digiti passing through the groove at its point of articulation with the ulna.

*Structure.* Similar to that of the other long bones.

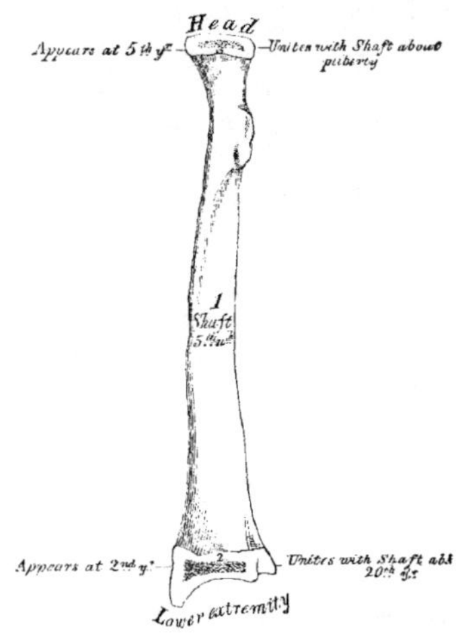


*93 – Plan of the development of the radius. By 3 centres.*

*Development* (fig. 93). By *three* centres: one for the shaft, and one for each extremity. That for the shaft, makes its appearance near the centre of the bone, soon after the development of the humerus commences. At birth, the shaft is ossified; but the ends of the bone are cartilaginous. About the end of the second year, ossification commences in the lower epiphysis; and about the fifth year, in the upper one. At the age of puberty, the upper epiphysis becomes joined to the shaft; the lower epiphysis becoming united about the twentieth year.

*Articulations.* With four bones; the humerus, ulna, scaphoid, and semilunar.

*Attachment of Muscles.* To the tuberosity, the Biceps; to the oblique ridge, the Supinator brevis, Flexor digitorum sublimis, and Flexor longus pollicis; to the shaft (its anterior surface), the Flexor longus pollicis and Pronator quadratus; (its posterior surface), the Extensor ossis metacarpi pollicis, and Extensor primi internodii pollicis; (its outer surface), the Pronator radii teres; and to the styloid process, the Supinator longus.

## THE HAND.

The Hand is subdivided into three segments, the Carpus or wrist, the Metacarpus or palm, and the Phalanges or fingers.

### Carpus.

The bones of the Carpus, eight in number, are arranged in two rows. Those of the upper row, enumerated from the radial to the ulnar side, are the scaphoid, semi-lunar, cuneiform, and pisiform; those of the lower row, enumerated in the same order, are the trapezium, trapezoid, magnum, and unciform.

### Common Characters of the Carpal Bones.

Each bone (excepting the pisiform) presents six surfaces. Of these, the *anterior* or *palmar*, and the *posterior* or *dorsal*, are rough, for ligamentous attachment, the dorsal surface being generally the broadest of the two. The *superior* and *inferior* are articular, the superior generally convex, the inferior concave; and the *internal* and *external* are also articular when in contact with contiguous bones, otherwise rough and tubercular. Their structure in all is similar, consisting within of cancellous tissue enclosed in a layer of compact bone. Each bone is also developed from a single centre of ossification.

## Bones of the Upper Row (Figs. 94, 95).

The *Scaphoid* is the largest bone of the first row. It has received its name from its fancied resemblance to a boat, being broad at one end, and narrowed like a prow at the opposite. It is situated at the upper and outer part of the carpus, its direction being from above downwards, outwards, and forwards. Its *superior surface* is convex, smooth, of triangular shape, and articulates with the lower end of the radius. Its *inferior surface*, directed downwards, outwards, and backwards, is smooth, convex, also triangular, and divided by a slight ridge into two parts, the external of which articulates with the trapezium, the inner with the trapezoid. Its *posterior* or *dorsal surface* presents a narrow, rough groove, which runs the entire breadth of the bone, and serves for the attachment of ligaments. The *anterior* or *palmar surface* is concave above, and elevated at its lower and outer part into a prominent rounded tubercle, which projects forwards from the front of the carpus, and gives attachment to the anterior annular ligament of the wrist. The *external surface* is rough and narrow, and gives attachment to the external lateral ligament of the wrist. The *internal surface* presents two articular facets: of these, the superior or smaller one is flattened, of semi-lunar form, and articulates with the semi-lunar; the inferior or larger is concave, forming, with the semi-lunar bone, a concavity for the head of the os magnum.

To ascertain to which hand this bone belongs, hold the convex radial articular surface upwards, and the dorsal surface backwards; the prominent tubercle will be directed to the side to which the bone belongs.

*Articulations.* With five bones; the radius above, trapezium and trapezoid below, os magnum and semi-lunar internally.

The *Semi-lunar* bone may be distinguished by its deep concavity and crescentic outline. It is situated in the centre of the upper range of the carpus, between the scaphoid and cuneiform. Its *superior surface*, convex, smooth, and quadrilateral in form, articulates with the radius. Its *inferior surface* is deeply concave, and of greater extent from before backwards, than transversely; it articulates with the head of the os magnum, and by a long narrow facet (separated by a ridge from the general surface) with the unciform bone. Its *anterior* or *palmar* and *posterior* or *dorsal surfaces* are rough, for the attachment of ligaments, the former being the broader, and of somewhat rounded form. The *external surface* presents a narrow, flattened, semi-lunar facet, for articulation with the scaphoid. The *internal surface* is marked by a smooth, quadrilateral facet, for articulation with the cuneiform.

To ascertain to which hand this bone belongs, hold it with the dorsal surface upwards, and the convex articular surface backwards; the quadrilateral articular facet will then point to the side to which the bone belongs.

*Articulations.* With five bones: the radius above, os magnum and unciform below, scaphoid and cuneiform on either side.

The *Cuneiform* (*l'Os Pyramidal*), may be distinguished by its pyramidal shape, and from having an oval-shaped, isolated facet, for articulation with the pisiform bone. It is situated at the upper and inner side of the carpus. The *superior surface* presents an internal, rough, non-articular portion; and an external or articular portion, which is convex, smooth, and separated from the lower end of the ulna by the inter-articular fibro-cartilage of the wrist. The *inferior surface*, directed outwards, is concave, sinuously curved, and smooth, for articulation with the unciform. Its *posterior* or *dorsal surface* is rough, for the attachment of ligaments. Its *anterior* or *palmar surface* presents, at its inner side, an oval-shaped facet, for articulation with the pisiform; and is rough externally, for ligamentous attachment. Its *external surface*, the base of the pyramid, is marked by a flat, quadrilateral, smooth facet, for articulation with the semi-lunar. The *internal surface*, the summit of the pyramid, is pointed and roughened, for the attachment of the internal lateral ligament of the wrist.

To ascertain to which hand this bone belongs, hold it so that the base is directed backwards, and the articular facet for the pisiform bone upwards; the concave articular facet will point to the side to which the bone belongs.

*Articulations.* With three bones: the semi-lunar externally, the pisiform in front, the unciform below, and with the triangular inter-articular fibro-cartilage which separates it from the lower end of the ulna.

The *Pisiform* bone may be known by its small size, and from its presenting a single articular facet. It is situated at the anterior and inner side of the carpus, is nearly circular in form, and presents on its *posterior surface* a smooth, oval facet, for articulation with the cuneiform bone. This facet approaches the superior, but not the inferior, border of the bone. Its *anterior* or *palmar surface* is rounded and rough, and gives attachment to the anterior annular ligament. The *outer* and *inner surfaces* are also rough, the former being convex, the latter usually concave.

To ascertain to which hand it belongs, hold the bone with its posterior or articular facet downwards, and the non-articular portion of the same surface backwards; the inner concave surface will then point to the side to which the bone belongs.

*Articulations.* With one bone, the cuneiform.

*Attachment of Muscles.* To two: the Flexor carpi ulnaris, and Abductor minimi digiti; and to the anterior annular ligament.

## Bones of the Lower Row (Figs. 94, 95).

The *Trapezium* is of very irregular form. It may be distinguished by a deep groove, for the tendon of the Flexor carpi radialis muscle. It is situated at the external and inferior part of the carpus, between the scaphoid and first metacarpal bone. The *superior surface*, concave and smooth, is directed

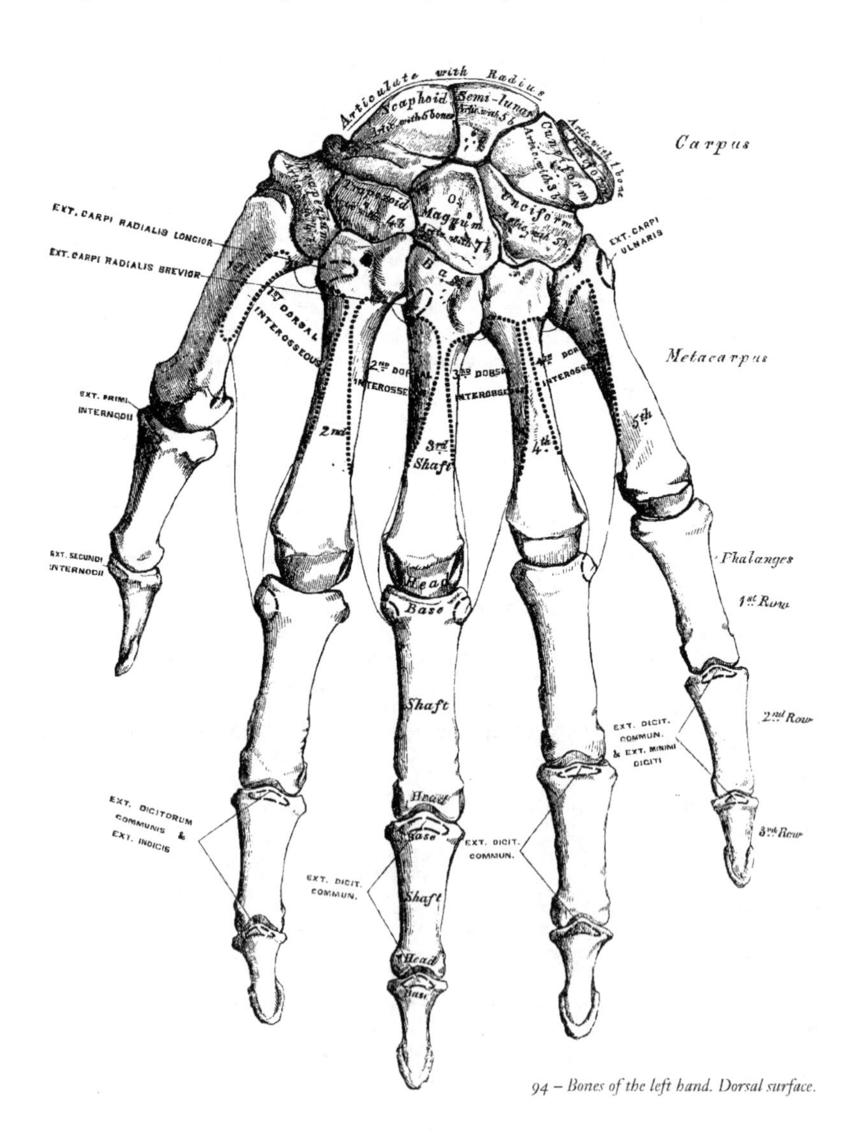


94 – *Bones of the left hand. Dorsal surface.*

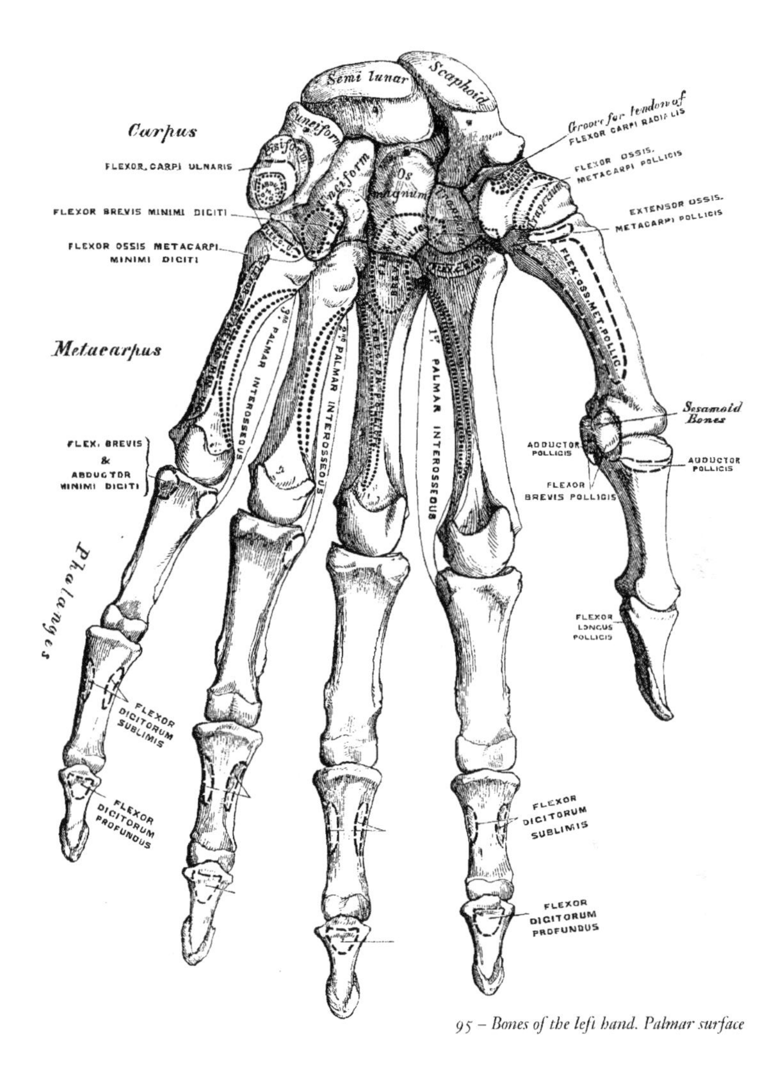


*95 – Bones of the left hand. Palmar surface*

upwards and inwards, and articulates with the scaphoid. Its *inferior surface*, directed downwards and outwards, is oval, concave from side to side, convex from before backwards, so as to form a saddle-shaped surface, for articulation with the base of the first metacarpal bone. The *anterior* or *palmar surface* is narrow and rough. At its upper part is a deep groove, running from above obliquely downwards and inwards; it transmits the tendon of the Flexor carpi radialis, and is bounded externally by a prominent ridge, the oblique ridge of the trapezium. This surface gives attachment to the Abductor pollicis, Flexor ossis metacarpi, and Flexor brevis pollicis muscles; and the anterior annular ligament. The *posterior* or *dorsal surface* is rough, and the *external surface* also broad and rough, for the attachment of ligaments. The *internal surface* presents two articular facets; the upper one, large and concave, articulates with the trapezoid; the lower one, narrow and flattened, with the base of the second metacarpal bone.

To ascertain to which hand it belongs, hold the bone with the grooved palmar surface upwards, and the external, broad, non-articular surface backwards; the saddle-shaped surface will then be directed to the side to which the bone belongs.

*Articulations.* With four bones: the scaphoid above, the trapezoid and second metacarpal bones internally, the first metacarpal below.

*Attachment of Muscles.* Abductor pollicis, Flexor ossis metacarpi, part of the Flexor brevis pollicis, and the anterior annular ligament.

The *Trapezoid* is the smallest bone in the second row. It may be known by its wedge-shaped form; its broad end occupying the dorsal, its narrow end the palmar surface of the hand. Its *superior surface*, quadrilateral in form, smooth and slightly concave, articulates with the scaphoid. The *inferior surface* articulates with the upper end of the second metacarpal bone; it is convex from side to side, concave from before backwards, and subdivided, by an elevated ridge, into two unequal lateral facets. The *posterior* or *dorsal*, and *anterior* or *palmar surfaces* are rough, for the attachment of ligaments; the former being the larger of the two. The *external surface*, convex and smooth, articulates with the trapezium. The *internal surface* is concave and smooth below, for articulation with the os magnum, rough above, for the attachment of an interosseous ligament.

To ascertain to which side this bone belongs, let the broad dorsal surface be held upwards, and its inferior concavo-convex surface forwards; the internal concave surface will then point to the side to which the bone belongs.

*Articulations.* With four bones; the scaphoid above, second metacarpal bone below, trapezium externally, os magnum internally.

*Attachment of Muscles.* Part of the Flexor brevis pollicis.

The *Os Magnum* is the largest bone of the carpus, and occupies the centre of the wrist. It presents, above, a rounded portion or head, which is received into the concavity formed by the scaphoid and semi-lunar bones; a constricted portion or neck; and, below, the body. Its *superior surface* is rounded, smooth, and articulates with the semi-lunar. Its *inferior surface* is divided by two ridges into three facets, for articulation with the second, third, and fourth metacarpal bones; that for the third (the middle facet) being the largest of the three. The *posterior* or *dorsal surface* is broad and rough; and the *anterior* or *palmar*, narrow, rounded, but also rough, for the attachment of ligaments. The *external surface* articulates with the trapezoid by a small facet at its anterior inferior angle, behind which is a rough depression for the attachment of an interosseous ligament. Above this is a deep and rough groove, which forms part of the neck, and serves for the attachment of ligaments, bounded superiorly by a smooth, convex surface, for articulation with the scaphoid. The *internal surface* articulates with the unciform by a smooth, concave, oblong facet, which occupies its posterior and superior parts; rough in front, for the attachment of an interosseous ligament.

To ascertain to which hand this bone belongs, the rounded head should be held upwards, and the broad dorsal surface forwards; the internal concave articular surface will point to its appropriate side.

*Articulations.* With seven bones: the scaphoid and semi-lunar above; the second, third, and fourth metacarpal below; the trapezoid on the radial side; and the unciform on the ulnar side.

*Attachment of Muscles.* Part of the Flexor brevis pollicis.

The *Unciform* bone may be readily distinguished by its wedge-shaped form, and the hook-like process that projects from its palmar surface. It is situated at the inner and lower angle of the carpus, with its base downwards, resting on the two inner metacarpal bones, and its apex directed upwards and outwards. Its *superior surface*, the apex of the wedge, is narrow, convex, smooth, and articulates with the semi-lunar. Its *inferior surface* articulates with the fourth and fifth metacarpal bones, the concave surface for each being separated by a ridge, which runs from before backwards. The *posterior* or *dorsal surface* is triangular and rough, for ligamentous attachment. The *anterior* or *palmar surface* presents, at its lower and inner side, a curved, hook-like process of bone, the unciform process, directed from the palmar surface forwards and outwards. It gives attachment, by its apex, to the annular ligament; by its inner surface, to the Flexor brevis minimi digiti, and the Flexor ossis metacarpi minimi digiti; and is grooved on its outer side, for the passage of the Flexor tendons into the palm of the hand. This is one of the four eminences on the front of the carpus, to which the anterior annular ligament is attached; the others being the pisiform internally, the oblique ridge of the trapezium, and the tuberosity of the scaphoid externally. The *internal surface* articulates with the cuneiform by an oblong surface, cut obliquely from above downwards and inwards. Its *external surface* articulates with the os magnum by its upper and posterior part, the remaining portion being rough, for the attachment of ligaments.

To ascertain to which hand it belongs, hold the apex of the bone upwards, and the broad dorsal surface backwards; the concavity of the unciform process will be directed to the side to which the bone belongs.

*Articulations.* With five bones: the semi-lunar above, the fourth and fifth metacarpal below, the cuneiform internally, the os magnum externally.

*Attachment of Muscles.* To two: the Flexor brevis minimi digiti, the flexor ossis metacarpi minimi digiti; and to the anterior annular ligament.

## The Metacarpus.

The Metacarpal bones are five in number: they are long cylindrical bones, presenting for examination a shaft, and two extremities.

## Common Characters of the Metacarpal Bones.

The *shaft* is prismoid in form, and curved longitudinally, so as to be convex in the longitudinal direction behind, concave in front. It presents three surfaces: two lateral, and one posterior. The *lateral surfaces* are concave, for the attachment of the Interossei muscles, and separated from one another by a prominent line. The *posterior* or *dorsal surface* is triangular, smooth, and flattened below, and covered, in the recent state, by the tendons of the Extensor muscles. In its upper half, it is divided by a ridge into two narrow lateral depressions, for the attachment of the Dorsal interossei muscles. This ridge bifurcates a little above the centre of the bone, and its branches run to the small tubercles on each side of the digital extremity.

The *carpal extremity*, or *base*, is of a cuboidal form, and broader behind than in front: it articulates, above, with the carpus; and, on each side, with the adjoining metacarpal bones; its *dorsal* and *palmar surfaces* being rough, for the attachment of tendons and ligaments.

The *digital extremity*, or *head*, presents an oblong surface, flattened at each side, for articulation with the first phalanx; it is broader and extends farther forwards in front than behind; and is longer in the antero-posterior than in the transverse diameter. On either side of the head is a deep depression, surmounted by a tubercle, for the attachment of the lateral ligament of the metacarpo-phalangeal joint. The *posterior surface*, broad and flat, supports the Extensor tendons; and the *anterior surface* presents a median groove, bounded on each side by a tubercle, for the passage of the Flexor tendons.

## Peculiar Metacarpal Bones.

The *metacarpal bone of the thumb* is shorter and wider than the rest, diverges to a greater degree from the carpus, and its *palmar surface* is directed inwards towards the palm. The *shaft* is flattened and broad on its dorsal aspect, and does not present the bifurcated ridge peculiar to the other metacarpal bones; concave from before backwards on its palmar surface. The *carpal extremity*, or *base*, presents a concavo-convex surface, for articulation with the trapezium, and has no lateral facets. The *digital extremity* is less convex than that of the other metacarpal bones, broader from side to side than from before backwards, and terminates anteriorly in a small articular eminence on each side, over which play two sesamoid bones.

The *metacarpal bone of the index-finger* is the longest, and its base the largest of the other four. Its *carpal extremity* is prolonged upwards and inwards; and its *dorsal* and *palmar surfaces* are rough, for the attachment of tendons and ligaments. It presents four articular facets: one at the end of the bone, which has an angular depression, for articulation with the trapezoid; on the radial side, a flat quadrilateral facet, for articulation with the trapezium; its ulnar side being prolonged upwards and inwards, to articulate, above, with the os magnum; internally, with the third metacarpal bone.

The *metacarpal bone of the middle-finger* is a little smaller than the preceding; it presents a pyramidal eminence on the radial side of its base (dorsal aspect), which extends upwards behind the os magnum. The carpal-articular facet is concave behind, flat and horizontal in front, and corresponds to the os magnum. On the radial side is a smooth concave facet, for articulation with the second metacarpal bone; and on the ulnar side two small oval facets, for articulation with the fourth metacarpal.

The *metacarpal bone of the ring-finger* is shorter and smaller than the preceding, and its base small and quadrilateral, its carpal surface presenting two facets, for articulation with the unciform and os magnum. On the radial side are two oval facets, for articulation with the third metacarpal bone; and on the ulnar side a single concave facet, for the fifth metacarpal.

The *metacarpal bone of the little-finger* may be distinguished by the concavo-convex form of its carpal surface, for articulation with the unciform, and from having only one lateral articular facet, which corresponds with the fourth metacarpal bone. On its ulnar side, is a prominent tubercle for the insertion of the tendon of the Extensor carpi ulnaris. The dorsal surface of the shaft is marked by an oblique ridge, which extends from near the ulnar side of the upper extremity, to the radial side of the lower. The outer division of this surface serves for the attachment of the fourth Dorsal interosseous muscle; the inner division is smooth, and covered by the Extensor tendons of the little finger.

*Articulations.* The first, with the trapezium; the second, with the trapezium, trapezoides, os magnum, and third metacarpal bones; the third, with the os magnum, and second and fourth metacarpal bones; the fourth, with the os magnum, unciform, and third and fifth metacarpal bones; and the fifth, with the unciform and fourth metacarpal.

*Attachment of Muscles.* To the metacarpal bone of the thumb, three: the Flexor ossis metacarpi pollicis, Extensor ossis metacarpi pollicis, and first Dorsal interosseous. To the second metacarpal bone, five: the Flexor carpi radialis, Extensor carpi radialis longior, first and second Dorsal interosseous, and first Palmar interosseous. To the third, five: the Extensor carpi radialis brevior, Flexor brevis pollicis, Adductor pollicis, and second and third Dorsal interosseous. To the fourth, three: the third and fourth Dorsal interosseous and second Palmar. To the fifth, four: the Extensor carpi ulnaris, Flexor carpi ulnaris, Flexor ossis metacarpi minimi digiti, and third Dorsal interosseous.

## Phalanges.

The Phalanges are the bones of the fingers; they are fourteen in number, three for each finger and two for the thumb. They are long bones, and present for examination a shaft, and two extremities. The *shaft* tapers from above downwards, is convex posteriorly, concave in front from above downwards, flat from side to side, and marked laterally by rough ridges, which give attachment to the fibrous sheaths of the Flexor tendons. The *metacarpal extremity* or *base*, in the first row, presents an oval concave articular surface, broader from side to side, than from before backwards; and the same extremity in the other two rows, a double concavity separated by a longitudinal median ridge, extending from before backwards. The *digital extremities* are smaller than the others, and terminate, in the first and second row, in two small lateral condyles, separated by a slight groove, the articular surface being prolonged farther forwards on the palmar, than on the dorsal surface, especially in the first row.

The *Ungual phalanges* are convex on their dorsal, flat on their palmar surfaces, they are recognised by their small size, and from their ungual extremity presenting, on its palmar aspect, a roughened elevated surface of a horse-shoe form, which serves to support the sensitive pulp of the finger.

*Articulations.* The first row with the metacarpal bones, and the second row of phalanges; the second row, with the first and third; the third, with the second row.

*Attachment of Muscles.* To the base of the first phalanx of the thumb, four muscles: the Extensor primi internodii pollicis, Flexor brevis pollicis, Abductor pollicis, Adductor pollicis. To the second phalanx, two: the Flexor longus pollicis, and the Extensor secundi internodii. To the base of the first phalanx of the index finger, the first Dorsal and the first Palmar interosseous; to that of the middle finger, the second and third Dorsal interosseous; to the ring finger, the fourth Dorsal and the second Palmar interosseous; and to that of the little finger, the third Palmar interosseous, the Flexor brevis minimi digiti, and Abductor minimi digiti. To the second phalanges, the Flexor sublimis digitorum, Extensor communis digitorum; and, in addition, the Extensor indicis, to the index finger; the Extensor minimi digiti, to the little finger. To the third phalanges, the Flexor profundus digitorum and Extensor communis digitorum.

## Development of the Hand.

The *Carpal bones* are each developed by a *single* centre; at birth they are all cartilaginous. Ossification proceeds in the following order (fig. 96); in the os magnum and unciform an ossific point appears during the first year, the former preceding the latter; in the cuneiform, at the third year; in the trapezium and semi-lunar, at the fifth year, the former preceding the latter; in the scaphoid, at the sixth year; in the trapezoid, during the eighth year; and in the pisiform, about the twelfth year.

The *Metacarpal bones* are each developed by *two* centres: one for the shaft, and one for the digital extremity, for the four inner metacarpal bones; one for the shaft and one for the base, for the

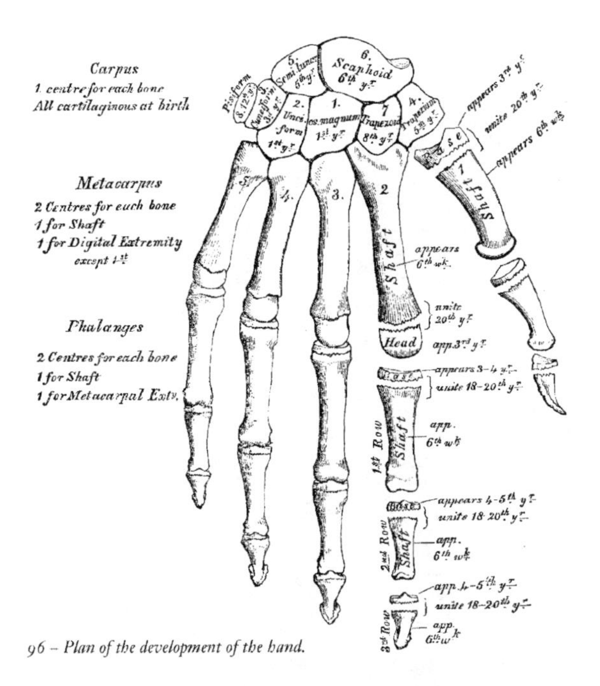


96 – Plan of the development of the hand.

metacarpal bone of the thumb, which, in this respect, resembles the phalanges. Ossification commences in the centre of the shaft about the sixth week, and gradually proceeds to either end of the bone; about the third year the digital extremities of the four inner metacarpal bones and the base of the first metacarpal, commence to ossify, and they unite about the twentieth year.

The *Phalanges* are each developed by *two* centres: one for the shaft and one for the base. Ossification commences in the shaft, in all three rows, at about the sixth week, and gradually involves the whole of the bone excepting the upper extremity. Ossification of the base commences in the first row between the third and fourth years, and a year later in those of the second and third row. The two centres become united in each row, between the eighteenth and twentieth years.

## Of the Lower Extremity.

The Lower Extremity consists of three segments, the *thigh*, *leg*, and *foot*, which correspond to the *arm*, *fore-arm*, and *hand* in the upper extremity. It is connected to the trunk through the os innominatum, or haunch, which is homologous with the shoulder.

## The Os Innominatum.

The *Os Innominatum* or nameless bone, so called from bearing no resemblance to any known object, is a large irregular-shaped bone, which, with its fellow of the opposite side, forms the sides and anterior wall of the pelvic cavity. In young subjects, it consists of three separate parts, which meet and form the large cup-like cavity, situated near the middle of the outer side of the bone; and, although in the adult these have become united, it is usual to describe the bone as divisible into three portions, the ilium, the ischium, and the pubes.

The *ilium*, so called from its supporting the flank (ilia), is the superior broad and expanded portion which runs upwards from the upper and back part of the acetabulum, and forms the prominence of the hip.

The *ischium* (ἰσχίον, the hip), is the inferior and strongest portion of the bone; it proceeds downwards from the acetabulum, expands into a large tuberosity, and then curving upwards, forms with the descending ramus of the pubes a large aperture, the obturator foramen.

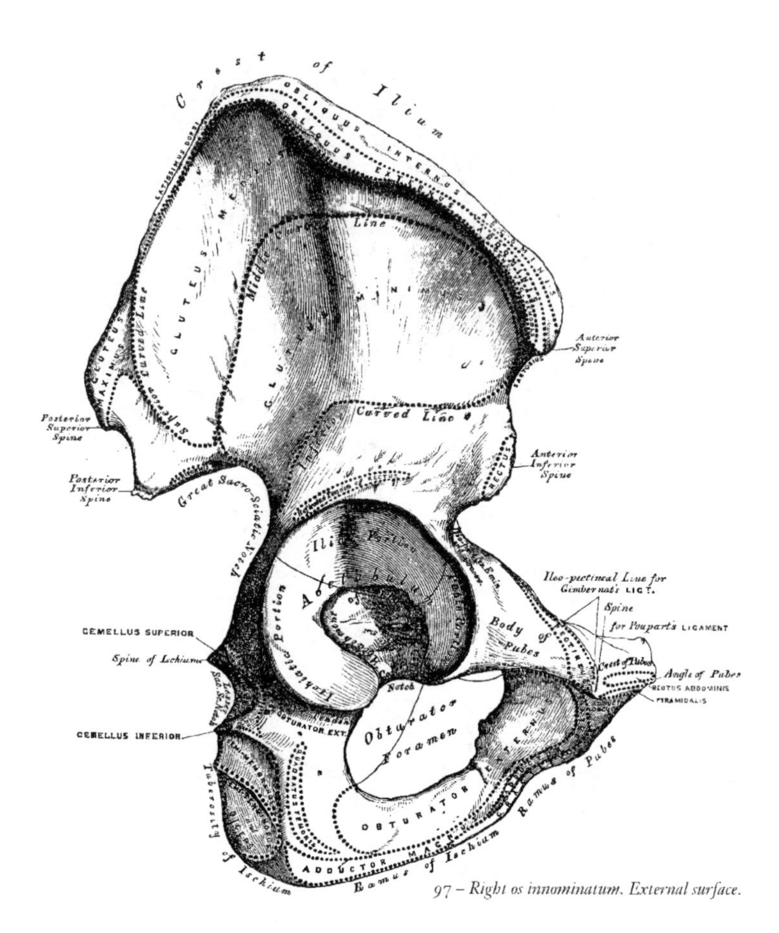


97 – *Right os innominatum. External surface.*

The *pubes* is that portion which runs horizontally inwards from the inner side of the acetabulum for about two inches, then makes a sudden bend, and descends to the same extent: it forms the front of the pelvis, supports the external organs of generation, and has received its name from being covered with hair.

The *Ilium* presents for examination two surfaces, an external and an internal, a crest, and two borders, an anterior and a posterior.

*External Surface* or *Dorsum of the Ilium* (fig. 97). The back part of this surface is directed backwards, downwards, and outwards; its front part forwards, downwards and outwards. It is smooth, convex in front, deeply concave behind; bounded above by the crest, below by the upper border of the acetabulum; in front and behind, by the anterior and posterior borders. This surface is crossed in an arched direction by three semicircular lines, the superior, middle, and inferior curved lines. The superior curved line, the shortest of the three, commences at the crest, about two inches in front of its posterior extremity; it is at first distinctly marked, but as it passes downwards and outwards to the upper part of the great sacro-sciatic notch, where it terminates, it becomes less marked, and is often altogether lost. The rough surface included between this line and the crest, affords attachment to part of the Gluteus maximus above, a few fibres of the Pyriformis below. The middle curved line, the longest of the three, commences at the crest, about an inch behind its anterior extremity, and, taking a curved direction downwards and backwards, terminates at the upper part of the great sacro-sciatic notch. The space between the middle and superior curved lines, and the crest, is concave, and affords attachment to the Gluteus medius muscle. Near the central part of this line may often be observed the orifice of a nutritious foramen. The inferior curved line, the least distinct of the three, commences in front at the upper part of the anterior inferior spinous process, and taking a curved direction backwards and downwards, terminates at the anterior part of the great sacro-sciatic notch. The surface of bone included between the middle and inferior curved lines, is concave from above downwards, convex from before backwards, and affords attachment to the Gluteus minimus muscle. Beneath the inferior curved line, and corresponding to the upper part of the acetabulum, is a smooth eminence (sometimes a depression), to which is attached the reflected tendon of the Rectus femoris muscle.

The *Internal Surface* (fig. 98) of the ilium is bounded above by the crest, below by a prominent line, the linea-ileo pectinea, and before and behind by the anterior and posterior borders. It presents anteriorly a large smooth concave surface called the *internal iliac fossa*, or *venter of the ilium*; it lodges the Iliacus muscle, and presents at its lower part the orifice of a nutritious canal. Behind the iliac fossa is a rough surface, divided into two portions, a superior and an inferior. The inferior, or auricular portion, so called from its resemblance to the external ear, is coated with cartilage in the recent state, and articulates with a similar shaped surface on the side of the sacrum. The superior portion is concave and rough for the attachment of the posterior sacro-iliac ligaments.

The crest of the ilium is convex in its general outline and sinuously curved, being bent inwards anteriorly, outwards posteriorly. It is longer in the female than in the male, very thick behind, and thinner at the centre than at the extremities. It terminates at either end in a prominent eminence, the anterior superior, and posterior superior spinous process. The surface of the crest is broad, and divided into an external lip, an internal lip, and an intermediate space. To the external lip is attached the Tensor vaginæ femoris, Obliquus externus abdominis, and Latissimus dorsi, and by its whole length the fascia lata; to the interspace between the lips, the Internal oblique; to the internal lip, the Transversalis, Quadratus lumborum, and Erector spinæ.

The anterior border of the ilium is concave. It presents two projections separated by a notch. Of these, the uppermost, situated at the junction of the crest and anterior border, is called the

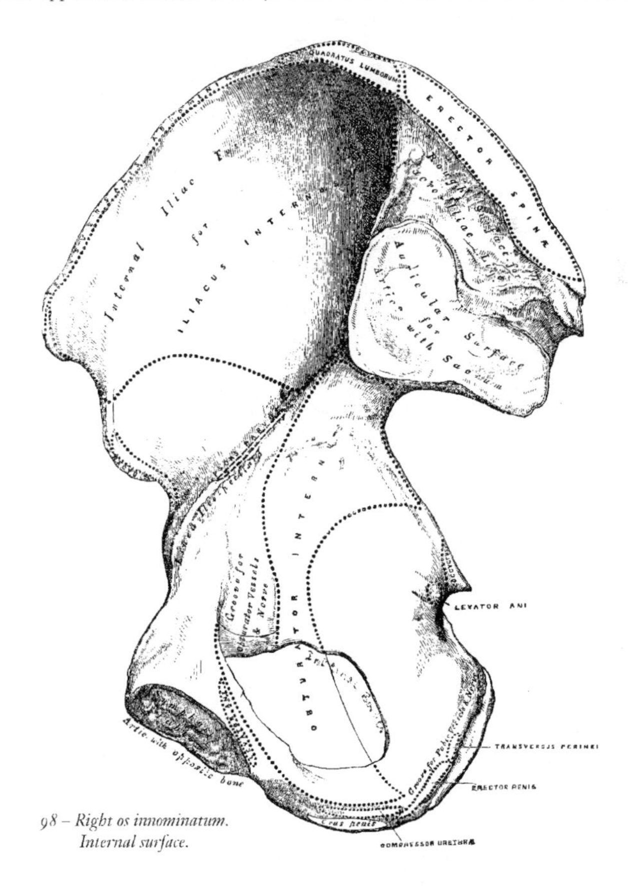


98 – *Right os innominatum. Internal surface.*

anterior superior spinous process of the ilium, the outer border of which gives attachment to the fascia lata, and the origin of the Tensor vaginæ femoris; its inner border, to the Iliacus internus; whilst its extremity affords attachment to Poupart's ligament, and the origin of the Sartorius. Beneath this eminence is a notch which gives attachment to the Sartorius muscle, and across which passes the external cutaneous nerve. Below the notch is the anterior inferior spinous process, which terminates in the upper lip of the acetabulum; it gives attachment to the straight tendon of the Rectus femoris muscle. On the inner side of the anterior inferior spinous process, is a broad shallow groove, over which passes the Iliacus muscle. The posterior border of the ilium, shorter than the anterior, also presents two projections separated by a notch, the posterior superior, and the posterior inferior spinous processes. The former corresponds with that portion of the posterior surface of the ilium, which serves for the attachment of the sacro-iliac ligaments; the latter, to the auricular portion which articulates with the sacrum. Below the posterior inferior spinous process is a deep notch, the great sacro-sciatic.

The *Ischium* forms the lower and back part of the os innominatum. It is divisible into a thick and solid portion, the body; and a thin ascending part, the ramus.

The *body*, somewhat triangular in form, presents three surfaces, external, internal, and posterior. The *external surface* corresponds to that portion of the acetabulum formed by the ischium; it is smooth and concave above, and forms a little more than two-fifths of that cavity; its outer margin is bounded by a prominent rim or lip, to which the cotyloid-fibro-cartilage is attached. Below the acetabulum, between it and the tuberosity, is a deep groove, along which the tendon of the Obturator externus muscle runs, as it passes outwards to be inserted into the digital fossa of the femur. The *internal surface* is smooth, concave, and forms the lateral boundary of the true pelvic cavity; it is broad above, and separated from the venter of the ilium by the linea-ileo-pectinea; narrow below; its posterior border is encroached upon a little below its centre, by the spine of the ischium, above and below which are the greater and lesser sacro-sciatic notches; in front it presents a sharp margin, which forms the outer boundary of the obturator foramen. This surface is perforated by two or three large vascular foramina, and affords attachment to part of the Obturator internus muscle. The *posterior surface* is quadrilateral in form, broad and smooth above, narrow below where it becomes continuous with the tuberosity; it is limited, in front, by the margin of the acetabulum; behind, by the front part of the great sacro-sciatic notch. This surface supports the Pyriformis, the two Gemelli, and the Obturator internus muscles, in their passage outwards to the great trochanter. The body of the ischium presents three borders, posterior, inferior, and internal. The *posterior border* presents, a little below the centre, a thin and pointed triangular eminence, the spine of the ischium, more or less elongated in different subjects. Its external surface gives attachment to the Gemellus superior; its internal surface, to the Coccygeus and Levator ani; whilst to the pointed extremity is connected the lesser sacro-sciatic ligament. Above the spine is a notch of large size, the great sacro-sciatic, converted into a foramen by the lesser sacro-sciatic ligament; it transmits the Pyriformis muscle, the gluteal vessels and nerve passing out of the pelvis above this muscle; the sciatic, and internal pudic vessels and nerve, and a small nerve to the Obturator internus muscle below it. Below the spine is a smaller notch, the lesser sacro-sciatic; it is smooth, coated with cartilage in the recent state, the surface of which presents numerous markings corresponding to the subdivisions of the tendon of the Obturator internus which winds over it. It is converted into a foramen by the sacro-sciatic ligaments, and transmits the tendon of the Obturator internus, the nerve which supplies this muscle, and the pudic vessels and nerve. The *inferior border* is thick and broad; at its point of junction with the posterior, is a large rough eminence upon which the body rests in sitting; it is called the tuberosity of the ischium. The *internal border* is thin, and forms the outer circumference of the obturator foramen.

The *tuberosity*, situated at the junction of the posterior and inferior borders, presents for examination an external lip, an internal lip, and an intermediate space. The external lip gives attachment to the Quadratus femoris, and part of the Adductor magnus muscles. The inner lip is bounded by a sharp ridge for the attachment of a falciform prolongation of the great sacro-sciatic ligament; presents a groove on the inner side of this for the lodgment of the internal pudic vessels and nerve; and, more anteriorly, has attached the Transversus perinei, Erector penis, and Compressor urethræ muscles. The intermediate surface presents four distinct impressions. Two of these, seen at the front part of the tuberosity, are rough, elongated, and separated from each other by a prominent ridge; the outer one gives attachment to the Adductor magnus, the inner one to the great sacro-sciatic ligament. Two, situated at the back part, are smooth, larger in size, and separated by an oblique ridge: from the upper and outer arises the Semi-membranosus; from the lower and inner, the Biceps and Semi-tendinosus. The uppermost part of the tuberosity gives attachment to the Gemellus inferior.

The *ramus* is the thin flattened part of the ischium, which ascends from the tuberosity upwards and inwards, and joins the ramus of the pubes, their point of junction being indicated in the adult by a rough eminence. Its outer surface is rough for the attachment of the Obturator externus muscle. Its inner surface forms part of the anterior wall of the pelvis. Its inner border is thick, rough, slightly everted, forms part of the outlet of the pelvis, and serves for the attachment of the crus-penis. Its outer border is thin and sharp, and forms part of the inner margin of the obturator foramen.

The *Pubes* forms the anterior part of the os innominatum; it is divisible into a horizontal ramus or body, and a perpendicular ramus.

The *body*, or *horizontal ramus*, presents for examination two extremities, an outer and an inner; and four surfaces. The *outer extremity*, the thickest part of the bone, forms one-fifth of the cavity of the acetabulum: it presents, above, a rough eminence, the ilio-pectineal, which serves to indicate the point of junction of the ilium and pubes. The *inner extremity* is the symphysis; it is oval, covered by eight or nine transverse ridges, or a series of nipple-like processes arranged in rows, separated by grooves; they serve for the attachment of the interarticular fibro-cartilage, placed between it and the opposite bone. The *upper surface*, triangular in form, wider externally than internally, is bounded behind by a sharp ridge, the pectineal line, or linea-ilio-pectinea, which, running outwards, marks the brim of the true pelvis. The surface of bone in front of the pubic portion of the linea-ilio-pectinea, serves for the attachment of the Pectineus muscle. This ridge terminates internally at a tubercle, which projects forwards, and is called the *spine* of the pubes. The portion of bone included between the spine and inner extremity of the pubes is called the *crest;* it serves for the attachment of the Rectus, Pyramidalis, and conjoined tendon of the Internal oblique and Transversalis. The point of junction of the crest with the symphysis is called the *angle of the pubes*. The *inferior surface* presents, externally, a broad and deep oblique groove, for the passage of the obturator vessels and nerve; and, internally, a sharp margin, which forms part of the circumference of the obturator foramen. Its *external surface*, flat and compressed, serves for the attachment of muscles. Its *internal surface*, convex from above downwards, concave from side to side, is smooth, and forms part of the anterior wall of the pelvis.

The *descending ramus* of the pubes passes outwards and downwards, becoming thinner and narrower as it descends, and joins with the ramus of the ischium. Its *external surface* is rough, for the attachment of muscles. Its *inner surface* is smooth. Its *inner border* is thick, rough, and everted, especially in females. In the male, it serves for the attachment of the crus penis. Its *outer border* forms part of the circumference of the obturator foramen.

The *cotyloid cavity*, or *acetabulum*, is a deep, cup-shaped, hemispherical depression; formed, internally, by the pubes; above, by the ilium; behind and below, by the ischium, a little less than two-fifths being formed by the ilium, a little more than two-fifths by the ischium, and the remaining fifth by the pubes. It is bounded by a prominent uneven rim, which is thick and strong above, and serves for the attachment of a fibro-cartilaginous structure, which contracts its orifice, and deepens the surface for articulation. It presents on its inner side a deep notch, the cotyloid notch, which transmits the nutrient vessels into the interior of the joint, and is continuous with a circular depression at the bottom of the cavity: this depression is perforated by numerous apertures, lodges a mass of fat, and its margins serve for the attachment of the ligamentum teres. The notch is converted, in the natural state, into a foramen by a dense ligamentous band which passes across it. Through this foramen, the nutrient vessels and nerves enter the joint.

The *obturator* or *thyroid foramen* is a large aperture, situated between the ischium and pubes. In the male it is large, of an oval form, its longest diameter being obliquely from above downwards; in the female smaller, and more triangular. It is bounded by a thin uneven margin, to which a strong membrane is attached; and presents, at its upper and outer part, a deep groove, which runs from the pelvis obliquely forwards, inwards, and downwards. This groove is converted into a foramen by the obturator membrane, and transmits the obturator vessels and nerve.

*Structure.* This bone consists of much cancellous tissue, especially where it is thick, enclosed between two layers of dense compact tissue. In the thinner parts of the bone, as at the bottom of the acetabulum, and centre of the iliac fossa, it is usually semi-transparent, and composed entirely of compact tissue.

*Development* (fig. 99). By *eight* centres: three primary—one for the ilium, one for the ischium, and one for the pubes; and *five* secondary—one for the crest of the ilium its whole length, one for the anterior inferior spinous process (said to occur more frequently in the male than the female), one for the tuberosity of the ischium, one for the symphysis pubis (more frequent in the female than the

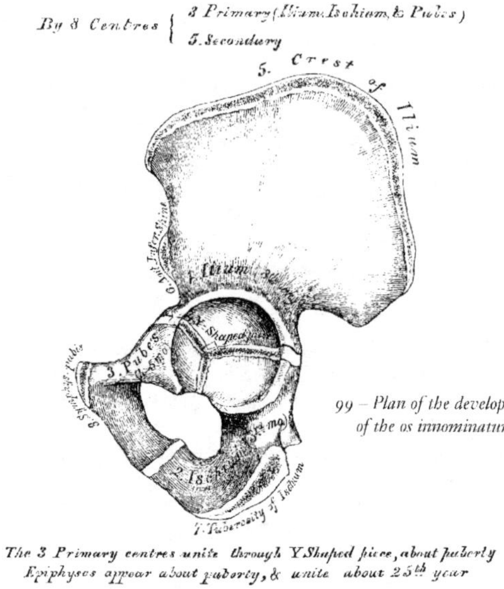

99 – *Plan of the development of the os innominatum.*

male), and one for the Y-shaped piece at the bottom of the acetabulum. These various centres appear in the following order: First, in the ilium, at the lower part of the bone immediately above the sciatic notch, at about the same period that the development of the vertebræ commences. Secondly, in the body of the ischium, at about the third month of fœtal life. Thirdly, in the body of the pubes, between the fourth and fifth months. At birth, the three primary centres are quite separate; the crest, the bottom of the acetabulum, and the rami of the ischium and pubes, being still cartilaginous. At about the sixth year, the rami of the pubes and ischium are almost completely ossified. About the thirteenth or fourteenth year, the three divisions of the bone have extended their growth into the bottom of the acetabulum, being separated from each other by a Y-shaped portion of cartilage, which now presents traces of ossification. The ilium and ischium then become joined, and lastly the pubes, through the intervention of the portion above-mentioned. At about the age of puberty, ossification takes place in each of the remaining portions, and they become joined to the rest of the bone about the twenty-fifth year.

*Articulations.* With its fellow of the opposite side, the sacrum and femur.

*Attachment of Muscles.* Ilium. To the outer lip of the crest, the Tensor vaginæ femoris, Obliquus externus abdominis, and Latissimus dorsi; to the internal lip, the Transversalis, Quadratus lumborum, and Erector spinæ; to the interspace between the lips, the Obliquus internus. To the outer surface of the ilium, the Gluteus maximus, Gluteus medius, Gluteus minimus, reflected tendon of Rectus, portion of Pyriformis; to the internal surface, the Iliacus; to the anterior border, the Sartorius and straight tendon of the Rectus. Ischium. To its outer surface, the Obturator externus; internal surface, Obturator internus and Levator ani. To the spine, the Gemellus superior, Levator ani, and Coccygeus. To the tuberosity, the Biceps, Semi-tendinosus, Semi-membranosus, Quadratus femoris, Adductor magnus, Gemellus inferior, Transversus perinæi, Erector penis. To the pubes, the Obliquus externus, Obliquus internus, Transversalis, Rectus, Pyramidalis, Psoas parvus, Pectineus, Adductor longus, Adductor brevis, Gracilis, Obturator externus and internus, Levator ani, Compressor urethræ, and occasionally a few fibres of the Accelerator urinæ.

## The Pelvis (Figs. 100, 101).

The pelvis, so called from its resemblance to a basin (πέλυξ), is stronger and more massively constructed than either of the other osseous cavities already considered; it is a bony ring, interposed between the lower end of the spine, which it supports, and the lower extremities, upon which it rests. It is composed of four bones—the two ossa innominata, which bound it on either side and in front; and the sacrum and coccyx, which complete it behind.

The pelvis is divided by a prominent line, the linea ileo pectinea, into a false and true pelvis.

The *false pelvis* is all that expanded portion of the pelvic cavity which is situated above the linea ileo pectinea. It is bounded on each side by the ossa ilii; in front it is incomplete, presenting a wide interval between the spinous processes of the ilia on either side, filled up in the recent state by the parietes of the abdomen; behind, in the middle line, is a deep notch. This broad shallow cavity is admirably adapted to support the intestines, and to transmit part of their weight to the anterior wall of the abdomen.

The *true pelvis* is all that part of the pelvic cavity which is situated beneath the linea ileo pectinea. It is smaller than the false pelvis, but its walls are more perfect. For convenience of description, it may be divided into a superior circumference or inlet, an inferior circumference or outlet, and a cavity.

The *superior circumference* forms the margin or brim of the pelvis, the included space being called the *inlet*. It is formed by the linea ileo pectinea, completed in front by the spine and crest of the pubes, and behind by the anterior margin of the base of the sacrum and sacro-vertebral angle.

The *inlet* of the pelvis is somewhat heart-shaped, obtusely pointed in front, diverging on either side, and encroached upon behind by the projection forwards of the promontory of the sacrum. It has three principal diameters: antero-posterior (sacro-pubic), transverse, and oblique. The antero-posterior extends from the sacro-vertebral angle to the symphysis pubis; its average measurement is four inches. The transverse extends across the greatest width of the inlet, from the middle of the brim on one side, to the same point on the opposite; its average measurement is five inches. The oblique extends from the margin of the pelvis, corresponding to the ileo-pectineal eminence on one side, to the sacro-iliac symphysis on the opposite side; its average measurement is also five inches.

The *cavity* of the true pelvis is bounded in front by the symphysis pubis, behind, by the concavity of the sacrum and coccyx, which, curving forwards above and below, contracts the inlet and outlet of the canal; and laterally it is bounded by a broad, smooth, quadrangular plate of bone, corresponding to the *inner surface* of the body of the ischium. The cavity is shallow in front, measuring at the symphysis an inch and a half in depth, three inches and a half in the middle, and four inches and a half posteriorly. From this description, it will be seen that the cavity of the pelvis is

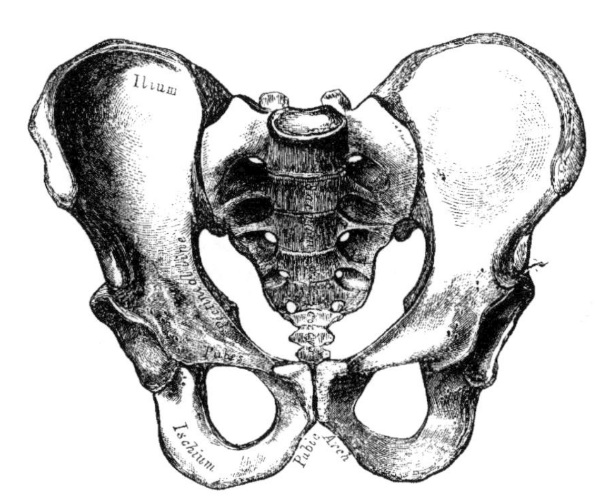


*100 – Male pelvis (adult).*

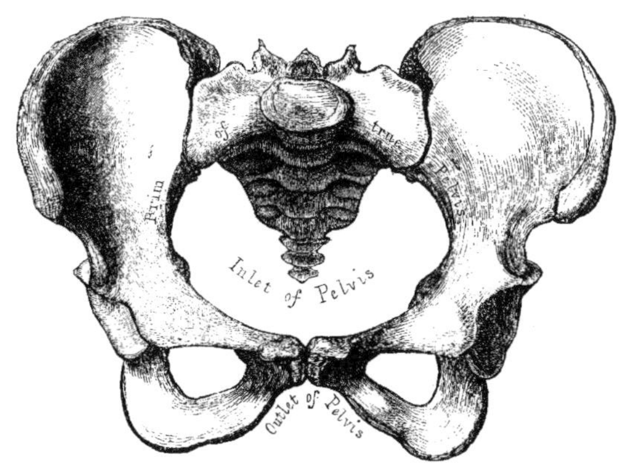


*101 – Female pelvis (adult).*

a short, curved canal, considerably deeper on its posterior than on its anterior wall, and broader in the middle than at either extremity, from the projection forwards of the sacro-coccygeal column above and below. This cavity contains, in the recent subject, the rectum, bladder, and part of the organs of generation. The rectum is placed at the back of the pelvis, and corresponds to the curve of the sacro-coccygeal column, the bladder in front, behind the symphysis pubis. In the female, the uterus and vagina occupy the interval between these parts.

The *lower circumference* of the pelvis is very irregular, and forms what is called the *outlet.* It is bounded by three prominent eminences: one posterior, formed by the point of the coccyx; and one on each side, the tuberosities of the ischia. These eminences are separated by three notches; one in front, the *public arch*, formed by the convergence of the rami of the ischia and pubes on each side. The other notches, one on each side, are formed by the sacrum and coccyx behind, the ischium in front, and the ilium above: they are called the *sacro-sciatic notches;* in the natural state they are converted into foramina by the lesser and greater sacro-sciatic ligaments.

The diameters of the outlet of the pelvis are two, antero-posterior and transverse. The *antero-posterior* extends from the tip of the coccyx to the lower part of the symphysis pubis; and the *transverse* from the posterior part of one ischiatic tuberosity, to the same point on the opposite side: the average measurement of both is four inches. The antero-posterior diameter varies with the length of the coccyx, and is capable of increase or diminution, on account of the mobility of this bone.

*Position of the Pelvis.* In the erect posture, the pelvis is placed obliquely with regard to the trunk of the body; the pelvic surface of the symphysis pubis looking upwards and backwards, the concavity of the sacrum and coccyx looking downwards and forwards; the base of the sacrum, in well-formed female bodies, being nearly four inches above the upper border of the symphysis pubis, and the apex of the coccyx a little more than half an inch above its lower border. This obliquity is much greater in the fœtus, and at an early period of life, than in the adult.

*Axes of the Pelvis* (fig. 102). The plane of the inlet of the true pelvis will be represented by a line drawn from the base of the sacrum to the upper margin of the symphysis pubis. A line carried at right angles with this at its middle, would correspond at one extremity with the umbilicus, and at the other with the middle of the coccyx; the axis of the inlet is therefore directed downwards and backwards. The axis of the outlet produced upwards, would touch the base of the sacrum; and is therefore directed downwards and forwards. The axis of the cavity is curved like the cavity itself: this curve corresponds to the concavity of the sacrum and coccyx, the extremities being indicated by the central points of the inlet and outlet. A knowledge of the direction of these axes serves to explain the course of the fœtus in its passage through the pelvis during parturition. It is also important to the surgeon at indicating the direction of the force required in the removal of calculi from the bladder, and as determining the direction in which instruments should be used in operations upon the pelvic viscera.

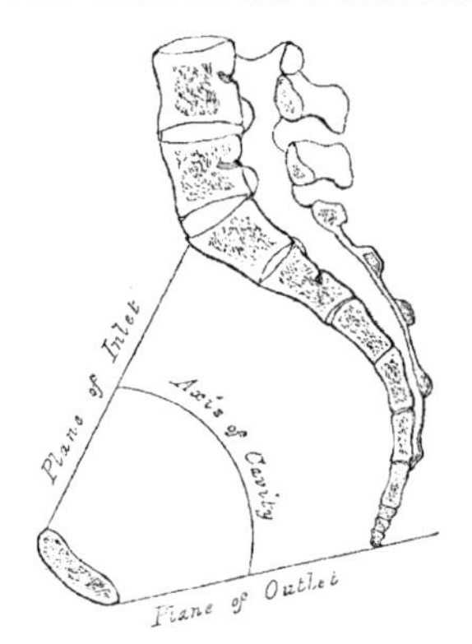


*102 – Vertical section of the pelvis, with lines indicating the axes of the pelvis.*

*Differences between the Male and Female Pelvis.* In the *male*, the bones are thicker and stronger, and the muscular eminences and impressions on their surfaces more strongly marked. The male pelvis is altogether more massive; its cavity is deeper and narrower, and the obturator foramina of larger size. In the *female*, the bones are lighter and more expanded, the muscular impressions on their surfaces are only slightly marked, and the pelvis generally is less massive in structure. The iliac fossæ are broad, and the spines of the ilia widely separated; hence the great prominence of the hips. The inlet and the outlet are larger; the cavity is more capacious, and the spines of the ischia project less into it. The promontory is less projecting, the sacrum wider and less curved,* and the coccyx more moveable. The arch of the pubes is wider, and its edges more everted. The tuberosities of the ischia and the acetabula are wider apart.

In the *fœtus*, and for several years after birth, the pelvis is small in proportion to that of the adult. The cavity is deep, and the projection of the sacro-vertebral angle less marked. The antero-

* It is not unusual to find the sacrum in the female presenting a considerable curve extending throughout its whole length.

posterior and transverse diameters are nearly equal. *About puberty*, the pelvis in both sexes presents the general characters of the adult male pelvis, but *after puberty* it acquires the sexual characters peculiar to it in adult life.

## OF THE THIGH.

The thigh is formed of a single bone, the femur.

### The Femur.

The Femur is the longest, largest, and strongest bone in the skeleton, and almost perfectly cylindrical in the greater part of its extent. In the erect posture, it is not vertical, being separated from its fellow above by a considerable interval which corresponds to the entire breadth of the pelvis, but gradually inclines downwards and inwards, so as to approach its fellow towards its lower part, for the purpose of bringing the knee-joint near the line of gravity of the body. The degree of this inclination varies in different persons, and is greater in the female than in the male, on account of the greater breadth of the pelvis. The femur, like other long bones, is divisible into a shaft, and two extremities.

The Upper Extremity presents for examination a head, a neck, and the greater and lesser trochanters.

The *head*, which is globular, and forms rather more than a hemisphere, is directed upwards, inwards, and a little forwards, the greater part of its convexity being above and in front. Its surface is smooth, coated with cartilage in the recent state, and presents, a little behind and below its centre, an ovoid depression, for the attachment of the ligamentum teres. The *neck* is a flattened pyramidal process of bone, which connects the head with the shaft. It varies in length and obliquity at various periods of life, and under different circumstances. Before puberty, it is directed obliquely, so as to form a gentle curve from the axis of the shaft. In the adult male, it forms an obtuse angle with the shaft, being directed upwards, inwards, and a little forwards. In the female, it approaches more nearly a right angle. Occasionally, in very old subjects, and more especially in those greatly debilitated, its direction becomes horizontal, so that the head sinks below the level of the trochanter, and its length diminishes to such a degree, that the head becomes almost contiguous with the shaft. The neck is flattened from before backwards, contracted in the middle, and broader at its outer extremity, where it is connected with the shaft, than at its summit, where it is attached to the head. It is much broader in the vertical than in the anterior posterior diameter, and much thicker below than above, on account of the greater amount of resistance required in sustaining the weight of the trunk. Its *anterior surface* is perforated by numerous vascular foramina. Its *posterior surface* is smooth, broader, and more concave than the anterior; and receives towards its outer side the attachment of the capsular ligament of the hip. Its *superior border* is short and thick, bounded externally by the great trochanter, and its surface perforated by large foramina. Its *inferior border*, long and narrow, curves a little backwards, to terminate at the lesser trochanter.

The Trochanters (*τρόχαω, to run* or *roll*) are prominent processes of bone which afford greater leverage to the muscles which rotate the thigh on its axis. They are two in number, the greater, and the lesser.

The *Great Trochanter* is a large irregular quadrilateral eminence, situated at the outer side of the neck, at its junction with the upper part of the shaft. It is directed a little outwards and backwards; and, in the adult, is about three quarters of an inch lower than the head. It presents for examination two surfaces, and four borders.

Its *external surface*, quadrilateral in form, is broad, rough, convex, and marked by a prominent diagonal line, which extends from the posterior superior to the anterior inferior angle: this line serves for the attachment of the tendon of the Gluteus medius. Above the line is a triangular surface, sometimes rough for part of the tendon of the same muscle, sometimes smooth for the interposition of a bursa between that tendon and the bone. Below and behind the diagonal line is a smooth triangular surface, over which the tendon of the Gluteus maximus muscle plays, a bursa being interposed. The *internal surface* is of much less extent than the external, and presents at its base a deep depression, the digital or trochanteric fossa, for the attachment of the tendon of the Obturator externus muscle.

The *superior border* is free; it is thick and irregular, and marked by impressions for the

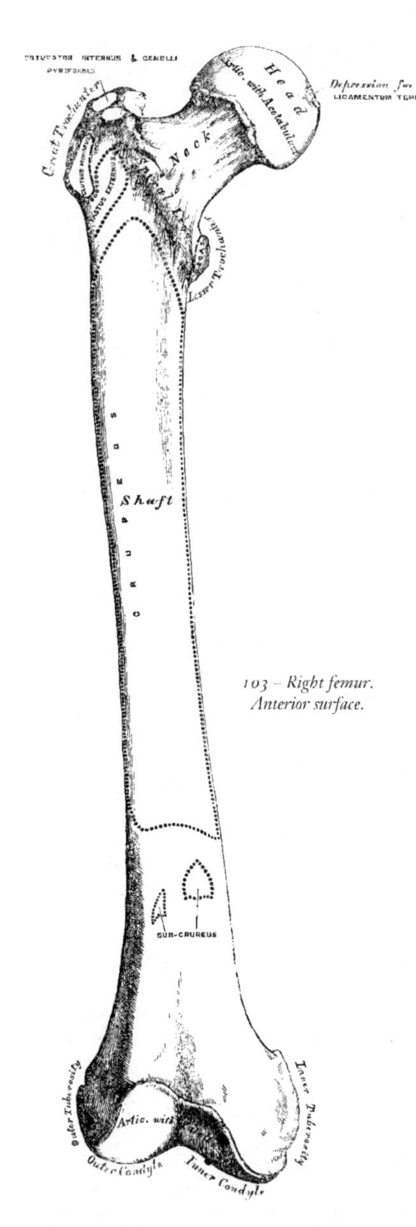


*103 – Right femur. Anterior surface.*

attachment of the Pyriformis behind, the Obturator internus and Gemelli in front. The *inferior border* corresponds to the point of junction of the base of the trochanter with the outer surface of the shaft; it is rough, prominent, slightly curved, and gives attachment to the upper part of the Vastus externus muscle. The *anterior border* is prominent, somewhat irregular, as well as the surface of bone immediately below it; it affords attachment by its outer part to the Gluteus minimus. The *posterior border* is very prominent, and appears as a free rounded edge, which forms the back part of the digital fossa.

The *Lesser Trochanter* is a conical eminence, which varies in size in different subjects; it projects from the lower and back part of the base of the neck. Its base is triangular, and connected with the adjacent parts of the bone by three well-marked borders: of these, the *superior* is continuous with the lower border of the neck; the *posterior*, with the posterior intertrochanteric line; and the *inferior*, with the middle bifurcation of the linea aspera. Its summit, which is directed inwards and backwards, is rough, and gives insertion to the tendon of the Psoas magnus. The Iliacus is inserted into the shaft below the lesser trochanter, between the Vastus internus in front, and the Pectineus behind. A well-marked prominence of variable size, which projects from the upper and front part of the neck, at its junction with the great trochanter, is called the *tubercle of the femur*; it is the point of meeting of three muscles, the Gluteus minimus externally, the Vastus externus below, and the tendon of the Obturator internus and Gemelli above. Running obliquely downwards and inwards from the tubercle, is the spiral line of the femur, or anterior intertrochanteric line; it winds round the inner side of the shaft, below the lesser trochanter, and terminates in the linea aspera; about two inches below this eminence. Its upper half is rough, and affords attachment to the capsular ligament of the hip-joint; its lower half is less prominent, and gives attachment to the upper part of the Vastus internus. The posterior intertrochanteric line is very prominent, and runs from the summit of the great trochanter downwards and inwards to the upper and back part of the lesser trochanter. Its upper half forms the posterior border of the great trochanter. A well-marked eminence commences about the middle of the posterior inter-trochanteric line, and passes vertically downwards for about two inches along the back part of the shaft: it is called the *linea quadrati*, and gives attachment to the Quadratus femoris, and a few fibres of the Adductor magnus muscles.

The *Shaft*, almost perfectly cylindrical in form, is a little broader above than in the centre, and somewhat flattened from before backwards below. It is slightly arched, so as to be convex in front; concave behind, where it is strengthened by a prominent longitudinal ridge, the linea aspera. It presents for examination three borders separating three surfaces. Of the three borders, one, the linea aspera, is posterior; the other two are placed laterally.

The *linea aspera* (fig. 104) is a prominent longitudinal ridge or crest, presenting on the middle third of the bone an external lip, an internal lip, and a rough intermediate space. A little above the centre of the shaft, this crest divides into three lines; the most external one becomes very rough, and is continued almost vertically upwards to the base of the great trochanter; the middle one, the least distinct, is continued to the base of the trochanter minor; and the internal one is lost above in the spiral line of the femur. Below, the linea aspera divides into two bifurcations, which enclose between them a triangular space (the popliteal space), upon which rests the popliteal artery. Of these two bifurcations, the outer branch is the most prominent, and descends to the summit of the outer condyle. The inner branch is less marked, presents a broad and shallow groove for the passage of the femoral artery, and terminates at a small tubercle at the summit of the internal condyle.

To the inner lip of the linea aspera, its whole length, is attached the Vastus internus; and to the whole length of the outer lip, the Vastus externus. The Adductor magnus is also attached to the whole length of the linea aspera, being connected with the outer lip above, and the inner lip below. Between the Vastus externus and the Adductor magnus are attached two muscles, viz., the Gluteus maximus above, and the short head of the Biceps below. Between the Adductor magnus and the Vastus internus, four muscles are attached: the Iliacus and Pectineus above (the latter to the middle division of the upper bifurcation); below these, the Adductor brevis and Adductor longus. The linea aspera is perforated a little below its centre by the nutritious canal, which is directed obliquely from below upwards.

The two lateral borders of the femur are only very slightly marked, the outer one extending from the anterior inferior angle of the great trochanter to the anterior extremity of the external condyle; the inner one passes from the spiral line, at a point opposite the trochanter minor, to the anterior extremity of the internal condyle. The internal border marks the limit of attachment of the Cruræus muscle internally.

The *anterior surface* includes that portion of the shaft which is situated between the two lateral borders. It is smooth, convex, broader above and below than in the centre, slightly twisted, so that its upper part is directed forwards and a little outwards, its lower part forwards and a little inwards. The upper three-fourths of this surface serve for the attachment of the Cruræus; the lower fourth is separated from this muscle by the intervention of the synovial membrane of the knee-joint, and affords attachment to the Sub-cruræus to a small extent. The *external surface* includes the portion of bone between the external border and the outer lip of the linea aspera; it is continuous, above, with the outer surface of the great trochanter; below, with the outer surface of the external condyle: to its upper three-fourths is attached the outer portion of

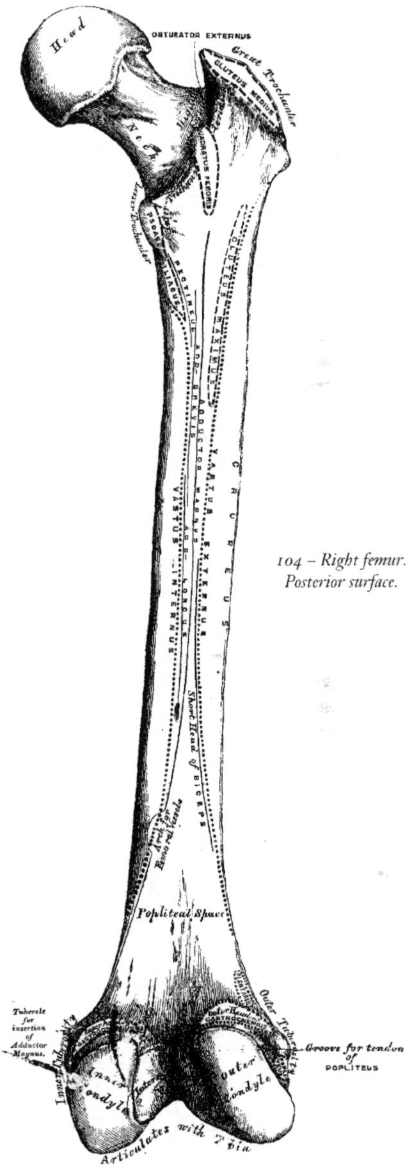

*104 – Right femur. Posterior surface.*

the Cruræus muscle. The *internal surface* includes the portion of bone between the internal border and the inner lip of the linea aspera; it is continuous, above, with the lower border of the neck; below, with the inner side of the internal condyle: it is covered by the Vastus internus muscle.

The *Lower Extremity*, larger than the upper, is of a cuboid form, flattened from before backwards, and divided by an interval presenting a smooth depression in front, and a notch of considerable size behind, into two large eminences, the condyles (*κόνδυλος, a knuckle*). The interval is called the *inter-condyloid notch*. The *external condyle* is the most prominent anteriorly, and is the broadest both in the antero-posterior and transverse diameters. The *internal condyle* is the narrowest, longest, and most prominent internally. This difference in the length of the two condyles is only observed when the bone is perpendicular, and depends upon the obliquity of the thigh-bones, in consequence of their separation above at the articulation with the pelvis. If the femur is held obliquely, the surfaces of the two condyles will be seen to be nearly horizontal. The two condyles are directly continuous in front, and form a smooth trochlear surface, the external border of which is more prominent, and ascends higher than the internal one. This surface articulates with the patella. It presents a median groove, which extends downwards and backwards to the inter-condyloid notch; and two lateral convexities, of which the external is the broader, more prominent, and prolonged farther upwards upon the front of the outer condyle. The inter-condyloid notch lodges the crucial ligaments; it is bounded laterally by the opposed surfaces of the two condyles, and in front by the lower end of the shaft.

*Outer Condyle.* The *outer surface* of the external condyle presents, a little behind its centre, an eminence, *the outer tuberosity;* it is less prominent than the inner tuberosity, and gives attachment to the external lateral ligament of the knee. Immediately beneath it is a groove which commences at a depression a little behind the centre of the lower border of this surface: the depression is for the tendon of origin of the Popliteus muscle; the groove in which this tendon is contained, is smooth, covered with cartilage in the recent state, and runs upwards and backwards to the posterior extremity of the condyle. The *inner surface* of the outer condyle forms one of the lateral boundaries of the inter-condyloid notch, and gives attachment, by its posterior part, to the anterior crucial ligament. The *inferior surface* is convex, smooth, and broader than that of the internal condyle. The posterior extremity is convex and smooth: just above the articular surface is a depression, for the tendon of the outer head of the Gastrocnemius.

*Inner Condyle.* The *inner surface* of the inner condyle presents a convex eminence, the *inner tuberosity*, rough, for the attachment of the internal lateral ligament. Above this tuberosity, at the termination of the inner bifurcation of the linea aspera, is a tubercle, for the insertion of the tendon of the Adductor magnus; and behind and beneath the tubercle a depression, for the tendon of the inner head of the Gastrocnemius. The *outer side* of the inner condyle forms one of the lateral boundaries of the inter-condyloid notch, and gives attachment, by its anterior part to the posterior crucial ligament. Its *inferior* or *articular surface* is convex, and presents a less extensive surface than the external condyle.

*Structure.* The shaft of the femur is a cylinder of compact tissue, hollowed by a large medullary canal. The cylinder is of great thickness and density in the middle third of the shaft, where the bone is narrowest, and the medullary canal well formed; but above and below this, the cylinder gradually becomes thinner, owing to a separation of the layers of the bone into cancelli, which project into the medullary canal, and finally obliterate it, so that the upper and lower ends of the shaft, and the articular extremities more especially, consist of cancellated tissue invested by a thin compact layer.

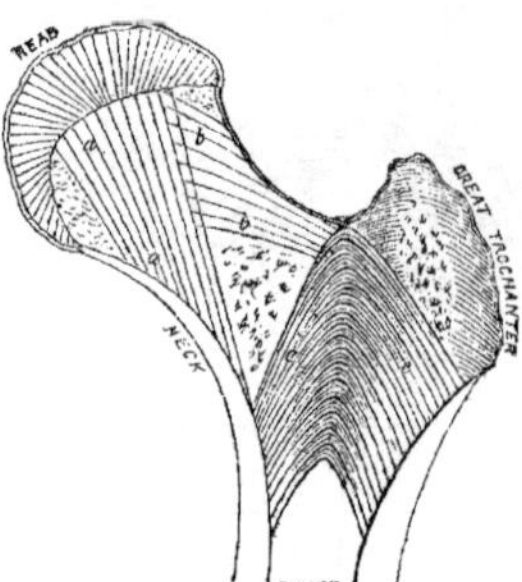


*105 – Diagram shewing the structure of the neck of the femur. (WARD).*

The arrangement of the cancelli in the ends of the femur is remarkable. In the upper end (fig. 105), they run in parallel columns *a a* from the summit of the head to the thick under wall of the neck, a series of transverse fibres *b b* decussate the parallel columns, and connect them to the thin upper wall of the neck. Another series of plates *c c* springs from the whole interior of the cylinder above the lesser trochanter; passing upwards, they converge to form a series of arches beneath the upper wall of the neck, near its junction with the great trochanter. This structure is admirably adapted to sustain, with the greatest mechanical advantage, concussion or weight transmitted from above, and serves an important office in strengthening a part especially liable to fracture.

In the lower end, the cancelli spring on all sides from the

inner surface of the cylinder, and descend in a perpendicular direction to the articular surface, the cancelli being strongest, and having a more decided perpendicular course, above the condyles.

*Articulations.* With three bones: the os innominatum, tibia, and patella.

*Development* (fig. 106). The femur is developed by *five* centres; one for the shaft, one for each extremity, and one for each trochanter. Of all the long bones, it is the first to show traces of ossification: this first commences in the shaft, at about the fifth week of fœtal life, the centres of ossification in the epiphyses appearing in the following order: First, in the lower end of the bone, at the ninth month of fœtal life; from this the condyles and tuberosities are formed; in the head, at the end of the first year after birth; in the great trochanter, during the fourth year; and in the lesser trochanter, between the thirteenth and fourteenth. The order in which the epiphyses are joined to the shaft, is the direct reverse of their appearance; their junction does not commence until after puberty, the lesser trochanter being first joined, then the greater, then the head, and, lastly, the inferior extremity (the first in which ossification commenced), which is not united until the twentieth year.

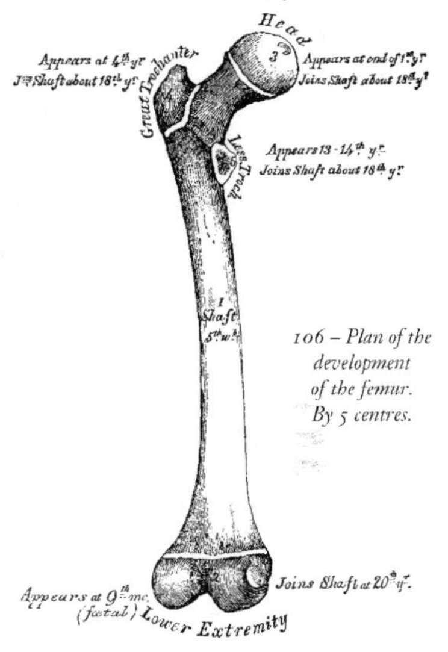


106 – *Plan of the development of the femur. By 5 centres.*

*Attachment of Muscles.* To the great trochanter: the Gluteus medius, Gluteus minimus, Pyriformis, Obturator internus, Obturator externus, Gemellus superior, Gemellus inferior, and Quadratus femoris. To the lesser trochanter: the Psoas magnus, and the Iliacus below it. To the shaft, its posterior surface: the Vastus externus, Gluteus maximus, short head of the Biceps, Vastus internus, Adductor magnus, Pectineus, Adductor brevis, and Adductor longus; to its anterior surface: the Cruræus, and Subcruræus. To the condyles: the Gastrocnemius, Plantaris, and Popliteus.

## THE LEG.

The Leg consists of three bones: the Patella, a large sesamoid bone, placed in front of the knee; and the Tibia, and Fibula.

### The Patella (Figs. 107, 108).

The *Patella* is a small, flat, triangular bone, situated at the anterior part of the knee-joint. It resembles the sesamoid bones, from being developed in the tendon of the Quadriceps extensor, and in its structure, being composed throughout of dense cancellous tissue; but it is generally regarded as analogous to the olecranon process of the ulna, which occasionally exists as a separate piece, connected to the shaft of the bone by a continuation of the tendon of the Triceps muscle.* It serves to protect the front of the joint, and increases the leverage of the Common extensor by making it act at a greater angle. It presents an anterior and posterior surface, three borders, a base, and an apex.

107 – *Right patella, anterior surface.*

The *anterior surface* is convex, perforated by small apertures, for the passage of nutrient vessels, and marked by numerous rough longitudinal striæ. This surface is covered, in the recent state, by an expansion from the tendon of the Quadriceps extensor, separated from the integument by a bursa, and gives attachment below to the ligamentum patellæ. The *posterior*

* Professor Owen states, that, 'in certain bats, there is a development of a sesamoid bone in the biceps brachii, which is the true homotype of the patella in the leg,' regarding the olecranon to be homologous, not with the patella, but with an extension of the upper end of the fibula above the knee-joint, which is met with in some animals. ('On the Nature of Limbs,' pp. 19, 24.)

*108 – Posterior surface.*

*surface* presents a smooth, oval-shaped, articular surface, covered with cartilage in the recent state, and divided into two facets by a vertical ridge, which descends from the superior towards the inferior angle of the bone. The ridge corresponds to the groove on the trochlear surface of the femur, and the two facets to the articular surfaces of the two condyles; the outer facet, for articulation with the outer condyle, being the broader and deeper, serves to indicate the leg to which the bone belongs. Below the articular surface is a rough, convex, non-articular depression, the lower half of which gives attachment to the ligamentum patellæ; the upper half being separated from the head of the tibia by adipose tissue.

Its *superior* and *lateral borders* give attachment to the tendon of the Quadriceps extensor; to the *superior border*, that portion of the tendon which is derived from the Rectus and Cruræus muscles; and to the *lateral borders*, the portion derived from the external and internal Vasti muscles.

The *base*, or *superior border*, is thick, directed upwards, and cut obliquely at the expense of its outer surface; it receives the attachment, as already mentioned, of part of the Quadriceps extensor tendon.

The *apex* is pointed, and gives attachment to the ligamentum patellæ.

*Structure.* It consists of dense cancellous tissue, covered by a thin compact lamina.

*Development.* By a single centre, which makes its appearance, according to Beclard, about the third year. In two instances, I have seen this bone cartilaginous throughout, at a much later period (six years). More rarely, the bone is developed by two centres, placed side by side.

*Articulations.* With the two condyles of the femur.

*Attachment of Muscles.* The Rectus, Cruræus, Vastus internus, and Vastus externus. These muscles joined at their insertion, constitute the Quadriceps extensor cruris.

## The Tibia.

The Tibia (so named from its resemblance to a flute or pipe) is situated at the front and inner side of the leg, and, excepting the femur, is the longest and largest bone in the skeleton. It is prismoid in form, expanded above, where it enters into formation with the knee joint, more slightly enlarged below. In the male, its direction is vertical, and parallel with the bone of the opposite side; but in the female it has a slight oblique direction downwards and outwards, to compensate for the oblique direction of the femur inwards. It presents for examination a shaft and two extremities.

The *Upper Extremity*, or head, is large and expanded on each side into two lateral eminences, the tuberosities. Superiorly, the tuberosities present two smooth concave surfaces, which articulate with the condyles of the femur; the internal articular surface is longer than the external, oval from before backwards, to articulate with the internal condyle; the external one being broader, flatter, and more circular, to articulate with the external condyle. Between the two articular surfaces, and nearer the posterior than the anterior aspect of the bone, is an eminence, the spinous process of the tibia, surmounted by a prominent tubercle on each side, which give attachment to the extremities of the semilunar fibro-cartilages; and in front and behind the spinous process, a rough depression for the attachment of the anterior and posterior crucial ligaments and the semilunar cartilages. Anteriorly the tuberosities are continuous with one another, presenting a large and somewhat flattened triangular surface, broad above, and perforated by large vascular foramina, narrow below, where it terminates in a prominent oblong elevation of large size, the tubercle of the tibia; the lower half of this tubercle is rough, for the attachment of the ligamentum patellæ; the upper half is a smooth facet corresponding, in the recent state, with a bursa which separates the ligament from the bone. Posteriorly, the tuberosities are separated from each other by a shallow depression, the popliteal notch, which gives attachment to the posterior crucial ligament. The posterior surface of the inner tuberosity presents a deep transverse groove, for the insertion of the tendon of the Semimembranosus; and the posterior surface of the outer one, a flat articular facet, nearly circular in form, directed downwards, backwards, and outwards, for articulation with the fibula. The lateral surfaces are convex and rough, the internal one, the most prominent, gives attachment to the internal lateral ligament.

The *Shaft* of the tibia is of a triangular prismoid form, broad above, gradually decreasing in size to the commencement of its lower fourth, its most slender part, where fracture most frequently occurs, and then enlarges again towards its lower extremity. It presents for examination three surfaces and three borders.

The *anterior border*, the most prominent of the three, is called the *crest of the tibia*, or in popular language, the *shin*; it commences above at the tubercle, and terminates below at the anterior margin of the inner malleolus. This border is very prominent in the upper two-thirds of its extent, smooth and rounded below. It presents a very flexuous course, being curved outwards above, and inwards below; it gives attachment to the deep fascia of the leg.

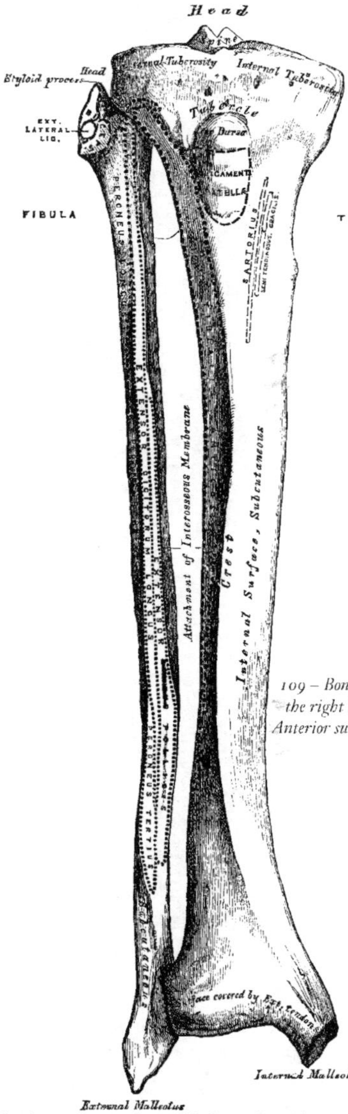

The *internal border* is smooth and rounded above and below, but more prominent in the centre; it commences at the back part of the inner tuberosity, and terminates at the posterior border of the internal malleolus; its upper third gives attachment to the internal lateral ligament of the knee, and to some fibres of the Popliteus muscle; its middle third, to some fibres of the Soleus and Flexor longus digitorum muscles.

The *external border* is thin and prominent, especially its central part, and gives attachment to the interosseous membrane; it commences above in front of the fibular articular facet, and bifurcates below, to form the boundaries of a triangular rough surface, for the attachment of the interosseous ligament, connecting the tibia and fibula.

The *internal surface* is smooth, convex, and broader above than below; its upper third, directed forwards and inwards, is covered by the aponeurosis derived from the tendon of the Sartorius, and by the tendons of the Gracilis and Semi-tendinosus, all of which are inserted nearly as far forwards as the anterior border; in the rest of its extent it is sub-cutaneous.

The *external surface* is narrower than the internal, its upper two-thirds present a shallow groove for the attachment of the Tibialis anticus muscle; its lower third is smooth, convex, curves gradually forwards to the anterior part of the bone, and is covered from within outwards by the tendons of the following muscles; Tibialis anticus, Extensor proprius pollicis, Extensor longus digitorum, Peroneus tertius.

The *posterior surface* (fig. 110) presents at its upper part a prominent ridge, the oblique line of the tibia, which extends from the back part of the articular facet for the fibula, obliquely downwards, to the internal border, at the junction of its upper and middle thirds. It marks the limit for the insertion of the Popliteus muscle, and serves for the attachment of the popliteal fascia, and part of the Soleus, Flexor longus digitorum, and Tibialis posticus muscles; the triangular concave surface, above, and to the inner side of this line, gives attachment to the Popliteus muscle. The middle third of the posterior surface is divided by a vertical ridge into two lateral halves; the ridge is well marked at its commencement at the oblique line, but becomes gradually indistinct below; the inner and broadest half gives attachment to the Flexor longus digitorum, the outer and narrowest, to part of the Tibialis posticus. The remaining part of the bone is covered by the Tibialis posticus, Flexor longus digitorum and Flexor longus pollicis muscles. Immediately below the oblique line is the medullary foramen, which is directed obliquely downwards.

*109 – Bones of the right leg. Anterior surface.*

The *Lower Extremity*, much smaller than the upper, is somewhat quadrilateral in form, and prolonged downwards, on its inner side, into a strong process, the internal malleolus. The *inferior surface* of the bone presents a quadrilateral smooth surface, for articulation with the astragalus;

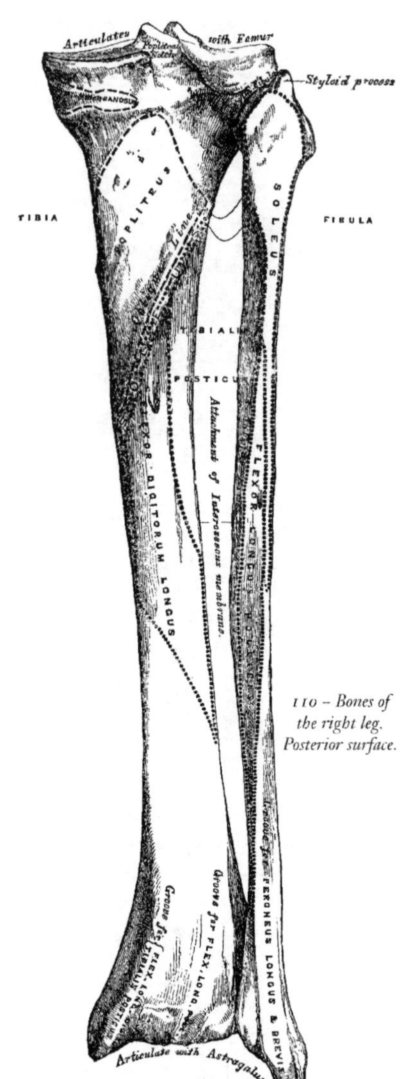


110 – *Bones of the right leg. Posterior surface.*

narrow internally, where it becomes continuous with the articular surface of the inner malleolus, broader externally, and traversed from before backwards by a slight elevation, separating two lateral depressions. The *anterior surface* is smooth and rounded above, and covered by the tendons of the Extensor muscles of the toes; its lower margin presents a rough transverse depression, for the attachment of the anterior ligament of the ankle joint. The *posterior surface* presents a superficial groove directed obliquely downwards and inwards, continuous with a similar groove on the posterior extremity of the astragalus; it serves for the passage of the tendon of the Flexor longus pollicis. The *external surface* presents a triangular rough depression, the lower part of which, in some bones, is smooth, covered with cartilage in the recent state and articulates with the fibula, the remaining part is rough for the attachment of the inferior interosseous ligament, which connects it with the fibula. This surface is bounded by two prominent ridges, continuous above with the interosseous ridge; they afford attachment to the anterior and posterior tibio-fibular ligaments. The *internal surface* is prolonged downwards to form a strong pyramidal-shaped process, flattened from without inwards, the inner malleolus; its *inner surface* is convex and subcutaneous. Its *outer surface*, smooth and slightly concave, deepens the articular surface for the astragalus. Its *anterior border* is rough, for the attachment of ligamentous fibres. Its *posterior border* presents a broad and deep groove, directed obliquely downwards and inwards; it is occasionally double, and transmits the tendons of the Tibialis posticus and Flexor longus digitorum muscles. Its *summit* is marked by a rough depression behind, for the attachment of the internal lateral ligament of the ankle joint.

*Structure.* Like that of the other long bones.

*Development.* By three centres (fig. 111): one for the shaft, and one for each extremity. Ossification commences in the centre of the shaft about the same time as in the femur, the fifth week, and gradually extends towards either extremity. The centre for the upper epiphysis appears at birth; it is flattened in form, and has a thin tongue-shaped process in front, which forms the tubercle. That for the lower epiphysis appears in the second year. The lower epiphysis joins the shaft at about the twentieth year, and the upper one about the twenty-fifth year. Two additional centres occasionally exist, one for the tongue-shaped process of the upper epiphysis, the tubercle, and one for the inner malleolus.

*Articulations.* With three bones: the femur, fibula, and astragalus.

*Attachment of Muscles.* To the inner tuberosity, the Semi-membranosus. To the outer tuberosity, the Tibialis anticus and Extensor longus digitorum: to the shaft; its internal surface, the Sartorius, Gracilis, and Semi-tendinosus: to its external surface, the Tibialis anticus: to its posterior surface, the Popliteus, Soleus, Flexor longus digitorum, and Tibialis posticus: to the tubercle, the ligamentum patellæ.

## The Fibula.

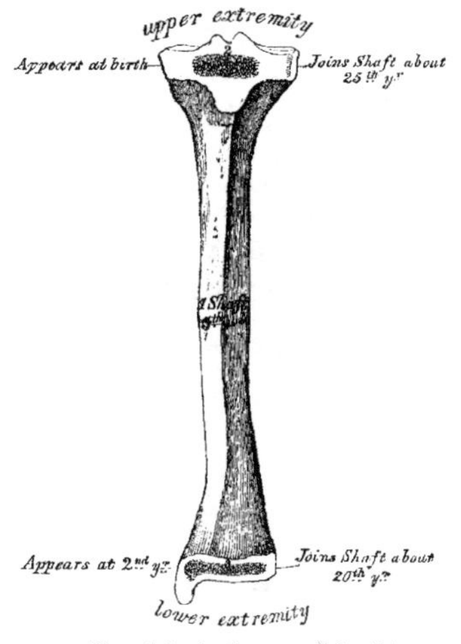


111 – *Plan of the development of the tibia. By 3 centres.*

The Fibula is situated at the outer side of the leg. It is the smaller of the two bones, and, in proportion to its length, the most slender of all the long bones; it is placed nearly parallel with the tibia, its upper extremity is small, placed below the level of the knee joint, and excluded from its formation; but the lower extremity inclines a little forwards, so as to be on a plane anterior to that of the upper end, projects below the tibia, and forms the outer ankle. It presents for examination a shaft and two extremities.

The *Upper Extremity* or *Head*, is of an irregular rounded form, presenting above a flattened articular facet, directed upwards and inwards, for articulation with a corresponding facet on the external tuberosity of the tibia. On the outer side is a thick and rough prominence, continued behind into a pointed eminence, the styloid process, which projects upwards from the posterior part of the head. The prominence above mentioned gives attachment to the tendon of the Biceps muscle, and to the long external lateral ligament of the knee, the ligament dividing this tendon into two parts. The summit of the styloid process gives attachment to the short external lateral ligament. The remaining part of the circumference of the head is rough, for the attachment, in front, of the anterior superior tibio-fibular ligament, and the upper and anterior part of the Peroneus longus; and behind, to the posterior superior tibio-fibular ligament, and the upper fibres of the outer head of the Soleus muscle.

The *Lower Extremity*, or *external malleolus*, is of a pyramidal form, somewhat flattened from without inwards, and is longer, and descends lower than the internal malleolus. Its *external surface* is convex, sub-cutaneous, and continuous with a triangular (also sub-cutaneous) surface on the outer side of the shaft. The *internal surface* presents in front a smooth triangular facet, broader above than below, convex from above downwards, which articulates with a corresponding surface on the outer side of the astragalus. Behind and beneath the articular surface is a rough depression, which gives attachment to the posterior fasciculus of the external lateral ligament of the ankle. The *anterior border* is thick and rough, and marked below by a depression for the attachment of the anterior fasciculus of the external lateral ligament. The *posterior border* is broad and marked by a shallow groove, for the passage of the tendons of the Peroneus longus and brevis muscles. Its *summit* is rounded, and gives attachment to the middle fasciculus of the external lateral ligament.

The *Shaft* presents three surfaces, and three borders. The *anterior border* commences above in front of the head, runs vertically downwards to a little below the middle of the bone, and then curving a little outwards, bifurcates below into two lines, which bound the triangular sub-cutaneous surface immediately above the outer side of the external malleolus. It gives attachment to an inter-muscular septum, which separates the muscles on the anterior surface from those on the external.

The *internal border* or *interosseous ridge*, is situated close to the inner side of the preceding, it runs nearly parallel with it in the upper third of its extent, but diverges from it so as to include a broader space in the lower two-thirds. It commences above just beneath the head of the bone (sometimes it is quite indistinct for about an inch below the head), and terminates below at the apex of a rough triangular surface immediately above the articular facet of the external malleolus. It serves for the attachment of the interosseous membrane, and separates the extensor muscles in front, from the flexor muscles behind. The portion of bone included between the anterior and interosseous lines, forms the anterior surface.

The *posterior border* is sharp and prominent; it commences above at the base of the styloid process, and terminates below in the posterior border of the outer malleolus. It is directed outwards above, backwards in the middle of its course, backwards and a little inwards below, and gives attachment to an aponeurosis which separates the muscles on the outer from those on the inner surface of the shaft. The portion of bone included between this line and the interosseous ridge, forms the internal surface. Its upper three-fourths are subdivided into two parts, an anterior and a posterior, by a very prominent ridge, the *oblique line of the fibula*, which commences above at the inner side of the head, and terminates by being continuous with the interosseous ridge at the lower fourth of the bone. It attaches an aponeurosis which separates the Tibialis posticus from the Soleus above, and the Flexor longus pollicis below. This ridge sometimes ceases just before approaching the interosseous ridge.

The *anterior surface* is the interval between the anterior and interosseous lines. It is extremely narrow and flat in the upper third of its extent; broader and grooved longitudinally in its lower third; it serves for the attachment of three muscles, the Extensor longus digitorum, Peroneus tertius, and Extensor longus pollicis.

The *external surface*, much broader than the preceding, is directed outwards in the upper two-thirds of its course, backwards in the lower third, where it is continuous with the posterior border of the external malleolus. This surface is completely occupied by the Peroneus longus and brevis muscles.

The *internal surface* is the interval between the interosseous ridge and the posterior border, and occupies nearly two-thirds of the circumference of the bone. Its upper three-fourths are divided into an anterior and a posterior portion by a very prominent ridge already mentioned, the oblique line of the fibula. The anterior portion is directed inwards, and is grooved for the attachment of the Tibialis posticus muscle. The posterior portion is continuous below with the rough triangular surface above the articular facet of the outer malleolus; it is directed backwards above, backwards and inwards at its middle, directly inwards below. Its upper fourth is rough, for the attachment of the Soleus muscle; its lower part presents a triangular rough surface, connected to the tibia by a strong interosseous ligament, and between these two points, the entire surface is covered by the fibres of origin of the Flexor longus pollicis muscle. At about the middle of this surface is the nutritious foramen, which is directed downwards.

In order to distinguish the side to which the bone belongs, hold it with the lower extremity downwards, and the broad groove for the Peronei tendons backwards, towards the holder, the triangular sub-cutaneous surface will then be directed to the side to which the bone belongs.

*Articulations.* With two bones; the tibia and astragalus.

*Development.* By three centres (fig. 112); one for the shaft, and one for each extremity. Ossification commences in the shaft about the sixth week of fœtal life, a little later than in the tibia, and extends gradually towards the extremities. At birth both ends are cartilaginous. Ossification commences in the lower end in the second year, and in the upper one about the fourth year. The lower epiphysis, the first in which ossification commences, becomes united to the shaft about the twentieth year, contrary to the law which appears to prevail with regard to the junction of the epiphyses with the shaft; the upper one is joined about the twenty-fifth year.

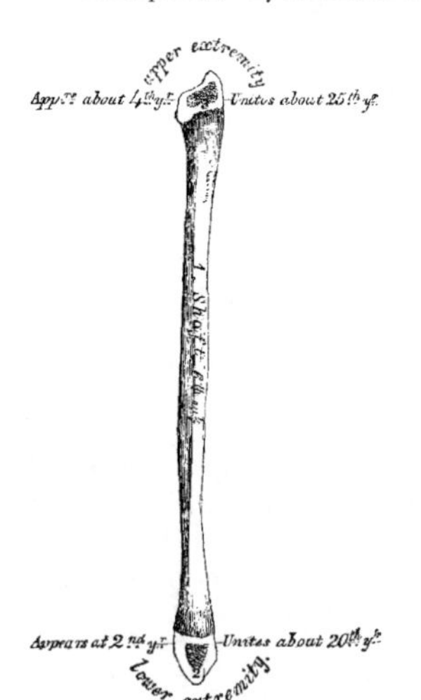


112 – *Plan of the development of the fibula. By 3 centres.*

*Attachment of Muscles.* To the head, the Biceps, Soleus, and Peroneus longus: to the shaft, its anterior surface, the Extensor longus digitorum, Peroneus tertius, and Extensor longus pollicis: to the internal surface, the Soleus, Tibialis posticus, and Flexor longus pollicis: to the external surface, the Peroneus longus and brevis.

## THE FOOT.

The Foot (figs. 113, 114) is the terminal part of the inferior extremity; it serves to support the body in the erect posture, and is an important instrument of locomotion. It consists of three divisions: the Tarsus, Metatarsus, and Phalanges.

### The Tarsus.

The bones of the Tarsus are seven in number; viz., the calcaneum, or os calcis, astragalus, cuboid, scaphoid, internal, middle, and external, cuneiform bones.

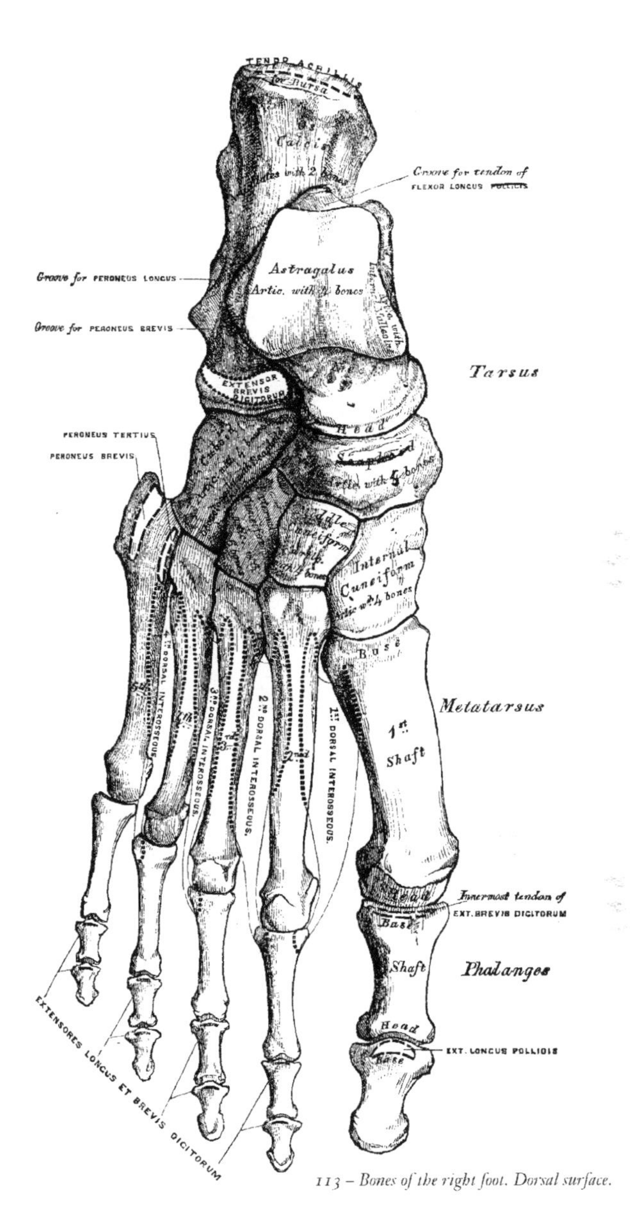


*113 – Bones of the right foot. Dorsal surface.*

## The Calcaneum.

The *Calcaneum*, or *Os Calcis*, is the largest and strongest of the tarsal bones. It is irregularly cuboidal in form, and situated at the lower and back part of the foot, serving to transmit the weight of the body to the ground, and forming a strong lever for the muscles of the calf. It presents for examination six surfaces; superior, inferior, external, internal, anterior, and posterior.

The *superior surface* is formed behind, of the upper aspect of that part of the os calcis which projects backwards to form the heel. It varies in length in different individuals; is convex from side to side, concave from before backwards, and corresponds above to a mass of adipose substance placed in front of the tendo Achillis. In the middle of this surface are two (sometimes three) articular facets, separated by a broad shallow groove, directed obliquely forwards and outwards, and rough for the attachment of the interosseous ligament connecting the astragalus and os calcis. Of these two articular surfaces, the *external* is the larger, and situated on the body of the bone; it is of an oblong form, wider behind than in front, and convex from before backwards. The *internal articular surface* is supported on a projecting process of bone, called the *lesser process* of the calcaneum (sustentaculum tali); it is of an oblong form, concave longitudinally, and sometimes subdivided into two, which differ in size and shape. More anteriorly is seen the upper surface of the *greater process*, marked by a rough depression for the attachment of numerous ligaments, and the Extensor brevis digitorum muscle.

The *inferior surface* is narrow, rough, uneven, wider behind than in front, and convex from side to side; it is bounded posteriorly by two tubercles, separated by a rough depression: the *external*, small, prominent, and rounded, gives attachment to part of the Abductor minimi digiti; the *internal*, broader and larger, for the support of the heel, gives attachment, by its prominent inner margin, to the Abductor pollicis, and in front to the Flexor brevis digitorum muscles; the depression between the tubercles attaches the Abductor minimi digiti, and plantar fascia. The rough surface in front of the tubercles gives attachment to the long plantar ligament; and to a prominent tubercle nearer the anterior part of the bone, as well as to a transverse groove in front of it, is attached the short plantar ligament.

The *external surface* is broad, flat, and almost subcutaneous, it presents near its centre a tubercle, for the attachment of the middle fasciculus of the external lateral ligament. Behind the tubercle is a broad smooth surface, giving attachment, at its upper and anterior part, to the external astragalo-calcanean ligament; and in front of the tubercle a narrow surface marked by two oblique grooves, separated by an elevated ridge: the *superior groove* transmits the tendon of the Peroneus brevis; the *inferior*, the tendon of the Peroneus longus; the intervening ridge gives attachment to a prolongation from the external annular ligament.

The *internal surface* presents a deep concavity, directed obliquely downwards and forwards, for the transmission of the plantar vessels and nerves and Flexor tendons into the sole of the foot; it affords attachment to part of the Flexor accessorius muscle. This surface presents an eminence of bone, the *lesser process*, which projects horizontally inwards from its upper and fore part. This process is concave above, and supports the anterior articular surface of the astragalus; below, it is convex, and grooved for the tendon of the Flexor longus pollicis. Its free margin is rough, for the attachment of ligaments.

The *anterior surface*, of a somewhat triangular form, is smooth, concavo-convex, and articulates with the cuboid. It is surmounted, on its outer side, by a rough prominence, which forms an important guide to the surgeon in the performance of Chopart's operation.

The *posterior surface* is rough, prominent, convex, and wider below than above. Its lower part is rough, for the attachment of the tendo Achillis; its upper part smooth, coated with cartilage, and corresponds to a bursa which separates this tendon from the bone.

*Articulations.* With two bones: the astragalus and cuboid.

*Attachment of Muscles.* Part of the Tibialis posticus, the tendo Achillis, Plantaris, Abductor pollicis, Abductor minimi digiti, Flexor brevis digitorum, Flexor accessorius, and Extensor brevis digitorum.

## The Cuboid.

The *Cuboid* bone is placed on the outer side of the foot, in front of the os calcis, and behind the fourth and fifth metatarsal bones. It is of a pyramidal shape, its base being directed upwards and inwards, its apex downwards and outwards. It may be distinguished from the other tarsal bones, by the existence of a deep groove on its under surface, for the tendon of the Peroneus longus muscle. It presents for examination six surfaces; three articular, and three non-articular: the non-articular surfaces are the superior, inferior, and external.

The *superior* or *dorsal surface*, directed upwards and outwards, is rough, for the attachment of numerous ligaments. The *inferior* or *plantar surface* presents in front a deep groove, which runs obliquely from without, forwards and inwards; it lodges the tendon of the Peroneus longus, and is bounded behind by a prominent ridge, terminating externally in an eminence, the tuberosity of the cuboid, the surface of which presents a convex facet, for articulation with the sesamoid bone of the tendon contained in the groove. The ridge and surface of bone behind it are rough, for the attachment of the long and short plantar ligaments. The *external surface*, the smallest and narrowest of the three, presents a deep notch, formed by the commencement of the peroneal groove.

The articular surfaces are the posterior, anterior, and internal. The *posterior surface* is smooth,

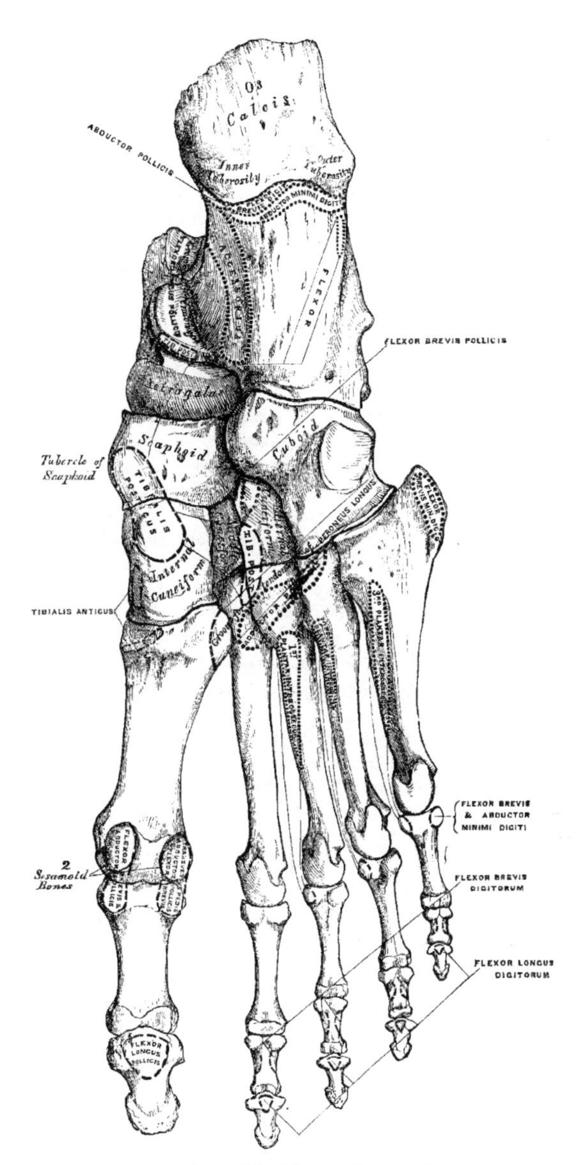


*114 – Bones of the right foot. Plantar surface.*

triangular, concavo-convex, for articulation with the anterior surface of the os calcis. The *anterior*, of smaller size, but also irregularly triangular, is divided by a vertical ridge into two facets; the inner facet, quadrilateral in form, articulates with the fourth metatarsal bone; the outer one larger and more triangular, articulates with the fifth metatarsal. The *internal surface* is broad, rough, irregularly quadrilateral, presenting at its middle and upper part a small oval facet, for articulation with the external cuneiform bone; and behind this (occasionally) a smaller facet, for articulation with the scaphoid; it is rough in the rest of its extent, for the attachment of strong interosseous ligaments.

To ascertain to which foot it belongs, hold the bone so that its under surface, marked by the peroneal groove, looks downwards, and the large concavo-convex articular surface backwards, towards the holder; the narrow non-articular surface marked by the commencement of the peroneal groove, will point to the side to which the bone belongs.

*Articulations.* With four bones: the os calcis, external cuneiform, and the fourth and fifth metatarsal bones, occasionally with the scaphoid.

*Attachment of Muscles.* Part of the Flexor brevis pollicis.

## The Astragalus.

The *Astragalus* (fig. 113), next to the os calcis, is the largest of the tarsal bones. It occupies the middle and upper part of the tarsus, supporting the tibia above, articulating with the malleoli on either side, resting below upon the os calcis, and joined in front to the scaphoid. This bone may easily be recognised by its large rounded head, the broad articular facet on its upper convex surface, and by the two articular facets separated by a deep groove on its under concave surface. It presents six surfaces for examination.

The *superior surface* presents, behind, a broad smooth trochlear surface, for articulation with the tibia; it is broader in front than behind, convex from before backwards, slightly concave from side to side. In front of the trochlea is the upper surface of the neck of the astragalus, rough for the attachment of ligaments. The *inferior surface* presents two articular facets separated by a deep groove. The groove runs obliquely forwards and outwards, becoming gradually broader and deeper in front: it corresponds with a similar groove upon the upper surface of the os calcis, and forms, when articulated with that bone, a canal, filled up in the recent state by the calcaneo-astragaloid interosseous ligament. Of the two articular facets, the posterior is the larger, of an oblong form, and deeply concave from side to side; the anterior, although nearly of equal length, is narrower, of an elongated oval form, convex longitudinally, and often subdivided into two by an elevated ridge; of these the posterior one articulates with the lesser process of the os calcis; the anterior one, with the upper surface of the calcaneo-scaphoid ligament. The *internal surface* presents at its upper part a pear-shaped articular facet for the inner malleolus, continuous above with the trochlear surface; below the articular surface is a rough depression, for the attachment of the deep portion of the internal lateral ligament. The *external surface* presents a large triangular facet, concave from above downwards, for articulation with the external malleolus; it is continuous above with the trochlear surface: and in front of it is a rough depression for the attachment of the anterior fasciculus of the external lateral ligament. The *anterior surface*, convex and rounded, forms the head of the astragalus; it is smooth, of an oval form, and directed obliquely inwards and downwards; it is continuous below with that part of the anterior facet on the under surface which rests upon the calcaneo-scaphoid ligament. The head is surrounded by a constricted portion, the neck of the astragalus. The *posterior surface* is narrow, and traversed by a groove, which runs obliquely downwards and inwards, and transmits the tendon of the Flexor longus Hallucis.

To ascertain to which foot it belongs, hold the bone with the broad articular surface upwards, and the rounded head forwards; the lateral triangular articular surface for the external malleolus will then point to the side to which the bone belongs.

*Articulations.* With four bones; tibia, fibula, os calcis, and scaphoid.

## The Scaphoid.

The *Scaphoid* or *Navicular* bone, so called from its fancied resemblance to a boat, is situated at the inner side of the tarsus, between the astragalus behind and the three cuneiform bones in front. This bone may be distinguished by its boat-like form, being concave behind, convex, and subdivided into three facets in front.

The *anterior surface*, of an oblong form, is convex from side to side, and subdivided by two ridges into three facets, for articulation with the three cuneiform bones. The *posterior surface* is oval, concave, broader externally than internally, and articulates with the rounded head of the astragalus. The *superior surface* is convex from side to side, and rough for the attachment of ligaments. The *inferior*, somewhat concave, irregular, and also rough for the attachment of ligaments. The *internal surface* presents a rounded tubercular eminence, the tuberosity of the scaphoid, which gives attachment to part of the tendon of the Tibialis posticus. The *external surface* is broad, rough, and irregular, for the attachment of ligamentous fibres, and occasionally presents a small facet for articulation with the cuboid bone.

To ascertain to which foot it belongs, hold the bone with the concave articular surface backwards, and the broad dorsal surface upwards; the broad external surface will point to the side to which the bone belongs.

*Articulations.* With four bones; astragalus and three cuneiform; occasionally also with the cuboid.

*Attachment of Muscles.* Part of the Tibialis posticus.

The Cuneiform Bones have received their name from their wedge-like shape. They form with the cuboid the most anterior row of the tarsus, being placed between the scaphoid behind, the three innermost metatarsal bones in front, and the cuboid externally. They are called the *first*, *second*, and *third*, counting from the inner to the outer side of the foot, and from their position, *internal*, *middle*, and *external*.

## The Internal Cuneiform.

The *Internal Cuneiform* is the largest of the three. It is situated at the inner side of the foot, between the scaphoid behind and the base of the first metatarsal in front. It may be distinguished by its large size, as compared with the other two, and from its more irregular wedge-like form. It presents for examination six surfaces.

The *internal surface* is subcutaneous, and forms part of the inner border of the foot; it is broad, quadrilateral, and presents at its anterior inferior angle a smooth oval facet, over which the tendon of the Tibialis anticus muscle glides; rough in the rest of its extent, for the attachment of ligaments. The *external surface* is concave, presenting, along its superior and posterior borders, a narrow surface for articulation with the middle cuneiform behind, and second metatarsal bone in front; in the rest of its extent, it is rough for the attachment of ligaments, and prominent below, where it forms part of the tuberosity. The *anterior surface*, reniform in shape, articulates with the metatarsal bone of the great toe. The *posterior surface* is triangular, concave, and articulates with the innermost and largest of the three facets on the anterior surface of the scaphoid. The *inferior* or *plantar surface* is rough, and presents a prominent tuberosity at its back part for the attachment of part of the tendon of the Tibialis posticus. It also gives attachment in front to part of the tendon of the Tibialis anticus. The *superior surface*, is the narrow pointed end of the wedge, which is directed upwards and outwards; it is rough for the attachment of ligaments.

To ascertain to which side it belongs, hold the bone so that its superior narrow edge looks upwards, and the long articular surface forwards; the external surface marked by its vertical and horizontal articular facets will point to the side to which it belongs.

*Articulations.* With four bones; scaphoid, middle cuneiform, and first and second metatarsal bones.

*Attachment of Muscles.* The Tibialis anticus and posticus.

## The Middle Cuneiform.

The *Middle Cuneiform*, the smallest of the three, is of very regular wedge-like form; the broad extremity being placed upwards, the narrow end downwards. It is situated between the other two bones of the same name, and corresponds to the scaphoid behind, and the second metatarsal in front.

The *anterior surface*, triangular in form, and narrower than the posterior, articulates with the base of the second metatarsal bone. The *posterior surface*, also triangular, articulates with the scaphoid. The *internal surface* presents an articular facet, running along the superior and posterior borders, for articulation with the internal cuneiform, and is rough below for the attachment of ligaments. The *external surface* presents posteriorly a smooth facet for articulation with the external cuneiform bone. The *superior surface* forms the base of the wedge; it is quadrilateral, broader behind than in front, and rough for the attachment of ligaments. The *inferior surface*, pointed and tubercular, is also rough for ligamentous attachment.

To ascertain to which foot the bone belongs, hold its superior or dorsal surface upwards, the broadest edge being towards the holder, and the smooth facet (limited to the posterior border) will point to the side to which it belongs.

*Articulations.* With four bones; scaphoid, internal and external cuneiform, and second metatarsal bone.

## The External Cuneiform.

The *External Cuneiform*, intermediate in size between the two preceding, is of a very regular wedge-like form, the broad extremity being placed upwards, the narrow end downwards. It occupies the centre of the front row of the tarsus between the middle cuneiform internally, the cuboid externally, the scaphoid behind, and the third metatarsal in front. It has six surfaces for examination.

The *anterior surface* triangular in form, articulates with the third metatarsal bone. The *posterior surface* articulates with the most external facet of the scaphoid, and is rough below for the attachment of ligamentous fibres. The *internal surface* presents two articular facets separated by a rough depression; the anterior one, situated at the superior angle of the bone, articulates with the outer side of the base of the second metatarsal bone; the posterior one skirts the posterior border, and articulates with the middle cuneiform; the rough depression between the two gives attachment to an interosseous ligament. The *external surface* also presents two articular facets, separated by a rough non-articular surface; the anterior facet, situated at the superior angle of the bone, is small, and articulates with the inner side of the base of the fourth metatarsal, the posterior, and larger one, articulates with the cuboid; the rough non-articular surface serves for the attachment of an interosseous ligament. The three facets for articulation with the three metatarsal bones are continuous with one another, and covered by a prolongation of the same cartilage; the facets for articulation with the middle cuneiform and scaphoid are also continuous, but that for articulation with the cuboid is usually separate. The *superior* or *dorsal surface*, of an oblong form, is rough for the attachment of ligaments. The *inferior* or *plantar surface* is an obtuse rounded margin, and serves for the attachment of part of the tendon of the Tibialis posticus, part of the Flexor brevis pollicis, and ligaments.

To ascertain to which side it belongs, hold the bone with the broad dorsal surface upwards, the prolonged edge backwards; the separate articular facet for the cuboid will point to the proper side.

*Articulations.* With six bones; the scaphoid, middle cuneiform, cuboid, and second, third, and fourth metatarsal bones.

*Attachment of Muscles.* Part of Tibialis posticus, and Flexor brevis pollicis.

## The Metatarsal Bones.

The Metatarsal bones are five in number; they are long bones, and subdivided into a shaft, and two extremities.

The *Shaft* is prismoid in form, tapers gradually from the tarsal to the phalangeal extremity, and is slightly curved longitudinally, so as to be concave below, slightly convex above.

The *Posterior Extremity*, or *Base*, is wedge-shaped, articulating by its terminal surface with the tarsal bones, and by its lateral surfaces with the contiguous bones; its dorsal and plantar surfaces being rough, for the attachment of ligaments.

The *Anterior Extremity*, or *Head*, presents a terminal rounded articular surface, oblong from above downwards, and extending further backwards below than above. Its sides are flattened, and present a depression, surmounted by a tubercle, for ligamentous attachment. Its under surface is grooved in the middle line, for the passage of the Flexor tendon, and marked on each side by an articular eminence continuous with the terminal articular surface.

## Peculiar Metatarsal Bones.

The *First* is remarkable for its great size, but is the shortest of all the metatarsal bones. The *shaft* is strong, and of well-marked prismoid form. The *posterior extremity* presents no lateral articular facets; its terminal articular surface is of large size, of semi-lunar form, and its circumference grooved for the tarso-metatarsal ligaments; its inferior angle presents a rough oval prominence, for the insertion of the tendon of the Peroneus longus. The *head* is of large size; on its plantar surface are two grooved facets, over which glide sesamoid bones; the facets are separated by a smooth elevated ridge.

The *Second* is the longest and largest of the remaining metatarsal bones; being prolonged backwards, into the recess formed between the three cuneiform bones. Its *tarsal extremity* is broad above, narrow and rough below. It presents four articular surfaces: one behind, of a triangular form, for articulation with the middle cuneiform; one at the upper part of its internal lateral surface, for articulation with the internal cuneiform; and two on its external lateral surface, a superior and an inferior, separated by a rough depression. Each of the latter articular surfaces is divided by a vertical ridge into two parts; the anterior segment of each facet articulates with the third metatarsal; the two posterior (sometimes continuous) with the external cuneiform.

The *Third* articulates behind, by means of a triangular smooth surface, with the external cuneiform; on its inner side, by two facets, with the second metatarsal; and on its outer side, by a single facet, with the third metatarsal. The latter facet is of circular form, and situated at the upper angle of the base.

The *Fourth* is smaller in size than the preceding; its *tarsal extremity* presents a terminal quadrilateral surface, for articulation with the cuboid; a smooth facet on the inner side, divided by a ridge into an anterior portion for articulation with the third metatarsal, and a posterior portion for articulation with the external cuneiform; on the outer side a single facet, for articulation with the fifth metatarsal.

The *Fifth* is recognised by the tubercular eminence on the outer side of its base: it articulates behind, by a triangular surface cut obliquely from without inwards with the cuboid; and internally, with the fourth metatarsal.

*Articulations.* Each bone articulates with the tarsal bones by one extremity, and by the other with the first row of phalanges. The number of tarsal bones with which each metatarsal articulates, is one for the first, three for the second, one for the third, two for the fourth, and one for the fifth.

*Attachment of Muscles.* To the first metatarsal bone, three: part of the Tibialis anticus, Peroneus longus, and First dorsal interosseous. To the second, three: the Adductor pollicis, and First and Second dorsal interosseous. To the third, four: the Adductor pollicis, Second and Third dorsal interosseous, and First plantar. To the fourth, four: the Adductor pollicis, Third and Fourth dorsal, and Second plantar interosseous. To the fifth, five: the Peroneus brevis, Peroneus tertius. Flexor brevis minimi digiti, Fourth dorsal, and Third plantar interosseous.

## Phalanges.

The *Phalanges* of the foot, both in number and general arrangement, resemble those in the hand; there being two in the great toe, and three in each of the other toes.

The phalanges of the *first row* resemble closely those of the hand. The *shaft* is compressed from side to side, convex above, concave below. The *posterior extremity* is concave; and the *anterior extremity* presents a trochlear surface, for articulation with the second phalanges.

The phalanges of the *second row* are remarkably small and short, but rather broader than those of the first row.

The *ungual* phalanges, in form, resemble those of the fingers, but they are smaller, flattened from above downwards, presenting a broad base for articulation with the second row, and an expanded extremity for the support of the nail and end of the toe.

*Articulations.* The first row, with the metatarsal bones, and second phalanges; the second of the great toe, with the first phalanx, and of the other toes, with the first and third phalanges; the third, with the second row.

*Attachment of Muscles.* To the first phalanges, great toe: innermost tendon of Extensor brevis digitorum, Abductor pollicis, Adductor pollicis, Flexor brevis pollicis, Transversus pedis. Second toe: First and Second dorsal interosseæ. Third toe: Third dorsal and First plantar interosseæ. Fourth toe: Fourth dorsal and Second plantar interosseæ. Fifth toe: Flexor brevis minimi digiti, Abductor minimi digiti, and Third plantar interosseous.—Second phalanges, great toe: Extensor longus pollicis, Flexor longus pollicis. Other toes: Flexor brevis digitorum, one slip from the Extensor brevis digitorum, and Extensor longus digitorum.—Third phalanges: two slips from the common tendon of the Extensor longus and Extensor brevis digitorum, and the Flexor longus digitorum.

## Development of the Foot (Fig. 115).

The Tarsal bones are each developed by a single centre, excepting the os calcis, which has an epiphysis for its posterior extremity. The centres make their appearance in the following order: in the os calcis, at the sixth month of fœtal life; in the astragalus, about the seventh month; in the cuboid, at the ninth month; external cuneiform, during the first year; internal cuneiform, in the

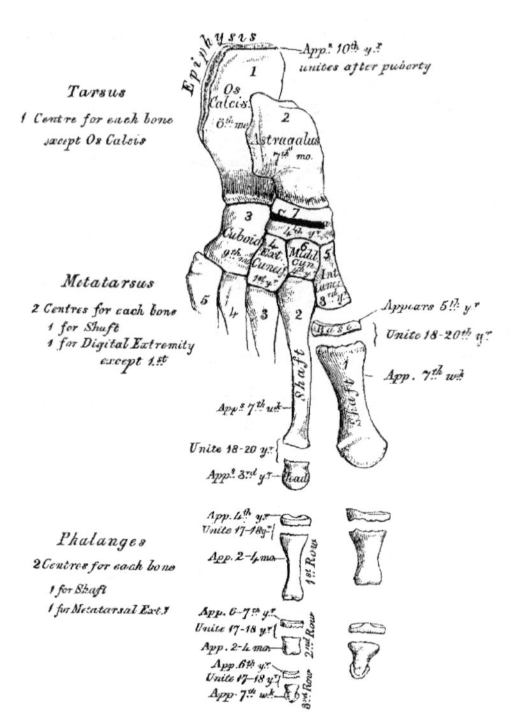


115 – *Plan of the development of the foot.*

third year; middle cuneiform, in the fourth year. The epiphysis for the posterior tuberosity of the os calcis appears at the tenth year, and unites with the rest of the bone soon after puberty.

The Metatarsal bones are each developed by *two* centres: one for the shaft, and one for the digital extremity in the four outer metatarsal; one for the shaft, and one for the base in the metatarsal bone of the great toe. Ossification commences in the centre of the shaft about the seventh week, and extends towards either extremity, and in the digital epiphyses about the third year; they become joined between the eighteenth and twentieth years.

The Phalanges are developed by *two* centres for each bone: one for the shaft, and one for the metatarsal extremity.

## Sesamoid Bones.

These are small rounded masses, cartilaginous in early life, osseous in the adult, which are developed in those tendons which exert a certain amount of pressure upon the parts over which they glide. It is said that they are more commonly found in the male than in the female, and in persons of an active muscular habit than in those that are weak and debilitated. They are invested throughout their whole surface by the fibrous tissue of the tendon in which they are found, excepting upon that side which lies in contact with the part over which they play, where they present a free articular facet. They may be divided into two kinds: those which glide over the articular surfaces of joints; those which play over the cartilaginous facets found on the surfaces of certain bones.

The sesamoid bones of the joints are, in the lower extremity, the patella, which is developed in the tendon of the Quadriceps extensor; two small sesamoid bones, found opposite the metatarso-phalangeal joint of the great toe in each foot, in the tendons of the Flexor brevis pollicis, and occasionally one in the metatarso-phalangeal joint of the second toe, the little toe, and, still more rarely, in the third and fourth toes.

In the upper extremity, there are two on the palmar surface of the metacarpo-phalangeal joint in the thumb, developed in the tendons of the Flexor brevis pollicis. Occasionally, one or two opposite the metacarpo-phalangeal articulations of the fore and little fingers, and, still more rarely, one opposite the same joints of the third and fourth fingers.

Those found in tendons which glide over certain bones, occupy the following situations. One in the tendon of the Peroneus longus, where it glides through the groove in the cuboid bone. One appears late in life in the tendon of the Tibialis anticus, opposite the smooth facet on the internal cuneiform bone. One in the tendon of the Tibialis posticus, opposite the inner side of the astragalus. One in the outer head of the Gastrocnemius, behind the outer condyle of the femur; and one in the Psoas and Iliacus, where they glide over the body of the pubes. Occasionally in the tendon of the Biceps, opposite the tuberosity of the radius; in the tendon of the Gluteus maximus, as it passes over the great trochanter; and in the tendons which wind round the inner and outer malleoli.

---

The author has to acknowledge valuable aid derived from the perusal of the works of Cloquet, Cruvelhier, Bourgery, and Boyer, especially of the latter. Reference has also been made to the following:—'Outlines of

Human Osteology,' by F. O. Ward. 'A Treatise on the Human Skeleton, and Observations on the Limbs of Vertebrate Animals,' by G. M. Humphry. Holden's 'Human Osteology.' Henle's 'Handbuch der Systematischen Anatomie des Menschen. Erster Band. Erste Abtheilung. Knochenlehre.' 'Osteological Memoirs (The Clavicle),' by Struthers. 'On the Archetype and Homologies of the Vertebrate Skeleton,' and 'On the Nature of Limbs,' by Owen.—Todd and Bowman's 'Physiological Anatomy,' and Kölliker's 'Manual of Human Microscopic Anatomy,' contain the most complete account of the structure and development of bone.—The development of the bones is minutely described in 'Quain's Anatomy,' edited by Sharpey and Ellis.—On the chemical analysis of bone, refer to 'Lehmann's Physiological Chemistry,' translated by Day; vol. iii., p. 12. 'Simon's Chemistry,' translated by Day; vol. ii., p. 396. A paper by Dr. Stark, 'On the Chemical Constitution of the Bones of the Vertebrated Animals' (Edinburgh Medical and Surgical Journal; vol. liii., p. 308); and Dr. Owen Rees' paper in the 21st vol. of the Medico-chirurgical Transactions.

# The Articulations.

The various bones of which the Skeleton consists are connected together at different parts of their surfaces, and such connection is designated by the name of *Joint* or *Articulation*. If the joint is *immoveable*, as between the cranial and most of the facial bones, their adjacent margins are applied in almost close contact, a thin layer of fibrous membrane, the *sutural ligament*, and, at the base of the skull, in certain situations, a thin layer of cartilage being interposed. Where slight movement is required, combined with great strength, the osseous surfaces are united by tough and elastic fibro-cartilages, as in the joints of the spine, the sacro-iliac, and inter-pubic articulation; but in the *moveable* joints, the bones forming the articulation are generally expanded for greater convenience of mutual connexion, covered by an elastic structure, called *cartilage*, held together by strong bands or capsules, of fibrous tissue, called *ligament*, and lined by a membrane, the *synovial membrane*, which secretes a fluid that lubricates the various parts of which the joint is formed, so that the structures which enter into the formation of a joint are bone, cartilage, fibro-cartilage, ligament, and synovial membrane.

*Bone* constitutes the fundamental element of all the joints. In the long bones, the extremities are the parts which form the articulations; they are generally somewhat enlarged, consisting of spongy cancellous tissue, with a thin coating of compact substance. In the flat bones, the articulations usually take place at the edges; and, in the short bones, by various parts of their surface. The layer of compact bone which forms the articular surface, and to which the cartilage is attached, is called the *articular lamella*. It is of a white colour, extremely dense, and varies in thickness. Its structure differs from ordinary bone tissue in this respect, that it contains no Haversian canals, and its lacunæ are much larger than in ordinary bone, and have no canaliculi. The vessels of the cancellous tissue, as they approach the articular lamella, turn back in loops, and do not perforate it; this layer is consequently more dense, and firmer than ordinary bone, and is evidently designed to form a firm and unyielding support for the articular cartilage.

*Cartilage*, is firm, opaque, of a pearly-white or bluish-white colour; in some varieties yellow; highly elastic, readily yielding to pressure, and recovering its shape when the force is removed; flexible, and possessed of considerable cohesive power. In man, that form of cartilage which constitutes the original framework of the body, and which in time becomes ossified throughout the greater part of its extent, is called *temporary cartilage*. But there is another form of cartilage employed in the construction of the body that is not prone to ossify, viz., *permanent cartilage*. This is found—1. In the joints, covering the ends of the bones (articular cartilage). 2. Forming a considerable part of the solid framework of the chest (costal cartilages). 3. Arranged in the form of plates or lamellæ, of greater or less thickness, which enter into the formation of the external ear, the nose, the eyelids, the Eustachian tube, the larynx, and the windpipe. They serve to maintain the shape of canals or passages, or to form tubes that require to be kept permanently open without the expenditure of vital force.

*Structure.* Cartilage consists either of a parenchyma of nucleated cells, or the cells are imbedded in an intercellular substance or matrix. The *cells*, or *cartilage corpuscles* are contained in hollow cavities or lacunæ in the intercellular substance, which appear to be lined by a firm, clear, or yellowish layer, the *cartilage capsule*. Under the influence of certain reagents, the cartilage cell shrinks up, and is separated from its capsule by a well marked interval. The cartilage cells are usually round or oblong, sometimes flattened or fusiform. Each contains a nucleus, furnished occasionally with one or two nucleoli. The nuclei vary from $\frac{1}{1400}$ to $\frac{1}{4000}$ of an inch; they sometimes contain fat globules, or appear converted into fat. The *intercellular substance* is either homogenous, or finely granular, or fibrous.

In *temporary cartilage*, the intercellular substance is not abundant; but the cartilage cells are numerous, and situated at nearly equal distances apart. The cells vary in shape and size, the majority being round or oval, and their nuclei are minutely granular. When ossification commences in it, the cells become arranged in clusters or rows, the ends of which are directed towards the ossifying part.

In *articular cartilage*, the intercellular substance is more abundant than in the former variety; it appears dim, like ground glass, and has a finely granular or homogenous aspect. The cells are oval or roundish, from $\frac{1}{1300}$ to $\frac{1}{900}$ of an inch, the nuclei small and commonly vesicular, and parent cells are frequently seen enclosing two or more younger cells. On the surface of the cartilage the cells are numerous, and disposed in isolated groups of two, three, or four, the groups being flattened, and lie with their planes parallel to the surface. In the interior, and nearer the bone, they are less numerous, and assume more or less of a linear direction, pointing towards the surface. This arrangement appears to be connected with a corresponding peculiarity of structure in the matrix, and serves to explain the disposition which this form of cartilage has to break in a direction perpendicular to the surface, the broken surface being to the eye striated in the same direction.

In the *costal cartilages* the intercellular substance is very abundant, finely mottled, and, in certain situations, presents a distinctly fibrous structure, the fibres being fine and parallel. This is most evident in advanced age. The cells, which are collected into groups, are larger than in any other cartilages of the body, being from $\frac{1}{650}$ to $\frac{1}{430}$ of an inch in diameter. Many contain two or more clear transparent nuclei, and some contain oil globules. Near the exterior of the cartilage the cells are flattened, and lie parallel with the surface; in the interior, the cells have a linear arrangement, the separate rows being turned in all directions.

The ensiform cartilage of the sternum, the cartilages of the nose, and the cartilages of the larynx and windpipe (excepting the epiglottis and cornicula laryngis) resemble the costal cartilages in their microscopic characters.

*Reticular cartilage.* The epiglottis, the cornicula laryngis, the cuneiform cartilages, the cartilage of the ear, of the eyelid, and of the Eustachian tube, are included in a separate class under the name of 'reticular,' 'yellow,' or 'spongy' cartilages. They are yellow, of a spongy texture throughout, more flexible than ordinary cartilage, and not prone to ossify. This variety of cartilage consists of an intercellular substance, composed of minute opaque fibres, which intersect each other in all directions, and are so arranged as to inclose numerous small oval spaces, in which the cartilage corpuscles are deposited.

*Articular cartilage* forms a thin incrustation upon the articular surfaces of bones, and is admirably adapted, by its elastic property, to break the force of concussions, and, by its smoothness, to afford perfect ease and freedom of movement between the bones. Where it covers the rounded ends of bones, upon which the greatest pressure is received, it is thick at the centre, and becomes gradually thinner towards the circumference: an opposite arrangement exists where it lines the corresponding cavities. On the articular surfaces of the short bones, as the carpus and tarsus, the cartilage is disposed in a layer of uniform thickness throughout. The attached surface of articular cartilage is closely adapted, by a rough, uneven surface, to the articular lamella; the free surface is smooth, polished, and partially covered by a perichondrium, prolonged from the periosteum, a short distance over the cartilage; in the fœtus, an extremely thin prolongation of synovial membrane may be traced over the surface of the cartilage, according to Toynbee, but, at a later period of life, this cannot be demonstrated. Articular cartilage in the adult does not contain blood-vessels; its nutrition being derived from the vessels of the synovial membrane which skirt the circumference of the cartilage, and from those of the adjacent bone, which are, however, separated from direct contact with the cartilage by means of the articular lamella. Mr. Toynbee has shown, that the minute vessels of the cancellous tissue, as they approach the articular lamella, dilate, and, forming arches, return into the substance of the bone. The vessels of the synovial membrane also advance forwards upon the circumference of the cartilage for a very short distance, and then return in loops; they are only found on the parts not subjected to pressure. In the fœtus, the vessels are said, by Toynbee, to advance for some distance upon the surface of the cartilage, beneath the synovial membrane; but Kölliker, from more recent examination, doubts this. Lymphatic vessels and nerves have not, as yet, been traced in its substance.

*Fibro-cartilage* is also employed in the construction of the joints, contributing to their strength and elasticity. It consists of a mixture of white fibrous and cartilaginous tissues in various proportions; it is to the first of these two constituents that its flexibility and toughness is chiefly owing, and to the latter its elasticity. The fibro-cartilages admit of arrangement into four groups, inter-articular, connecting, circumferential, and stratiform.

The *inter-articular fibro-cartilages* (*menisci*) are flattened fibro-cartilaginous plates, of a round, oval, or sickle-like form, interposed between the articular cartilages of certain joints. They are free on both surfaces, thinner toward their centre than at their circumference, and held in position by their extremities being connected to the surrounding ligaments. The synovial membrane of the joint is prolonged over them a short distance from their attached margin. They are found in the temporo-maxillary, sterno-clavicular, acromio-clavicular, wrist and knee joints. These cartilages are usually found in those joints most exposed to violent concussions, and subject to frequent movement. Their use is—to maintain the apposition of the opposed surfaces in their various motions; to increase the depth of the articular surface, and give ease to the gliding movement; to moderate the effects of great pressure, and deaden the intensity of the shocks to which they may be submitted.

The *connecting fibro-cartilages* are interposed between the bony surfaces of those joints which admit of only slight mobility, as between the bodies of the vertebræ, and the symphysis of the pubes; they exist in the form of discs, intimately adherent to the opposed surfaces, being composed of concentric rings of fibrous tissue, with cartilaginous laminæ interposed, the former tissue predominating towards the circumference, the latter towards the centre.

The *circumferential fibro-cartilages* consist of a rim of fibro-cartilage, which surrounds the margin of some of the articular cavities, as the cotyloid cavity of the hip, and the glenoid cavity of the shoulder; they serve to deepen the articular surface and protect the edges of the bone.

The *stratiform fibro-cartilages* are those which form a thin layer in the osseous grooves, through which the tendons of certain muscles glide.

*Ligaments* are found in nearly all the moveable articulations; they consist of bands of various forms, serving to connect together the articular extremities of bones, and composed mainly of bundles of *white fibrous tissue* placed parallel with, or closely interlaced with, one another, and presenting a white, shining, silvery aspect. Ligament is pliant and flexible, so as to allow of the most perfect freedom of movement, but strong, tough, and inextensile, so as not readily to yield under the most severely applied force; it is, consequently, admirably adapted to serve as the connecting medium between the bones. Some ligaments consist entirely of *yellow elastic tissue*, as the ligamenta subflava, which connect together the adjacent arches of the vertebræ, and the ligamentum nuchæ.

*Synovial Membrane* is a thin, delicate membrane, arranged in the form of a short wide tube, attached by its open ends to the margins of the articular extremities of the bones, and covering the inner surface of the various ligaments which connect the articulating surfaces. It resembles the serous membranes in structure, but differs in the nature of its secretion, which is thick, viscid, and glairy, like the white of egg; and hence termed *synovia*. The synovial membranes found in the body admit of subdivision into three kinds, articular, bursal, and vaginal.

The *articular synovial membranes* are found in all the freely moveable joints. In the fœtus, this membrane is said, by Toynbee, to be continued over the surface of the cartilages; but in the adult it is wanting, excepting at their circumference, upon which it encroaches for a short distance: it then invests the inner surface of the capsular or other ligaments enclosing the joint, and is reflected over the surface of any tendons passing through its cavity, as the tendon of the Popliteus in the knee, and the tendon of the Biceps in the shoulder. In most of the joints, the synovial membrane is thrown into folds, which project into the cavity. Some of these folds contain large masses of fat. These are especially distinct in the hip and the knee. Others are flattened folds, subdivided at their margins into fringe-like processes, the vessels of which have a convoluted arrangement. The latter generally project from the synovial membrane near the margin of the cartilage, and lie flat upon its surface. They consist of connective tissue, covered with epithelium, and contain fat cells in variable quantity, and, more rarely, isolated cartilage cells. They are found in most of the bursal and vaginal, as well as in the articular synovial membranes, and were described, by Clopton Havers, as mucilaginous glands, and as the source of the synovial secretion. Under certain diseased conditions, similar processes are found covering the entire surface of the synovial membrane, forming a mass of pedunculated fibro-fatty growths, which project into the joint.

The *bursæ* are found interposed between surfaces which move upon each other, producing friction, as in the gliding of a tendon, or of the integument over projecting bony surfaces. They admit of subdivision into two kinds, the *bursæ mucosæ*, and the *synovial bursæ*. The former are large, simple, or irregular cavities in the subcutaneous areolar tissue, enclosing a clear viscid fluid. They are found in various situations, as between the integument and front of the patella, over the olecranon, the malleoli, and other prominent parts. The *synovial bursæ* are found interposed between muscles or tendons as they play over projecting bony surfaces, as between the Glutei muscles and surface of the great trochanter. They consist of a thin wall of connective tissue, partially covered by epithelium, and contain a viscid fluid. Where one of these exists in the neighbourhood of a joint, it usually communicates with its cavity, as is generally the case with the bursa between the tendon of the Psoas and Iliacus, and the capsular ligament of the hip, or the one interposed between the under surface of the Subscapularis and the neck of the scapula.

The *vaginal synovial membranes* (synovial sheaths) serve to facilitate the gliding of tendons in the osseo-fibrous canals through which they pass. The membrane is here arranged in the form of a sheath, one layer of which adheres to the wall of the canal, and the other is reflected upon the outer surface of the contained tendon; the space between the two free surfaces of the membrane, being partially filled with synovia. These sheaths are chiefly found surrounding the tendons of the Flexor and Extensor muscles of the fingers and toes, as they pass through the osseo-fibrous canals in the hand or foot.

*Synovia* is a transparent, yellowish-white, or slightly reddish fluid, viscid like the white of egg, having an alkaline reaction, and slightly saline taste. It consists, according to Frerichs, in the ox, of 94.85 water, 0.56 mucus and epithelium, 0.07 fat, 3.51 albumen and extractive matter, and 0.99 salts.

The Articulations are divided into three classes: *Synarthrosis*, or immoveable; *Amphiarthrosis*, or mixed; and *Diarthrosis*, or moveable.

## 1. Synarthrosis. Immoveable Articulations.

*Synarthrosis* (*σύν, with, ἄρθρον, a joint*), or *Immoveable Joints*, include all those articulations in which the surfaces of the bones are in almost direct contact, not separated by an intervening synovial cavity, and immoveably connected with each other, as between the bones of the cranium and face, excepting the lower jaw. The varieties of synarthrosis are three in number: Sutura, Schindylesis, and Gomphosis.

*Sutura* (a seam). Where the articulating surfaces are connected by a series of processes and indentations interlocked together, it is termed *sutura vera;* of which there are three varities: sutura dentata, serrata and limbosa. The surfaces of the bones are not in direct contact, being separated by a layer of membrane continuous externally with the pericranium, internally with the dura mater. The *sutura dentata* (*dens*, a tooth) is so called from the tooth-like form of the projecting articular processes, as in the suture between the parietal bones. In the *sutura serrata*, (*serra*, a saw), the edges of the two bones forming the articulation are serrated like the teeth of a fine saw, as between the two portions of the frontal bone. In the *sutura limbosa* (*limbus*, a selvage), besides the dentated processes, there is a certain degree of bevelling of the articular surfaces, so that the bones overlap one another, as in the suture between the parietal and frontal bones. Where the articulation is formed by roughened surfaces placed in apposition with one another, it is termed the *false suture*, *sutura notha*, of which there are two kinds: the *sutura squamosa* (*squama*, a scale), formed by the overlapping of two contiguous bones by broad bevelled margins, as in the temporo-parietal (squamous) suture; and the *sutura harmonia* (*ἀρεῖν, to adapt*), where there is simple apposition of two contiguous rough bony surfaces, as in the articulation between the two superior maxillary bones, or of the horizontal plates of the palate.

*Schindylesis* (*σχινδύλησις*, a fissure) is that form of articulation in which a thin plate of bone is received into a cleft or fissure formed by the separation of two laminæ of another, as in the articulation of the rostrum of the sphenoid, and perpendicular plate of the ethmoid with the vomer, or in the reception of the latter in the fissure between the superior maxillary and palate bones.

*Gomphosis* (*γόμφος, a nail*) is an articulation formed by the insertion of a conical process into a socket, as a nail is driven into a board; and is illustrated in the articulation of the teeth in the alveoli of the maxillary bones.

## 2. Amphiarthrosis. Mixed Articulations.

*Amphiarthrosis* (*ἀμφὶ, 'on all sides,' ἄρθρον, a joint*), or *Mixed Articulation*. In this form of articulation, the contiguous osseous surfaces are connected together by broad flattened discs of fibro-cartilage, which adhere to the ends of both bones, as in the articulation between the bodies of the vertebræ, and first two pieces of the sternum; or the articulating surfaces are covered with fibro-cartilage, partially lined by synovial membrane, and connected together by external ligaments, as in the sacro-iliac and pubic symphyses; both these forms being capable of limited motion in every direction. The former resemble the synarthrodial joints in the continuity of their surfaces, and absence of synovial sac; the latter, the diarthrodial. These joints occasionally become obliterated in old age: this is frequently the case in the inter-pubic articulation, and occasionally in the intervertebral and sacro-iliac.

## 3. Diarthrosis. Moveable Articulations.

*Diarthrosis* (*διὰ, through, ἄρθρον, a joint*). This form of articulation includes the greater number of the joints in the body, mobility being their distinguishing character. They are formed by the approximation of two contiguous bony surfaces, covered with cartilage, connected by ligaments, and lined by synovial membrane. The varieties of joints in this class, have been determined by the kind of motion permitted in each; they are four in number: Arthrodia, Enarthrosis, Ginglymus, Diarthrosis Rotatorius.

*Arthrodia* is that form of joint which admits of a gliding movement; it is formed by the approximation of plane surfaces, or one slightly concave, the other slightly convex; the amount of motion between them being limited by the ligaments, or osseous processes, surrounding the articulation; as in the articular processes of the vertebræ, temporo-maxillary, sterno and acromio-clavicular, inferior radio-ulnar, carpal, carpo-metacarpal, superior tibio-fibular, tarsal, and tarso-metatarsal articulations.

*Enarthrosis* is that form of joint which is capable of motion in all directions. It is formed by the reception of a globular head into a deep cup-like cavity (hence the name 'ball and socket'), the parts being kept in apposition by a capsular ligament strengthened by accessory ligamentous bands. Examples of this form of articulation are found in the hip and shoulder.

*Ginglymus, Hinge-joint* (*γιγγλυμὸς, a hinge*). In this form of joint, the articular surfaces are moulded to each other in such a manner, as to permit motion only in two directions, forwards and backwards, the extent of motion at the same time being considerable. The articular surfaces are connected together by strong lateral ligaments, which form their chief bond of union. The most perfect forms of ginglymi are the elbow and ankle; the knee is less perfect, as it allows a slight degree of rotation in certain positions of the limb: there are also the metatarso-phalangeal and phalangeal joints in the lower extremity, and the metacarpo-phalangeal and phalangeal joints in the upper extremity.

*Diarthrosis rotatorius* (Lateral Ginglymus). Where the movement is limited to rotation, the joint is formed by a pivot-like process turning within a ring, or the ring on the pivot, the ring being formed partly of bone, partly of ligament. In the articulation of the odontoid process of the axis with the atlas, the ring is formed in front by the anterior arch of the atlas; behind, by the transverse ligament; here the ring rotates round the odontoid process. In the superior radio-ulnar articulation, the ring is formed partly by the lesser sigmoid cavity of the ulna; in the rest of its extent, by the orbicular ligament; here, the head of the radius rotates within the ring.

Subjoined, in a tabular form, are the names, distinctive characters, and examples of the different kinds of articulations.

- *Synarthrosis*, or immoveable joint. Surfaces separated by fibrous membrane, no intervening synovial cavity, and immoveably connected with each other. Example: bones of cranium and face (except lower jaw).
  - *Sutura*. Articulation by processes and indentations interlocked together.
    - *Sutura vera* (true) articulate by indented borders.
      - *Dentata*, having tooth-like processes. Inter-parietal suture.
      - *Serrata*, having serrated edges, like the teeth of a saw. Inter-frontal suture.
      - *Limbosa*, having bevelled margins, and dentated processes. Fronto-parietal suture.
    - *Sutura notha* (false) articulate by rough surfaces.
      - *Squamosa*, formed by thin bevelled margins overlapping each other. Temporo-parietal suture.
      - *Harmonia*, formed by the apposition of contiguous rough surfaces. Inter-maxillary suture.
  - *Schindylesis*. Articulation formed by the reception of a thin plate of bone into a fissure of another. Rostrum of sphenoid with vomer.
  - *Gomphosis*. An articulation formed by the insertion of a conical process into a socket. Tooth in socket.

| | |
|---|---|
| *Amphiarthrosis*, Mixed Articulation. | 1. Surfaces connected by fibro-cartilage, not separated by synovial membrane, and having limited motion. Bodies of vertebræ.<br>2. Surfaces covered by fibro-cartilage; lined by a partial synovial membrane. Sacro-iliac and pubic symphyses. |
| *Diarthrosis*, Moveable Joint. | *Arthrodia*. Gliding joint; articulation by plane surfaces, which glide upon each other. As in sterno and acromio-clavicular articulations.<br>*Enarthrosis*. Ball and socket joint; capable of motion in all directions. Articulation by a globular head received into a cup-like cavity. As in hip and shoulder joints.<br>*Ginglymus*. Hinge joint; motion limited to two directions, forwards and backwards. Articular surfaces fitted together so as to permit of movement in one plane. As in the elbow, ankle, and knee.<br>*Diarthrosis rotatorius*. Articulation by a pivot process turning within a ring, or ring around a pivot. As in superior radio-ulnar articulation, and atlo-axoid joint. |

### The Kinds of Movement admitted in joints.

The movements admissible in joints may be divided into four kinds, gliding, angular movement, circumduction, and rotation.

Gliding movement is the most simple kind of motion that can take place in a joint, one surface gliding over another. It is common to all moveable joints; but in some, as in the articulations of the carpus and tarsus, is the only motion permitted. This movement is not confined to plane surfaces, but may exist between any two contiguous surfaces, of whatever form, limited by the ligaments which enclose the articulation.

Angular movement occurs only between the long bones, and may take place in four directions, forwards or backwards, constituting flexion and extension, or inwards and outwards, which constitutes abduction and adduction. Flexion and extension are confined to the strictly ginglymoid or hinge joints. Abduction and adduction, combined with flexion and extension, are met with only in the most moveable joints; as in the hip, shoulder, and metacarpal joint of the thumb, and partially in the wrist and ankle.

Circumduction is that limited degree of motion which takes place between the head of a bone and its articular cavity, whilst the extremity and sides of a limb are made to circumscribe a conical space, the base of which corresponds with the inferior extremity of the limb, the apex to the articular cavity; and is best seen in the shoulder and hip joints.

Rotation is the movement of a bone upon its own axis, the bone retaining the same relative situation with respect to the adjacent parts; as in the articulation between the atlas and axis, where the odontoid process serves as a pivot around which the atlas turns; or in the rotation of the radius upon the humerus, and also in the hip and shoulder.

The articulations may be arranged into those of the trunk, those of the upper extremity, and those of the lower extremity.

## ARTICULATIONS OF THE TRUNK.

These may divided into the following groups, viz.:—

1. Of the vertebral column.
2. Of the atlas with the axis.
3. Of the atlas with the occipital bone.
4. Of the axis with the occipital bone.
5. Of the lower jaw.
6. Of the ribs with the vertebræ.
7. Of the cartilages of the ribs with the sternum, and with each other.
8. Of the sternum.
9. Of the vertebral column with the pelvis.
10. Of the pelvis.

## 1. Articulations of the Vertebral Column.

The different segments of the spine are connected together by ligaments, which admit of the same arrangement as the vertebræ. They may be divided into five sets. 1. Those connecting the *bodies* of the vertebræ. 2. Those connecting the *laminæ*. 3. Those connecting the *articular processes*. 4. The ligaments connecting the *spinous processes*. 5. Those of the *transverse processes*.

The articulation of the *bodies* of the vertebræ with each other, form a series of amphiarthrodial joints; whilst those between the *articular processes* form a series of arthrodial joints.

### 1. The Ligaments of the Bodies are,

Anterior Common Ligament. Posterior Common Ligament.

Intervertebral Substance.

The *Anterior Common Ligament* (fig. 116) is a broad and strong band of ligamentous fibres, which extends along the front surface of the bodies of the vertebræ from the axis to the sacrum. It is broader below than above, thicker in the dorsal than in the cervical or lumbar regions, it is also somewhat thicker opposite the front of the body of each vertebra, than opposite the intervertebral substance. It is attached, above, to the body of the axis by a pointed process, which is connected with the tendon of origin of the Longus colli muscle; and, extends down as far as the upper bone of the

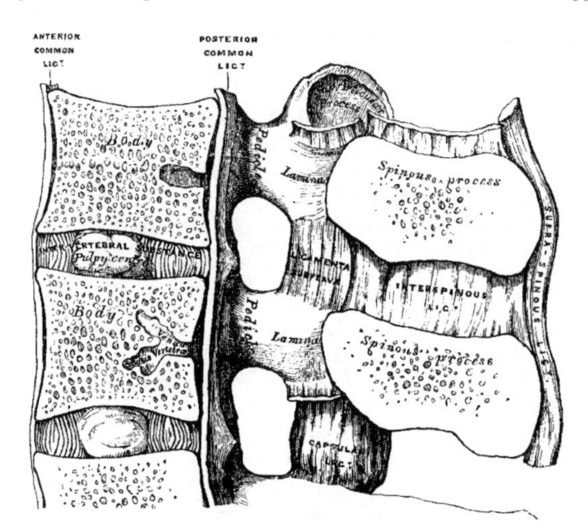


*116 – Vertical section of two vertebræ and their ligaments, from the lumbar region.*

sacrum. It consists of dense longitudinal fibres, which are intimately adherent to the intervertebral substance and prominent margins of the vertebræ; but less closely with the middle of the bodies. In the latter situation the fibres are exceedingly thick, and serve to fill up the concavities on their front surface, and to make the anterior surface of the spine more even. This ligament is composed of several layers of fibres, which vary in length but are closely interlaced with each other. The most superficial or longest fibres extend between four or five vertebræ. A second subjacent set extend between two or three vertebræ; whilst a third set, the shortest and deepest, extend from one vertebra to the next. At the sides of the bodies, this ligament consists of few short fibres, which pass from one vertebra to the next, separated from the median portion by large oval apertures, for the passage of vessels.

The *Posterior Common Ligament* is situated within the spinal canal, and extends along the posterior surface of the bodies of the vertebræ, from the body of the axis above, where it is continuous with the occipito-axoid ligament, to the sacrum below. It is broader at the upper than at the lower part of the spine, and thicker in the dorsal than in the cervical or lumbar regions. In the situation of the intervertebral substance and contiguous margins of the vertebræ, where the ligament is more intimately adherent, it is broad, and presents a series of dentations with intervening concave margins; but it is narrow and thick over the centre of the bodies, from which it is separated by the *venæ basis vertebræ*. This ligament is composed of smooth, shining, longitudinal fibres, denser and more compact than those of the anterior ligament, and composed of a superficial layer occupying the interval between three or four vertebræ, and of a deeper layer, which extends between one vertebra and the next adjacent to it. It is separated from the dura mater of the spinal cord by some loose filamentous tissue, very liable to serous infiltration.

The *Intervertebral Substance* (fig. 116) is a lenticular disc of fibro-cartilage, interposed between the adjacent surfaces of the bodies of the vertebræ, from the axis to the sacrum, forming the chief bond of connexion between these bones. These discs vary in shape, size, and thickness, in different parts of the spine. In shape they accurately correspond with the surfaces of the bodies between which they are placed, being oval in the cervical and lumbar regions, circular in the dorsal. Their *size* is greatest in the lumbar region. In *thickness* they vary not only in the different regions of the spine, but in different parts of the same region: thus, they are uniformly thick in the lumbar region; thickest, in front, in the cervical and lumbar regions which are convex forwards; and behind, to a slight extent, in the dorsal region. They thus contribute, in a great measure, to the curvatures of the spine in the neck and loins; whilst the concavity of the dorsal region is chiefly due to the shape of the bodies of the vertebræ. The intervertebral discs form about one-fourth of the spinal column, exclusive of the first two vertebræ; they are not equally distributed, however, between the various bones; the dorsal portion on the spine having, in proportion to its length, a much smaller quantity than in the cervical and lumbar regions, which necessarily gives to the latter parts greater pliancy and freedom of movement. The intervertebral discs are adherent, by their surfaces, to the adjacent parts of the bodies of the vertebræ; and by their circumference are closely connected in front to the anterior, and behind to the posterior common ligament; whilst, in the dorsal region, they are connected laterally to the heads of those ribs which articulate with two vertebræ, by means of the inter-articular ligament; they, consequently, form part of the articular cavities in which the heads of these bones are received.

The intervertebral substance is composed, at its circumference, of laminæ of fibrous tissue and fibro-cartilage; and, at its centre, of a soft, elastic, pulpy matter. The laminæ are arranged concentrically one within the other, with their edges turned towards the corresponding surfaces of the vertebræ, and consist of alternate plates of fibrous tissue and fibro-cartilage. These plates are not quite vertical in their direction, those near the circumference being curved outwards and closely approximated; whilst those nearest the centre curve in the opposite direction, and are somewhat more widely separated. The fibres of which each plate is composed, are directed, for the most part, obliquely from above downwards; the fibres of an adjacent plate have an exactly opposite arrangement, varying in their direction in every layer; whilst in some few they are horizontal. This laminar arrangement belongs to about the outer half of each disc, the central part being occupied by a soft, pulpy, highly elastic substance, of a yellowish colour, which rises up considerably above the surrounding level, when the disc is divided horizontally. This substance presents no concentric arrangement, and consists of white fibrous tissue, having interspersed cells of variable shape and size. The pulpy matter, which is especially well developed in the lumbar region, is separated from immediate contact with the vertebræ, by the interposition of thin plates of cartilage.

## 2. Ligaments connecting the Laminæ.

### Ligamenta Subflava.

The *Ligamenta Subflava* are interposed between the laminæ of the interior from the axis to the sacrum. They are most distinct when seen from the interior of the spinal canal; when viewed from the outer surface, they appear short, being overlapped by the laminæ. Each ligament consists of two lateral portions, which commence on each side at the root of either articular process, and pass backwards to the point where the laminæ converge to form the spinous process, where their margins are thickest, and separated by a slight interval, filled up with areolar tissue. These ligaments consist of yellow elastic tissue, the fibres of which, almost perpendicular in direction, are attached to the anterior surface of the margin of the lamina above, and to the posterior surface, as well as to the margin of the lamina below. In the cervical region, they are thin in texture, but very broad and long; they become thicker in the dorsal region; and in the lumbar acquire very considerable thickness. Their highly elastic property serves to preserve the upright posture, and to counteract the efforts of the flexor muscles of the spine. These ligaments do not exist between the occiput and atlas, or between the atlas and axis.

## 3. Ligaments connecting the Articular Process.

Capsular.

The *Capsular Ligaments* are thin and loose ligamentous sacs, attached to the contiguous margins of the articulating processes of each vertebra, through the greater part of their circumference, and completed internally by the ligamenta subflava. They are longer and more loose in the cervical than in the dorsal or lumbar regions. The capsular ligaments are lined on their inner surface by synovial membrane.

## 4. Ligaments connecting the Spinous Processes.

Inter-spinous. Supra-spinous.

The *Inter-spinous Ligaments*, thin and membranous, are interposed between the spinous processes in the dorsal and lumbar regions. Each ligament extends from the root to near the summit of each spinous process, and connects together their adjacent margins. They are narrow and elongated in the dorsal region, broader, quadrilateral in form, and thicker in the lumbar region.

The *Supra-spinous Ligament* is a strong fibrous cord, which connects together the apices of the spinous processes from the seventh cervical to the spine of the sacrum. It is thicker and broader in the lumbar than in the dorsal region, and intimately blended, in both situations, with the neighbouring aponeuroses. The most superficial fibres of this ligament connect three or four vertebræ; those deeper seated pass between two or three vertebræ; whilst the deepest connect the contiguous extremities of neighbouring vertebræ.

## 5. Ligaments connecting the Transverse Processes.

Inter-transverse.

The *Inter-transverse Ligaments* consist of a few thin scattered fibres, interposed between the transverse processes. They are generally wanting in the cervical region; in the dorsal, they are rounded cords; in the lumbar region, thin and membranous.

*Actions.* The movements permitted in the spinal column are, Flexion, Extension, Lateral movement, Circumduction, and Rotation.

In *Flexion*, or movement of the spine forwards, the anterior common ligament is relaxed, and the intervertebral substances are compressed in front; the posterior common ligament, the ligamenta subflava, and the inter- and supra-spinous ligaments, are stretched, as well as the posterior fibres of the intervertebral discs. The interspaces between the laminæ are widened, and the inferior articular processes glide upwards, upon the articular processes of the vertebræ below. Flexion is the most extensive of all the movements of the spine.

In *Extension*, or movements of the spine backwards, an exactly opposite disposition of the parts takes place. This movement is not extensive, being limited by the anterior common ligament, and by the approximation of the spinous processes.

Flexion and extension are most free in the lower part of the lumbar, and in the cervical regions; extension in the latter region being greater than flexion, the reverse of which exists in the lumbar region. These movements are least free in the middle and upper part of the back.

In *Lateral Movement*, the sides of the intervertebral discs are compressed, the extent of motion being limited by the resistance offered by the surrounding ligaments, and by the approximation of the transverse processes. This movement may take place in any part of the spine, but is most free in the neck and loins.

*Circumduction* is very limited, and is produced merely by a succession of the preceding movements.

*Rotation* is produced by the twisting of the intervertebral substances; this, although only slight between any two vertebræ, produces great extent of movement, when it takes place in the whole length of the spine, the front of the column being turned to one or the other side. This movement takes place only to a slight extent in the neck, but is more free in the lower part of the dorsal and lumbar regions.

It is thus seen, that the *cervical region* enjoys the greatest extent of each variety of movement, flexion and extension being very free; lateral movement and rotation, although less extensive than the former, being greater than in any other region. In the *dorsal region*, especially at its upper part, the movements are most limited: flexion, extension, and lateral motion taking place only to a slight extent. In the *lumbar region*, all the movements are very free.

## 2. Articulation of the Atlas with the Axis.

The articulation of the anterior arch of the atlas with the odontoid process forms a lateral ginglymoid joint, whilst that between the articulating processes of the two bones forms a double arthrodia. The ligaments of this articulation are, the

| | |
|---|---|
| Two Anterior Atlo-Axoid. | Transverse. |
| Posterior Atlo-Axoid. | Two Capsular. |

Of the *Two Anterior Atlo-Axoid Ligaments* (fig. 117), the most superficial is a rounded cord, situated in the middle line; attached, above, to the tubercle on the anterior arch of the atlas; below, to the base of the odontoid process and body of the axis. The deeper ligament is a membranous layer, attached, above, to the lower border of the anterior arch of the atlas; below, to the base of the odontoid process, and body of the axis. These ligaments are in relation, in front, with the Recti antici majores.

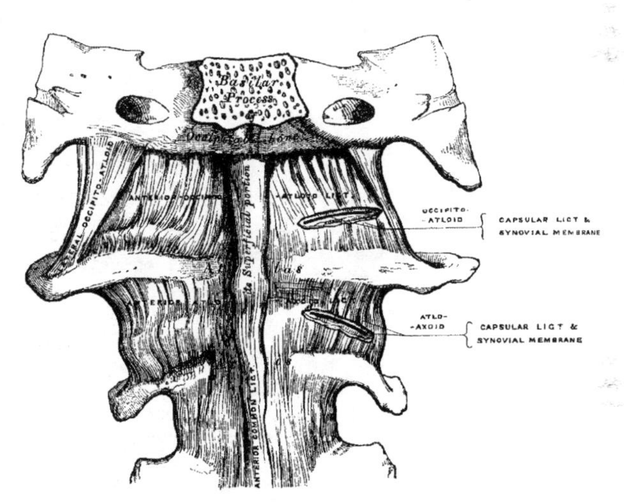


117 – *Occipito–atloid and atlo–axoid ligaments. Front view.*

The *Posterior Atlo-Axoid Ligament* (fig. 118) is a broad and thin membranous layer, attached, above, to the lower border of the posterior arch of the atlas; below, to the upper edge of the laminæ of the axis. This ligament supplies the place of the ligamenta subflava, and is in relation, behind, with the Inferior oblique muscles.

The *Transverse Ligament* (figs. 119, 120) is a thick and strong ligamentous band, which arches across the ring of the atlas, and serves to retain the odontoid process in firm connection with its anterior arch. This ligament is flattened from before backwards, broader and thicker in the middle than at either extremity, and firmly attached on each side of the atlas to a small tubercle on the inner surface of each of its lateral masses. As it crosses the odontoid process, a small fasciculus is derived from its upper and lower borders; the former passing upwards, to be inserted into the basilar process of the occipital bone; the latter, downwards, to be attached to the root of the odontoid process: hence, this ligament has received the name of *cruciform*. The transverse ligament divides the ring of the atlas into two unequal parts: of these, the posterior and larger serves for the transmission of the cord and its membranes: the anterior and smaller serving to retain the odontoid process in its position. The lower border of the space between the atlas and transverse ligament being smaller than the upper (on account of the transverse ligament embracing firmly the narrow neck of the odontoid process), this process is retained in firm connection with the atlas when all the other ligaments have been divided.

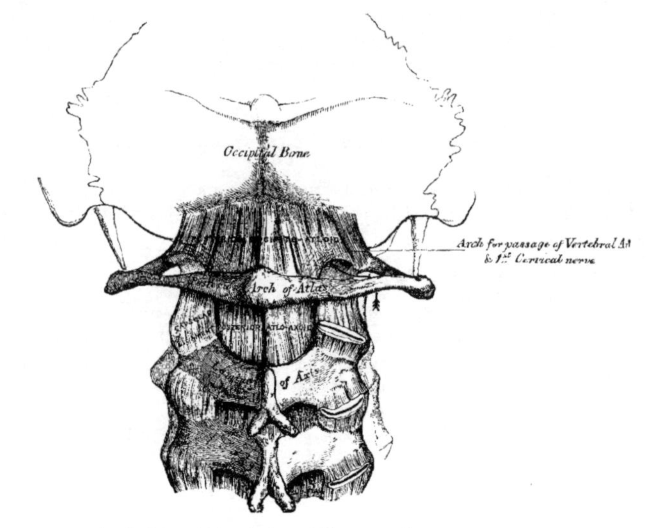

118 – *Occipito-atloid and atlo-axoid ligaments. Posterior view.*

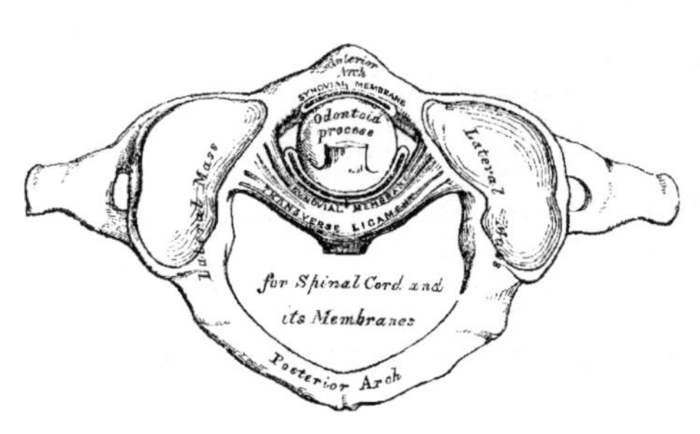

119 – *Articulation between odontoid process and atlas.*

The *Capsular Ligaments* are two thin and loose capsules, connecting the articular surfaces of the atlas and axis, the fibres being strongest on the anterior and external part of the articulation.

There are *four Synovial Membranes* in this articulation. One lining the inner surface of each of the capsular ligaments: one between the anterior surface of the odontoid process and anterior arch of the atlas: and one between the posterior surface of the odontoid process and the transverse ligament. The latter often communicates with those between the condyles of the occipital bone and the articular surfaces of the atlas.

*Actions.* This joint is capable of great mobility, and allows the rotation of the atlas, and, with it, of the cranium upon the axis, the extent of rotation being limited by the odontoid ligaments.

## Articulation of the Spine with the Cranium.

The ligaments connecting the spine with the cranium may be divided into two sets: Those connecting the occipital bone with the atlas; Those connecting the occipital bone with the axis.

## 3. Articulation of the Atlas with the Occipital Bone.

This articulation is a double arthrodia. Its ligaments are the

Two Anterior Occipito-Atloid.
Posterior Occipito-Atloid.
Two Lateral Occipito-Atloid.
Two Capsular.

Of the *Two Anterior Ligaments* (fig. 117), the most superficial is a strong narrow, rounded cord, attached, above, to the basilar process of the occiput; below, to the tubercle on the anterior arch of the atlas: the deeper ligament is a broad and thin membranous layer, which passes between the anterior margin of the foramen magnum above, and the whole length of the upper border of the anterior arch of the atlas below. This ligament is in relation, in front, with the Recti antici minores; behind, with the odontoid ligaments.

The *Posterior Occipito-Atloid Ligament* (fig. 118) is a very broad but thin membranous lamina, intimately blended with the dura mater. It is connected above, to the posterior margin of the foramen magnum; below, to the upper border of the posterior arch of the atlas. This ligament is incomplete at each side, and forms, with the superior intervertebral notch, an opening for the passage of the vertebral artery and sub-occipital nerve. It is in relation, behind, with the Recti postici minores and Obliqui superiores; in front, with the dura mater of the spinal canal, to which it is intimately adherent.

The *Lateral Ligaments* are strong fibrous bands, directed obliquely upwards and inwards, attached, above, to the jugular process of the occipital bone; below, to the base of the transverse process of the atlas.

The *Capsular Ligaments* surround the condyles of the occipital bone, and connect them with the articular surfaces of the atlas; they consist of thin and loose capsules, which enclose the synovial membrane of the articulation. The synovial membranes between the occipital bone and atlas communicate occasionally with that between the posterior surface of the odontoid process and transverse ligament.

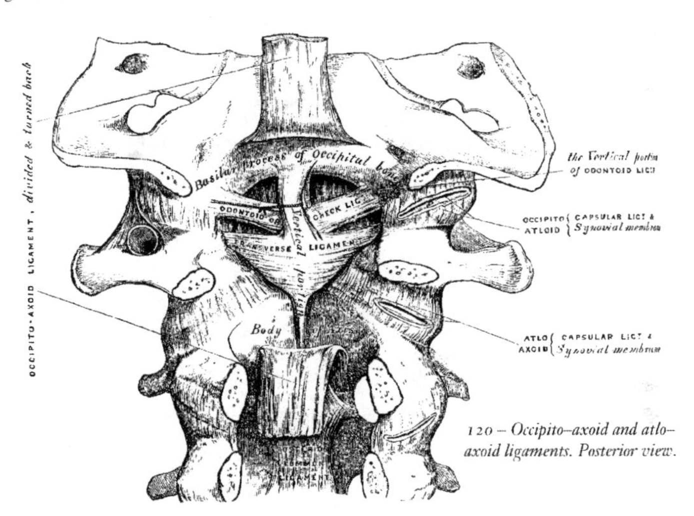


120 – *Occipito–axoid and atlo–axoid ligaments. Posterior view.*

*Actions.* The movements permitted in this joint are flexion and extension, which give rise to the ordinary forward or backward nodding of the head, besides slight lateral motion to one or the other side. When either of these actions is carried beyond a slight extent, the whole of the cervical portion of the spine assists in its production.

## 4. Articulation of the Axis with the Occipital Bone.

Occipito-Axoid. Three Odontoid.

To expose these ligaments, the spinal canal should be laid open by removing the posterior arch of the atlas, the laminæ and spinous process of the axis, and that portion of the occipital bone behind the foramen magnum, as seen in fig. 120.

The *Occipito-Axoid Ligament* (Apparatus ligamentosus colli) is situated at the upper part of the front surface of the spinal canal. It is a broad and strong ligamentous band, which covers the odontoid process and its ligaments, and appears to be a prolongation upwards of the posterior common ligament of the spine. It is attached, below, to the posterior surface of the body of the axis, and becoming expanded as it ascends, is inserted into the basilar groove of the occipital bone, in front of the foramen magnum.

*Relations.* By its anterior surface, it is intimately connected with the transverse ligament; by its posterior surface with the dura mater. By dividing this ligament transversely across, and turning its ends aside, the transverse and odontoid ligaments are exposed.

The *Odontoid* or *Check Ligaments* are strong rounded fibrous cords, which arise one on either side of the apex of the odontoid process, and passing obliquely upwards and outwards, are inserted into the rough depressions on the inner side of the condyles of the occipital bone. In the triangular interval left between these ligaments and the margin of the foramen magnum, a third strong ligamentous band (ligamentum suspensorium) may be seen, which passes almost perpendicularly from the apex of the odontoid process to the anterior margin of the foramen, being intimately blended with the anterior occipito-atloid ligament, and upper fasciculus of the transverse ligament of the atlas.

*Actions.* The odontoid ligaments serve to limit the extent to which rotation of the cranium may be carried; hence they have received the name of *check ligaments.*

## 5. Temporo-Maxillary Articulation.

This articulation is a double arthrodia. The parts entering into its formation are, on each side, the anterior part of the glenoid cavity of the temporal bone and the eminentia articularis above; with the condyle of the lower jaw below. The ligaments are the following.

| | |
|---|---|
| External Lateral. | Stylo-maxillary. |
| Internal Lateral. | Capsular. |

[Inter-articular Fibro-cartilage.]

The *External Lateral Ligament* (fig. 121) is a short, thin, and narrow fasciculus, attached above to the outer surface of the zygoma and to the rough tubercle on its lower border; below, to the outer surface and posterior border of the neck of the lower jaw. This ligament is broader above than below; its fibres are placed parallel with one another, and directed obliquely downwards and backwards. Externally, it is covered by the parotid gland and by the integument. Internally, it is in relation with the inter-articular fibro-cartilage and the synovial membranes.

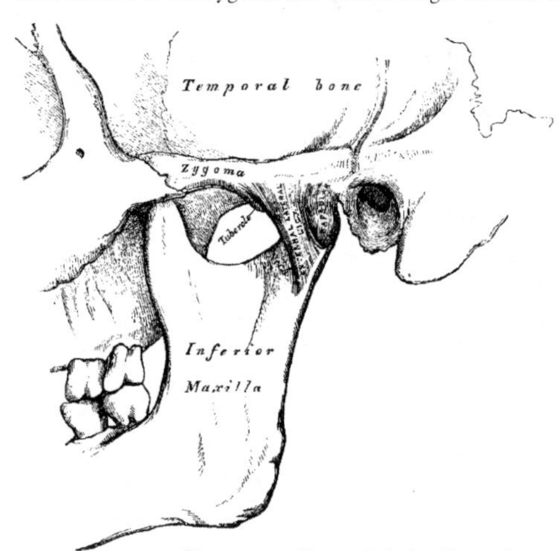

121 – *Temporo-maxillary articulation. External view.*

The *Internal Lateral Ligament* (fig. 122) is a long, thin, and loose band, attached above to the spinous process of the sphenoid bone, and becoming broader as it descends, is inserted into the inner margin of the dental foramen. Its outer surface is in relation above with the External pterygoid muscle; lower down it is separated from the neck of the condyle by the internal maxillary artery; and still more inferiorly the inferior dental vessels and nerve separate it from the ramus of the jaw. Internally it is in relation with the Internal pterygoid.

The *Stylo-maxillary Ligament* is a thin aponeurotic cord, which extends from near the apex of the styloid process of the temporal bone, to the angle and posterior border of the ramus of the lower jaw, between the Masseter and Internal pterygoid muscles. This ligament separates the parotid from the sub-maxillary gland, and has attached to its inner side, part of the fibres of origin of the Stylo-glossus muscle. Although usually classed among the ligaments of the jaw, it can only be considered as an accessory in the articulation.

The *Capsular Ligament* consists of a thin and loose ligamentous capsule, attached above to the circumference of the glenoid cavity and the articular surface immediately in front; below, to the neck of the condyle of the lower jaw. It consists of a few, thin scattered fibres, and can hardly be considered as a distinct ligament; it is thickest at the back part of the articulation.

The *Inter-articular fibro-cartilage* (fig. 123) is a thin plate of an oval form, placed horizontally between the condyle of the jaw and the glenoid cavity. Its upper surface is concave from before backwards, and a little convex transversely, to accommodate itself to the form of the glenoid cavity. Its under surface, where it is in contact with the condyle, is concave. Its circumference is connected externally to the external lateral ligament; internally, to the capsular ligament; and in front to the tendon of the External pterygoid muscle. It is thicker at its circumference, especially behind, than at its centre, where it is sometimes perforated. The fibres of which it is composed have a concentric arrangement, more apparent at the circumference than at the centre. Its surfaces are smooth, and divide the joint into two cavities, each of which is furnished with a separate synovial membrane. When the fibro-cartilage is perforated, the synovial membranes are continuous with one another.

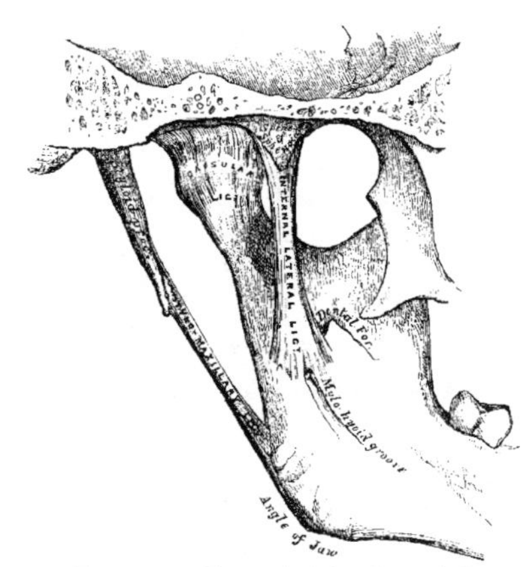


122 – *Temporo-maxillary articulation. Internal view.*

The *Synovial Membranes*, two in number, are placed one above, and the other below the fibro-cartilage. The upper one, the larger and looser of the two, is continued from the margin of the cartilage covering the glenoid cavity and eminentia articularis, over the upper surface of the fibro-cartilage. The lower one is interposed between the under surface of the fibro-cartilage and the condyle of the jaw, being prolonged downwards a little further behind than in front.

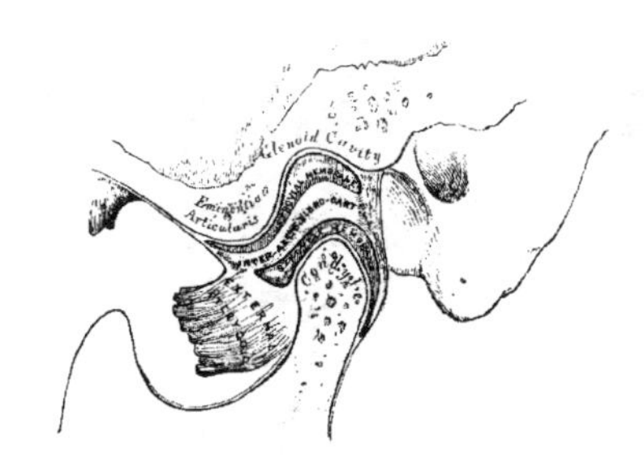


123 – *Vertical section of temporo–maxillary articulation.*

The *Nerves* of this joint are derived from the auriculo-temporal, and masseteric branches of the inferior maxillary.

*Actions.* The movements permitted in this articulation are very extensive. Thus, the jaw may be depressed or elevated, or it may be carried forwards or backwards, or from side to side. It is by the alternation of these movements performed in succession, that a kind of rotatory movement of the lower jaw upon the upper takes place, which materially assists in the mastication of the food.

If the movement of depression is carried only to a slight extent, the condyles remain in the glenoid cavities, their anterior part descending only to a slight extent; but if depression is considerable, the condyles glide from the glenoid fossæ on to the eminentia articularis, carrying with them the inter-articular fibro-cartilages. When this movement is carried to too great an extent, as, for instance, during a convulsive yawn, dislocation of the condyle into the zygomatic fossa occurs; the inter-articular cartilage being carried forwards, and the capsular ligament ruptured. When the jaw is elevated, the condyles and fibro-cartilages are carried backwards into their original position. When the jaw is carried forwards or backwards, a horizontal gliding movement of the fibro-cartilages and condyles upon the glenoid cavities takes place in the antero-posterior direction; whilst in the movement from side to side, this occurs in the lateral direction.

## 6. Articulation of the Ribs with the Vertebræ.

The articulation of the ribs with the vertebral column, may be divided into two sets. 1. Those which connect the heads of the ribs with the bodies of the vertebræ; 2. Those which connect the neck and tubercle of the ribs with the transverse processes.

## I. Articulation between the Heads of the Ribs and the Bodies of the Vertebræ.

These constitute a series of angular ginglymoid joints, formed by the articulation of the heads of the ribs with the cavities on the contiguous margins of the bodies of the dorsal vertebræ, connected together by the following ligaments:—

Anterior Costo-vertebral or Stellate.
Capsular.
Inter-articular.

The *Anterior Costo-vertebral* or *Stellate Ligament* (fig. 124) connects the anterior part of the head of each rib, with the sides of the bodies of the vertebræ, and the intervening intervertebral disc. It consists of three flat bundles of ligamentous fibres, which radiate from the anterior part of the head of the rib. The superior fasciculus passes upwards to be connected with the body of the vertebra above; the inferior one descends to the body of the vertebra below; and the middle one, the smallest and least distinct, passes horizontally inwards to be attached to the intervertebral substance.

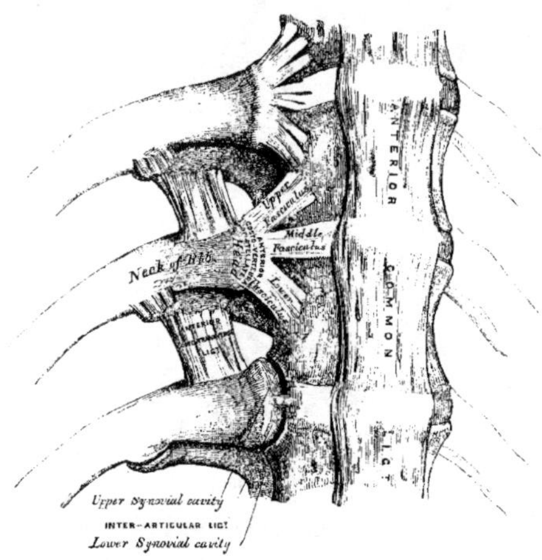

124 – *Costo-vertebral and costo-transverse articulations. Anterior view.*

*Relations.* In front, with the thoracic ganglia of the sympathetic, the pleura, and, on the right side, with the vena azygos major; behind, with the inter-articular ligament and synovial membranes.

In the first rib, which articulates with a single vertebra only, this ligament does not present a distinct division into three fasciculi; its superior fibres, however, pass to be attached to the body of the last cervical vertebra, as well as to the body of the vertebra with which the rib articulates. In the eleventh and twelfth ribs, which also articulate with a single vertebra, the same division does not exist; but the upper fibres of the ligament, in each case, are connected with the vertebra above as well as to that with which the ribs articulate.

The *Capsular Ligament* is a thin and loose ligamentous bag, which surrounds the joint between the head of the rib and the articular cavity formed by the junction of the vertebræ. It is very thin, firmly connected with the anterior ligament and most distinct at the upper and lower parts of the articulation.

The *Inter-articular Ligament* is situated in the interior of the joint. It consist of a short band of fibres, flattened from above downwards, attached by one extremity to the sharp crest on the head of the rib, and by the other to the intervertebral disc. It divides the joint into two cavities, which have no communication with one another, and are each lined by a separate synovial membrane. In the first, eleventh, and twelfth ribs, the inter-articular ligament does not exist; consequently, there is but one synovial membrane.

*Actions.* The movements permitted in these articulations, are limited to elevation, depression, and slightly forwards and backwards. This movement varies, however, very much in its extent in different ribs. The first rib is almost entirely immoveable, excepting in deep inspiration. The movement of the second rib is also not very extensive. In the other ribs, their mobility increases successively to the last two, which are very moveable. The ribs are generally more moveable in the female than in the male.

## 2. Articulation between the Neck and Tubercle of the Ribs with the Transverse Processes.

The ligaments connecting these parts, are—

Anterior Costo-Transverse.
Middle Costo-Transverse (Interosseous).
Posterior Costo-Transverse.
Capsular.

The *Anterior Costo-Transverse Ligament* (fig. 125) is a broad and strong band of fibres, attached below, to the sharp crest on the upper border of the neck of each rib, and passing obliquely upwards and outwards, to the lower border of the transverse process immediately above. It is broader below than above, broader and thinner between the lower ribs than between the upper, and more distinct in front than behind. This ligament is in relation, in front, with the intercostal vessels and nerves; behind, with the Longissimus dorsi. Its *internal border* completes an aperture formed between it and the articular processes, through which pass the posterior branches of the intercostal vessels and nerves. Its *external border* is continuous with a thin aponeurosis, which covers the External intercostal muscle.

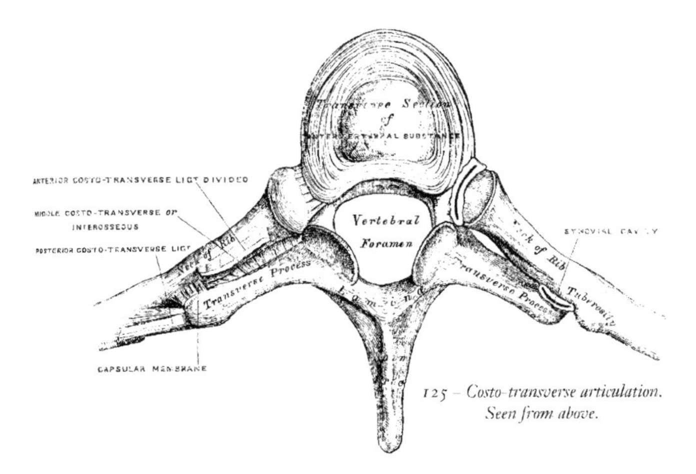


125 – *Costo-transverse articulation. Seen from above.*

The *first* and *last ribs* have no anterior costo-transverse ligament.

The *Middle Costo-Transverse* or *Interosseous Ligament* consists of short, but strong, fibres, which pass between the rough surface on the posterior part of the neck of each rib, and the anterior surface of the adjacent transverse process. In order fully to expose this ligament, a horizontal section should be made across the transverse process and corresponding part of the rib; or the rib may be forcibly separated from the transverse process, and its fibres torn asunder.

In the *eleventh* and *twelfth ribs*, this ligament is quite rudimentary.

The *Posterior Costo-Transverse Ligament* is a short, but thick and strong fasciculus, which passes obliquely from the summit of the transverse process to the rough non-articular portion of the tubercle of the rib. This ligament is shorter and more oblique in the upper, than in the lower ribs. Those corresponding to the superior ribs ascend, and those of the inferior ones slightly descend.
In the *eleventh* and *twelfth ribs*, this ligament is wanting.

The articular portion of the tubercle of the rib, and adjacent transverse process form an arthrodial joint, provided with a thin *capsular ligament* attached to the circumference of the articulating surfaces, and enclosing a small *synovial membrane*.
In the *eleventh* and *twelfth ribs*, this ligament is wanting.

*Actions.* The movement permitted in these joints, is limited to a slight gliding motion of the articular surfaces one upon the other.

## 7. Articulation of the Cartilages of the Ribs with the Sternum.

The articulation of the cartilages of the true ribs with the sternum are arthrodial joints. The ligaments connecting them are—

Anterior Costo-Sternal.
Posterior Costo-Sternal.
Capsular.

The *Anterior Costo-Sternal Ligament* (fig. 126) is a broad and thin membranous band that radiates from the inner extremity of the cartilages of the true ribs, to the anterior surface of the sternum. It is composed of fasciculi, which pass in different directions. The *superior* fasciculi ascend obliquely, the *inferior* pass obliquely downwards, and the *middle* fasciculi horizontally. The superficial fibres of this ligament are the longest; they intermingle with the fibres of the ligaments above and below them, with those of the opposite side, and with the tendinous fibres of origin of the Pectoralis major; forming a thick fibrous membrane, which covers the surface of the sternum. This is more distinct at the lower than at the upper part.

The *Posterior Costo-Sternal Ligament*, less thick and distinct than the anterior, is composed of fibres which radiate from the posterior surface of the sternal end of the cartilages of the true ribs, to the posterior surface of the sternum, becoming blended with the periosteum.

The *Capsular Ligament* surrounds the joints formed between the cartilages of the true ribs and the sternum. It is very thin, intimately blended with the anterior and posterior ligaments, and strengthened at the upper and lower part of the articulation by a few fibres, which pass from the cartilage to the side of the sternum. These ligaments protect the synovial membranes.

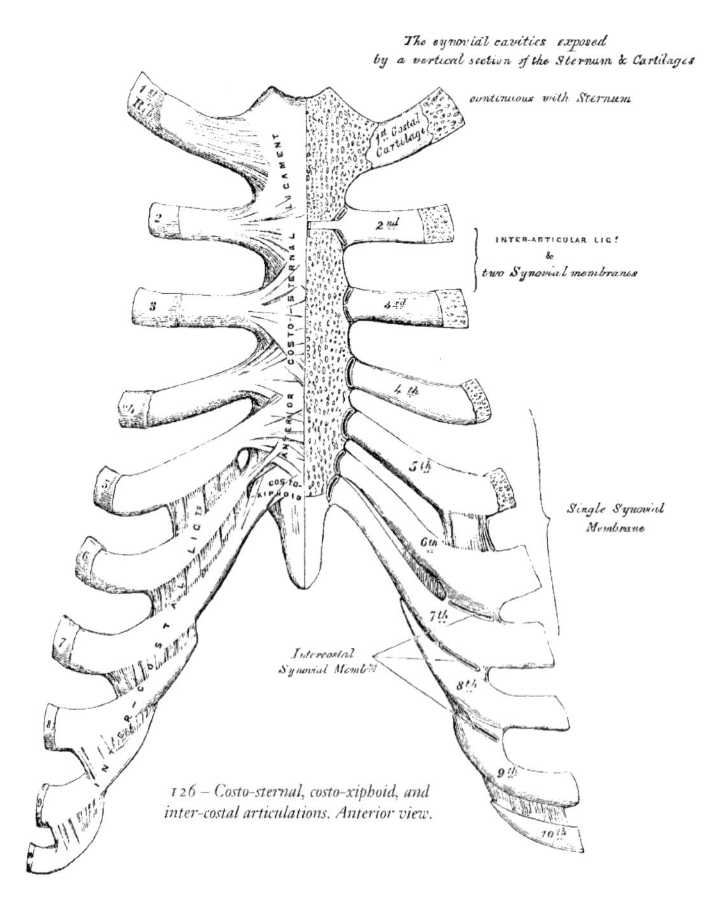


*126 – Costo-sternal, costo-xiphoid, and inter-costal articulations. Anterior view.*

*Synovial Membranes.* The cartilage of the *first rib* is directly continuous with the sternum, the synovial membrane being absent. The cartilage of the *second rib* is connected with the sternum by means of an inter-articular ligament, attached by one extremity to the cartilage of the second rib, and by the other extremity to the cartilage which unites the first and second pieces of the sternum. This articulation is provided with two synovial membranes. That of the third rib has also two synovial membranes; and that of the fourth, fifth, sixth, and seventh, each a single synovial membrane. Thus there are *eight* synovial cavities in the articulations between the costal cartilages of the true ribs and the sternum. They may be demonstrated by removing a thin section from the anterior surface of the sternum and cartilages, as seen in the figure. After middle life, the articular surfaces lose their polish, become roughened, and the synovial membranes appear to be wanting. In old age, the articulations do not exist, the cartilages of most of the ribs becoming firmly united to the sternum. The cartilage of the *seventh rib*, and occasionally also that of the *sixth*, is connected to the anterior surface of the ensiform appendix, by a band of ligamentous fibres, which varies in length and breadth in different subjects. It is called the *costo-xiphoid ligament.*

*Actions.* The movements which are permitted in the costo-sternal articulations, are limited to elevation and depression; and these only to a slight extent.

## Articulation of the Cartilages of the Ribs with each other.

The cartilages of the sixth, seventh, and eighth ribs articulate, by their lower borders, with the corresponding margin of the adjoining cartilages, by means of a small, smooth, oblong-shaped facet. Each articulation is enclosed in a thin *capsular ligament*, lined by *synovial membrane*, and strengthened externally and internally by ligamentous fibres (intercostal ligaments), which pass from one cartilage to the other. Sometimes the cartilage of the fifth rib, more rarely that of the ninth, articulates, by its lower border, with the adjoining cartilage by a small oval facet; more frequently they are connected together by a few ligamentous fibres. Occasionally, the articular surfaces above mentioned are found wanting.

## Articulation of the Ribs with their Cartilages.

The outer extremity of each costal cartilage is received into a depression in the sternal end of the ribs, and held together by the periosteum.

## 8. Ligaments of the Sternum.

The first and second pieces of the Sternum are united by a layer of cartilage which rarely ossifies, except at an advanced period of life. These two segments are connected by an anterior and posterior ligament.

The *anterior sternal ligament* consists of a layer of fibres, having a longitudinal direction; it blends with the fibres of the anterior costo-sternal ligaments on both sides, and with the aponeurosis of origin of the Pectoralis major. This ligament is rough, irregular, and much thicker at the lower than at the upper part of this bone.

The *posterior sternal ligament* is disposed in a somewhat similar manner on the posterior surface of the articulation.

## 9. Articulation of the Pelvis with the Spine.

The ligaments connecting the last lumbar vertebra with the sacrum are similar to those which connect the segments of the spine with each other, viz.:—1. The continuation downwards of the anterior and posterior common ligaments. 2. The inter-vertebral substance connecting the flattened oval surfaces of the two bones, thus forming an amphiarthrodial joint. 3. Ligamenta subflava, connecting the arch of the last lumbar vertebra with the posterior border of the sacral canal. 4. Capsular ligaments connecting the articulating processes and forming a double anthrodia. 5. Inter- and supra-spinous ligaments.

The two proper ligaments connecting the pelvis with the spine are the lumbo-sacral and lumbo-iliac.

The *Lumbo-sacral Ligament* (fig. 127) is a short, thick, triangular fasciculus, connected above to the lower and front part of the transverse process of the last lumbar vertebra, and passing obliquely outwards, is attached below to the lateral surface of the base of the sacrum, becoming blended with the anterior sacro-iliac ligament. This ligament is in relation in front with the Psoas muscle.

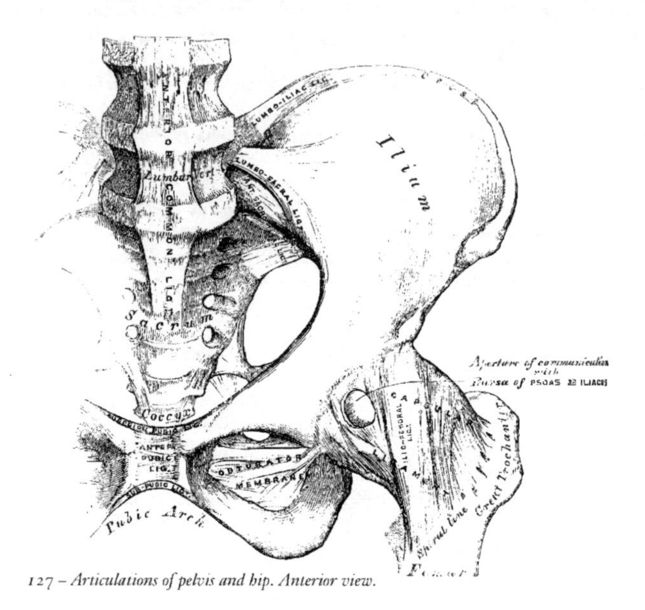


*127 – Articulations of pelvis and hip. Anterior view.*

The *Lumbo-iliac Ligament* (fig. 127) passes horizontally outwards from the apex of the transverse process of the last lumbar vertebra, to that portion of the crest of the ilium immediately in front of the sacro-iliac articulation. It is of a triangular form, thick and narrow internally, broad and thinner externally. It is in relation, in front, with the Psoas muscle; behind, with the muscles occupying the vertebral groove; above, with the Quadratus lumborum.

## 10. Articulations of the Pelvis.

The Ligaments connecting the bones of the pelvis with each other may be divided into four groups. 1. Those connecting the sacrum and ilium. 2. Those passing between the sacrum and ischium. 3. Those connecting the sacrum and coccyx. 4. Those between the two pubic bones.

### 1. Articulation of the Sacrum and Ilium.

The sacro-iliac articulation is an amphiarthrodial joint, formed between the lateral surfaces of the sacrum and ilium. The anterior or auricular portion of each articular surface is covered with a thin plate of cartilage, thicker on the sacrum than on the ilium. The surfaces of these cartilages in the adult are rough and irregular, and separated from one another by a soft yellow pulpy substance. At an early period of life, occasionally in the adult, and in the female during pregnancy, they are smooth and lined by a delicate synovial membrane. The ligaments connecting these surfaces are the anterior and posterior sacro-iliac.

The *Anterior Sacro-iliac Ligament* consists of numerous thin ligamentous bands, which connect the anterior surfaces of the sacrum and ilium.

The *Posterior Sacro-Iliac* (fig. 128) is a strong interosseous ligament, situated in the deep depression between the sacrum and ilium behind, and forming the chief bond of connexion between these bones. It consists of numerous strong fasciculi, which pass between the bones in various directions. Three of these are of large size; the *two superior*, nearly horizontal in direction, arise from the first and second transverse tubercles on the posterior surface of the sacrum, and are inserted into the

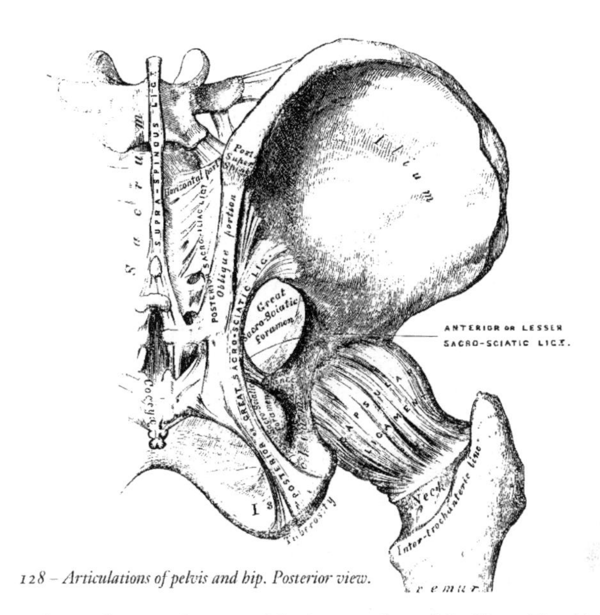


*128 – Articulations of pelvis and hip. Posterior view.*

rough uneven surface at the posterior part of the inner surface of the ilium. The third fasciculus, oblique in direction, is attached by one extremity to the third or fourth transverse tubercle on the posterior surface of the sacrum, and by the other to the posterior superior spine of the ilium; it is sometimes called the *oblique sacro-iliac ligament.*

## 2. Articulation of the Sacrum and Ischium.

The Great Sacro-Sciatic (Posterior).
The Lesser Sacro-Sciatic (Anterior).

The *Great* or *Posterior Sacro-Sciatic Ligament* is situated at the lower and back part of the pelvis. It is thin, flat, and triangular in form; narrower in the middle than at the extremities; attached by its broad base to the posterior inferior spine of the ilium, to the third and fourth transverse tubercles on the sacrum, and to the lower part of the lateral margin of that bone and the coccyx; passing obliquely downwards, outwards, and forwards, it becomes narrow and thick; and at its insertion into the inner margin of the tuberosity of the ischium, it increases in breadth, and is prolonged forwards along the inner margin of the ramus forming the falciform ligament. The free concave edge of this ligament has attached to it the obturator fascia, with which it forms a kind of groove, protecting the internal pudic vessels and nerve. One of its surfaces is turned towards the perinæum, the other towards the Obturator internus muscle.

The *posterior surface* of this ligament gives origin, by its whole extent, to fibres of the Gluteus maximus. Its *anterior surface* is united to the lesser sacro-sciatic ligament. Its *superior border* forms the lower boundary of the lesser sacro-sciatic foramen. Its *lower border* forms part of the boundary of the perinæum. This ligament is pierced by the coccygeal branch of the sciatic artery.

The *Lesser* or *Anterior Sacro-Sciatic Ligament*, much shorter and smaller than the preceding, is thin, triangular in form, attached by its apex to the spine of the ischium, and internally, by its broad base, to the lateral margin of the sacrum and coccyx, anterior to the attachment of the great sacro-sciatic ligament, with which its fibres are intermingled.

It is in relation, *anteriorly*, with the Coccygeus muscle; *posteriorly*, it is covered by the posterior ligament, and crossed by the pudic vessels and nerve. Its *superior border* forms the lower boundary of the great sacro-sciatic foramen. Its *inferior border*, part of the lesser sacro-sciatic foramen.

These two ligaments convert the sacro-sciatic notches into foramina. The *superior* or *great* sacro-sciatic foramen is bounded, in front and above, by the posterior border of the os innominatum; behind, by the great sacro-sciatic ligament; and below, by the lesser ligament. It is partially filled up, in the recent state, by the Pyriformis muscle. Above this muscle, the gluteal vessels and nerve emerge from the pelvis; and below it, the ischiatic vessels and nerves, the internal pudic vessels and nerve, and the nerve to the Obturator internus. The *inferior* or *lesser* sacro-sciatic foramen is bounded, in front, by the tuber ischii; above, by the spine and lesser ligament; behind, by the greater ligament. It transmits the tendon of the Obturator internus muscle, its nerve, and the pudic vessels and nerve.

## 3. Articulation of the Sacrum and Coccyx.

This articulation is an amphiarthrodial joint, formed between the oval surface on the summit of the sacrum, and the base of the coccyx. It is analogous to the joints between the bodies of the vertebræ, and is connected by similar ligaments. They are the—

Anterior Sacro-Coccygeal.
Posterior Sacro-Coccygeal.
Inter-articular Fibro-Cartilage.

The *Anterior Sacro-Coccygeal Ligament* consists of a few irregular fibres, which descend from the anterior surface of the sacrum to the front of the coccyx, becoming blended with the periosteum.

The *Posterior Sacro-Coccygeal Ligament* is a flat band of ligamentous fibres, of a pearly tint, which arises from the margin of the lower orifice of the sacral canal, and descends to be inserted into the posterior surface of the coccyx. This ligament completes the lower and back part of the sacral canal. Its superficial fibres are much longer then the deep-seated; the latter extend from the apex of the sacrum to the upper cornua of the coccyx. This ligament is in relation in front with the arachnoid membrane of the sacral canal, a portion of the sacrum, and almost the whole of the posterior surface of the coccyx; behind with the Gluteus maximus.

An *Inter-articular Fibro-Cartilage* is interposed between the contiguous surfaces of the sacrum and coccyx; it differs from that interposed between the bodies of the vertebræ, in being thinner, and its central part more firm in texture. It is somewhat thicker in front and behind, than at the sides. Occasionally a synovial membrane is found where the coccyx is freely movable, which is more especially the case during pregnancy.

The different segments of the coccyx are connected together by an extension downwards of the anterior and posterior sacro-coccygeal ligaments, a thin annular disc of fibro-cartilage being interposed between each of the bones. In the adult male, all the pieces become ossified; but in the female, this does not commonly occur until a later period of life. The separate segments of the coccyx are first united, and at a more advanced age the joint between the sacrum and the coccyx.

*Actions.* The movements which take place between the sacrum and coccyx, and between the different pieces of the latter bone, are slightly forwards and backwards; they are very limited. Their mobility increases during pregnancy.

## 4. Articulation of the Pubes.

The articulation between the pubic bones is an amphiarthrodial joint, formed by the junction of the two oval surfaces which has received the name of the *symphysis*. The ligaments of this articulation are the

Anterior Pubic. Posterior Pubic.
Superior Pubic. Sub-Pubic.
Inter-articular Fibro-Cartilage.

The *Anterior Pubic Ligament* consists of several superimposed layers, which pass across the front of the articulation. The superficial fibres pass obliquely from one bone to the other, decussating and forming an interlacement with the fibres of the aponeurosis of the External oblique muscle. The deep fibres pass transversely across the symphysis, and are blended with the inter-articular fibro-cartilage.

The *Posterior Pubic Ligament* consists of a few thin, scattered fibres, which unite the two pubic bones posteriorly.

The *Superior Pubic Ligament* is a band of fibres, which connects together the two pubic bones superiorly.

The *Sub-Pubic Ligament* is a thick, triangular arch of ligamentous fibres connecting together the two pubic bones below, and forming the upper boundary of the pubic arch. Above, it is blended with the inter-articular fibro-cartilage; laterally, with the rami of the pubes. Its fibres are of a yellowish colour, closely connected, and have an arched direction.

The *Inter-articular Fibro-Cartilage* consists of two oval-shaped plates, one covering the surface of each symphysis pubis. They vary in thickness is different subjects, and project somewhat beyond the level of the bones, especially behind. The outer surface of each plate is firmly connected to the bone by a series of nipple-like processes, which accurately fit within corresponding depressions on the osseous surface. Their opposed surfaces are connected, in the greater part of their extent, by an intermediate fibrous elastic-tissue; and by their circumference to the various ligaments surrounding the joint. An interspace is left between the plates at the upper and back part of the articulation, when the fibrous-tissue is deficient, and the surface of the fibro-cartilage is lined by epithelium. This space is found at all periods of life, both in the male and female; but it is larger in the latter, especially during pregnancy, and after parturition. It is most frequently limited to the upper and back part of the joint but it occasionally reaches to the front, and may extend the entire length of the cartilages. This structure may be easily demonstrated, by making a vertical section of the symphysis pubis near its posterior surface.

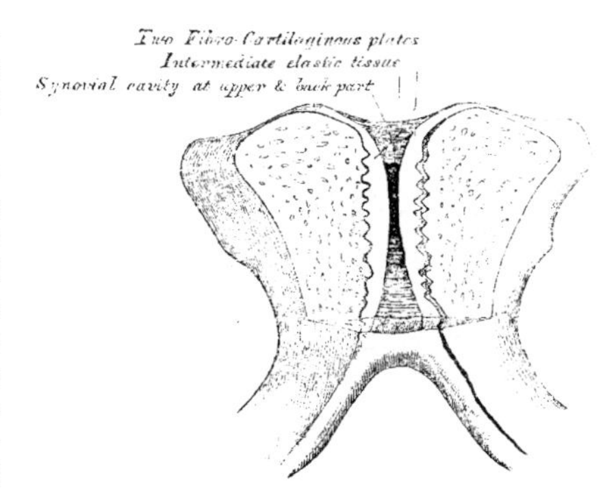


*129 – Vertical section of the symphysis pubis. Made near its posterior surface.*

The *Obturator Ligament* is a dense membranous layer, consisting of fibres which interlace in various directions. It is attached to the circumference of the obturator foramen, which it closes completely, except at its upper and outer part, where a small oval canal is left for the passage of the obturator vessels and nerve. It is in relation, in front, with the Obturator externus; behind, with the Obturator internus; both of which muscles are in part attached to it.

## ARTICULATIONS OF THE UPPER EXTREMITY.

The articulations of the Upper Extremity may be arranged into the following groups:—1. Sterno-clavicular articulation. 2. Scapulo-clavicular articulation. 3. Ligaments of the Scapula. 4. Shoulder-joint. 5. Elbow-joint. 6. Radio-ulnar articulation. 7. Wrist-joint. 8. Articulations of the Carpal bones. 9. Carpo-metacarpal articulations. 10. Metacarpo-phalangeal articulations. 11. Articulations of the Phalanges.

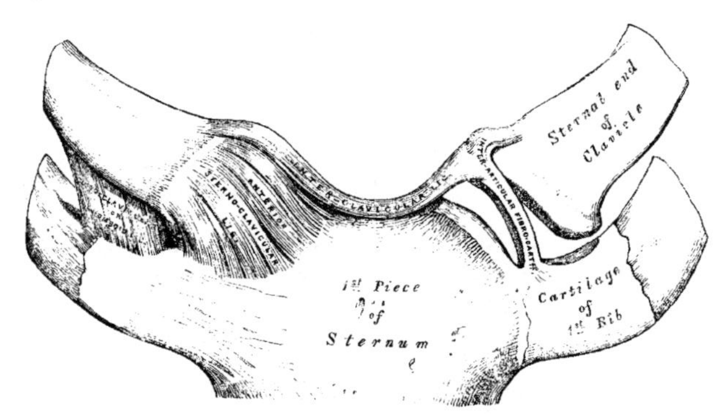


*130 – Sterno-clavicular articulation. Anterior view.*

### 1. Sterno-Clavicular Articulation.

The *Sterno-Clavicular* is an arthrodial joint. The parts entering into its formation are the sternal end of the clavicle, the upper and lateral part of the first piece of the sternum, and the cartilage of the first rib. The articular surface of the clavicle is much longer than that of the sternum, and invested with a layer of cartilage,* which is considerably thicker than that on the latter bone. The ligaments of this joint are the

| | |
|---|---|
| Anterior Sterno-Clavicular. | Inter-Clavicular. |
| Posterior Sterno-Clavicular. | Costo-Clavicular (rhomboid). |

Inter-Articular Fibro-Cartilage.

The *Anterior Sterno-Clavicular Ligament* is a broad band of ligamentous fibres, which covers the anterior surface of the articulation, being attached, above, to the upper and front part of the inner extremity of the clavicle; and, passing obliquely downwards and inwards, is attached, below, to the front and upper part of the first piece of the sternum. This ligament is covered in front by the sternal portion of the Sterno-cleido-mastoid and the integument; behind, it is in relation with the inter-articular fibro-cartilage and the two synovial membranes.

The *Posterior Sterno-Clavicular Ligament* is a broad band of fibres, which covers the posterior surface of the articulation, being attached, above, to the posterior part of the inner extremity of the clavicle; and, passing obliquely downwards and inwards, is connected, below, to the posterior and upper part of the sternum. It is in relation, in front, with the inter-articular fibro-cartilage and synovial membranes; behind, with the Sterno-hyoid and Sterno-thyroid muscles.

The *Inter-Clavicular Ligament* is a flattened ligamentous band, which varies considerably in form and size in different individuals; it passes from the upper part of the inner extremity of one clavicle to the other, and is closely attached to the upper margin of the sternum. It is in relation, in front, with the integument; behind with the Sterno-thyroid muscles.

The *Costo-Clavicular Ligament* (*rhomboid*) is a short, flat, and strong band of ligamentous fibres of a rhomboid form, attached, below, to the upper and inner part of the cartilage of the first rib; and ascending obliquely backwards and outwards, is attached, above, to the rhomboid depression on the under surface of the clavicle. It is in relation, in front, with the tendon of origin of the Subclavius; behind, with the subclavian vein.

The *Inter-articular Fibro-Cartilage* is a flat and nearly circular disc, interposed between the articulating surfaces of the sternum and clavicle. It is attached above, to the upper and posterior border of the clavicle; below, to the cartilage of the first rib, at its junction with the sternum; and by its circumference to the anterior and posterior sterno-clavicular ligaments. It is thicker at the circumference, especially its upper and back part, than at its centre, or below. It divides the joint into two cavities, each of which is furnished with a separate synovial membrane; when the fibro-cartilage is perforated, which not unfrequently occurs, the synovial membranes communicate.

Of the *two Synovial Membranes* found in this articulation, one is reflected from the sternal end of the clavicle, over the adjacent surface of the fibro-cartilage, and cartilage of the first rib; the other is placed between the articular surface of the sternum and adjacent surface of the fibro-cartilage; the latter is the more loose of the two. They seldom contain much synovia.

*Actions.* This articulation is the centre of the movements of the shoulder, and admits of motion in nearly every direction—upwards, downwards, backwards forwards, as well as circumduction; the sternal end of the clavicle and the inter-articular cartilage gliding on the articular surface of the sternum.

## 2. Scapulo-Clavicular Articulation.

The *Scapulo-Clavicular* is an arthrodial joint, formed between the outer extremity of the clavicle, and the upper edge of the acromian process of the scapula. Its ligaments are the

Superior Acromio-Clavicular.
Inferior Acromio-Clavicular.
Coraco-Clavicular { Trapezoid and Conoid.
Inter-articular Fibro-Cartilage.

The *Superior Acromio-Clavicular Ligament* is a broad band of fibres, of a quadrilateral form, which

* According to Bruch, the sternal end of the clavicle is covered by a tissue which is more fibrous than cartilaginous in structure.

covers the superior part of the articulation, extending between the upper part of the outer end of the clavicle, and the adjoining part of the acromion. It is composed of parallel fibres, which interlace, with the aponeurosis of the Trapezius and Deltoid muscles; below, it is in contact with the inter-articular fibro-cartilage and synovial membranes.

The *Inferior Acromio-Clavicular Ligament*, somewhat thinner than the preceding, covers the under part of the articulation, and is attached to the adjoining surfaces of the two bones. It is in relation, above, with the inter-articular fibro-cartilage (when it exists) and the synovial membranes; below, with the tendon of the Supra-spinatus. These two ligaments are continuous with each other in front and behind, and form a complete capsule around the joint.

The *Coraco-Clavicular Ligament* serves to connect the clavicle with the coracoid process of the scapula. It consists of two fasciculi, called the trapezoid and conoid ligaments.

The *trapezoid ligament*, the anterior and external fasciculus, is a broad, thin, quadrilateral-shaped band of fibres, placed obliquely between the coracoid process and the clavicle. It is attached, below, to the upper surface of the coracoid process; above, to the oblique line on the under surface of the clavicle. Its anterior border is free; its posterior border is joined with the conoid ligament, the two forming by their junction a projecting angle.

The *conoid ligament*, the posterior and internal fasciculus, is a dense band of fibres, conical in form, the base being turned upwards, the summit downwards. It is attached by its apex to a rough depression at the base of the coracoid process, internal to the preceding; above, by its expanded base, to the conoid tubercle on the under surface of the clavicle, and into a line proceeding internally from it for half an inch. These ligaments are in relation, in front, with the Subclavius; behind, with the Trapezius: they serve to limit rotation of the scapula forwards and backwards.

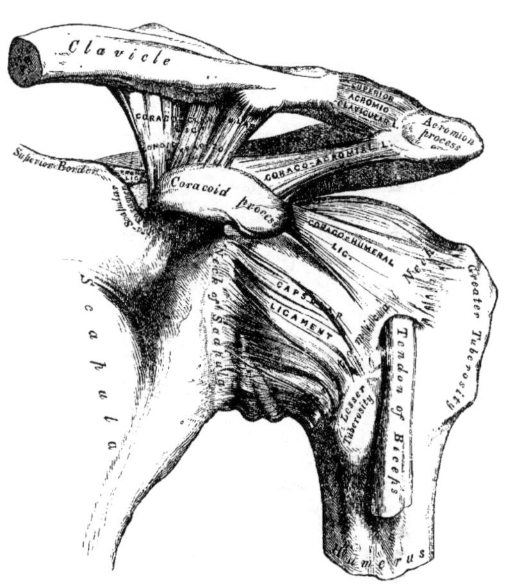


131 – *The left shoulder-joint, scapulo-clavicular articulations, and proper ligaments of scapula.*

The *Inter-articular Fibro-Cartilage* is most frequently absent in this articulation. When it exists, it generally only partially separates the articular surfaces, and occupies the upper part of the articulation. More rarely, it completely separates the joint into two cavities.

There are *two Synovial Membranes* where a complete inter-articular cartilage exists; more frequently, there is only one synovial membrane.

*Actions.* The movements of this articulation are of two kinds. 1. A gliding motion of the articular end of the clavicle on the acromion. 2. Rotation of the scapula forwards and backwards upon the clavicle, the extent of this rotation being limited by the two portions of the coraco-clavicular ligament.

## 3. Proper Ligaments of the Scapula.

The proper ligaments of the scapula are, the

Coraco-acromial. Transverse.

The *Coraco-acromial Ligament* is a broad, thin, flat band, of a triangular shape, extended transversely above the upper part of the shoulder-joint, between the coracoid and acromion processes. It is attached, by its apex, to the summit of the acromion just in front of the articular surface for the clavicle; and by its broad base, to the whole length of the outer border of the coracoid process. Its posterior fibres are directed obliquely backwards and outwards, its anterior fibres transversely. This ligament completes the vault formed by the coracoid and acromion processes for the protection of the head of the humerus. It is in relation, above, with the clavicle and under surface of the Deltoid; below, with the tendon of the Supra-spinatus muscle, a bursa being interposed. Its anterior border is continuous with a dense cellular lamina that passes beneath the Deltoid upon the tendons of the Supra- and Infra-spinati muscles.

The *Transverse* or *Coracoid Ligament* converts the supra-scapular notch into a foramen. It is a thin and flat fasciculus, narrower at the middle than at the extremities, attached, by one end, to the base of the coracoid process, and by the other, to the inner extremity of the scapular notch. The supra-scapular nerve passes through the foramen; its accompanying vessels above it.

## 4. Shoulder-Joint.

The Shoulder is an enarthrodial or ball and socket joint. The bones entering into its formation, are the large globular head of the humerus, which is received into the shallow glenoid cavity of the scapula, an arrangement which permits of very considerable movement, whilst the joint itself is protected against displacement by the strong ligaments and tendons which surround it, and above by a arched vault, formed by the under surface of the coracoid and acromion processes, and the coraco-acromial ligament. The articular surfaces are covered by a layer of cartilage: that on the head of the humerus is thicker at the centre than at the circumference, the reverse being observed in the glenoid cavity. Its ligaments are, the

Capsular. Coraco-humeral.
Glenoid.

The *Capsular Ligament* completely encircles the articulation; being attached above, to the circumference of the glenoid cavity beyond the glenoid ligament; below, to the anatomical neck of the humerus, approaching nearer to the articular cartilage above, than in the rest of its extent. It is thicker above than below, remarkably loose and lax, and much larger and longer than is necessary to keep the bones in contact, allowing them to be separated from each other more than an inch, an evident provision for that extreme freedom of movement which is peculiar to this articulation. Its external surface is strengthened, above, by the Supra-spinatus; above and internally, by the coraco-humeral ligament; below, by the long head of the Triceps; externally, by the tendons of the Infra-spinatus and Teres minor; and internally, by the tendon of the Subscapularis. The capsular ligament usually presents three openings: one at its inner side, below the coracoid process, partially filled up by the tendon of the Subscapularis; it establishes a communication between the synovial membrane of the joint and a bursa beneath the tendon of that muscle; a second, not constant, at its outer part, where a communication sometimes exists between the joint and a bursal sac belonging to the Infra-spinatus muscle. The third is seen in the lower border of the ligament, between the two tuberosities, for the passage of the tendon of the Biceps muscle.

The *Coraco-humeral* or *Accessory Ligament*, is a broad band which strengthens the upper and inner part of the capsular ligament. It arises from the outer border of the coracoid process, and descends obliquely downwards and outwards to the front of the great tuberosity of the humerus, being blended with the tendon of the Supra-spinatus muscle. This ligament is intimately united to the capsular in the greater part of its extent.

The *Glenoid Ligament* is a firm fibrous band attached round the margin of the glenoid cavity. It is triangular on section, the thickest portion being fixed to the circumference of the cavity, the free edge being thin and sharp. It is continuous above with the long tendon of the Biceps muscle, which bifurcates at the upper part of the cavity into two fasciculi, which encircle its margin, and unite at its lower part. This ligament deepens the cavity for articulation, and protects the edges of the bone. It is lined by the synovial membrane.

The *Synovial Membrane* lines the margin of the glenoid cavity and the fibro-cartilaginous rim surrounding it; it is then reflected over the internal surface of the capsular ligament, covers the lower part and sides of the neck of the humerus, and is continued a short distance over the cartilage covering the head of this bone. The long tendon of the Biceps muscle which passes through the joint, is enclosed in a tubular sheath of synovial membrane, which is reflected upon it at the point where it perforates the capsule, and is continued around it as far as the summit of the glenoid cavity. The tendon of the Biceps is thus enabled to traverse the articulation, but is not contained in the interior of the synovial cavity. The synovial membrane communicates with a large bursal sac beneath the tendon of the Subscapularis, by an opening at the inner side of the capsular ligament; it also occasionally communicates with another bursal sac, beneath the tendon of the Infra-spinatus, through an orifice at its outer part. A third bursal sac, which does not communicate with the joint, is placed between the under surface of the deltoid and the outer surface of the capsule.

The Muscles in relation with the joint are, above, the Supra-spinatus; below, the long head of the Triceps; internally, the Subscapularis; externally, the Infra-spinatus, and Teres minor; within, the long tendon of the Biceps. The Deltoid is placed most externally, and covers the articulation on its outer side, and in front and behind.

The Arteries supplying the joint, are articular branches of the anterior and posterior circumflex, and supra-scapular.

The Nerves are derived from the circumflex and supra-scapular.

*Actions.* The shoulder-joint is capable of movement in almost any direction, forwards, backwards, abduction, adduction, circumduction, and rotation.

## 5. Elbow-Joint.

The *Elbow* is a *ginglymoid* or hinge joint. The bones entering into its formation are the trochlear surface of the humerus, which is received in the greater sigmoid cavity of the ulna, and admits of the movements peculiar to this joint; those of flexion and extension, whilst the cup-shaped depression of the head of the radius articulates with the radial tuberosity of the humerus, its circumference with the lesser sigmoid cavity of the ulna, allowing of the movement of rotation of the radius on the ulna, the chief action of the superior radio-ulnar articulation. The articular surfaces are covered with a thin layer of cartilage, and connected together by the following ligaments:—

| | |
|---|---|
| Anterior Ligament. | Internal Lateral. |
| Posterior Ligament. | External Lateral. |

The *Anterior Ligament* (fig. 132) is a broad and thin fibrous layer, which covers the anterior surface of the joint. It is attached to the front of the humerus immediately above the coronoid fossa; below, to the anterior surface of the coronoid process of the ulna and orbicular ligament, being continuous on each side with the lateral ligaments. Its superficial or oblique fibres pass from the inner condyle of the humerus outwards to the orbicular ligament. The middle fibres, vertical in direction, pass from the upper part of the coronoid depression, and become blended with the preceding. A third, or transverse set, intersect these at right angles. This ligament is in relation in front, with the Brachialis anticus; behind, with the synovial membrane.

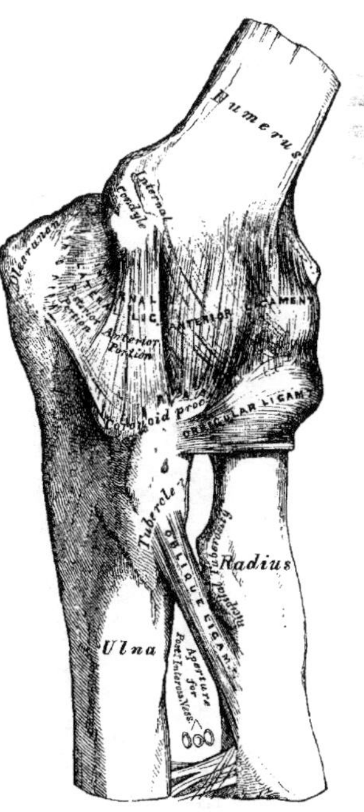

*132 – Left elbow-joint, showing anterior and internal ligaments.*

The *Posterior Ligament* is a thin and loose membranous fold, attached, above, to the lower end of the humerus, immediately above the olecranon depression; below, to the margin of the olecranon. The superficial or transverse fibres pass between the adjacent margins of the olecranon fossa. The deeper portion consists of vertical fibres, which pass from the upper part of the olecranon fossa to the margin of the olecranon. This ligament is in relation, behind, with the tendon of the Triceps and Anconeus; in front, with the synovial membrane.

The *Internal Lateral Ligament* is a thick triangular band of ligamentous fibres, consisting of two distinct portions, an anterior and posterior. The *anterior portion*, directed obliquely forwards, is attached, above, by its apex, to the front part of the internal condyle of the humerus; and, below, by its broad base, to the inner margin of the coronoid process. The *posterior portion*, also of triangular form, is attached, above, by its apex, to the lower and back part of the internal condyle; below, to the inner margin of the olecranon. This ligament is in relation, internally, with the Triceps and Flexor carpi ulnaris muscles, and the ulnar nerve.

The *External Lateral Ligament* (fig. 133) is a short and narrow fibrous fasciculus, less distinct than the internal, attached, above, to the external condyle of the humerus; below, to the orbicular ligament, some of its most posterior fibres passing over that ligament, to be inserted into the outer margin of the ulna. This ligament is intimately blended with the tendon of origin of the Supinator brevis muscle.

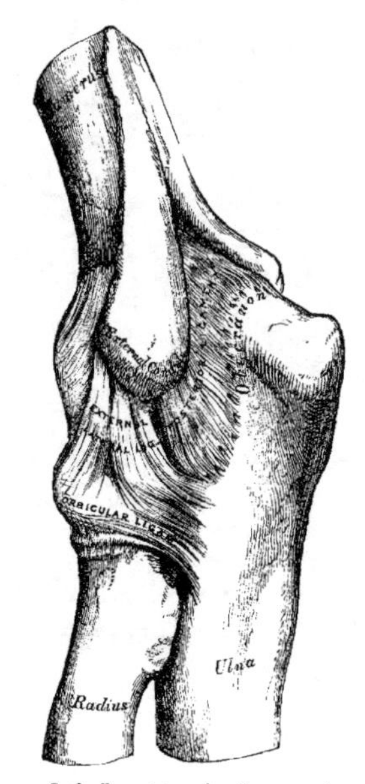

133 – *Left elbow-joint, shewing posterior and external ligaments.*

The *Synovial Membrane* is very extensive. It covers the margin of the articular surface of the humerus, and lines the coronoid and olecranon depressions on that bone; from these points, it is reflected over the anterior, posterior, and lateral ligaments; and forms a pouch between the lesser sigmoid cavity, the internal surface of the annular ligament, and the circumference of the head of the radius.

The *Muscles* in relation with the joint are, in front, the Brachialis anticus; behind, the Triceps and Anconeus; externally, the Supinator brevis, and the common tendon of origin of the Extensor muscles; internally, the common tendon of origin of the Flexor muscles, the Flexor carpi ulnaris, and ulnar nerve.

The *Arteries* supplying the joint are derived from the communicating branches between the superior profunda, inferior profunda, and anastomotic branches of the brachial, with the anterior, posterior and interosseous recurrent branches of the ulnar, and the recurrent branch of the radial. These vessels form a complete chain of inosculation around the joint.

The *Nerves* are derived from the ulnar, as it passes between the internal condyle and the olecranon; and a few filaments from the musculo-cutaneous.

*Actions.* The elbow is one of the most perfect hinge-joints in the body; its movements are, consequently, limited to flexion and extension, the exact apposition of the articular surfaces preventing the least lateral motion.

## 6. Radio-Ulnar Articulations.

The articulation of the radius with the ulna is effected by ligaments, which connect together both extremities as well as the shafts of these bones. They may, consequently, be subdivided into three sets:—1, the superior radio-ulnar; 2, the middle radio-ulnar; and, 3, the inferior radio-ulnar articulations.

### 1. Superior Radio-Ulnar Articulation.

This articulation is a lateral ginglymoid joint. The bones entering into its formation are the inner side of the circumference of the head of the radius, which rotates within the lesser sigmoid cavity of the ulna. These surfaces are covered with cartilage, and invested with a duplicature of synovial membrane, continuous with that which lines the elbow-joint. Its only ligament is

#### The Annular or Orbicular.

The *Orbicular Ligament* (fig. 133) is a strong flat band of ligamentous fibres, which surrounds the head of the radius, and retains it in firm connection with the lesser sigmoid cavity of the ulna. It forms about three-fourths of a fibrous ring attached by each end to the extremities of the sigmoid cavity, and is broader at the upper part of its circumference than below, by which means the head of the radius is more securely held in its position. Its *outer surface* is strengthened by the external lateral ligament of the elbow, and affords partial origin to the Supinator brevis muscle. Its *inner surface* is smooth, and lined by synovial membrane.

*Actions.* The movement which takes place in this articulation is limited to rotation of the head of the radius within the orbicular ligament, and upon the lesser sigmoid cavity of the ulna; rotation forwards being called *pronation;* rotation backward, *supination.*

## 2. Middle Radio-Ulnar Articulation.

The interval between the shafts of the radius and ulna is occupied by two ligaments.

Oblique. Interosseous.

The *Oblique* or *Round Ligament* (fig. 132) is a small round fibrous cord, which extends obliquely downwards and outwards, from the tubercle of the ulna at the base of the coronoid process, to the radius a little below the bicipital tuberosity. Its fibres run in the opposite direction to those of the interosseous ligament; and it appears to be placed as a substitute for it in the upper part of the interosseous interval. This ligament is sometimes wanting.

The *Interosseous Membrane* is a broad and thin plane of aponeurotic fibres, descending obliquely downwards and inwards, from the interosseous ridge on the radius to that on the ulna. It is deficient above, commencing about an inch beneath the tubercle of the radius; is broader in the middle than at either extremity; and presents an oval aperture just above its lower margin for the passage of the anterior interosseous vessels to the back of the forearm. This ligament serves to connect the bones, and to increase the extent of surface for the attachment of the deep muscles. Between its upper border and the oblique ligament an interval exists, through which the posterior interosseous vessels pass. Two or three fibrous bands are occasionally found on the posterior surface of this membrane, which descend obliquely from the ulna towards the radius, and which have consequently a direction contrary to that of the other fibres. It is in relation, in *front*, by its upper three-fourths (radial margin) with the Flexor longus pollicis (ulnar margin), with the Flexor profundus digitorum, lying upon the interval between which are the anterior interosseous vessels and nerve, by its lower fourth with the Pronator quadratus; *behind*, with the Supinator brevis, Extensor ossis metacarpi pollicis, Extensor primi internodii pollicis, Extensor secundi internodii pollicis, Extensor indicis; and, near the wrist, with the anterior interosseous artery and posterior interosseous nerve.

## 3. Inferior Radio-Ulnar Articulation.

This is a lateral ginglymoid joint, formed by the head of the ulna being received into the sigmoid cavity at the inner side of the lower end of the radius. The articular surfaces are covered by a thin layer of cartilage, and connected together by the following ligaments.

Anterior radio-ulnar.
Posterior radio-ulnar.
Triangular Inter-articular Fibro-cartilage.

The *Anterior Radio-ulnar Ligament* (fig. 134) is a narrow band of fibres, extending from the anterior margin of the sigmoid cavity of the radius to the anterior surface of the head of the ulna.

The *Posterior Radio-ulnar Ligament* (fig. 135) extends between similar points on the posterior surface of the articulation.

The *Triangular Fibro-cartilage* (fig. 136) is placed transversely beneath the head of

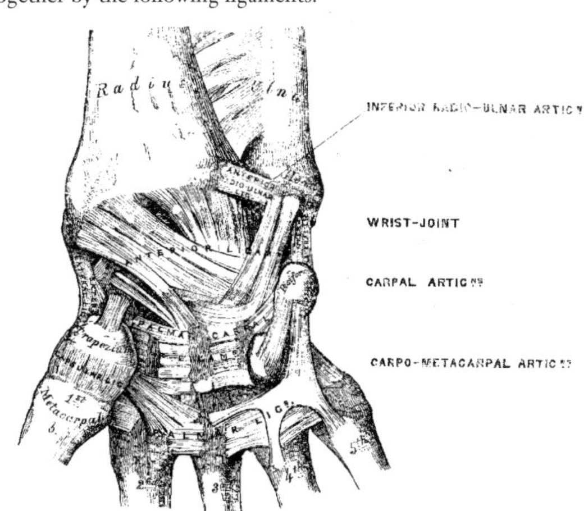


*134 – Ligaments of wrist and hand. Anterior view.*

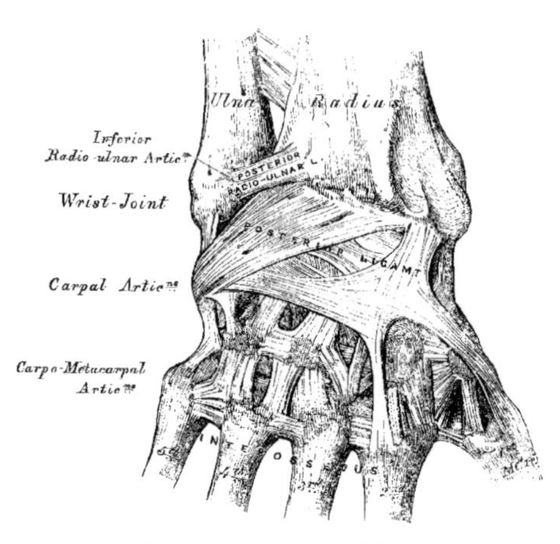


135 – Ligaments of wrist and hand. Posterior view.

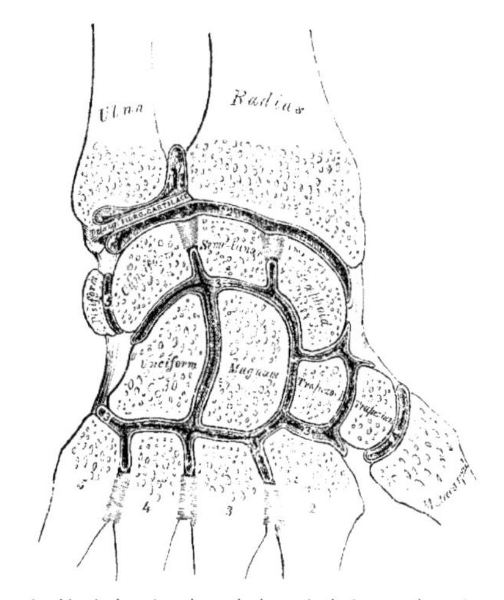


136 – Vertical section through the articulations at the wrist, showing the five synovial membranes.

the ulna, binding the lower end of this bone and the radius firmly together. Its circumference is thicker than its centre, which is thin and occasionally perforated. It is attached by its apex to a depression which separates the styloid process of the ulna from the head of that bone; and, by its base, which is thin, to the prominent edge of the radius, which separates the sigmoid cavity from the carpal articulating surface. Its margins are united to the ligaments of the wrist joint. Its *upper surface*, smooth and concave, is contiguous with the head of the ulna; its *under surface*, also concave and smooth, with the cuneiform bone. Both surfaces are lined by a synovial membrane: the upper surface, by one peculiar to the radio-ulnar articulation; the under surface, by the synovial membrane of the wrist.

The *Synovial Membrane* of this articulation has been called, from its extreme looseness, the *membrana sacciformis;* it covers the margin of the articular surface of the head of the ulna, and where reflected from this bone on to the radius, forms a very loose *cul-de-sac;* from the radius, it is continued over the upper surface of the fibro-cartilage. The quantity of synovia which it contains is usually considerable. When the fibro-cartilage is perforated, the synovial membrane is continuous with that which lines the wrist.

*Actions.* The movement which occurs in the inferior radio-ulnar articulation is just the reverse of that which takes place between the two bones above; it is limited to rotation of the radius round the head of the ulna; rotation forwards being termed *pronation*, rotation backwards *supination*. In pronation, the sigmoid cavity glides forward on the articular edge of the ulna; in supination, it rolls in the opposite direction, the extent of these movements being limited by the anterior and posterior ligaments.

## 7. Wrist Joint.

The *Wrist* presents most of the characters of an enarthrodial joint. The parts entering into its formation are, the lower end of the radius, and under surface of the triangular inter-articular fibro-cartilage, above; and the scaphoid, semilunar, and cuneiform bones below. The articular surfaces of the radius and inter-articular fibro-cartilage form a transversely elliptical concave surface. The radius is subdivided into two parts by a line extending from before backwards; and these, together with the inter-articular cartilage, form three facets, one for each carpal bone. The three carpal bones are connected together, and form a convex surface, which is received into the concavity above mentioned. All the bony surfaces of the articulation are covered with cartilage, and connected together by the following ligaments.

| | |
|---|---|
| External Lateral. | Anterior. |
| Internal Lateral. | Posterior. |

The *External Lateral Ligament* extends from the summit of the styloid process of the radius to the outer side of the scaphoid, some of its fibres being prolonged to the trapezium and annular ligament.

The *Internal Lateral Ligament* is a rounded cord, attached, above, to the extremity of the styloid process of the ulna; below, it divides into two fasciculi, which are attached, one to the inner side of the cuneiform bone, the other to the pisiform bone and annular ligament.

The *Anterior Ligament* is a broad membranous band, consisting of three fasciculi, attached, above, to the anterior margin of the lower end of the radius, its styloid process, and the ulna; its fibres pass downwards and inwards, to be inserted into the palmar surface of the scaphoid, semilunar, and cuneiform bones. This ligament is perforated by numerous apertures for the passage of vessels, and is in relation, in front, with the tendons of the Flexor profundus digitorum and Flexor longus pollicis; behind, with the synovial membrane of the wrist-joint.

The *Posterior Ligament*, less thick and strong than the anterior, is attached, above, to the posterior border of the lower end of the radius; its fibres descend obliquely downwards and inwards to be attached to the dorsal surface of the scaphoid, semilunar, and cuneiform bones, its fibres being continuous with those of the dorsal carpal ligaments. This ligament is in relation, behind, with the extensor tendons of the fingers; in front, with the synovial membrane of the wrist.

The *Synovial Membrane* lines the under surface of the triangular inter-articular fibro-cartilage above; and is reflected on the inner surface of the ligaments above mentioned.

*Relations.* The wrist-joint is covered in front by the flexor, and behind by the extensor tendons; it is also in relation with the radial and ulnar arteries.

The Arteries supplying the joint are the anterior and posterior carpal branches of the radial and ulnar, the anterior and posterior interosseous, and some ascending branches from the deep palmar arch.

The Nerves are derived from the ulnar.

*Actions.* The movements permitted in this joint are flexion, extension, abduction, adduction, and circumduction. It is totally incapable of rotation, one of the characteristic movements in true enarthrodial joints.

## 8. Articulations of the Carpus.

These articulations may be subdivided into three sets.

1. The articulation of the first row of carpal bones.
2. The articulation of the second row of carpal bones.
3. The articulation of the two rows with each other.

### 1. Articulation of the First Row of Carpal Bones.

These are arthrodial joints. The articular surfaces are covered with cartilage, and connected together by the following ligaments.

Two Dorsal. Two Palmar.
Two Interosseous.

The *Dorsal Ligaments*, two in number, are placed transversely behind the bones of the first row; they connect the scaphoid and semilunar, and the semilunar and cuneiform.

The *Palmar Ligaments*, also two in number, connect the scaphoid and semilunar, and the semilunar and cuneiform bones; they are less strong than the dorsal, and placed very deep under the anterior ligament of the wrist.

The *Interosseous Ligaments* (fig. 136) are two narrow bundles of fibrous tissue, connecting the semilunar bone, on one side with the scaphoid, on the other with the cuneiform. They close the upper part of the interspaces between the scaphoid, semilunar, and cuneiform bones, their upper surfaces being smooth, and lined by the synovial membrane of the wrist-joint.

The articulation of the pisiform with the cuneiform is provided with a separate synovial membrane, protected by a thin capsular ligament. There are also two strong fibrous fasciculi, which connect this bone to the unciform, and base of the fifth metacarpal bone.

## 2. Articulation of the Second Row of Carpal Bones.

These are arthrodial joints, the articular surfaces are covered with cartilage, and connected by the following ligaments.

Three Dorsal. Three Palmar.
Two Interosseous.

The *three Dorsal Ligaments* extend transversely from one bone to another on the dorsal surface, connecting the trapezium with the trapezoid, the trapezoid with the os magnum, and the os magnum with the unciform.

The *three Palmar Ligaments* have a similar arrangement on the palmar surface.

The *two Interosseous Ligaments*, much thicker than those of the first row, are placed one on each side of the os magnum, connecting it with the trapezoid externally, and the unciform internally. The former is less distinct than the latter.

## 3. Articulation of the Two Rows of Carpal Bones with each other.

The articulation between the two rows of the carpus consists of an enarthrodial joint in the middle, formed by the reception of the os magnum into a cavity formed by the scaphoid and semilunar bones, and of an arthrodial joint on each side, the outer one formed by the articulation of the scaphoid with the trapezium and trapezoid, the internal one by the articulation of the cuneiform and unciform. The articular surfaces are covered by a thin layer of cartilage, and connected by the following ligaments.

Anterior or Palmar. External Lateral.
Posterior or Dorsal. Internal Lateral.

The *Anterior* or *Palmar Ligaments* consist of short fibres, which pass obliquely between the bones of the first and second row on the palmar surface.

The *Posterior* or *Dorsal Ligaments* have a similar arrangement on the dorsal surface of the carpus.

The *Lateral Ligaments* are very short; they are placed, one on the radial, the other on the ulnar side of the carpus; the former, the stronger and more distinct, connecting the scaphoid and trapezium bones, the latter the cuneiform and unciform: they are continuous with the lateral ligaments of the wrist-joint.

There are *two Synovial Membranes* found in the articulation of the carpal bones with each other. The first of these, the more extensive, lines the under surface of the scaphoid, semilunar, and cuneiform bones, sending upwards two prolongations between their contiguous surfaces; it is then reflected over the bones of the second row, and sends down three prolongations between them, which line their contiguous surfaces, and invest the carpal extremities of the four outer metacarpal bones. The second is the synovial membrane between the pisiform and cuneiform bones.

*Actions.* The partial movement which takes place between the bones of each row is very inconsiderable; the movement between the two rows is more marked but limited chiefly to flexion and extension.

## 9. Carpo-Metacarpal Articulations.

### Articulation of the Metacarpal Bone of the Thumb with the Trapezium.

This is an enarthrodial joint. Its ligaments are a capsular and synovial membrane. The *capsular ligament* is a thick but loose capsule, which passes from the circumference of the upper extremity of the metacarpal bone, to the rough edge bounding the articular surface of the trapezium; it is thickest externally and behind, and lined by a separate *synovial membrane*.

## Articulation of the Metacarpal Bones of the Fingers with the Carpus.

The joints formed between the carpus and four inner metacarpal bones, are connected together by dorsal, palmar, and interosseous ligaments.

The *Dorsal Ligaments*, the strongest and most distinct, connect the carpal and metacarpal bones on their dorsal surface. The second metacarpal bone receives two fasciculi, one from the trapezium, the other from the trapezoid; the third metacarpal receives one from the os magnum; the fourth two, one from the os magnum, and one from the unciform; the fifth receives a single fasciculus from the unciform bone.

The *Palmar Ligaments* have a somewhat similar arrangement on the palmar surface, with the exception of the third metacarpal, which has three ligaments, an external one from the trapezium, situated above the sheath of the tendon of the Flexor carpi radialis; a middle one, from the os magnum; and an internal one, from the unciform.

The *Interosseous Ligaments* consist of short thick fibres, which are limited to one part of the carpo-metacarpal articulation; they connect the contiguous inferior angles of the os magnum and unciform, with the adjacent surfaces of the third and fourth metacarpal bones.

The *Synovial Membrane* is a continuation of that between the two rows of carpal bones. Occasionally, the articulation of the unciform with the fourth and fifth metacarpal bones, has a separate synovial membrane.

The Synovial Membranes of the wrist (fig. 136) are thus seen to be five in number. The *first*, the membrana sacciformis, lining the lower end of the ulna, the sigmoid cavity of the radius, and upper surface of the triangular inter-articular fibro-cartilage. The *second* lines the lower end of the radius and inter-articular fibro-cartilage above, and the scaphoid, semilunar, and cuneiform bones below. The *third*, the most extensive, covers the contiguous surfaces of the two rows of carpal bones, and, passing between the bones of the second range, lines the carpal extremities of the four inner metacarpal bones. The *fourth* lines the adjacent surfaces of the trapezium and metacarpal bone of the thumb. And the *fifth* the adjacent surfaces of the cuneiform and pisiform bones.

*Actions.* The movement permitted in the carpo-metacarpal articulations is limited to a slight gliding of the articular surfaces upon each other, the extent of which varies in the different joints. Thus the articulation of the metacarpal bone of the thumb with the trapezium is most moveable, then the fifth metacarpal, and then the fourth. The second and third are almost immoveable. In the articulation of the metacarpal bone of the thumb with the trapezium, the movements permitted are flexion, extension, adduction, abduction, and circumduction.

## Articulations of the Metacarpal Bones with each other.

The carpal extremities of the metacarpal bones articulate with one another at each side by small surfaces covered with cartilage, and connected together by dorsal, palmar, and interosseous ligaments.

The *Dorsal* and *Palmar Ligaments* pass transversely from one bone to another on the dorsal and palmar surfaces. The *Interosseous Ligaments* pass between their contiguous surfaces, just beneath their lateral articular facets.

The *Synovial Membrane* lining the lateral facets, is a reflection from that between the two rows of carpal bones.

The digital extremities of the metacarpal bones are connected together by a narrow fibrous band, the transverse ligament, which passes transversely across their under-surfaces, and is blended with the ligaments of the metacarpo-phalangeal articulations. Its *anterior surface* presents four grooves for the passage of the flexor tendons. Its *posterior surface* blends with the ligaments of the metacarpo-phalangeal articulation.

## 10. Metacarpo-phalangeal Articulations (Fig. 137).

These articulations are of the ginglymoid kind, formed by the reception of the rounded head of the metacarpal bone, into a superficial cavity in the extremity of the first phalanx. They are connected by the following ligaments:

Anterior. Two Lateral.

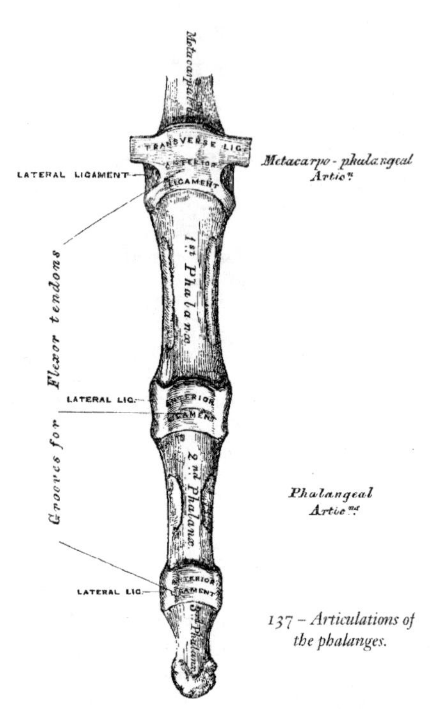


137 – *Articulations of the phalanges.*

The *Anterior Ligaments* are thick, dense, and fibro-cartilaginous in texture. Each is placed on the palmar surface of the joint, in the interval between the lateral ligaments, to which they are connected; they are loosely united on the metacarpal bone, but very firmly to the base of the first phalanges. Their palmar surface is intimately blended with the transverse ligament, each ligament forming with it a groove for the passage of the flexor tendons, the sheath surrounding which is connected to it at each side. By their internal surface, they form part of the articular surface for the head of the metacarpal bone, and are lined by a synovial membrane.

The *Lateral Ligaments* are strong rounded cords, placed one on each side of the joint, each being attached by one extremity to the tubercle on the side of the head of the metacarpal bone, and by the other, to the contiguous extremity of the phalanx.

The Posterior Ligament is supplied by the extensor tendon placed over the back of each joint.

*Actions.* The movements which occur in these joints are flexion, extension, adduction, abduction, and circumduction; the lateral movements are very limited.

### 11. Articulations of the Phalanges.

These are ginglymoid joints, connected by the following ligaments:

Anterior. Two Lateral.

The arrangement of these ligaments is similar to those in the metacarpo-phalangeal articulations; the extensor tendon supplies the place of a posterior ligament.

*Actions.* The only movements permitted in the phalangeal joints are flexion and extension; these movements are more extensive between the first and second phalanges than between the second and third. The movement of flexion is very extensive, but extension is limited by the anterior and lateral ligaments.

## ARTICULATIONS OF THE LOWER EXTREMITY.

The articulations of the Lower Extremity comprise the following groups. 1. The hip joint. 2. The knee joint. 3. The articulations between the tibia and fibula. 4. The ankle joint. 5. The articulations of the tarsus. 6. The tarso-metatarsal articulations. 7. The metatarso-phalangeal articulations. 8. The articulation of the phalanges.

### 1. Hip Joint (Fig. 138).

This articulation is an enarthrodial, or ball and socket joint, formed by the reception of the head of the femur into the cup-shaped cavity of the acetabulum. The articulating surfaces are covered with cartilage, that on the head of the femur being thicker at the centre than at the circumference, and covering the entire surface with the exception of a depression just below its centre for the ligamentum teres; that covering the acetabulum is much thinner at the centre than at the circumference,

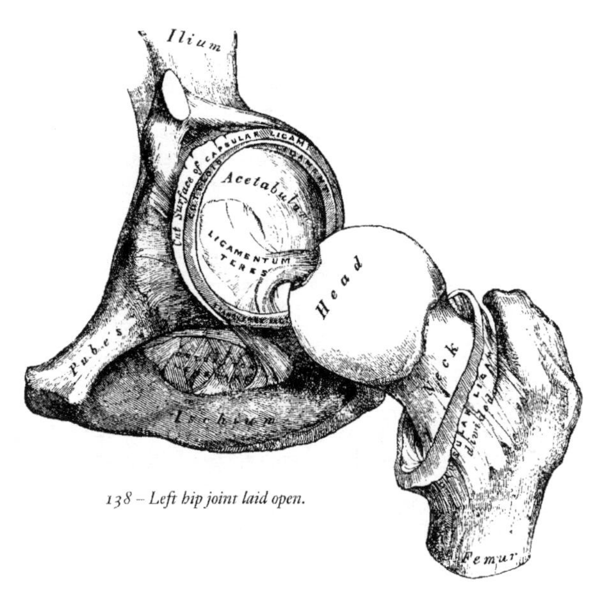


138 – *Left hip joint laid open.*

being deficient in the situation of the circular depression at the bottom of this cavity. The ligaments of the joint are the

| | |
|---|---|
| Capsular. | Teres. |
| Ilio-femoral. | Cotyloid. |
| Transverse. | |

The *Capsular Ligament* is a strong, dense, ligamentous capsule, embracing the margin of the acetabulum above, and surrounding the neck of the femur below. Its *upper circumference* is attached to the acetabulum two or three lines external to the cotyloid ligament; but opposite the notch where the margin of this cavity is deficient, it is connected with the transverse ligament, and by a few fibres to the edge of the obturator foramen. Its *lower circumference* surrounds the neck of the femur, being attached, in front, to the spiral or anterior inter-trochanteric line; above, to the base of the neck; behind, to the middle of the neck of the bone, about half an inch from the posterior inter-trochanteric line. It is much thicker at the upper and fore part of the joint where the greatest amount of resistance is required, than below, where it is thin, loose, and longer than in any other situation. Its external surface is rough, covered by numerous muscles, and separated in front from the Psoas and Iliacus by a synovial bursa, which not unfrequently communicates by a circular aperture with the cavity of the joint. It differs from the capsular ligament of the shoulder, in being much less loose and lax, and in not being perforated for the passage of a tendon.

The *Ilio-femoral Ligament* (fig. 127) is an accessory band of fibres, extending obliquely across the front of the joint: it is intimately connected with the capsular ligament, and serves to strengthen it in this situation. It is attached, above, to the anterior inferior spine of the ilium; below, to the anterior inter-trochanteric line.

The *Ligamentum Teres* is a triangular band of fibres, implanted, by its apex, into the depression a little behind and below the centre of the head of the femur: and by its broad base, which consists of two bundles of fibres, into the margins of the notch at the bottom of the acetabulum, becoming blended with the transverse ligament. It is formed of a bundle of fibres, the thickness and strength of which is very variable, surrounded by a tubular sheath of synovial membrane. Sometimes, the synovial fold only exists, or the ligament may be altogether absent. The use of the round ligament is to check rotation outwards, when combined with flexion; it thus assists in preventing dislocation of the head of the femur forwards and outwards, an accident likely to occur from the necessary mechanism of the joint, if not provided against by this ligament and the thick anterior part of the capsule.*

* See an interesting paper, 'On the Use of the Round Ligament of the Hip-joint,' by Dr. J. Struthers. *Edinburgh Medical Journal*, 1858.

The *Cotyloid Ligament* is a fibro-cartilaginous rim attached to the margin of the acetabulum, the cavity of which it deepens; at the same time it protects the edges of the bone, and fills up the inequalities on its surface. It is prismoid in form, its base being attached to the margin of the acetabulum, its opposite edge being free and sharp; whilst its two surfaces are invested by synovial membrane, the external one being in contact with the capsular ligament, the internal one being inclined inwards so as to narrow the acetabulum and embrace the cartilaginous surface of the head of the femur. It is much thicker above and behind than below and in front, and consists of close compact fibres, which arise from different points of the circumference of the acetabulum, and interlace with each other at very acute angles.

The *Transverse Ligament* is a strong flattened band of fibres, which crosses the notch at the lower part of the acetabulum, and converts it into a foramen. It is continuous at each side with the cotyloid ligament. An interval is left beneath the ligament for the passage of nutrient vessels to the joint.

The *Synovial Membrane* is very extensive. Commencing at the margin of the cartilaginous surface of the head of the femur, it covers all that portion of the neck which is contained within the joint; from this point it is reflected on the internal surface of the capsular ligament, covers both surfaces of the cotyloid ligament, and the mass of fat contained in the fossa at the bottom of this cavity, and is prolonged in the form of a tubular sheath around the ligamentum teres, as far as the head of the femur.

The Muscles in relation with the joint are, in front, the Psoas and Iliacas, separated from the capsular ligament by a synovial bursa; above, the short head of the Rectus and Gluteus minimus, the latter being closely adherent to it; internally, the Obturator externus and Pectineus; behind, the Pyriformis, Gemellus superior, Obturator internus, Gemellus inferior, Obturator externus, and Quadratus femoris.

The Arteries supplying it are derived from the obturator, sciatic, internal circumflex, and gluteal.

The Nerves are articular branches from the sacral plexus, great sciatic, obturator, and accessory obturator nerves.

*Actions.* The movements of the hip, like all enarthrodial joints, are very extensive; they are, flexion, extension, adduction, abduction, circumduction, and rotation.

## 2. Knee-Joint.

The knee is a ginglymoid, or hinge-joint; the bones entering into its formation are, the condyles of the femur above, the head of the tibia below, and the patella in front. The articular surfaces are covered with cartilage, and connected together by ligaments, some of which are placed on the exterior of the joint, whilst others occupy its interior.

| *External Ligaments.* | *Internal Ligaments.* |
|---|---|
| Anterior, or Ligamentum Patellæ. | Anterior, or External Crucial. |
| Posterior, or Ligamentum Posticum Winslowii. | Posterior, or Internal Crucial. |
| Internal Lateral. | Two Semilunar Fibro-cartilages. |
| Two External Lateral. | Transverse. |
| Capsular. | Coronary. |
| | Ligamentum mucosum. |
| | Ligamenta alaria. |

The *Anterior Ligament*, or *Ligamentum Patellæ* (fig. 139), is that portion of the common tendon of the extensor muscles of the thigh which is continued from the patella to the tubercle of the tibia, supplying the place of an anterior ligament. It is a strong, flat, ligamentous band, about three inches in length, attached, above, to the apex of the patella and the rough depression on its posterior surface; below, to the lower part of the tuberosity of the tibia; its superficial fibres being continuous across the front of the patella with those of the tendon of the Quadriceps extensor. Two synovial bursæ are connected with this ligament and the patella; one is interposed between the patella and the skin covering its anterior surface; the other, of small size, between the ligamentum patellæ and the upper part of the tuberosity of the tibia. The posterior surface of this ligament is separated above from the knee-joint by a large mass of adipose tissue; its lateral margins are continuous with the aponeuroses derived from the Vasti muscles.

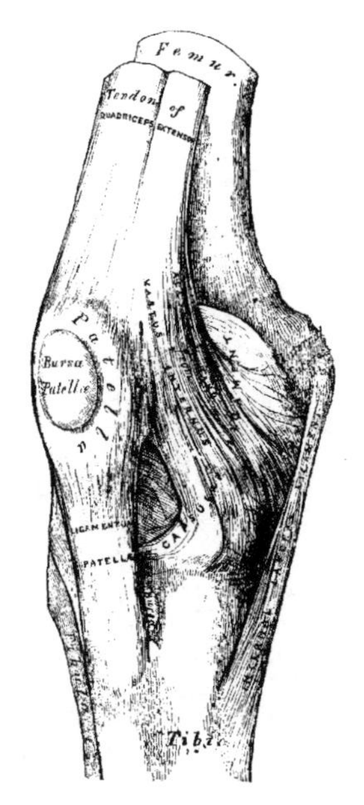


139 – *Right knee-joint. Anterior view.*

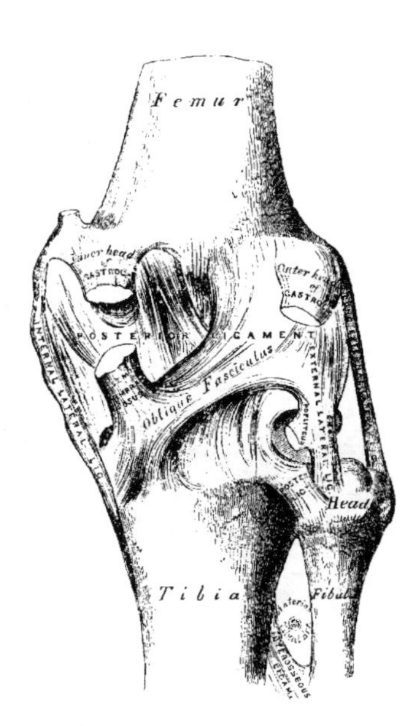


140 – *Right knee-joint. Posterior view.*

The *Posterior Ligament, Ligamentum Posticum Winslowii* (fig. 140), is a broad, flat, fibrous band, which covers over the whole of the back part of the joint. It consists of two lateral portions, formed chiefly of vertical fibres, which arise above from the condyles of the femur, and are connected below with the back part of the head of the tibia, being closely united with the tendons of the Gastrocnemius, Plantaris, and Popliteus muscles; the central portion is formed of fasciculi obliquely directed and separated from one another by apertures for the passage of vessels. The strongest of these fasciculi is derived from the tendon of the Semi-membranosus; it passes from the back part of the inner tuberosity of the tibia, obliquely upwards and outwards to the back part of the outer condyle of the femur. The posterior ligament forms part of the floor of the popliteal space, and upon it rests the popliteal artery.

The *Internal Lateral Ligament* is a broad, flat, membranous band, thicker behind than in front, and situated nearer to the back than the front of the joint. It is attached, above, to the inner tuberosity of the femur; below, to the inner tuberosity and inner surface of the shaft of the tibia, to the extent of about two inches. It is crossed, at its lower part, by the aponeurosis of the Sartorius, and the tendon of the Gracilis and Semi-tendinosus muscles, a synovial bursa being interposed. Its *deep surface* covers the anterior portion of the tendon of the Semi-membranosus, the synovial membrane of the joint, and the inferior internal articular artery; it is intimately adherent to the internal semilunar fibro-cartilage.

The *Long External Lateral Ligament* is a strong, rounded, fibrous cord, situated nearer to the back than the front of the joint. It is attached, above, to the outer condyle of the femur; below, to the outer part of the head of the fibula. Its *outer surface* is covered by the tendon of the Biceps, which divides into two parts, separated by the ligament, at its insertion. It has, passing beneath it the tendon of the Popliteus muscle, and the inferior external articular vessels and nerve.

The *Short External Lateral Ligament* is an accessory bundle of fibres, placed behind and parallel with the preceding; attached, above, to the lower part of the outer condyle of the femur; below, to the summit of the styloid process of the fibula. This ligament is intimately connected with the capsular ligament, and has, passing beneath it, the tendon of the Popliteus muscle.

The *Capsular Ligament* consists of an exceedingly thin, but strong, fibrous membrane, which fills in the intervals left by the preceding ligaments. It is attached to the femur immediately above its articular surface; below, to the upper border and sides of the patella, the margins of the head of the tibia and inter-articular cartilages, and is continuous behind with the posterior ligament. This membrane is strengthened by fibrous expansions, derived from the fascia lata, from the Vasti and Cruræus muscles, and from the Biceps, Sartorius, and tendon of the Semi-membranosus.

The *Crucial* are two interosseous ligaments of considerable strength, situated in the interior of the joint, nearer its posterior than its anterior part. They are called *crucial*, because they cross each other, somewhat like the lines of the letter X; and have received the names *anterior* and *posterior*, from the position of their attachment to the tibia.

The *Anterior* or *External Crucial Ligament* (fig. 141), smaller than the posterior, is attached to the inner side of the depression in front of the spine of the tibia, being blended with the anterior

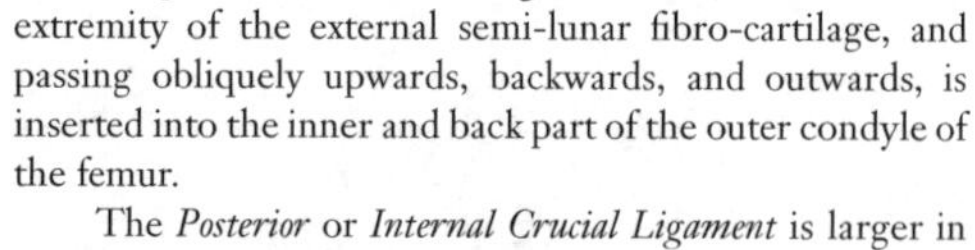

extremity of the external semi-lunar fibro-cartilage, and passing obliquely upwards, backwards, and outwards, is inserted into the inner and back part of the outer condyle of the femur.

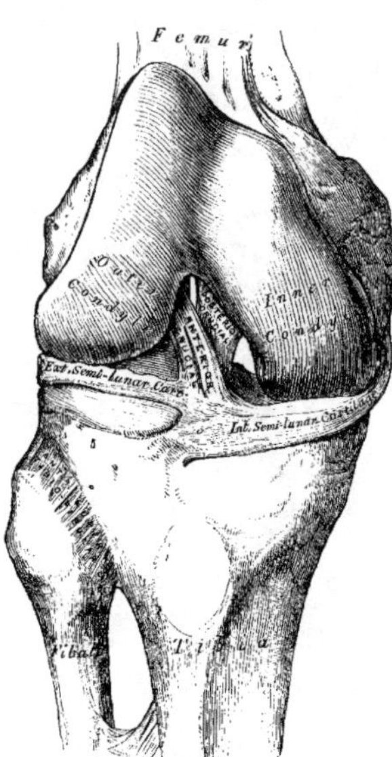


141 – *Right knee-joint. Shewing internal ligaments.*

The *Posterior* or *Internal Crucial Ligament* is larger in size, but less oblique in its direction than the anterior. It is attached to the back part of the depression behind the spine of the tibia, and to the posterior extremity of the external semi-lunar fibro-cartilage; passing upwards, forwards, and inwards, it is inserted into the outer and fore part of the inner condyle of the femur. As it crosses the anterior crucial ligament, a fasciculus is given off from it, which blends with its posterior part. It is in relation, in front, with the anterior crucial ligament; behind, with the ligamentum posticum Winslowii.

The *Semi-lunar Fibro-Cartilages* (fig. 142) are two crescentic lamellæ attached to the margins of the head of the tibia, serving to deepen its surface for articulation with the condyles of the femur. The circumference of each cartilage is thick and convex; the inner free border, thin and concave. Their upper surfaces are concave, and in relation with the condyles of the femur; their lower surfaces are flat, and rest upon the head of the tibia. Each cartilage covers nearly the outer two-thirds of the corresponding articular surface of the tibia, the inner third being uncovered; both surfaces are smooth, and invested by synovial membrane.

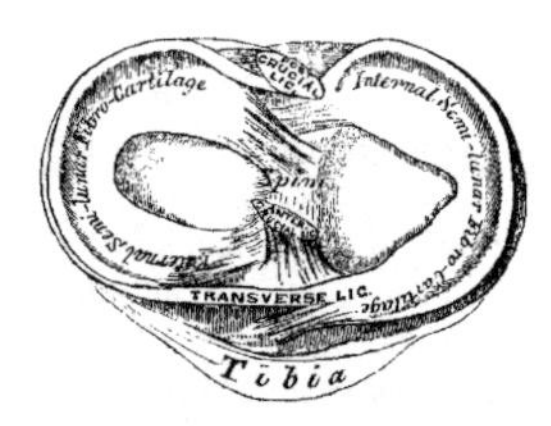


142 – *Head of tibia, with semi-lunar cartilages, etc. Seen from above. Right side.*

The *Internal Semi-lunar Fibro-Cartilage* is nearly semicircular in form, a little elongated from before backwards, and broader behind than in front; its convex border is united to the internal lateral ligament, and to the head of the tibia, by means of the coronary ligaments; its anterior extremity, thin and pointed, is firmly implanted into a depression in front of the inner articular surface of the tibia; its posterior extremity to the depression behind the spine, between the attachment of the external cartilage and posterior crucial ligament.

The *External Semi-lunar Fibro-Cartilage* forms nearly an entire circle, covering a larger portion of the articular surface than the internal one. It is grooved on its outer side, for the tendon of the Popliteus muscle. Its circumference is held in connexion with the head of the tibia, by means of the coronary ligaments; and its two extremities are firmly implanted in the depressions in front and behind the spine of the tibia. These extremities, at their insertion, are interposed between the attachments of the internal cartilage. The external semi-lunar fibro-cartilage gives off from its *anterior border* a fasciculus, which forms the transverse ligament. By its *anterior extremity*, it is continuous with the anterior crucial ligament. Its *posterior extremity* divides into three slips; two of these pass upwards and forwards, and are inserted into the outer side of the inner condyle, one in front, the other behind the posterior crucial ligament; the third fasciculus is inserted into the back part of the anterior crucial ligament.

The *Transverse Ligament* is a band of fibres, which passes transversely between the anterior convex margin of the external semi-lunar cartilage, to the anterior extremity of the internal cartilage; its thickness varies considerably in different subjects.

The *Coronary Ligaments* consist of numerous short fibrous bands, which connect the convex border of the semi-lunar cartilages with the circumference of the head of the tibia, and with the other ligaments surrounding the joint.

The *Synovial Membrane* of the knee-joint is the largest and most extensive in the body. Commencing at the upper border of the patella, it forms a large *cul-de-sac* beneath the Extensor tendon of the thigh: this is sometimes replaced by a synovial bursa interposed between the tendon and the front of the femur, which in some subjects communicates with the synovial membrane of the knee-joint, by an orifice of variable size. On each side of the patella, the synovial membrane extends beneath the aponeurosis of the Vasti muscles, and more especially beneath that of the Vastus internus; and, beneath the patella, it is separated from the anterior ligament by a considerable quantity of adipose tissue. In this situation, it sends off a triangular-shaped prolongation, containing a few ligamentous fibres which extends from the anterior part of the joint below the patella, to the front of the inter-condyloid notch. This fold has been termed the *ligamentum mucosum*. The *ligamenta alaria* consist of two fringe-like folds, which extend from the sides of the ligamentum mucosum, upwards and outwards, to the sides of the patella. The synovial membrane invests the semi-lunar fibro-cartilages, and on the back part of the external one forms a *cul-de-sac* between the groove on its surface and the tendon of the Popliteus; it is continued to the articular surface of the tibia; surrounds the crucial ligaments, and the inner surface of the ligaments which enclose the joint; lastly, it approaches the condyles of the femur, and from them is continued on to the lower part of the front of the shaft. The pouch of synovial membrane between the Extensor tendon and front of the femur is supported during the movements of the knee, by a small muscle, the Sub-cruræus, which is inserted into it.

The Arteries supplying the joint are derived from the anastomotic branch of the femoral, articular branches of the popliteal, and recurrent branch of the anterior tibial.

The Nerves are derived from the obturator, anterior crural, and external and internal popliteal.

*Actions.* The chief movements of this joint are flexion and extension; but it is also capable of performing some slight rotatory movement. During flexion, the articular surfaces of the tibia, covered by their inter-articular cartilages, glide backwards upon the condyles of the femur, the lateral, posterior, and crucial ligaments are relaxed, the ligamentum patellæ is put upon the stretch, the patella filling up the vacuity in front of the joint between the femur and tibia. In extension, the tibia and inter-articular cartilages glide forwards upon the femur; all the ligaments are stretched, with the exception of the ligamentum patellæ, which is relaxed, and admits of considerable lateral movement. The movement of rotation is permitted when the knee is semi-flexed, rotation outwards being most extensive.

## 3. Articulation between the Tibia and Fibula.

The articulations between the tibia and fibula are effected by ligaments which connect both extremities, as well as the shaft of these bones. They may, consequently, be subdivided into three sets. 1. The Superior Tibio-Fibular articulation. 2. The Middle Tibio-Fibular articulation. 3. The Inferior Tibio-Fibular articulation.

### 1. Superior Tibio-Fibular Articulation.

This articulation is an arthrodial joint. The contiguous surfaces of the bones present two flat oval surfaces covered with cartilage, and connected together by the following ligaments.

Anterior Superior Tibio-Fibular.
Posterior Superior Tibio-Fibular.

The *Anterior Superior Ligament* (fig. 141) consists of two or three broad and flat bands, which pass obliquely upwards and inwards, from the head of the fibula to the outer tuberosity of the tibia.

The *Posterior Superior Ligament* is a single thick and broad band, which passes from the back part of the head of the fibula to the back part of the outer tuberosity of the tibia. It is covered in by the tendon of the Popliteus muscle.

A *Synovial Membrane* lines this articulation. It is occasionally continuous with that of the knee joint at its upper and back part.

## 2. Middle Tibio-Fibular Articulation.

An interosseous membrane extends between the contiguous margins of the tibia and fibula, and separates the muscles on the front from those on the back of the leg. It consists of a thin aponeurotic lamina composed of oblique fibres, which pass between the interosseous ridges on the two bones. It is broader above than below, and presents at its upper part a large oval aperture for the passage of the anterior tibial artery forwards to the anterior aspect of the leg; and at its lower part, an opening for the passage of the anterior peroneal vessels. It is continuous below with the inferior interosseous ligament; and is perforated in numerous parts for the passage of small vessels. It is in relation in front with the Tibialis anticus, Extensor longus digitorum, Extensor proprius pollicis, Peroneus tertius, and the anterior tibial vessels and nerve; behind, with the Tibialis posticus and Flexor longus pollicis.

## 3. Inferior Tibio-Fibular Articulation.

This articulation is formed by the rough convex surface at the inner side of the lower end of the fibula, being connected with a similar rough surface on the outer side of the tibia. Below, to the extent of about two lines, these surfaces are smooth and covered with cartilage, which is continuous with that of the ankle-joint. Its ligaments are—

Inferior Interosseous.
Anterior Inferior Tibio-fibular.
Posterior Inferior Tibio-fibular.
Transverse.

The *Inferior Interosseous Ligament* consists of numerous short, strong fibrous bands, which pass between the contiguous rough surfaces of the tibia and fibula, constituting the chief bond of union between these bones. It is continuous, above, with the interosseous membrane.

The *Anterior Inferior Ligament* (fig. 144) is a flat triangular band of fibres, broader below than above, which extends obliquely downwards and outwards between the adjacent margins of the tibia and fibula on the front aspect of the articulation. It is in relation, in front, with the Peroneus tertius, the aponeurosis of the leg, and the integument; behind, with the inferior interosseous ligament, and lies in contact with the cartilage covering the astragalus.

The *Posterior Inferior Ligament*, smaller than the preceding, is disposed in a similar manner on the posterior surface of the articulation.

The *Transverse Ligament* is a long narrow band of ligamentous fibres, continuous with the preceding, passing transversely across the back of the joint, from the external malleolus to the tibia, a short distance from its malleolar process. This ligament projects below the margins of the bones, and forms part of the articulating surface for the astragalus.

The *Synovial Membrane* lining the articular surfaces is derived from that of the ankle-joint.

*Actions.* The movement permitted in these articulations is limited to a very slight gliding of the articular surfaces one upon another.

## 4. Ankle Joint.

The *Ankle* is a ginglymoid or hinge joint. The bones entering into its formation are the lower extremity of the tibia and its malleolus, and the malleolus of the fibula, above, which, united, form an arch, in which is received the upper convex surface of the astragalus and its two lateral facets. These surfaces are covered with cartilage, and connected together by the following ligaments—

Anterior.
Internal Lateral.
External Lateral.

The *Anterior Ligament* (fig. 143) is a broad, thin membranous layer, attached above, to the margin of the articular surface of the tibia; below, to the margin of the astragalus, in front of its articular surface. It is in relation, in front, with the extensor tendons of the toes, the tendons of the Tibialis anticus and Peroneus tertius, and the anterior tibial vessels and nerve; behind, it lies in contact with the synovial membrane.

The *Internal Lateral* or *Deltoid Ligament* consists of two layers, superficial and deep. The *superficial layer* is a strong, flat, triangular band, attached, above, to the apex and anterior and posterior borders of the inner malleolus. The most anterior fibres pass forwards to be inserted into

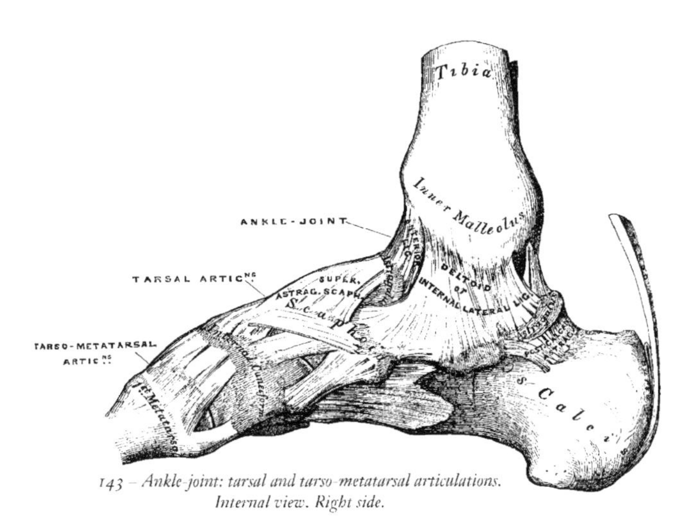

143 – *Ankle-joint: tarsal and tarso-metatarsal articulations. Internal view. Right side.*

the scaphoid; the middle descend almost perpendicularly to be inserted into the os calcis; and the posterior fibres pass backwards and outwards to be attached to the inner side of the astragalus. The *deeper layer* consists of a short, thick, and strong fasciculus, which passes from the apex of the malleolus to the inner surface of the astragalus, below the articular surface. This ligament is covered in by the tendons of the Tibialis posticus and Flexor longus digitorum muscles.

The *External Lateral Ligament* (fig. 144) consists of three fasciculi, taking different directions, and separated by distinct intervals.

The *anterior fasciculus*, the shortest of the three, passes from the anterior margin of the summit of the external malleolus, downwards and forwards, to the astragalus, in front of its external articular facet.

The *posterior fasciculus*, the most deeply seated, passes from the depression at the inner and back part of the external malleolus to the astragalus, behind its external malleolar facet. Its fibres are almost horizontal in direction.

The *middle fasciculus*, the longest of the three, is a narrow rounded cord, passing from the apex of the external malleolus downwards and slightly backwards to the middle of the outer side of the os calcis. It is covered by the tendons of the Peroneus longus and brevis. There is no posterior ligament, its place being supplied by the transverse ligament of the tibia and fibula.

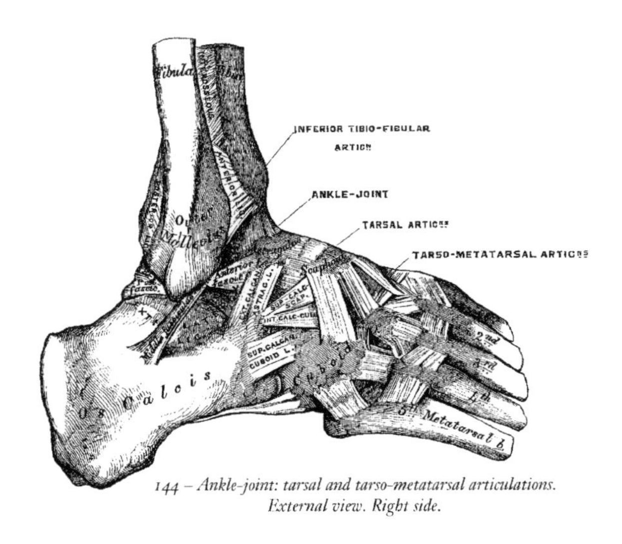

144 – *Ankle-joint: tarsal and tarso-metatarsal articulations. External view. Right side.*

The *Synovial Membrane* invests the inner surface of the ligaments, and sends a duplicature upwards between the lower extremities of the tibia and fibula for a short distance.

*Relations.* The tendons, vessels, and nerves in connection with the joint are, in front, from within outwards, the Tibialis anticus, Extensor proprius pollicis, anterior tibial vessels, anterior tibial nerve, Extensor communis digitorum, and Peroneus tertius; behind, from within outwards, Tibialis posticus, Flexor longus digitorum, posterior tibial vessels, posterior tibial nerve, Flexor longus pollicis, and, in the groove behind the external malleolus, the tendons of the Peroneus longus and brevis.

The *Arteries* supplying the joint are derived from the malleolar branches of the anterior tibial and peroneal.

The *Nerves* are derived from the anterior tibial.

*Actions.* The movements of the joint are limited to flexion and extension. There is no lateral motion.

## 5. Articulations of the Tarsus.

These articulations may be subdivided into three sets: 1. The articulation of the first row of tarsal bones. 2. The articulation of the second row of tarsal bones. 3. The articulations of the two rows with each other.

### 1. Articulation of the First Row of Tarsal Bones.

The articulation between the astragalus and os calcis is an arthrodial joint, connected together by three ligaments:—

External Calcaneo-Astragaloid. Posterior Calcaneo-Astragaloid.
Interosseous.

The *External Calcaneo-Astragaloid Ligament* (fig. 144) is a short, strong fasciculus, passing from the outer surface of the astragalus, immediately beneath its external malleolar facet, to the outer edge of the os calcis. It is placed in front of the middle fasciculus of the external lateral ligament of the ankle-joint, with the fibres of which it is parallel.

The *Posterior Calcaneo-Astragaloid Ligament* (fig. 143) connects the posterior extremity of the astragalus with the upper contiguous surface of the os calcis; it is a short narrow band, the fibres of which are directed obliquely backwards and inwards.

The *Interosseous Ligament* forms the chief bond of union between these bones. It consists of numerous vertical and oblique fibres, attached, by one extremity, to the groove between the articulating surfaces of the astragalus; by the other, to a corresponding depression on the upper surface of the os calcis. It is very thick and strong, being at least an inch in breadth from side to side, and serves to unite the os calcis and astragalus solidly together.

The *Synovial Membranes* (fig. 146) are two in number: one for the posterior calcaneo-astragaloid articulation; a second for the anterior calcaneo-astragaloid joint. The latter synovial membrane is continued forwards between the contiguous surfaces of the astragalus and scaphoid bones.

### 2. Articulations of the Second Row of Tarsal Bones.

The articulations between the scaphoid, cuboid, and three cuneiform are effected by the following ligaments:—

Dorsal. Plantar.
Interosseous.

The *Dorsal Ligaments* are small bands of parallel fibres, which pass from each bone to the neighbouring bones with which it articulates.

The *Plantar Ligaments* have the same arrangement on the plantar surface.

The *Interosseous Ligaments* are four in number. They consist of strong transverse fibres, which pass between the rough non-articular surfaces of adjoining bones. There is one between the sides of the scaphoid and cuboid; a second between the internal and middle cuneiform bones; a third between the middle and external cuneiform; and a fourth between the external cuneiform and cuboid. The scaphoid and cuboid, when in contact, present each a small articulating facet, covered with cartilage, and lined either by a separate synovial membrane, or by an offset from the common tarsal synovial membrane.

### 3. Articulations of the Two Rows of the Tarsus with each other.

These articulations may be conveniently divided into three sets. 1. The articulation of the os calcis with the cuboid. 2. The os calcis with the scaphoid. 3. The astragalus with the scaphoid.

1. The ligaments connecting the os calcis with the cuboid, are four in number:—

Dorsal. { Superior Calcaneo-Cuboid. / Internal Calcaneo-Cuboid (Interosseous).

Plantar. { Long Calcaneo-Cuboid. / Short Calcaneo-Cuboid.

The *Superior Calcaneo-Cuboid Ligament* (fig. 144) is a thin and narrow fasciculus; which passes between the contiguous surfaces of the os calcis and cuboid, on the dorsal surface of the joint.

The *Internal Calcaneo-Cuboid (Interosseous) Ligament* (fig. 144) is a short, but thick and strong, band of fibres, arising from the os calcis, in the deep groove which intervenes between it and the astragalus; being closely blended, at its origin, with the superior calcaneo-scaphoid ligament. It is inserted into the inner side of the cuboid bone. This ligament forms one of the chief bonds of union between the first and second row of the tarsus.

The *Long Calcaneo-Cuboid* (fig. 145), the most superficial of the two plantar ligaments, is the longest of all the ligaments of the tarsus, being attached to the under surface of the os calcis, from near the tuberosities, as far forwards as the anterior tubercle; its fibres pass forwards to be attached to the ridge on the under surface of the cuboid bone, the more superficial fibres being continued onwards to the bases of the second, third, and fourth metatarsal bones. This ligament crosses the groove on the under surface of the cuboid bone, converting it into a canal for the passage of the tendon of the Peroneus longus.

The *Short Calcaneo-Cuboid Ligament* lies nearer to the bones than the preceding, from which it is separated by a little areolar adipose tissue. It is exceedingly broad, about an inch in length, and extends from the tubercle and the depression in front of it on the fore part of the under surface of the os calcis, to the inferior surface of the cuboid bone behind the peroneal groove. A synovial membrane is found in this articulation.

2. The ligaments connecting the os calcis with the scaphoid, are two in number:—

Superior Calcaneo-Scaphoid.
Inferior Calcaneo-Scaphoid.

The *Superior Calcaneo-Scaphoid* (fig. 144) arises, as already mentioned, with the internal calcaneo-cuboid in the deep groove between the astragalus and os calcis; it passes forward from the inner side of the anterior extremity of the os calcis to the outer side of the scaphoid bone. These two ligaments resemble the letter Y, being blended together behind, but separated in front.

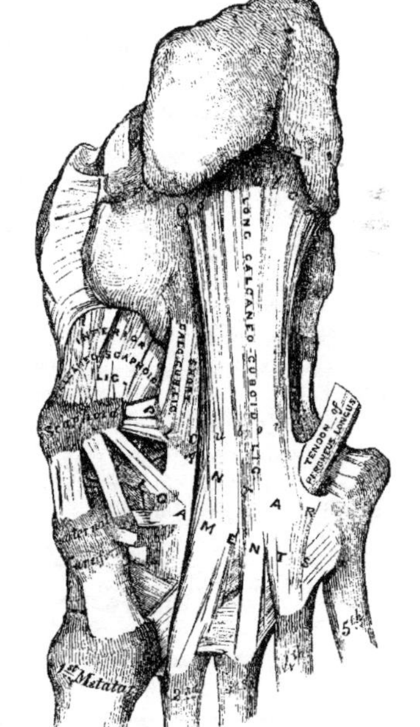

*145 – Ligaments of plantar surface of the foot.*

The *Inferior Calcaneo-Scaphoid* (fig. 145) is by far the largest and strongest of the two ligaments of this articulation; it is a broad and thick band of ligamentous fibres, which passes forwards and inwards from the anterior and inner extremity of the os calcis, to the under surface of the scaphoid bone. This ligament not only serves to connect the os calcis and scaphoid, but supports the head of the astragalus, forming part of the articular cavity in which it is received. Its *upper surface* is lined by the synovial membrane continued from the anterior calcaneo-astragaloid articulation. Its *under surface* is in contact with the tendon of the Tibialis posticus muscle.

3. The articulation between the astragalus and scaphoid is an enarthrodial joint; the rounded head of the astragalus being received into the concavity formed by the posterior surface of the scaphoid, the anterior articulating surface of the calcaneum, and the upper surface of the calcaneo-scaphoid ligament, which fills up the triangular interval between these bones. The only ligament of this joint is the superior astragalo-scaphoid, a broad band of ligamentous fibres, which passes obliquely forwards from the neck of the astragalus, to the superior surface of the scaphoid bone. It is thin and weak in texture, and covered by the Extense tendons. The inferior calcaneo-scaphoid supplies the place of an inferior ligament.

The *Synovial Membrane* which lines the joint, is continued forwards from the anterior calcaneo-astragaloid articulation. This articulation permits of considerable mobility; but its feebleness is such as to occasionally allow of dislocation of the astragalus.

The *Synovial Membranes* (fig. 146) found in the articulations of the tarsus, are four in number: *one* for the posterior calcaneo-astragaloid articulation; a *second* for the anterior calcaneo-astragaloid and astragalo-scaphoid articulations; a *third* for the calcaneo-cuboid articulation; and a *fourth* for the articulations between the scaphoid and the three cuneiform, the three cuneiform with each other, the external cuneiform with the cuboid, and the middle and external cuneiform with the bases of the second and third metatarsal bones. The prolongation which lines the metatarsal bones, passes forwards between the external and middle cuneiform bones. A small synovial membrane is sometimes found between the contiguous surfaces of the scaphoid and cuboid bones.

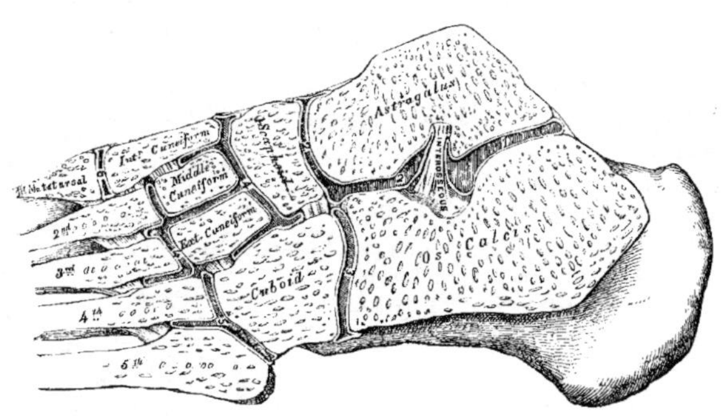


*146 – Oblique section of the articulations of the tarsus and metatarsus. Shewing the six synovial membranes.*

*Actions.* The movements permitted between the bones of the first row, the astragalus and os calcis, are limited to a gliding upon each other from before backwards, and from side to side. The gliding movement which takes place between the bones of the second row, is very slight, the articulation between the scaphoid and cuneiform bones being more moveable than those of the cuneiform with each other and with the cuboid. The movement which takes place between the two rows, is more extensive, and consists in a sort of rotation, by means of which the sole of the foot may be slightly flexed, and extended, or carried inwards and outwards.

## 6. Tarso-Metatarsal Articulations.

These are arthrodial joints. The bones entering into their formation are the internal, middle, external cuneiform, and cuboid, which articulate with the metatarsal bones of the five toes. The metatarsal bone of the great toe articulates with the internal cuneiform; that of the second is deeply wedged in between the internal and external cuneiform, resting against the middle cuneiform, and being the most strongly articulated of all the metatarsal bones; the third metatarsal articulates with the extremity of the external cuneiform; the fourth with the cuboid and external cuneiform; and the fifth with the cuboid. The articular surfaces are covered with cartilage, lined by synovial membrane, and connected together by the following ligaments.

Dorsal. Plantar.
Interosseous.

The *Dorsal Ligaments* consist of strong, flat, fibrous bands, which connect the tarsal with the metatarsal bones. The first metatarsal is connected to the internal cuneiform by a single broad, thin, fibrous band; the second has three dorsal ligaments, one from each cuneiform bone; the third has one from the external cuneiform; and the fourth and fifth have one each from the cuboid.

The *Plantar Ligaments* consist of longitudinal and oblique fibrous bands connecting the tarsal and metatarsal bones, But disposed with less regularity than on the dorsal surface. Those for the first and second metatarsal are the most strongly marked; the second and third receive strong fibrous bands, which pass obliquely across from the internal cuneiform; the plantar ligaments of the fourth and fifth consist of a few scanty fibres derived from the cuboid.

The *Interosseous Ligaments* are three in number: internal, middle, and external: The *internal* one passes from the outer extremity of the internal cuneiform, to the adjacent angle of the second metatarsal. The *middle* one, less strong than the preceding, connects the external cuneiform with the adjacent angle of the second metatarsal. The *external* interosseous ligament connects the outer angle of the external cuneiform with the adjacent side of the third metatarsal.

The *Synovial Membranes* of these articulations are three in number: one for the metatarsal bone of the great toe, with the internal cuneiform: one for the second and third metatarsal bones, with the middle and external cuneiform; this is continuous with the great tarsal synovial membrane: and one for the fourth and fifth metatarsal bones with the cuboid. The synovial membranes of the tarsus and metatarsus are thus seen to be six in number (fig. 146).

## Articulations of the Metatarsal Bones with each other.

The bases of the metatarsal bones, except the first, are connected together by dorsal, plantar, and interosseous ligaments. The *dorsal* and *plantar ligaments* pass from one metatarsal bone to another. The *interosseous ligaments* lie deeply between the rough non-articular portions of their lateral surfaces. The articular surfaces are covered with cartilage, and provided with synovial membrane, continued forwards from the tarso-metatarsal joints. The digital extremities of the metatarsal bones are united by the transverse metatarsal ligament. It connects the great toe with the rest of the metatarsal bones; in this respect it differs from the transverse ligament in the hand.

*Actions.* The movement permitted in the tarsal ends of the metatarsal bones is limited to a slight gliding of the articular surfaces upon one another; considerable motion, however, takes place in their digital extremities.

## Metatarso-Phalangeal Articulations.

The heads of the metatarsal bones are connected with the concave articular surfaces of the first phalanges by the following ligaments—

Anterior or Plantar. Two Lateral.

They are arranged precisely similar to the corresponding parts in the hand. The expansion of the extensor tendon supplies the place of a posterior ligament.

*Actions.* The movements permitted in the metatarso-phalangeal articulations are flexion, extension, abducation, and adduction.

## Articulation of the Phalanges.

The ligaments of these articulations are similar to those found in the hand; each pair of phalanges being connected by an anterior or plantar and two lateral ligaments, and their articular surfaces lined by synovial membrane. Their actions are also similar.

---

For further information on this subject, the Student is referred to Cruvelhier's 'Anatomie Descriptive'; to Mr. Humphry's able work on the 'Human Skeleton, including the Joints'; and to Arnold's 'Tabulæ Anatomicæ,' Fascic. 4 Pars. 2. Icones articulorum et ligamentorum. On the textures composing the Joints refer to Todd and Bowman's 'Physiological Anatomy', and Kölliker's 'Manual of Human Microscopic Anatomy'.

# The Muscles and Fasciæ.*

The Muscles are the active organs of locomotion. They are formed of bundles of reddish fibres, consisting chemically of fibrine, and endowed with the property of contractility. Two kinds of muscular tissue are found in the animal body, viz., that of voluntary or animal life, and that of involuntary or organic life.

The *muscles of animal life* (*striped muscles*) are capable of being either exerted or controlled by the efforts of the will. They are composed of bundles of fibres enclosed in a delicate web of areolar tissue. Each bundle consists of numerous smaller ones, enclosed in a similar fibro-areolar covering, and these again of primitive fasciculi.

The *primitive fasciculi* consist of a number of filaments, inclosed in a tubular sheath of transparent, elastic, and apparently homogenous membrane, named by Bowman the 'Sarcolemma.' The primitive fasciculi are cylindriform or prismatic. Their breadth varies in man from $\frac{1}{200}$ to $\frac{1}{500}$ of an inch, the average of the majority being about $\frac{1}{400}$; their length is not always in proportion to the length of the muscle, but depends on the arrangement of the tendons. This form of muscular fibre is especially characterised by being apparently marked with very fine, dark, parallel lines or *striæ*, which pass transversely round them, in curved or wavy parallel lines, from $\frac{1}{10000}$ to $\frac{1}{12000}$ of an inch apart. Other striæ pass longitudinally over the tubes, indicating the direction of the primitive fibrils of which the primitive fasciculus is composed. They are less distinct than the former.

The *primitive fibrils* constitute the proper contractile tissue of the muscle. Each fibril is cylindriform, somewhat flattened, about $\frac{1}{18000}$ of an inch in thickness, and marked by transverse striæ placed at the same distance from each other as the striæ on the surface of the fasciculus. Each fibril apparently consists of a single row of minute particles (named 'sarcous elements' by Bowman), connected together like a string of beads. Closer examination, however, shews that the elementary particles are little masses of pellucid substance, having a rectangular outline, and appearing dark in the centre. These appearances would favour the suggestion that the elementary particles of which the fibrils are composed, are possibly nucleated cells, cohering in a linear series, the transverse marks between them corresponding to their line of junction. Kölliker, however, considers 'the sarcous elements as artificial products, occasioned by the breaking up of the fibrils at the parts where they are thinner.'

This form of muscular fibre composes the whole of the voluntary muscles, all the muscles of the ear, those of the larynx, pharynx, tongue, and upper half of the œsophagus, the heart, and the walls of the large veins at the point where they open into it.

The *muscles of organic life* (*unstriped muscles*) consist of flattened bands, or of elongated, spindle-shaped fibres, flattened, of a pale colour, from $\frac{1}{4700}$ to $\frac{1}{3100}$ of an inch broad, homogeneous in texture, having a finely mottled aspect, which sometimes appears granular, the granules being occasionally arranged in a linear series, so as to present a striated appearance. Each fibre contains a cylindrical rod-shaped nucleus, which sometimes appears as a narrow, continuous, dark streak. The fibres are united into bundles, which are connected together by areolar tissue and elastic fibres. This form of muscular tissue occurs either scattered in the areolar tissue, or exists in the form of a muscular membrane, the bundles being arranged parallel, or forming a close interlacement, crossing each other at various angles. The muscular fibre of organic life is found in the alimentary canal, forming the muscular coat of the digestive tube from the middle of the œsophagus to the internal sphincter of the anus; in the posterior wall of the trachea, and in the bronchi; in the ducts of the submaxillary glands; in the gall bladder and common bile duct; in the calyces and pelvis of the kidney; in the

* The Muscles and Fasciæ are described conjointly, in order that the student may consider the arrangement of the latter in his dissection of the former. It is rare for the student of anatomy in this country to have the opportunity of dissecting the fasciæ separately; and it is for this reason, as well as from the close connexion that exists between the muscles and their investing aponeuroses, that they are considered together. Some general observations are first made on the anatomy of the muscles and fasciæ, the special description being given in connexion with the different regions.

ureters and bladder; and, scantily, in the urethra. In the female, in the vagina, the uterus, Fallopian tubes, and broad ligaments. In the male, it is met with in the scrotum, the epididymis, the vas deferens, vesiculæ seminales, the prostate; and, in the cavernous bodies, in both sexes. It is found also in the coats of all arteries, in most veins, and lymphatic vessels; in the iris and ciliary muscle, and in the skin.

*Blood-vessels* are distributed in considerable abundance to the muscular tissue. In the voluntary muscles the capillaries, which are of extremely minute size, form narrow, oblong meshes, which run in the direction of the fibres.

The *lymphatic vessels* in muscles are few in number, and appear to exist only in the largest muscles. The *nerves* of voluntary muscles are of large size. The larger branches pass between the fasciculi, and, subdividing, unite to form primary plexuses; from these, finer bundles, or, single nerve tubes, pass between the muscular fibres, and, forming loops, return to the plexus.

Each muscle is invested externally by a thin cellular layer, forming what is called its *sheath*, which not only covers its outer surface, but penetrates into its interior in the intervals between the fasciculi, surrounding these, and serving as a bond of connection between them.

The muscles are connected with the bones, cartilages, ligaments and skin, either directly or through the intervention of fibrous structures, called tendons or aponeuroses. Where a muscle is attached to bone or cartilage, the fibres terminate in blunt extremities upon the periosteum or perichondrium, and do not come into direct relation with the osseous or cartilaginous tissue. Where muscles are connected with the skin, they either lie as a flattened layer beneath it, or are connected with its areolar tissue by larger or smaller bundles of fibres, as in the muscles of the face.

The muscles vary considerably in their form. In the limbs, they are of considerable length, especially the more superficial ones, the deep ones being generally broad; they surround the bones, and form an important protection to the various joints. In the trunk, they are broad, flattened, and expanded, forming the parietes of the cavities which they enclose; hence the reason of the terms, *long, broad, short*, etc., used in the description of a muscle.

There is considerable variation in the arrangement of the fibres of certain muscles, to the tendons to which they are attached. In some, the fibres are arranged longitudinally, and terminate at either end in a narrow tendon. If the fibres converge, like the plumes of a pen, to one side of a tendon, which runs the entire length of a muscle, it is said to be *penniform*, as the Peronei; or, if they converge to both sides of a tendon, they are called *bipenniform*, as the Rectus femoris; if they converge from a broad surface to a narrow tendinous point, they are then said to be *radiated*, as the Temporal and Glutei muscles.

Their size presents considerable variation: the Gastrocnemius forms the chief bulk of the back of the leg, and the fibres of the Sartorius are nearly two feet in length, whilst the Stapedius, a small muscle of the internal ear, weighs about a grain, and its fibres are not more than two lines in length. In each case, however, they are admirably adapted to execute the various movements they are required to perform.

The names applied to the various muscles have been derived: 1, from their situation, as the Tibialis, Radialis, Ulnaris, Peroneus; 2, from their direction, as the Rectus abdominis, Obliqui capitis, Transversalis; 3, from their uses, as Flexors, Extensors, Abductors, etc.; 4, from their shape, as the Deltoid, Trapezius, Rhomboideus; 5, from the number of their divisions, as the Biceps (from having two heads), the Triceps (from having three heads); 6, from their points of attachment, as the Sterno-cleido-mastoid, Sterno-hyoid, Sterno-thyroid.

In the description of a muscle, the term *origin* is meant to imply its more fixed or central attachment; and the term *insertion*, the moveable point upon which the force of the muscle is directed: this holds true, however, for only a very small number of muscles, such as those of the face, which are attached by one extremity to the bone, and by the other to the moveable integument; in the greater number, the muscle can be made to act from either extremity.

In the dissection of the muscles, the student should pay especial attention to the exact *origin, insertion*, and *actions* of each, and its more important *relations* with surrounding parts. An accurate knowledge of the points of attachment of the muscles is of great importance in the determination of their action. By a knowledge of the action of the muscles, the surgeon is able at once to explain the causes of displacement in the various forms of fracture, or the causes which produce distortion in various forms of deformities, and, consequently, to adopt appropriate treatment in each case. The relations, also, of some of the muscles, especially those in immediate apposition with the larger blood-vessels, and the surface-markings they produce, should be especially remembered, as they form most useful guides to the surgeon in the application of a ligature to these vessels.

*Tendons* are white, glistening, fibrous cords, varying in length and thickness, sometimes round, sometimes flattened, of considerable strength, and only slightly elastic. They consist almost entirely of white fibrous tissue, the fibrils of which have an undulating course parallel with each other, and firmly united together. They are very sparingly supplied with blood-vessels, the smaller tendons presenting in their interior not a trace of them. Nerves also are not present in the smaller tendons; but the larger ones, as the tendo Achillis, receive nerves which accompany the nutrient vessels. The tendons consist principally of a substance which yields gelatine.

*Aponeuroses* are fibrous membranes, of a pearly-white colour, iridescent, glistening, and similar in structure to the tendons. They are destitute of nerves, and the thicker ones only sparingly supplied with blood-vessels.

The tendons and aponeuroses are connected, on the one hand, with the muscles; and, on the other hand, with the moveable structures, as the bones, cartilages, ligaments, fibrous membranes (for instance, the sclerotic), and the synovial membranes (subcruræus, subanconeus). Where the muscular fibres are continuous in a direct line, with those of the tendon or aponeurosis, the two are directly continuous, the muscular fibre being distinguishable from that of the tendon only by its striation. But where the muscular fibre joins the tendon or aponeurosis at an oblique angle, the former terminates, according to Kölliker, in rounded extremities, which are received into corresponding depressions on the surface of the latter, the connective tissue between the fibres being continuous with that of the tendon. The latter mode of attachment occurs in all the penniform and semi-penniform muscles, and in those muscles the tendons of which commence in a membranous form, as the Gastrocnemius and Soleus.

The Fasciæ (*fascia*, a bandage) are fibro-areolar or aponeurotic laminæ, of variable thickness and strength, found in all regions of the body, investing the softer and more delicate organs. The fasciæ have been subdivided, from the structure which they present, into two groups, fibro-areolar or superficial fasciæ, and aponeurotic or deep fasciæ.

The *fibro-areolar fascia* is found immediately beneath the integument over almost the entire surface of the body, and is generally known as the *superficial fascia*. It connects the skin with the deep or aponeurotic fascia, and consists of fibro-areolar tissue, containing in its meshes pellicles of fat in varying quantity. In the eyelids and scrotum, where adipose tissue is rarely deposited, this tissue is very liable to serous infiltration. The superficial fascia varies in thickness in different parts of the body: in the groin it is so thick as to be capable of being subdivided into several laminæ, but in the palm of the hand it is of extreme thinness, and intimately adherent to the integument. The superficial fascia is capable of separation into two or more layers, between which are found the superficial vessels and nerves, and superficial lymphatic glands; as the superficial epigastric vessels in the abdominal region, the radial and ulnar veins in the forearm, the saphenous veins in the leg and thigh, as well as in certain situations cutaneous muscles, as the Platysma myoides in the neck, Orbicularis palpebrarum around the eyelids. It is most distinct at the lower part of the abdomen, the scrotum, perinæum, and in the extremities; is very thin in those regions where muscular fibres are inserted into the integument, as on the side of the neck, the face, and around the margin of the anus, and almost entirely wanting in the palms of the hands and soles of the feet, where the integument is adherent to the subjacent aponeurosis. The superficial fascia connects the skin to the subjacent parts, serves as a soft nidus, for the passage of vessels and nerves to the integument, and retains the warmth of the body, from the adipose tissue contained in its areolæ being a bad conductor of caloric.

The *aponeurotic* or *deep fascia* is a dense inelastic and unyielding fibrous membrane, forming sheaths for the muscles, and affording them broad surfaces for attachment, it consists of shining tendinous fibres, placed parallel with one another, and connected together by other fibres disposed in a reticular manner. It is usually exposed on the removal of the superficial fascia, forming a strong investment, which not only binds down collectively the muscles in each region, but gives a separate sheath to each, as well as to the vessels and nerves. The fasciæ are thick in unprotected situations, as on the outer side of a limb, and thinner on the inner side. Aponeurotic fasciæ are divided into two classes, aponeuroses of insertion, and aponeuroses of investment.

The *aponeuroses of insertion* serve for the insertion of muscles. Some of these are formed by the expansion of a tendon into an aponeurosis, as, for instance, the tendon of the Sartorius; others do not originate in tendons, as the aponeuroses of the abdominal muscles.

The *aponeuroses of investment* form a sheath for the entire limb, as well as for each individual muscle. Many aponeuroses, however, serve both for investment and insertion. Thus the aponeurosis given off from the tendon of the Biceps brachialis near its insertion is continuous with, and partly forms, the investing fascia of the forearm, and gives origin to the muscles in this region. The deep fasciæ assist the muscles in their action, by the degree of tension and pressure they make upon their surface; and, in certain situations, this is increased and regulated by muscular action, as, for instance, by the Tensor vaginæ femoris and Gluteus maximus in the thigh, by the Biceps in the leg, and Palmaris longus in the hand. In the limbs, the fasciæ not only invest the entire limb, but give off septa, which separate the various muscles, and are attached beneath to the periosteum; these prolongations of fasciæ are usually spoken of as intermuscular septa.

The Muscles and Fasciæ may be arranged, according to the general division of the body, into, 1. Those of the head, face, and neck. 2. Those of the trunk. 3. Those of the upper extremity. 4. Those of the lower extremity.

## MUSCLES AND FASCIÆ OF THE HEAD AND FACE.

The Muscles of the Head and Face consist of ten groups, arranged according to the region in which they are situated.

1. Cranial Region.
2. Auricular Region.
3. Palpebral Region.
4. Orbital Region.
5. Nasal Region.
6. Superior Maxillary Region.
7. Inferior Maxillary Region.
8. Inter-Maxillary Region.
9. Temporo-Maxillary Region.
10. Pterygo-Maxillary Region.

The Muscles contained in each of these groups are the following.

1. *Epicranial Region.*

Occipito-frontalis.

2. *Auricular Region.*

Attollens aurem.
Attrahens aurem.
Retrahens aurem.

3. *Palpebral Region.*

Orbicularis palpebrarum.
Corrugator supercilii.
Tensor tarsi.

4. *Orbital Region.*

Levator palpebræ.
Rectus superior.
Rectus inferior.
Rectus internus.
Rectus externus.
Obliquus superior.
Obliquus inferior.

5. *Nasal Region.*

Pyramidalis nasi.
Levator labii superioris alæque nasi.
Dilator naris posterior.
Dilator naris anterior.
Compressor naris.
Compressor narium minor.
Depressor alæ nasi.

6. *Superior Maxillary Region.*

Levator labii superioris.
Levator anguli oris.
Zygomaticus major.
Zygomaticus minor.

7. *Inferior Maxillary Region.*

Levator labii inferioris.
Depressor labii inferioris.
Depressor anguli oris.

8. *Inter-Maxillary Region.*

Buccinator.
Risorius.
Orbicularis oris.

9. *Temporo-Maxillary Region.*

Masseter.
Temporal.

10. *Pterygo-Maxillary Region.*

Pterygoideus externus.
Pterygoideus internus.

### 1. Epicranial Region—Occipito-Frontalis.

*Dissection* (fig. 147). The head being shaved, and a block placed beneath the back of the neck, make a vertical incision through the skin from before backwards, commencing at the root of the nose in front, and terminating behind at the occipital protuberance; make a second incision in a horizontal direction along the forehead and round the side of the head, from the anterior to the posterior extremity of the preceding. Raise the skin in front from the subjacent muscle from below upwards; this must be done with extreme care, on account of their intimate union. The tendon of the muscle is best avoided by removing the integument from the outer surface of the vessels and nerves which lie between the two.

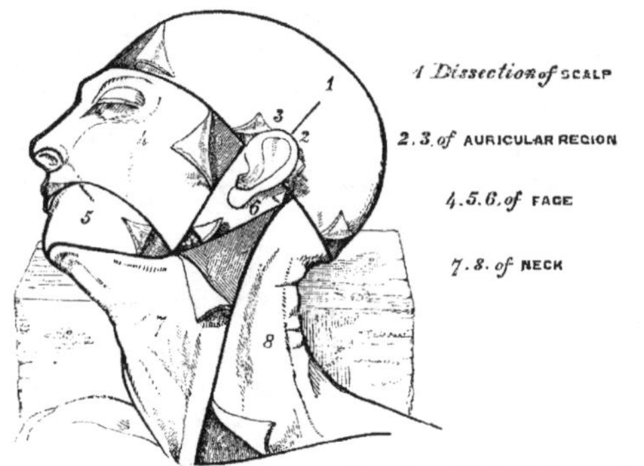

147 – *Dissection of the head, face, and neck.*

The *superficial fascia* in the epicranial region is a firm, dense layer, intimately adherent to the integument, and to the Occipito-frontalis and its tendinous aponeurosis; it is continuous, behind, with the superficial fascia at the back part of the neck; and, laterally, is continued over the temporal aponeurosis: it contains between its layers the small muscles of the auricle, and the superficial temporal vessels and nerves.

The *Occipito-frontalis* (fig. 148) is a broad musculo-fibrous layer, which covers the whole of one side of the vertex of the skull, from the occiput to the eyebrow. It consists of two muscular slips, separated by an intervening tendinous aponeurosis. The *occipital portion*, thin, quadrilateral in form, and about an inch and a half in length, arises from the outer two-thirds of the superior curved line of the occipital bone, and from the mastoid portion of the temporal. Its fibres of origin are tendinous, but they soon become muscular, and ascend in a parallel direction to terminate in the tendinous aponeurosis. The *frontal portion* is thin, of a quadrilateral form, and intimately adherent to the skin. It is broader, its fibres are longer, and their structure paler than the occipital portion. Its internal fibres are continuous with those of the Pyramidalis nasi. Its

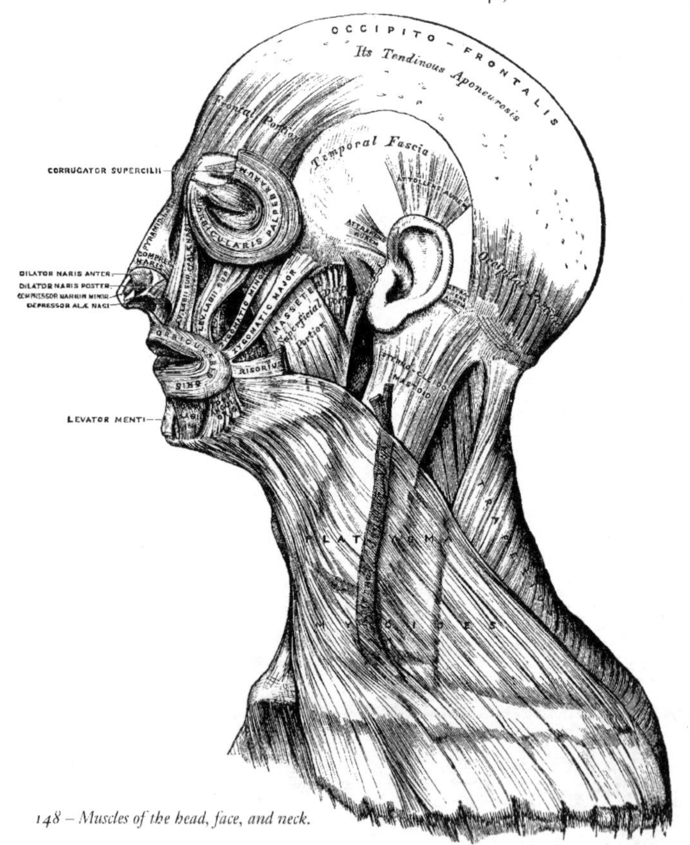

148 – *Muscles of the head, face, and neck.*

middle fibres become blended with the Corrugator supercilii and Orbicularis: and the outer fibres are also blended with the latter muscle over the external angular process. From this attachment, the fibres are directed upwards and join the aponeurosis below the coronal suture. The inner margins of the two frontal portions of the muscle are joined together for some distance above the root of the nose; but between the occipital portions there is a considerable but variable interval.

The *aponeurosis* covers the upper part of the vertex of the skull, being continuous across the middle line with the aponeurosis of the opposite muscle. Behind, it is attached, in the interval between the occipital origins, to the occipital protuberance and superior curved lines above the attachment of the Trapezius; in front, it forms a short angular prolongation between the frontal portions; and on each side, it has connected with it the Attollens and Attrahens aurem muscles; in this situation it loses its aponeurotic character, and is continued over the temporal fascia to the zygoma by a layer of laminated areolar tissue. This aponeurosis is closely connected to the integument by a dense fibro-cellular tissue, which contains much granular fat, and in which ramify the numerous vessels and nerves of the integument; it is loosely connected with the pericranium by a quantity of loose cellular tissue, which allows of a considerable degree of movement of the integument.

*Nerves.* The Occipito-frontalis is supplied (frontal portion) by the supra-orbital and facial nerves; (occipital portion) by the posterior auricular branch of the facial and sometimes by the small occipital.

*Actions.* The frontal portion of the muscle raises the eyebrows and the skin over the root of the nose; at the same time throwing the integument of the forehead into transverse wrinkles, a predominant expression in the emotions of delight. By bringing alternately into action the occipital and frontal portions, the entire scalp may be moved from before backwards.

## 2. Auricular Region (Fig. 148).

Attollens Aurem. Attrahens Aurem.
Retrahens Aurem.

These three small muscles are placed immediately beneath the skin around the external ear. In man, in whom the external ear is almost immoveable, they are rudimentary. They are the analogues of large and important muscles in some of the mammalia.

*Dissection.* This requires considerable care, and should be performed in the following manner. To expose the Attollens aurem; draw the pinna or broad part of the ear downwards, when a tense band will be felt beneath the skin, passing from the side of the head to the upper part of the concha; by dividing the skin over the tendon, in a direction from below upwards, and then reflecting it on each side, the muscle is exposed. To bring into view the Attrahens aurem, draw the helix backwards by means of a hook, when the muscle will be made tense, and may be exposed in a similar manner to the preceding. To expose the Retrahens aurem, draw the pinna forwards, when the muscle being made tense may be felt beneath the skin, at its insertion into the back part of the concha, and may be exposed in the same manner as the other muscles.

The *Attollens Aurem,* the largest of the three, is thin, and fan-shaped; it arises from the aponeurosis of the Occipito-frontalis, its fibres converge to be inserted by a thin, flattened tendon into the upper part of the cranial surface of the pinna.

*Relations. Externally*, with the integument; *internally*, with the Temporal aponeurosis.

The *Attrahens Aurem,* the smallest of the three, is thin, fan-shaped, and its fibres pale and indistinct. It arises from the lateral edge of the aponeurosis of the Occipito-frontalis; its fibres converge to be inserted into a projection on the front of the helix.

*Relations. Externally*, with the skin; *internally*, with the temporal fascia, which separates it from the temporal artery and vein.

The *Retrahens Aurem* consists of two or three fleshy fasciculi, which arise from the mastoid portion of the temporal bone by short aponeurotic fibres. They are inserted into the lower part of the cranial surface of the concha.

*Relations. Externally*, with the integument; *internally*, with the mastoid portion of the temporal bone.

*Nerves.* The Attollens aurem is supplied by the small occipital; the Attrahens aurem, by the facial and auriculo-temporal branch of the inferior maxillary; and the Retrahens aurem, by the posterior auricular branch of the facial.

*Actions.* In man, these muscles posses very little action; the Attollens aurem slightly raises the ear; the Attrahens aurem draws it forwards and upwards; and the Retrahens aurem draws it backwards.

## 3. Palpebral Region (Fig. 148).

| | |
|---|---|
| Orbicularis Palpebrarum. | Levator Palpebræ. |
| Corrugator Supercilii. | Tensor Tarsi. |

*Dissection* (fig. 147). In order to expose the muscles of the face, continue the longtudinal incision made in the dissection of the Occipito-frontalis, down the median line of the face to the tip of the nose, and from this point onwards to the upper lip; another incision should be carried along the margin of the lip to the angle of the mouth, and transversely across the face to the angle of the jaw. The integument should also be divided by an incision made in front of the external ear, from the angle of the jaw, upwards, to the transverse incision made in exposing the Occipito-frontalis. These incisions include a square-shaped flap which should be carefully removed in the direction marked in the figure, as the muscles at some points are intimately adherent to the integument.

The *Orbicularis Palpebrarum* is a sphincter muscle which surrounds the circumference of the orbit and eyelids. It arises from the internal angular process of the frontal bone, from the nasal process of the superior maxillary in front of the lachrymal groove, and from the anterior surface and borders of a short tendon, the tendo palpebrarum, placed at the inner angle of the orbit. From this origin, the fibres are directed outwards, forming a broad, thin, and flat layer, which cover the eyelids, surrounds the circumference of the orbit, and spreads out over the temple, and downwards on the cheek, becoming blended with the Occipito-frontalis and Corrugator supercilii. The palpebral portion (ciliaris) of the Orbicularis is thin and pale; it arises from the bifurcation of the tendo palpebrarum, and forms a series of concentric curves, which are united on the outer side of the eyelids at an acute angle by a cellular raphe, some being inserted into the external tarsal ligament and malar bone. The orbicular portion (orbicularis latus) is thicker, of a reddish colour, its fibres well developed, forming a complete ellipse.

The *tendo palpebrarum* (tendo oculi) is a short tendon, about two lines in length and one in breadth, attached to the nasal process of the superior maxillary bone in front of the lachrymal groove. Crossing the lachrymal sac, it divides into two parts, each division being attached to the inner extremity of the corresponding tarsal cartilage. As the tendon crosses the lachrymal sac, a strong aponeurotic lamina is given off from its posterior surface, which expands over the sac, and is attached to the ridge on the lachrymal bone. This is the reflected aponeurosis of the tendo palpebrarum.

*Relations.* By its *superficial surface*, with the integument. By its *deep surface*, above, with the Occipito-frontalis and Corrugator supercilii, with which it is intimately blended, and with the supra-orbital vessels and nerve; below, it covers the lachrymal sac, and the origin of the Levator labii superioris, and the Levator labii superioris alæque nasi muscles. *Internally*, it is occasionally blended with the Pyramidalis nasi. *Externally*, it lies on the temporal fascia. On the eyelids, it is separated from the conjunctiva by a fibrous membrane and the tarsal cartilages.

The *Corrugator Supercilii* is a small, narrow, pyramidal muscle, placed at the inner extremity of the eyebrow beneath the Occipito-frontalis and Orbicularis palpebrarum muscles. It arises from the inner extremity of the superciliary ridge; its fibres pass upwards and outwards, to be inserted into the under surface of the orbicularis, opposite the middle of the orbital arch.

*Relations.* By its *anterior surface*, with the Occipito-frontalis and Orbicularis palpebrarum muscles. By its *posterior surface*, with the frontal bone and supra-trochlear nerve.

The *Levator Palpebræ* will be described with the muscles of the orbital region.

The *Tensor Tarsi* is a small thin muscle, about three lines in breadth and six in length, situated at the inner side of the orbit, behind the tendo oculi. It arises from the crest and adjacent part of the orbital surface of the lachrymal bone, and, passing across the lachrymal sac, divides into two slips, which cover the lachrymal canals, and are inserted into the tarsal cartilages near the puncta lachrymalia. Its fibres appear to be continuous with those of the palpebral portion of the Orbicularis; it is occasionally very indistinct.

*Nerves.* The Orbicularis palpebrarum and Corrugator supercilii are supplied by the facial and supra-orbital nerves; the Tensor tarsi by the facial.

*Actions.* The Orbicularis palpebrarum is the sphincter muscle of the eyelids. The palpebral portion acts involuntarily in closing the lids, and independently of the orbicular portion, which is subject to the will. When the entire muscle is brought into action, the integument of the forehead, temple, and cheek is drawn inwards towards the inner angle of the eye, and the eyelids are firmly closed. The Levator palpebræ is the direct antagonist of this muscle; it raises the upper eyelid, and exposes the globe. The Corrugator supercilii draws the eyebrow downwards and inwards, producing the vertical wrinkles of the forehead. This muscle may be regarded as the principal agent in the expression of grief. The Tensor tarsi draws the eyelids and the extremities of the lachrymal canals inwards, and compresses them against the surface of the globe of the eye; thus placing them in the most favourable situation for receiving the tears. It serves, also, to compress the lachrymal sac.

## 4. Orbital Region (Fig. 149).

Levator Palpebræ. Rectus Internus.
Rectus Superior. Rectus Externus.
Rectus Inferior. Obliquus Superior.
Obliquus Inferior.

*Dissection.* To open the cavity of the orbit, the skull-cap and brain should be first removed; then saw through the frontal bone at the inner extremity of the supra-orbital ridge, and externally at its junction with the malar. The thin roof of the orbit should then be comminuted by a few slight blows with the hammer, and the superciliary portion of the frontal bone driven forwards by a smart stroke; but it must not be removed. The several fragments may then be detached, when the periosteum of the orbit will be exposed: this being removed, together with the fat which fills the cavity of the orbit, the several muscles of this region can be examined. To facilitate their dissection, the globe of the eye should be distended; this may be effected by puncturing the optic nerve near the eyeball, with a curved needle, and pushing it onwards into the globe. Through this aperture the point of a blow-pipe should be inserted, and a little air forced into the cavity of the eyeball; then apply a ligature round the nerve, so as to prevent the air escaping. The globe should now be drawn forwards, when the muscles will be put upon the stretch.

The *Levator Palpebræ Superioris* is thin, flat, and triangular in shape. It arises from the under surface of the lesser wing of the sphenoid, immediately in front of the optic foramen; and is inserted, by a broad aponeurosis, into the upper border of the superior tarsal cartilage. At its origin, it is narrow and tendinous; but soon becomes broad and fleshy, and finally terminates in a broad aponeurosis.

*Relations.* By its *upper surface*, with the frontal nerve and artery, the periosteum of the orbit; and, in front, with the inner surface of the broad tarsal ligament. By its *under surface*, with the Superior rectus; and, in the lid, with the conjunctiva. A small branch of the third nerve enters its under surface.

The *Rectus Superior*, the thinnest and narrowest of the four Recti, arises from the upper margin of the optic foramen, beneath the Levator palpebræ, and Superior oblique, and from the fibrous sheath of the optic nerve; and is inserted, by a tendinous expansion, into the sclerotic coat, about three or four lines from the margin of the cornea.

*Relations.* By its *upper surface*, with the Levator palpebræ. By its *under surface*, with the optic nerve, the ophthalmic artery, and nasal nerve; and, in front, with the tendon of the Superior oblique, and the globe of the eye.

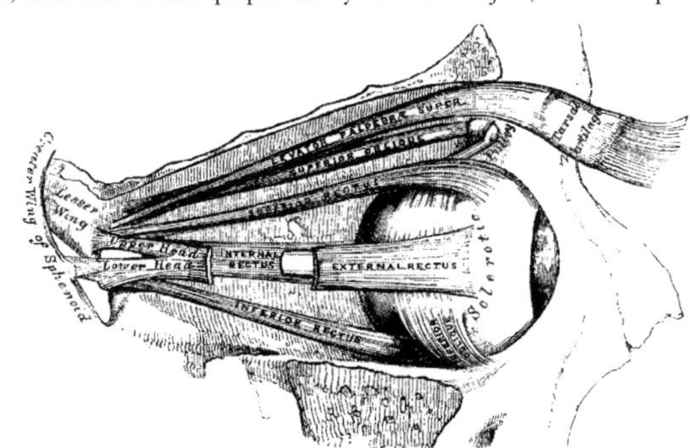


149 – *Muscles of the right orbit.*

The *Inferior* and *Internal Recti* arise by a common tendon (the ligament of Zinn), which is attached round the circumference of the optic foramen, except at its upper and outer part. The External rectus has two heads: the upper one arises from the outer margin of the optic foramen, immediately beneath the Superior rectus; the lower head, partly from the ligament of Zinn, and partly from a small pointed process of bone on the lower margin of the sphenoidal fissure. Each muscle passes forward in the position implied by its name, to be inserted, by a tendinous expansion, into the sclerotic coat, about three or four lines from the margin of the cornea. Between the two heads of the External rectus is a narrow interval, through which pass the third, nasal branch of the fifth, and sixth nerves, and the ophthalmic vein. Although

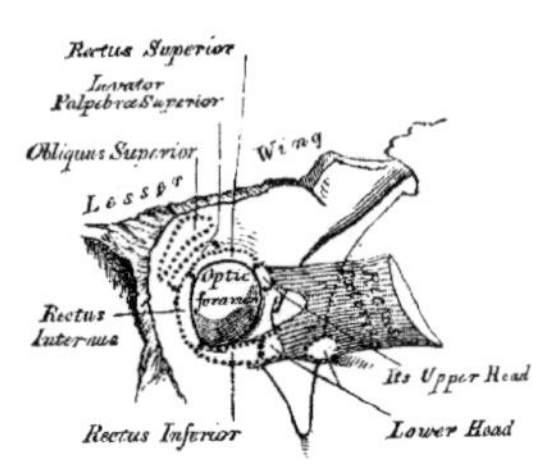


*150 – The relative position and attachment of the muscles of the left eyeball.*

nearly all these muscles present a common origin and are inserted in a similar manner in the sclerotic coat, there are certain differences to be observed in them, as regards their length and breadth. The Internal rectus is the broadest; the External, the longest; and the Superior, the thinnest and narrowest.

The *Superior Oblique* is a fusiform muscle, placed at the upper and inner side of the orbit, internal to the Levator palpebræ. It arises about a line above the inner margin of the optic foramen, and, passing forwards to the inner angle of the orbit, terminates in a rounded tendon, which passes through a fibro-cartilaginous ring, attached by fibrous tissue to a depression beneath the internal angular process of the frontal bone, the contiguous surfaces of the tendon and ring being lined by a delicate synovial membrane, and enclosed in a thin fibrous investment. The tendon is reflected backwards and outwards beneath the Superior rectus to the outer part of the globe of the eye, and is inserted into the sclerotic coat, midway between the cornea and entrance of the optic nerve, the insertion of the muscle lying between the Superior and External recti.

*Relations.* By its *upper surface*, with the periosteum covering the roof of the orbit, and the fourth nerve. By its *under surface*, with the nasal nerve, and the upper border of the Internal rectus.

The *Inferior Oblique* is a thin, narrow muscle, placed near the anterior margin of the orbit. It arises from a depression in the orbital plate of the superior maxillary bone, external to the lachrymal groove. Passing outwards and backwards beneath the Inferior rectus, and between the eyeball and the External rectus, it is inserted into the outer part of the sclerotic coat between the Superior and External rectus, and near the tendon of insertion of the Superior oblique.

*Relations.* By its *upper surface*, with the globe of the eye, and with the Inferior rectus. By its *under surface*, with the periosteum covering the floor of the orbit, and with the External rectus. Its borders look forwards and backwards; the posterior one receives a branch of the third nerve.

*Nerves.* The Levator palpebræ, Inferior oblique, and all the recti excepting the External, are supplied by the third nerve; the Superior oblique, by the fourth; the External rectus, by the sixth.

*Actions.* The Levator palpebræ raises the upper eyelid, and is the direct antagonist of the Orbicularis palpebrarum. The four Recti muscles are attached in such a manner to the globe of the eye, that, acting singly, they will turn it either upwards, downwards, inwards, or outwards, as expressed by their names. If any two Recti act together, they carry the globe of the eye in the diagonal of these directions, viz., upwards and inwards, upwards and outwards, downwards and inwards, or downwards and outwards. The movement of circumduction, as in turning the eyes round a room, is performed by the alternate action of the four Recti. By some anatomists, these muscles have been considered the chief agents in adjusting the sight at different distances, by compressing the globe, and so lengthening its antero-posterior diameter. The Oblique muscles rotate the eyeball on its *antero-posterior axis*, this kind of movement being required, for the correct viewing of an object, when the head is moved laterally, as from shoulder to shoulder, in order that the picture may fall in all respects on the same part of the retina.*

*Surgical Anatomy.* The position and exact point of insertion of the tendons of the Internal and External recti muscles into the globe, should be carefully examined from the front of the eyeball, as the surgeon is often required to divide one or the other muscle for the cure of strabismus. In convergent strabismus, which is the most common form of the disease, the eye is turned inwards, requiring the division of the Internal rectus. In the divergent form, which is more rare, the eye is turned outwards, the External rectus being especially implicated. The deformity produced in either case is considerable, and is easily remedied by division of one or the other muscle. This operation is readily effected by having the lids well separated by retractors held by an assistant; the eyeball being drawn outwards, the conjunctiva should be raised by a pair of forceps, and divided immediately beneath the lower border of the tendon of the Internal rectus, a little behind its insertion into the sclerotic; the submucous areolar tissue is then divided, and into the small aperture thus made, a blunt hook is passed upwards between the muscle and the globe, and the tendon of the muscle and conjunctiva covering it, divided by a pair of blunt-pointed scissors. Or the tendon may be divided by a sub-conjunctival incision, one blade of the scissors being passed upwards between the tendon and the conjunctiva, and the other between the tendon and sclerotic. The student, when dissecting these muscles, should remove on one side of the subject the conjunctiva from the front of the eye, in order see more accurately the position of these tendons, and on the opposite side the operation may be performed.

* 'On the Oblique Muscles of the Eye in Man and Vertebrate Animals,' by John Struthers, M.D. '*Anatomical and Physiological Observations.*'

## 5. Nasal Region (Fig. 148).

Pyramidalis Nasi.
Levator Labii Superioris Alæque Nasi.
Dilator Naris Posterior.
Dilator Naris Anterior.
Compressor Naris.
Compressor Narium Minor.
Depressor Alæ Nasi.

The *Pyramidalis Nasi* is a small pyramidal slip, prolonged downwards from the Occipito-frontalis upon the side of the nose, where it becomes tendinous, and blends with the Compressor naris. As the two muscles descend, they diverge, leaving an angular interval between them.

*Relations.* By its *upper surface*, with the skin. By its *under surface*, with the frontal and nasal bones.

The *Levator Labii Superioris Alæque Nasi* is a thin triangular muscle, placed by the side of the nose, and extending between the inner margin of the orbit and upper lip. It arises by a pointed extremity from the upper part of the nasal process of the superior maxillary bone, and passing obliquely downwards and outwards divides into two slips, one of which is inserted into the cartilage of the ala of the nose; the other is prolonged into the upper lip, becoming blended with the Orbicularis and Levator labii proprius.

*Relations.* In front, with the integument; and with a small part of the Orbicularis palpebrarum above.

Lying upon the superior maxillary bone, beneath this muscle, is a longitudial muscular fasciculus about an inch in length. It is attached by one end near the origin of the Compressor naris, and by the other to the nasal process about an inch above it; it was described by Albinus as the Musculus 'anomalus,' and by Santorini as the 'Rhomboideus.'

The *Dilator naris posterior* is a small muscle, which is placed partly beneath the proper elevator of the nose and lip. It arises from the margin of the nasal notch of the superior maxilla, and from the sesamoid cartilages, and is inserted into the skin near the margin of the nostril.

The *Dilator naris anterior* is a thin, delicate fasciculus, passing from the cartilage of the ala of the nose to the integument near its margin. This muscle is situated in front of the preceding.

The *Compressor Naris* is a small, thin, triangular muscle, arising by its apex from the superior maxillary bone, above and a little external to the incisive fossa; its fibres proceed upwards and inwards, expanding into a thin aponeurosis which is attached to the fibro-cartilage of the nose, and is continuous on the bridge of the nose with that of the muscle of the opposite side, and with the aponeurosis of the Pyramidalis nasi.

The *Compressor Narium Minor* is a small muscle, attached by one end to the alar cartilage, and by the other to the integument at the end of the nose.

The *Depressor Alæ Nasi* is a short, radiated muscle, arising from the incisive fossa of the superior maxilla; its fibres ascend to be inserted into the septum, and back part of the ala of the nose. This muscle lies between the mucous membrane and muscular structure of the lip.

*Nerves.* All the muscles of this group are supplied by the facial nerve.

*Actions.* The Pyramidalis nasi draws down the inner angle of the eyebrow; by some anatomists it is also considered as an elevator of the ala, and, consequently a dilator of the nose. The Levator labii superioris alæque nasi draws upwards the upper lip and ala of the nose; its most important action is upon the nose, which it dilates to a considerable extent. The action of this muscle produces a marked influence over the countenance, and is the principal agent in the expression of contempt. The two Dilatores nasi enlarge the aperture of the nose, and the Compressor naris appears to act as a dilator of the nose rather than as a constrictor. The Depressor alæ nasi is a direct antagonist of the preceding muscles, drawing the ala of the nose downwards, and thereby constricting the aperture of the nares.

## 6. Superior Maxillary Region (Fig. 148).

| | |
|---|---|
| Levator Labii Superioris. | Zygomaticus major. |
| Levator Anguli Oris. | Zygomaticus minor. |

The *Levator Labii Superioris* is a thin muscle of a quadrilateral form. It arises from the lower margin of the orbit immediately above the infra-orbital foramen, some of its fibres being attached to the superior maxilla, some to the malar bone; its fibres converge to be inserted into the muscular substance of the upper lip.

*Relations.* By its *superficial surface*, with the lower segment of the Orbicularis palpebrarum; below, it is sub-cutaneous. By its *deep surface*, it conceals the origin of the Compressor naris and Levator anguli oris muscles, and the infra-orbital vessels and nerves, as they escape from the infra-orbital foramen.

The *Levator Anguli Oris* arises from the canine fossa, immediately below the infra-orbital foramen; its fibres incline downwards and a little outwards, to be inserted into the angle of the mouth, its fibres intermingling with those of the Zygomatici, the Depressor anguli oris, and the Orbicularis.

*Relations.* By its *superficial surface*, with the Levator labii superioris and the infra-orbital vessels and nerves. By its *deep surface*, with the superior maxilla, the Buccinator, and the mucous membrane.

The *Zygomaticus major* is a slender fasciculus, which arises from the malar bone, in front of the zygomatic suture, and, descending obliquely downwards and inwards, is inserted into the angle of the mouth, where it blends with the fibres of the Orbicularis and Depressor anguli oris.

*Relations.* By its *superficial surface*, with the subcutaneous adipose tissue. By its *deep surface*, with the malar bone, the Masseter and Buccinator muscles.

The *Zygomaticus minor* arises from the malar bone, immediately behind the maxillary suture, and, passing downwards and inwards, is continuous with the outer margin of the Levator labii superioris. It lies in front of the preceding.

*Relations.* By its *superficial surface*, with the integument and the Orbicularis palpebrarum above. By its *deep surface*, with the Levator anguli oris.

*Nerves.* This group of muscles is supplied by the facial nerve.

*Actions.* The Levator labii superioris is the proper elevator of the upper lip, carrying it at the same time a little outwards. The Levator anguli oris raises the angle of the mouth and draws it inwards; whilst the Zygomatici raise the upper lip and draw it somewhat outwards, as in laughing.

## 7. Inferior Maxillary Region (Fig. 148).

Levator Labii Inferioris (Levator menti).
Depressor Labii Inferioris (Quadratus menti).
Depressor Anguli Oris (Triangularis menti).

*Dissection.* The Muscles in this region may be dissected by making a vertical incision through the integument from the margin of the lower lip to the chin: a second incision should then be carried along the margin of the lower jaw as far as the angle, and the integument carefully removed in the direction shewn in fig. 147.

The *Levator Labii Inferioris* (*Levator menti*) is to be dissected by everting the lower lip and raising the mucous membrane. It is a small conical fasciculus. placed on the side of the frænum of the lower lip. It arises from the incisive fossa, external to the symphysis of the lower jaw: its fibres descend to be inserted into the integument of the chin.

*Relations.* On its *inner surface*, with the mucous membrane; in the *median line*, it is blended with the muscle of the opposite side; and on its *outer side*, with the Depressor labii inferioris.

The *Depressor Labii Inferioris* (*Quadratus menti*) is a small quadrilateral muscle, situated at the outer side of the preceding. It arises from the external oblique line of the lower jaw, between the symphysis and mental foramen, and passes obliquely upwards and inwards, to be inserted into the integument of the lower lip, its fibres blending with the Orbicularis, and with those of its fellow of the opposite side. It is continuous with the fibres of the Platysma at its origin. This muscle contains much yellow fat intermingled with its fibres.

*Relations.* By its *superficial surface*, with part of the Depressor anguli oris and with the integument, to which it is closely connected. By its *deep surface*, with the mental vessels and nerves, the mucous membrane of the lower lip, the labial glands, and the Levator menti, with which it is intimately united.

The *Depressor Anguli Oris* is triangular in shape, arising, by its broad base from the external oblique line of the lower jaw; its fibres pass upwards, to be inserted, by a narrow fasciculus, into the angle of the mouth. It is continuous with the Platysma at its origin, and with the Orbicularis and Risorius at its insertion.

*Relations.* By its *superficial surface*, with the integument. By its *deep surface*, with the Depressor labii inferioris and Buccinator.

*Nerves.* This group of muscles is supplied by the facial nerve.

*Actions.* The Levator labii inferioris raises the lower lip, and protrudes it forwards; at the same time it wrinkles the integument of the chin. The Depressor labii inferioris draws the lower lip directly downwards and a little outwards. The Depressor anguli oris depresses the angle of the mouth, being the great antagonist to the Levator anguli oris and Zygomaticus major: acting with these muscles, it will draw the angle of the mouth directly backwards.

## 8. Inter-Maxillary Region.

Orbicularis Oris. Buccinator. Risorius.

*Dissection.* The dissection of these muscles may be considerably facilitated by filling the cavity of the mouth with tow, so as to distend the cheeks and lips; the mouth should then be closed by a few stitches, and the integument carefully removed from the surface.

The *Orbicularis Oris* is a sphincter muscle, elliptic in form, composed of concentric fibres, which surround the orifice of the mouth. It consists of two thick semicircular planes of muscular fibre, which surround the oral aperture, and interlace on either side with those of the Buccinator and other muscles inserted into this part. On the free margin of the lips the muscular fibres are continued uninterruptedly from one lip to the other, around the corner of the mouth, forming a roundish fasciculus of fine pale fibres closely approximated. To the outer part of each segment some special fibres are added, by which the lips are connected directly with the maxillary bones and septum of the nose. The additional fibres for the upper segment consist of four bands, two of which (Accessorii orbicularis superioris) arise from the alveolar border of the superior maxilla, opposite the incisor teeth, and arching outwards on each side, are continuous at the angles of the mouth with the other muscles inserted into this part. The two remaining muscular slips, called the Naso-labialis, connect the upper lip to the septum of the nose: as they descend from the septum, an interval is left between them, which corresponds to that left by the divergence of the accessory portions of the Orbicularis above described. It is this interval which forms the depression seen on the surface of the skin beneath the septum of the nose.

The additional fibres for the lower segment (Accessorii orbicularis inferior) arise from the inferior maxilla, externally to the Levator labii inferioris; arching outwards to the angles of the mouth, they join the Buccinator and the other muscles attached to this part.

*Relations.* By its *superficial surface*, with the integument to which it is closely connected. By its *deep surface*, with the buccal mucous membrane, the labial glands, and coronary vessels. By its *outer circumference*, it is blended with the numerous muscles which converge to the mouth from various parts of the face. Its *inner circumference* is free, and covered by mucous membrane.

The *Buccinator* is a broad, thin muscle, quadrilateral in form, occupying the interval between the jaws at the side of the face. It arises from the outer surface of the alveolar processes of the upper and lower jaws, corresponding to the last three molar teeth; and, behind, from the anterior border of the pterygo-maxillary ligament. The fibres converge towards the angle of the mouth, where the central ones intersect each other, those from below being continuous with the upper segment of the Orbicularis oris; and those from above, with the inferior segment; but the highest and lowest fibres continue forward uninterruptedly into the corresponding segment of the lip, without decussation.

*Relations.* By its *superficial surface*, behind, with a large mass of fat, which separates it from the ramus of the lower jaw, the Masseter, and a small portion of the Temporal muscle; anteriorly, with the Zygomatici, Risorius, Levator anguli oris, Depressor anguli oris, and Stenon's duct, which pierces it opposite the second molar tooth of the upper jaw; the facial artery and vein cross it from below upwards; it is also crossed by the branches of the facial and buccal nerves. By its *internal surface*, with the buccal glands and mucous membrane of the mouth.

The *pterygo-maxillary ligament* separates the Buccinator muscle from the Superior constrictor of the pharynx. It is a tendinous band, attached by one extremity to the apex of the internal pterygoid plate, and by the other to the posterior extremity of the internal oblique line of the lower jaw. Its *inner surface* corresponds to the cavity of the mouth, and is lined by mucous membrane. Its *outer surface* is separated from the ramus of the jaw by a quantity of adipose tissue. Its *posteri border* gives attachment to the Superior constrictor of the pharynx; its *anterior border*, to the fibres of the Buccinator.

The *Risorius* (*Santorini*) consists of a narrow bundle of fibres, which arises in the fascia over the Masseter muscle, and passing horizontally forwards, is inserted into the angle of the mouth, joining with the fibres of the Depressor anguli oris. It is placed superficial to the Platysma, and is broadest at its outer extremity. This muscle varies much in its size and form.

*Nerves.* The Orbicularis oris is supplied by the facial, the Buccinator by the facial and buccal branch of the inferior maxillary nerve.

*Actions.* The Orbicularis oris is the direct antagonist of all those muscles which converge to the lips from the various parts of the face, its action producing the direct closure of the lips; and its forcible action throwing the integument into wrinkles, on account of the firm connection between the latter and the surface of the muscle. The Buccinators contract and compress the cheeks, so that, during the process of mastication, the food is kept under the immediate pressure of the teeth.

## 9. Temporo-Maxillary Region (Fig. 151).

Masseter. Temporal.

The Masseter has been already exposed by the removal of the integument from the side of the face (fig. 148).

The *Masseter* is a short thick muscle, somewhat quadrilateral in form, consisting of two portions, superficial and deep. The *superficial portion*, the largest, arises by a thick tendinous aponeurosis from the malar process of the superior maxilla, and from the anterior two-thirds of the lower border of the zygomatic arch: its fibres pass downwards and backwards, to be inserted into the angle and lower half of the ramus of the jaw. The *deep portion* is much smaller, and more muscular in texture; it arises from the posterior third of the lower border and whole of the inner surface of the zygomatic arch; its fibres pass downwards and forwards to be inserted into the upper half of the ramus and outer surface of the coronoid process of the jaw. The deep portion of the muscle is partly concealed, in front, by the superficial portion; behind, it is covered by the parotid gland. The fibres of the two portions are united at their insertion.

*Relations.* By its *superficial surface*, with the integument; above, with the Orbicularis palpebrarum and Zygomatici; and has passing across it, transversely, Stenon's duct, the branches of the facial nerve, and the transverse facial vessels. By its *deep surface*, with the ramus of the jaw, and the Buccinator, from which it is separated by a mass of fat. Its *posterior margin* is overlapped by the parotid gland. Its *anterior margin* projects over the Buccinator muscle.

The *temporal fascia* is seen, at this stage of the dissection, covering-in the Temporal muscle. It is a strong aponeurotic investment, affording attachment, by its inner surface, to the superficial fibres of this muscle. Above, it is a single layer, attached to the entire extent of the temporal ridge; but below, where it is attached to the zygoma, it consists of two layers, one of which is inserted into the outer, and the other to the inner border of the zygomatic arch. A small quantity of fat, the orbital branch of the temporal artery, and a filament from the orbital branch of the superior maxillary nerve, are contained between these two layers. It is covered, on its outer surface, by the aponeurosis of the Occipito-frontalis, the Orbicularis palpebrarum, and Attollens and Attrahens aurem muscles; the temporal vessels and nerves cross it from below upwards.

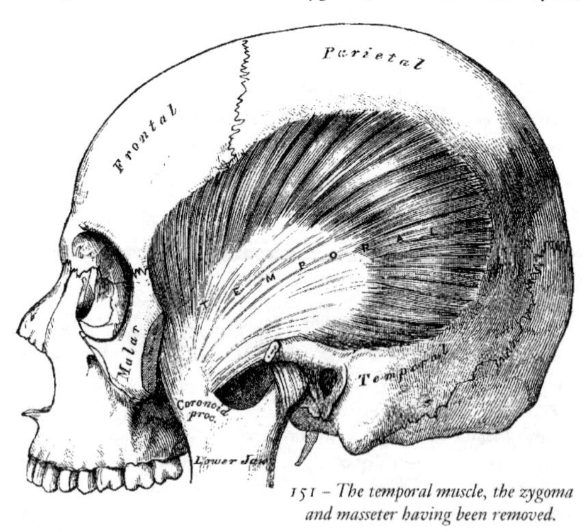

*151 – The temporal muscle, the zygoma and masseter having been removed.*

*Dissection.* In order to expose the Temporal muscle, this fascia should be removed: this may be effected by separating it at its attachment along the upper border of the zygoma and dissecting it upwards from the surface of the muscle. The zygomatic arch should then be divided, in front, at its junction with the malar bone; and, behind, near the external auditory meatus, and drawn downwards with the Masseter, which should be detached from its insertion into the ramus and angle of the jaw. The whole extent of the Temporal muscle is then exposed.

The *Temporal* is a broad radiating muscle, situated at the side of the head, and occupying the entire extent of the temporal fossa. It arises from the whole of the temporal fossa, which extends from the external angular process of the frontal in front, to the mastoid portion of the temporal behind; and from the curved line on the frontal and parietal bones above, to the pterygoid ridge on the great wing of the sphenoid below. It is also attached to the inner surface of the temporal fascia. Its fibres converge as they descend, and terminate in an aponeurosis, the fibres of which, radiated at its commencement, converge into a thick and flat tendon, which is inserted into the inner surface, apex, and anterior border of the coronoid process of the jaw, nearly as far forwards as the last molar tooth.

*Relations.* By its *superficial surface*, with the integument, the temporal fascia, aponeurosis of the Occipito-frontalis, the Attollens and Attrahens aurem muscles, the temporal vessels and nerves, the zygoma and Masseter. By its *deep surface*, with the temporal fossa, the External pterygoid and part of the Buccinator muscles, the internal maxillary artery, its deep temporal branches, and the temporal nerves.

*Nerves.* Both muscles are supplied by the inferior maxillary nerve.

## 10. Pterygo-Maxillary Region.

Internal Pterygoid. External Pterygoid.

*Dissection.* The Temporal muscle having been examined, the muscles in the pterygo-maxillary region may be exposed by sawing through the base of the coronoid process, and drawing it upwards, together with the Temporal muscle, which should be detached from the surface of the temporal fossa. Divide the ramus of the jaw just below the condyle, and also, by a transverse incision extending across the commencement of its lower third, just above the dental foramen; remove the fragment, and the Pterygoid muscles will be exposed.

The *Internal Pterygoid* is a thick quadrilateral muscle, and resembles the Masseter in form, structure, and in the direction of its fibres. It arises from the pterygoid fossa, its fibres being attached to the inner surface of the external pterygoid plate, and to the grooved surface of the tuberosity of the palate bone; its fibres pass downwards, outwards, and backwards, to be inserted, by strong tendinous laminæ, into the lower and back part of the inner side of the ramus and angle of the lower jaw, as high as the dental foramen.

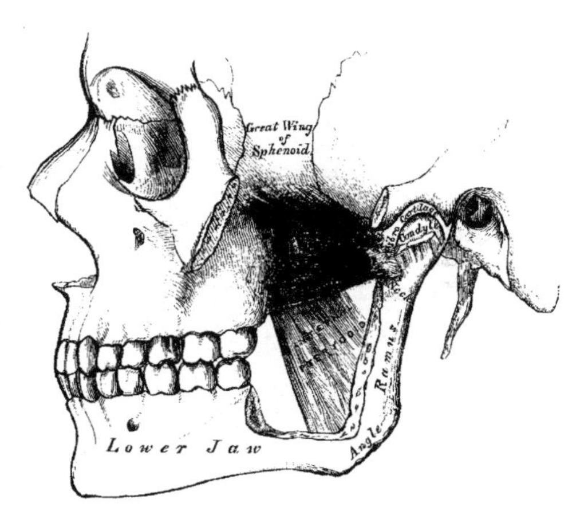

152 – *The pterygoid muscles; the zygomatic arch and a portion of the ramus of the jaw having been removed.*

*Relations.* By its *external surface*, with the ramus of the lower jaw, from which it is separated, at its upper part, by the External pterygoid, the internal lateral ligament, the internal maxillary artery, and the dental vessels and nerve. By its *internal surface*, with the Tensor palati, being separated from the Superior constrictor of the pharynx by a cellular interval.

The *External Pterygoid* is a short thick muscle, somewhat conical in form, and extends almost horizontally between the zygomatic fossa and the condyle of the jaw. It arises from the pterygoid ridge on the great wing of the sphenoid and the portion of bone included between it and the base of the pterygoid process from the outer surface of the external pterygoid plate; and from the tuberosity of the palate and superior maxillary bones. Its fibres pass horizontally backwards and outwards, to be inserted into a depression in front of the neck of the condyle of the lower jaw, and into the corresponding part of the inter-articular fibro-cartilage. This muscle, at its origin, appears to consist of two portions separated by a slight interval; hence the terms upper and lower head sometimes used in the description of the muscle.

*Relations.* By its *external surface*, with the ramus of the lower jaw, the internal maxillary artery, which crosses it, the tendon of the Temporal muscle, and the Masseter. By its *internal surface*, it rests against the upper part of the Internal pterygoid, the internal lateral ligament, the middle meningeal artery, and inferior maxillary nerve; by its *upper border* it is in relation with the temporal and masseteric branches of the inferior maxillary nerve.

*Nerves.* These muscles are supplied by the inferior maxillary nerve.

*Actions.* The Temporal, Masseter, and Internal pterygoid raise the lower jaw against the upper with great force. The two latter muscles, from the obliquity in the direction of their fibres, assist the External pterygoid in drawing the lower jaw forwards upon the upper, the jaw being drawn back again by the deep fibres of the Masseter, and posterior fibres of the Temporal. The External pterygoid muscles are the direct agents in the trituration of the food, drawing the lower jaw directly forwards, so as to make the lower teeth project beyond the upper. If the muscle of one side acts, the corresponding side of the jaw is drawn forwards, and the other condyle remaining fixed, the symphysis deviates to the opposite side. The alternation of these movements on the two sides, produces trituration.

## MUSCLES AND FASCIÆ OF THE NECK.

The muscles of the Neck may be arranged into groups, corresponding with the region in which they are situated.

These groups are nine in number:—

1. Superficial Region.
2. Depressors of the Os Hyoides and Larynx.
3. Elevators of the Os Hyoides and Larynx.
4. Muscles of the Tongue.
5. Muscles of the Pharynx.
6. Muscles of the Soft Palate.
7. Muscles of the Anterior Vertebral Region.
8. Muscles of the Lateral Vertebral Region.
9. Muscles of the Larynx.

1. *Superficial Region.*

Platysma myoides.
Sterno-cleido-mastoid.

*Infra-hyoid Region.*

2. *Depressors of the Os Hyoides and Larynx.*

Sterno-hyoid.
Sterno-thyroid.
Thyro-hyoid.
Omo-hyoid.

*Supra-hyoid Region.*

3. *Elevators of the Os Hyoides and Larynx.*

Digastric.
Stylo-hyoid.
Mylo-hyoid.
Genio-hyoid.

*Lingual Region.*

4. *Muscles of the Tongue.*

Genio-hyo-glossus.
Hyo-glossus.
Lingualis.
Stylo-glossus.
Palato-glossus.

5. *Muscles of the Pharynx.*

Constrictor inferior.
Constrictor medius.
Constrictor superior.
Stylo-pharyngeus.
Palato-pharyngeus.

6. *Muscles of the Soft Palate.*

Levator palati.
Tensor palati.
Azygos uvulæ.
Palato-glossus.
Palato-pharyngeus.

7. *Muscles of the Anterior Vertebral Region.*

Rectus capitis anticus major.
Rectus capitis anticus minor.
Rectus lateralis.
Longus colli.

8. *Muscles of the Lateral Vertebral Region.*

Scalenus anticus.
Scalenus medius.
Scalenus posticus.

9. *Muscles of the Larynx.*

Included in the description of the Larynx.

## 1. Superficial Cervical Region.

Platysma Myoides. Sterno-Cleido-Mastoid.

*Dissection.* A block having been placed at the back of the neck, and the face turned to the side opposite to that to be dissected, so as to place the parts upon the stretch, two transverse incisions are to be made: one from the chin, along the margin of the lower jaw, to the mastoid process; and the other along the upper border of the clavicle. These are to be connected by an oblique incision made in the course of the Sterno-mastoid muscle, from the mastoid process to the sternum; the two flaps of integument having been removed in the direction shewn in fig. 147, the superficial fascia will be exposed.

The *superficial cervical fascia* is exposed on the removal of the integument from the side of the neck; it is an extremely thin aponeurotic lamina, which is hardly demonstrable as a separate membrane. Beneath it is found the Platysma-myoides muscle, the external jugular vein, and some superficial branches of the cervical plexus of nerves.

The *Platysma Myoides* (fig. 148) is a broad thin plane of muscular fibres, placed immediately beneath the skin on each side of the neck. It arises from the clavicle and acromion, and from the fascia covering the upper part of the Pectoral, Deltoid, and Trapezius muscles; its fibres proceed obliquely upwards and inwards along the side of the neck, to be inserted into the lower jaw beneath the external oblique line, some fibres passing forwards to the angle of the mouth, and others becoming lost in the cellular tissue of the face. The most anterior fibres interlace, in front of the jaw, with the fibres of the muscle of the opposite side; those next in order become blended with the Depressor labii inferioris and the Depressor anguli oris; others are prolonged upon the side of the cheek, and interlace, near the angle of the mouth, with the muscles in this situation, and may occasionally be traced to the Zygomatic muscles, or to the margin of the Orbicularis palpebrarum. Beneath the Platysma, the external jugular vein may be seen descending from the angle of the jaw to the clavicle. It is essential to remember the direction of the fibres of the Platysma, in connection with the operation of bleeding from this vessel; for if the point of the lancet is introduced in the direction of the muscular fibres, the orifice made will be filled up by the contraction of the muscle, and blood will not flow; but if the incision is made in a direction opposite to the course of the fibres, they will retract, and expose the orifice in the vein, and so facilitate the flow of blood.

*Relations.* By its *external surface*, with the integument to which it is united closely below, but more loosely above. By its *internal surface*, below the clavicle which it covers, with the Pectoralis major, Deltoid, and Trapezius. In the *neck*, with the external and anterior jugular veins, the deep cervical fascia, the superficial cervical plexus, the Sterno-mastoid, Sterno-hyoid, Omo-hyoid, and Digastric muscles. In front of the Sterno-mastoid, it covers the sheath of the carotid vessels; and behind it, the Scaleni muscles and the nerves of the brachial plexus. On the *face*, it is in relation with the parotid gland, the facial artery and vein, and the Masseter and Buccinator muscles.

The *deep cervical fascia* is exposed on the removal of the Platysma myoides. It is a strong fibrous layer, which invests the muscles of the neck, and encloses the vessels and nerves. It commences, as an extremely thin layer, at the back part of the neck, where it is attached to the spinous processes of the cervical vertebræ, and to the ligamentum nuchæ; and, passing forwards to the posterior border of the Sterno-mastoid muscle, divides into two layers, one of which passes in front, and the other behind it. These join again at its anterior border; and being continued forwards to the front of the neck, blend with the fascia of the opposite side. The superficial layer of the deep cervical fascia (that which passes in front of the Sterno-mastoid), if traced upwards, is found to pass across the parotid gland and Masseter muscle, forming the parotid and masseteric fasciæ, and is attached to the lower border of the zygoma, and more anteriorly to the lower border of the body of the jaw; if the same layer is traced downwards, it is seen to pass to the upper border of the clavicle and sternum, being pierced just above the former bone for the external jugular vein. In the middle line of the neck, the fascia is thin above, and connected to the hyoid bone; but it becomes thicker below, and divides, just below the thyroid gland, into two layers, the more superficial of which is attached to the upper border of the sternum and interclavicular ligament; the deeper and stronger layer is connected to the posterior border of that bone, covering in the Sterno-hyoid and Sterno-thyroid muscles. Between these two layers is a little areolar tissue and fat, and occasionally a small lymphatic gland. The deep layer of the cervical fascia (that which lies behind the posterior surface of the Sterno-mastoid) sends numerous prolongations which invest the muscles and vessels of the neck; if traced upwards, a process of this fascia, of extreme density, passes behind and to the inner side of the parotid gland, and is attached to the base of the styloid process and angle of the lower jaw, forming the stylo-maxillary ligament; if traced downwards and outwards, it will be found to enclose the posterior belly of the Omo-hyoid muscle, binding it down by a distinct process, which descends to be inserted into the clavicle and cartilage of the first rib. The deep layer of the cervical fascia also assists in forming the sheath which encloses the common carotid artery, internal jugular vein, and pneumogastric nerve. There are fibrous septa intervening between each of these parts, which, however, are included together in one common investment. More internally, a thin layer is continued across the trachea and thyroid gland beneath the Sterno-thyroid muscles; and at the root of the neck this may be traced, over the large vessels, to be continuous with the fibrous layer of the pericardium.

The *Sterno-Cleido-Mastoid* (fig. 153) is a large thick muscle, which passes obliquely across the side of the neck, being enclosed between the two layers of the deep cervical fascia. It is thick and narrow at its central part, but is broader and thinner at each extremity. It arises, by two heads, from the sternum and clavicle. The *sternal portion* arises by a rounded fasciculus, tendinous in front, fleshy behind, from the upper and anterior part of the first piece of the sternum, and is directed upwards and backwards. The *clavicular portion* arises from the inner third of the superior border of the clavicle, being composed of fleshy and aponeurotic fibres; it is directed almost vertically upwards. These two portions are separated from one another, at their origin, by a triangular cellular interval; but become gradually blended, below the middle of the neck, into a thick rounded muscle, which is inserted, by a strong tendon, into the outer surface of the mastoid process, from the apex to its superior border, and by a thin aponeurosis into the outer two-thirds of the superior curved line of the occipital bone. This muscle varies much in its extent of attachment to the clavicle: in one case it may be as narrow as the sternal portion; in another, as much as three inches in breadth. When the clavicular origin is broad, it is occasionally subdivided into numerous slips, separated by narrow intervals. More rarely, the corresponding margins of the Sterno-mastoid and Trapezius have been found in contact. In the application of a ligature to the third part of the subclavian artery, it will be necessary, where the muscles have an arrangement similar to that above mentioned, to divide a portion of one or of both, in order to facilitate the operation.

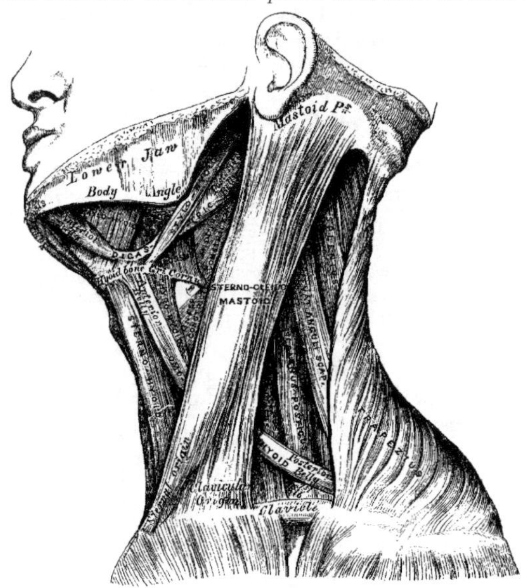


*153 – Muscles of the neck, and boundaries of the triangles.*

This muscle divides the quadrilateral space at the side of the neck into two triangles, an anterior and a posterior. The boundaries of the *anterior* triangle being, in front, the median line of the neck; above, the lower border of the body of the jaw, and an imaginary line drawn from the angle of the jaw to the mastoid process; behind, the anterior border of the Sterno-mastoid muscle. The boundaries of the *posterior* triangle are, in front, the posterior border of the Sterno-mastoid; below, the upper border of the clavicle; behind, the anterior margin of the Trapezius.

The anterior edge of the muscle forms a very prominent ridge beneath the skin, which it is important to notice, as it forms a guide to the surgeon in making the necessary incisions for ligature of the common carotid artery, and the œsophagotomy.

*Relations.* By its *superficial surface*, with the integument and Platysma, from which it is separated by the external jugular vein, the superficial branches of the cervical plexus, and the anterior layer of the deep cervical fascia. By its *deep surface*, it rests on the sterno-clavicular articulation, the deep layer of the cervical fascia, the Sterno-hyoid, Sterno-thyroid, Omo-hyoid, the posterior belly of the Digastric, Levator anguli scapulæ, the Splenius and Scaleni muscles. Below, with the lower part of the common carotid artery, internal jugular vein, pneumogastric, descendens noni, and communicans noni nerves, and with the deep lymphatic glands; with the spinal accessory nerve, which pierces its upper third, the cervical plexus, the occipital artery, and a part of the parotid gland.

*Nerves.* The Platysma-myoides is supplied by the facial and superficial cervical nerves. The Sterno-cleido-mastoid by the spinal accessory and deep branches of the cervical plexus.

*Actions.* The Platysma-myoides produces a slight wrinkling of the surface of the skin of the neck, in a vertical direction, when the entire muscle is brought into action. Its anterior portion, the thickest part of the muscle, depresses the lower jaw; it also serves to draw down the lower lip and angle of the mouth on each side, being one of the chief agents in the expression of melancholy. The Sterno-mastoid muscles, when both are brought into action, serve to depress the head upon the neck, and the neck upon the chest. Either muscle, acting singly, flexes the head, and (combined with the Splenius) draws it towards the shoulder of the same side, and rotates it so as to carry the face towards the opposite side.

*Surgical Anatomy.* The relations of the sternal and clavicular parts of the Sterno-mastoid should be carefully examined, as the surgeon is sometimes required to divide one or both portions of the muscle in *wry neck.* One variety of this distortion is produced by spasmodic contraction or rigidity of the Sterno-mastoid; the head being carried down towards the shoulder of the same side, and the face turned to the opposite side, and fixed in that position. When all other remedies for the relief of this disease have failed, subcutaneous division of the muscle is resorted to. This is performed by introducing a long narrow bistoury beneath it, about half an inch above its origin, and dividing it from behind forwards whilst the muscle is put well upon the stretch. There is seldom any difficulty in dividing the sternal portion. In dividing the clavicular portion care must be taken to avoid wounding the external jugular vein, which runs parallel with the posterior border of the muscle in this situation.

## 2. Infra-Hyoid Region (Figs. 153, 154).

### Depressors of the Os Hyoides and Larynx.

| | |
|---|---|
| Sterno-Hyoid. | Thyro-Hyoid. |
| Sterno-Thyroid. | Omo-Hyoid. |

*Dissection.* The muscles in this region may be exposed by removing the deep fascia from the front of the neck. In order to see the entire extent of the Omo-hyoid, it is necessary to divide the Sterno-mastoid at its centre, and turn its ends aside, and to detach the Trapezius from the clavicle and scapula, if this muscle has been previously dissected; but not otherwise.

The *Sterno-Hyoid* is a thin, narrow, riband-like muscle, which arises from the inner extremity of the clavicle, and the upper and posterior part of the first piece of the sternum; and, passing upwards and

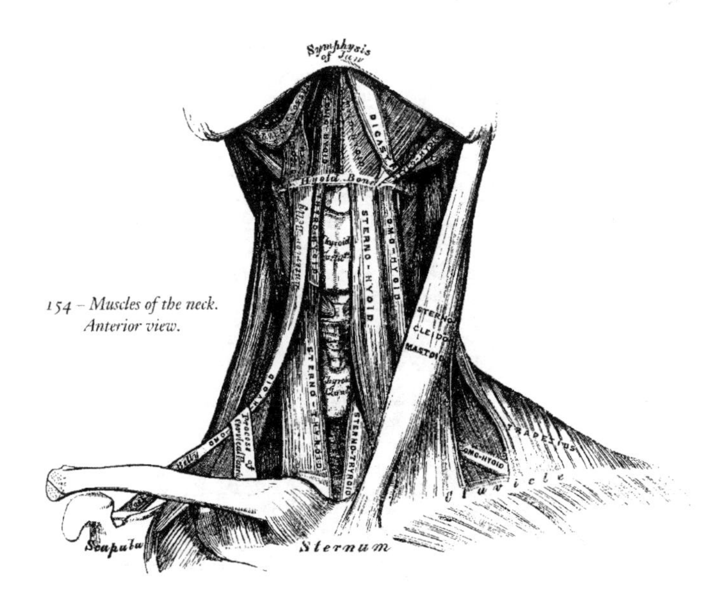


154 – *Muscles of the neck. Anterior view.*

inwards, is inserted, by short tendinous fibres, into the lower border of the body of the os hyoides. This muscle is separated, below, from its fellow by a considerable interval; but they approach one another in the middle of their course, and again diverge as they ascend. It often presents, immediately above its origin, a transverse tendinous intersection, analogous to those in the Rectus abdominis.

*Variations.* This muscle sometimes arises from the inner extremity of the clavicle, and the posterior sterno-clavicular ligament; or from the sternum and this ligament; from either bone alone, or from all these parts; and occasionally has a fasciculus connected with the cartilage of the first rib.

*Relations.* By its *superficial surface*, below, with the sternum, sternal end of the clavicle, and the Sterno-mastoid; and above, with the Platysma and deep cervical fascia. By its *deep surface*, with the Sterno-thyroid, Crico-thyroid, and Thyro-hyoid muscles, the thyroid gland, the superior thyroid vessels, the crico-thyroid and thyro-hyoid membranes.

The *Sterno-Thyroid* is situated beneath the preceding muscle, but is shorter and wider than it. It arises from the posterior surface of the first bone of the sternum, below the origin of the Sterno-hyoid, and occasionally from the edge of the cartilage of the first rib; and is inserted into the oblique line on the side of the ala of the thyroid cartilage. This muscle is in close contact with its fellow at the lower part of the neck; and is frequently traversed by a transverse or oblique tendinous intersection, analogous to those in the Rectus abdominis.

*Variations.* This muscle is sometimes continuous with the Thyro-hyoid and Inferior constrictor of the pharynx; and a lateral prolongation from it sometimes passes as far as the os hyoides.

*Relations.* By its *anterior surface*, with the Sterno-hyoid, Omo-hyoid, and Sterno-mastoid. By its *posterior surface*, from below upwards, with the trachea vena innominata, common carotid (and on the right side the arteria innominata, the thyroid gland and its vessels, and the lower part of the larynx. The middle thyroid vein lies along its inner border, an important relation to be remembered in the operation of tracheotomy.

The *Thyro-Hyoid* is a small quadrilateral muscle, appearing like a continuation of the Sterno-thyroid. It arises from the oblique line on the side of the thyroid cartilage, and passes vertically upwards to be inserted into the lower border of the body, and greater cornu of the hyoid bone.

*Relations.* By its *external surface*, with the Sterno-hyoid and Omo-hyoid muscles. By its *internal surface*, with the thyroid cartilage, the thyro-hyoid membrane, and the superior laryngeal vessels and nerve.

The *Omo-hyoid* passes across the side of the neck, from the scapula to the hyoid bone. It consists of two fleshy bellies, united by a central tendon. It arises from the upper border of the scapula, and occasionally from the transverse ligament which crosses the supra-scapular notch; its extent of attachment to the scapula varying from a few lines to an inch. From this origin, the posterior belly forms a flat, narrow fasciculus, which inclines forwards across the lower part of the neck; behind the Sterno-mastoid muscle, where it becomes tendinous, it changes its direction, forming an obtuse angle, and ascends almost vertically upwards, close to the outer border of the Sterno-hyoid, to be inserted into the lower border of the body of the os hyoides, just external to the insertion of the Sterno-hyoid. The tendon of this muscle, which much varies in its length and form in different subjects, is held in its position by a process of the deep cervical fascia, which includes it in a sheath, and is prolonged down, to be attached to the cartilage of the first rib. It is by this means that the angular form of the muscle is maintained.

This muscle subdivides each of the two large triangles at the side of the neck into two smaller triangles. The two posterior ones being the *posterior superior* or *sub-occipital*, and the *posterior inferior* or *subclavian*; the two anterior, the *anterior superior* or *superior carotid*, and the *anterior inferior* or *inferior carotid* triangle.

*Relations.* By its *superficial surface*, with the Trapezius, Subclavius, the clavicle, the Sterno-mastoid, deep cervical fascia, Platysma, and integument. By its *deep surface*, with the Scaleni, brachial plexus, sheath of the common carotid artery, and internal jugular vein, the descendens noni nerve, Sterno-thyroid and Thyro-hyoid muscles.

*Nerves.* The Thyro-hyoid is supplied by the hypo-glossal; the other muscles of this group by branches from the loop of communication between the descendens and communicans noni.

*Actions.* These muscles depress the larynx and hyoid bone, after they have been drawn up with the pharynx in the act of deglutition. The Omo-hyoid muscles not only depress the hyoid bone, but carry it backwards, and to one or the other side. These muscles also are tensors of the cervical fascia.

The Thyro-hyoid may act as an elevator of the thyroid cartilage, when the hyoid bone ascends drawing upwards the thyroid cartilage behind the os hyoides.

## 3. Supra-Hyoid Region (Figs. 153, 154).

### Elevators of the Os Hyoides—Depressors of the Lower Jaw.

| | |
|---|---|
| Digastric. | Mylo-Hyoid. |
| Stylo-Hyoid. | Genio-Hyoid. |

*Dissection.* To dissect these muscles, a block should be placed beneath the back of the neck, and the head drawn backwards, and retained in that position. On the removal of the deep fascia, the muscles are at once exposed.

The *Digastric* consists of two fleshy bellies united by an intermediate rounded tendon. It is a small muscle, situated below the side of the body of the lower jaw, and extending, in a curved form, from the side of the head to the symphysis of the jaw. The *posterior belly*, longer than the anterior, arises from the digastric groove on the inner side of the mastoid process of the temporal bone, and passes downwards, forwards, and inwards. The *anterior belly*, being reflected upwards and forwards, is inserted into a depression on the inner side of the lower border of the jaw, close to the symphysis. The tendon of the muscle perforates the Stylo-hyoid, and is held in connection with the side of the body of the hyoid bone by an aponeurotic loop, lined by a synovial membrane. A broad aponeurotic layer is given off from the tendon of the Digastric on each side, which is attached to the body and great cornu of the hyoid bone: this is termed the *supra-hyoid aponeurosis*. It forms a strong layer of fascia between the anterior portion of the two muscles, and a firm investment for the other muscles of the supra-hyoid region which lie beneath it.

The Digastric muscle divides the anterior superior triangle of the neck into two smaller triangles; the upper, or sub-maxillary, being bounded, above, by the lower jaw, and mastoid process; below, by the two bellies of the Digastric muscle: the lower, or superior carotid triangle, being bounded, above, by the posterior belly of the Digastric; behind, by the Sterno-mastoid; below, by the Omo-hyoid.

*Relations.* By its *superficial surface*, with the Platysma, Sterno- and Trachelo-mastoid, part of the Stylo-hyoid muscle, and the parotid and sub-maxillary glands. By its *deep surface*, the anterior belly lies on the Mylo-hyoid; the posterior belly on the Stylo-glossus, Stylo-pharyngeus, and Hyo-glossus muscles, the external carotid and its lingual and facial branches, the internal carotid, internal jugular vein, and hypoglossal nerve.

The *Stylo-Hyoid* is a small, slender muscle, lying in front of, and above, the posterior belly of the Digastric. It arises from the middle of the outer surface of the styloid process; and, passing downwards and forwards, is inserted into the body of the hyoid bone, just at its junction with the greater cornu, and immediately above the Omo-hyoid. This muscle is perforated, near its insertion, by the tendon of the Digastric.

*Relations.* The same as the posterior belly of the Digastric.

The Digastric and Stylo-hyoid should be removed, in order to expose the next muscle.

The *Mylo-Hyoid* is a flat triangular muscle, situated immediately beneath the anterior belly of the Digastric, and forming, with its fellow of the opposite side, a muscular floor for the cavity of the mouth. It arises from the whole length of the mylo-hyoid ridge, from the symphysis in front, to the last molàr tooth behind. The posterior fibres pass obliquely forwards, to be inserted into the body of the os hyoides. The middle and anterior fibres are inserted into a median fibrous raphe, where they join at an angle with the fibres of the opposite muscle. This median raphe is sometimes wanting; the muscular fibres of the two sides are then directly continuous with one another.

*Relations.* By its *cutaneous surface*, with the Platysma, the anterior belly of the Digastric, the supra-hyoid fascia, the submaxillary gland, and submental vessels. By its *deep* or *superior surface*, with the Genio-hyoid, part of the Hyo-glossus, and Stylo-glossus muscles, the lingual and gustatory nerves, the sublingual gland, and the buccal mucous membrane. Wharton's duct curves round its posterior border in its passage to the mouth.

*Dissection.* The Mylo-hyoid should now be removed, in order to expose the muscles which lie beneath; this is effected by detaching it from its attachments to the hyoid bone and jaw, and separating it by a vertical incision from its fellow of the opposite side.

The *Genio-Hyoid* is a narrow slender muscle, situated immediately beneath the inner border of the preceding. It arises from the inferior genial tubercle on the inner side of the symphysis of the jaw, and passes downwards and backwards, to be inserted into the anterior surface of the body of the os hyoides. This muscle lies in close contact with its fellow of the opposite side, and increases slightly in breadth as it descends.

*Relations.* It is covered by the Mylo-hyoid, and lies on the Genio-hyo-glossus.

*Nerves.* The Digastric is supplied, its anterior belly, by the mylo-hyoid branch of the inferior dental; its posterior belly, by the facial; the Stylo-hyoid, by the facial; the Mylo-hyoid, by the mylo-hyoid branch of the inferior dental; the Genio-hyoid, by the hypoglossal.

*Actions.* This group of muscles performs two very important actions. They raise the hyoid bone, and with it the base of the tongue, during the act of deglutition; or, when the hyoid bone is fixed by its depressors and those of the larynx, they depress the lower jaw. During the first act of deglutition, when the mass is being driven from the mouth into the pharynx, the hyoid bone, and with it the tongue, is carried upwards and forwards by the anterior belly of the Digastric, the Mylo-hyoid, and Genio-hyoid muscles. In the second act, when the mass is passing through the pharynx, the direct elevation of the hyoid bone takes place by the combined action of all the muscles; and after the food has passed, the hyoid bone is carried upwards and backwards by the posterior belly of the Digastric and Stylo-hyoid muscles, which assists in preventing the return of the morsel into the cavity of the mouth.

## 4. Lingual Region.

Genio-Hyo-Glossus. Lingualis.
Hyo-Glossus. Stylo-Glossus.
Palato-Glossus.

*Dissection.* After completing the dissection of the preceding muscles, saw through the lower jaw just external to the symphysis. The tongue should then be drawn forwards with a hook, and its muscles, which are thus put on the stretch, may be examined.

The *Genio-Hyo-Glossus* has received its name from its triple attachment to the chin, hyoid bone, and tongue; it is a thin, flat, triangular muscle, placed vertically in the middle line, its apex corresponding with its point of attachment to the lower jaw, its base with its insertion into the tongue and hyoid bone. It arises by a short tendon from the superior genial tubercle on the inner side of the symphysis of the chin, immediately above the Genio-hyoid; from this point, the muscle spreads out in a fan-like form, the inferior fibres passing downwards, to be inserted into the upper part of the body of the hyoid bone, a few being continued into the side of the pharynx; the middle fibres passing backwards, and the superior one upwards and forwards, to be attached to the whole length of the under surface of the tongue, from the base to the apex.

*Relations.* By its *internal surface*, it is in contact with its fellow of the opposite side, from which it is separated, at the back part of the tongue, by a fibro-cellular structure, which extends forwards through the middle of the organ. By its *external surface*, with the Lingualis, Hyo-glossus, and Stylo-glossus, the lingual artery and hypoglossal nerve, the gustatory nerve, and sublingual gland. By its *upper border*, with the mucous membrane of the floor of the mouth. By its *lower border*, with the Genio-hyoid.

The *Hyo-Glossus* is a thin, flat, quadrilateral muscle, arising from the side of the body, the lesser cornu, and whole length of the greater cornu of the hyoid bone, and passing almost vertically upwards, is inserted into the side of the tongue between the Stylo-glossus and Lingualis. Those fibres of this muscle which arise from the body, are directed upwards and backwards, overlapping those from the greater cornu, which are directed obliquely forwards. Those from the lesser cornu extend forwards and outwards along the side of the tongue, under cover of the portion arising from the body.

The difference in the direction of the fibres of this muscle, and their separate origin from different segments of the hyoid bone, led Albinus and other anatomists to describe it as three muscles, under the names of the Basio-glossus, the Cerato-glossus, and the Chondro-glossus.

*Relations.* By its *external surface*, with the Digastric, the Stylo-hyoid, Stylo-glossus, and Mylo-hyoid muscles, the gustatory and hypoglossal nerves, Wharton's duct, and the sublingual gland. By its *deep surface*, with the Genio-hyo-glossus, Lingualis, and the Middle constrictor, the lingual vessels, and the glosso-pharyngeal nerve.

The *Lingualis* is a longitudinal band of muscular fibres, situated on the under surface of the tongue, lying in the interval between the Hyo-glossus and the Genio-hyo-glossus, and extending from the base to the apex of that organ. Posteriorly, some of its fibres are lost in the base of the tongue, and others are attached to the hyoid bone. It blends with the fibres of the Stylo-glossus, in front of the Hyo-glossus, and is continued forwards as far as the apex of the tongue. It is in relation, by its under surface, with the ranine artery.

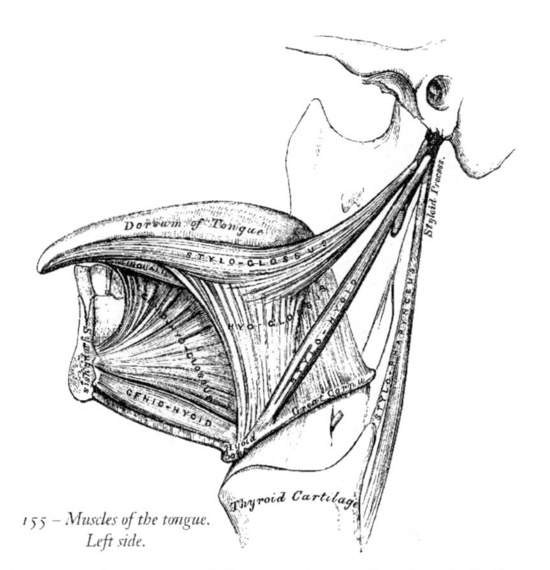


155 – *Muscles of the tongue. Left side.*

The *Stylo-Glossus*, the shortest and smallest of the three styloid muscles, arises from the anterior and outer side of the styloid process, near its centre, and from the stylo-maxillary ligament, to which its fibres, in most cases, are attached by a thin aponeurosis. Passing downwards and forwards, so as to become nearly horizontal in its direction, it divides upon the side of the tongue into two portions: one longitudinal, which is inserted along the side of the tongue, blending with the fibres of the Lingualis, in front of the Hyo-glossus; the other, oblique, which overlaps the Hyo-glossus muscle, and decussates with its fibres.

*Relations.* By its *external surface*, from above downwards, with the parotid gland, the Internal pterygoid muscle, the sublingual gland, the gustatory nerve, and the mucous membrane of the mouth. By its *internal surface*, with the tonsil, the Superior constrictor, and the Hyo-glossus muscle.

The *Palato-Glossus*, or *Constrictor Isthmi Faucium*, although one of the muscles of the tongue, serving to draw its base upwards during the act of deglutition, is more nearly associated with the soft palate, both in its situation and function; it will, consequently, be described with that group of muscles.

*Nerves.* The Palato-glossus is supplied by the palatine branches of Meckel's ganglion; the Lingualis, by the chorda tympani; the remaining muscles of this group, by the hypoglossal.

*Actions.* The movements of the tongue, although numerous and complicated, may easily be understood by carefully considering the direction of the fibres of the muscles of this organ. The *Genio-hyo-glossi*, by means of their posterior and inferior fibres, draw upwards the hyoid bone, bringing it and the base of the tongue forwards, so as to protrude the apex from the mouth. The anterior fibres will restore it to its original position by retracting the organ within the mouth. The whole length of these two muscles acting along the middle line of the tongue will draw it downwards, so as to make it concave from before backwards, forming a channel along which fluids may pass towards the pharynx, as in sucking. The *Hyo-glossi* muscles draw down the sides of the tongue, so as to render it convex from side to side. The *Linguales*, by drawing downwards the centre and apex of the tongue, render it convex from before backwards. The *Palato-glossi* draw the base of the tongue upwards, and the Stylo-glossi upwards and backwards.

## 5. Pharyngeal Region.

Constrictor Inferior. Constrictor Superior.
Constrictor Medius. Stylo-pharyngeus.
Palato-pharyngeus.

*Dissection* (fig. 156). In order to examine the muscles of the pharynx, cut through the trachea and œsophagus just above the sternum, and draw them upwards by dividing the loose areolar tissue connecting the pharynx with the front of the vertebral column. The parts being drawn well forwards, the edge of the saw should be applied immediately behind the styloid processes, and the base of the skull sawn through from below upwards. The pharynx and mouth should then be stuffed with tow, in order to distend its cavity and render the muscles tense and easier of dissection.

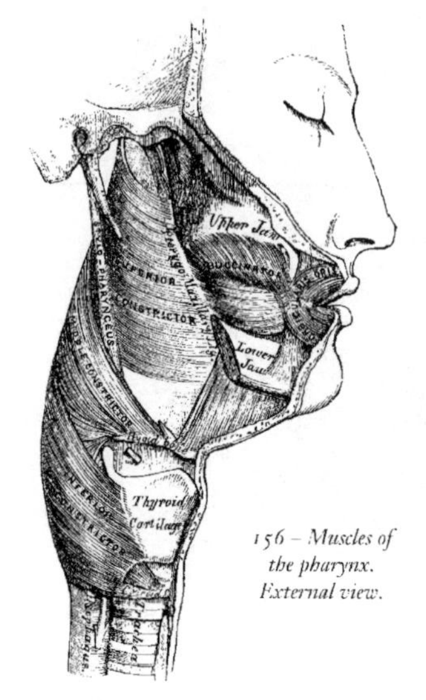


156 – *Muscles of the pharynx. External view.*

The *Inferior Constrictor*, the most superficial and thickest of the three constrictors, arises from the side of the cricoid and thyroid cartilages. To the cricoid cartilage it is attached in the interval between the crico-thyroid, in front, and the articular faces for the thyroid cartilage behind. To the thyroid cartilage it is attached to the oblique line on the side of the great ala, the cartilaginous surface behind it, nearly as far as its posterior border, and to the inferior cornu. From these attachments, the fibres spread backwards and inwards, to be inserted into the fibrous raphe in the posterior median line of the pharynx. The inferior fibres are horizontal, and continuous with the fibres of the œsophagus; the rest ascend, increasing in obliquity, and overlap the Middle constrictor. The superior laryngeal nerve passes near the upper border, and the inferior, or recurrent laryngeal, beneath the lower border of this muscle, previous to their entering the larynx.

*Relations.* It is covered by a dense cellular membrane which surrounds the entire pharynx. *Behind*, it lies on the vertebral column and the Longus colli. *Laterally*, it is in relation with the thyroid gland, the common carotid artery, and the Sterno-thyroid muscle. By its *internal surface*, with the Middle constrictor, the Stylo-pharyngeus, Palato-pharyngeus, and the mucous membrane of the pharynx.

The *Middle Constrictor* is a flattened, fan-shaped muscle, smaller than the preceding, and situated on a plane anterior to it. It arises from the whole length of the greater cornu of the hyoid bone, from the lesser cornu, and from the stylo-hyoid ligament. The fibres diverge from their origin; the lower ones descending beneath the Inferior constrictor, the middle fibres passing transversely, and the upper fibres ascending to cover in the Superior constrictor. It is inserted into the posterior median fibrous raphe, blending in the middle line with the fibres of the opposite muscle.

*Relations.* This muscle is separated from the Superior constrictor by the glosso-pharyngeal nerve and the Stylo-pharyngeus muscle; and from the Inferior constrictor, by the superior laryngeal nerve. *Behind*, it lies on the vertebral column, the Longus colli, and the Rectus anticus major. *On each side* it is in relation with the carotid vessels, the pharyngeal plexus, and some lymphatic glands. Near its origin, it is covered by the Hyo-glossus, from which it is separated by the lingual artery. It covers in the Superior constrictor, the Stylo-pharyngeus, the Palato-pharyngeus, and the mucous membrane.

The *Superior Constrictor* is a quadrilateral muscle, thinner and paler than the other constrictors, and situated at the upper part of the pharynx. It arises from the lower third of the margin of the internal pterygoid plate and its hamular process, from the contiguous portion of the palate bone and the reflected tendon of the Tensor palati muscle, from the pterygo-maxillary ligament, from the alveolar process above the posterior extremity of the mylo-hyoid ridge, and by a few fibres from the side of the tongue in connexion with the Genio-hyo-glossus. From these points, the fibres curve backwards, to be inserted into the median raphe, being also prolonged by means of a fibrous aponeurosis to the pharyngeal spine on the basilar process of the occipital bone. Its superior fibres arch beneath the Levator palati and the Eustachian tube, the interval between the upper border of the muscle and the basilar process being deficient in muscular fibres, and closed by fibrous membrane.

*Relations.* By its *outer surface*, with the vertebral column, the carotid vessels, the internal jugular vein, the three divisions of the eighth and ninth nerves, the Middle constrictor which overlaps it, and the Stylo-pharyngeus. It covers the Palato-pharyngeus and the tonsil, and is lined by mucous membrane.

The *Stylo-pharyngeus* is a long, slender muscle, round above, broad and thin below. It arises from the inner side of the base of the styloid process, passes downwards along the side of the pharynx between the Superior and Middle constrictors, and spreading out beneath the mucous membrane, some of its fibres are lost in the constrictor muscles, and others joining with the Palato-pharyngeus, are inserted into the upper border of the thyroid cartilage. The glosso-pharyngeal nerve runs on the outer side of this muscle, and crosses over it in passing forward to the tongue.

*Relations. Externally*, with the Stylo-glossus muscle, the external carotid artery, the parotid gland, and the Middle constrictor. *Internally*, with the internal carotid, the internal jugular vein, the Superior constrictor, Palato-pharyngeus and mucous membrane.

*Nerves.* The muscles of this group are supplied by branches from the pharyngeal plexus and glosso-pharyngeal nerve; and the Inferior constrictor, by an additional branch from the external laryngeal nerve.

*Actions.* When deglutition is about to be performed, the pharynx is drawn upwards and dilated in different directions, to receive the morsel propelled into it from the mouth. The Stylo-pharyngei, which are much farther removed from one another at their origin than at their insertion, draw upwards and outwards the sides of this cavity, the breadth of the pharynx in the antero-posterior direction being increased, by the larynx and tongue being carried forwards in their ascent. As soon as the morsel is received in the pharynx, the elevator muscles relax, the bag descends, and the Constrictors contract upon the morsel, and convey it gradually downwards into the œsophagus. The pharynx also exerts an important influence in the modulation of the voice, especially in the production of the higher tones.

## 6. Palatal Region.

Levator Palati. Azygos Uvulæ.
Tensor Palati. Palato-glossus.
Palato-pharyngeus.

*Dissection* (fig. 157). Lay open the pharynx from behind, by a vertical incision extending from its upper to its lower part, and partially divide the occipital attachment by a transverse incision on each side of the vertical one; the posterior surface of the soft palate is then exposed. Having fixed the uvula so as to make it tense, the mucous membrane and glands should be carefully removed from the posterior surface of the soft palate, and the muscles of this part are at once exposed.

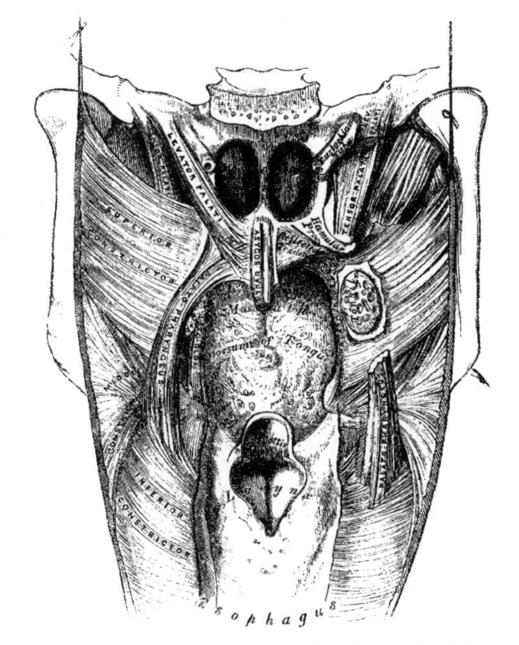

*157 – Muscles of the soft palate. The pharynx being laid open from behind.*

The *Levator Palati* is a long, thick, rounded muscle, placed on the outer side of the posterior aperture of the nares. It arises from the under surface of the apex of the petrous portion of the temporal bone, and from the adjoining cartilaginous portion of the Eustachian tube; after passing into the pharynx, above the upper concave margin of the Superior constrictor, it descends obliquely downwards and inwards, its fibres spreading out in the posterior surface of the soft palate as far as the middle line, where they blend with those of the opposite side.

*Relations. Externally*, with the Tensor palati and Superior constrictor. *Internally*, it is lined by the mucous membrane of the pharynx. *Posteriorly*, with the mucous lining of the soft palate. This muscle must be removed and the pterygoid attachment of the Superior constrictor dissected away, in order to expose the next muscle.

The *Circumflexus* or *Tensor Palati* is a broad, thin, riband-like muscle, placed on the outer side of the preceding, and consisting of a vertical and a horizontal portion. The vertical portion arises by a broad, thin, and flat lamella from the scaphoid fossa at the base of the internal pterygoid plate, its origin extending as far back as the spine of the sphenoid; it also arises from the anterior aspect of the cartilaginous portion of the Eustachian tube; descending vertically between the internal pterygoid plate and the inner surface of the Internal pterygoid muscle; it terminates in a tendon which winds round the hamular process, being retained in this situation by a tendon of origin of the Internal pterygoid muscle, and lubricated by a bursa. The tendon or horizontal portion then passes horizontally inwards, and expands into a broad aponeurosis on the anterior surface of the soft palate, which unites in the median line with the aponeurosis of the opposite muscle, the fibres being attached in front to the transverse ridge on the posterior border of the horizontal portion of the palate bone.

*Relations. Externally*, with the Internal pterygoid. *Internally*, with the Levator palati, from which it is separated by the Superior constrictor, and the internal pterygoid plate. In the soft palate its aponeurotic expansion is anterior to that of the Levator palati, being covered by mucous membrane.

The *Azygos Uvulæ* is not a single muscle as implied by its name, but a pair of narrow cylindrical fleshy fasciculi, placed side by side in the median line of the soft palate. Each muscle arises from the posterior nasal spine of the palate bone, and from the contiguous tendinous aponeurosis of the soft palate, and descends to be inserted into the uvula.

*Relations. Anteriorly*, with the tendinous expansion of the Levatores palati; *behind*, with the mucous membrane.

The two next muscles are exposed by removing the mucous membrane which covers the pillars of the soft palate throughout nearly their whole extent.

The *Palato-Glossus* (*Constrictor Isthmi Faucium*) is a small fleshy fasciculus, narrower in the middle than at either extremity, forming, with the mucous membrane covering its surface, the anterior pillar of the soft palate. It arises from the anterior surface of the soft palate on each side of the uvula, and passing forwards and outwards in front of the tonsil, is inserted into the side and dorsum of the tongue, where it blends with the fibres of the Stylo-glossus muscle. In the soft palate, the fibres of this muscle are continuous with those of the opposite side.

The *Palato-Pharyngeus* is a long fleshy fasciculus, narrower in the middle than at either extremity, forming, with the mucous membrane covering its surface, the posterior pillar of the soft palate. It is separated from the preceding by an angular interval, in which the tonsil is lodged. It arises from the soft palate by an expanded fasciculus, which is divided into two parts by the Levator palati. The *anterior fasciculus*, the thickest, enters the soft palate between the Levator and Tensor, and joins in the middle line the corresponding part of the opposite muscle; the *posterior fasciculus* lies in contact with the mucous membrane, and also joins with the corresponding muscle in the middle line. Passing outwards and downwards behind the tonsil, it joins the Stylo-pharyngeus, and is inserted with it into the posterior border of the thyroid cartilage, some of its fibres being lost on the side of the pharynx, and others passing across the middle line posteriorly, to decussate with the muscle of the opposite side.

*Relations.* In the soft palate, its *anterior* and *posterior surfaces* are covered by mucous membrane, from which it is separated by a layer of palatine glands. By its *superior border*, it is in relation with the Levator palati. Where it forms the posterior pillar of the fauces, it is covered by mucous membrane, excepting on its outer surface. In the *pharynx*, it lies between the mucous membrane and the constrictor muscles.

*Nerves.* The Tensor palati is supplied by a branch from the otic ganglion; the Levator palati, and Azygos uvulæ, by the facial, through the connection of its trunk with the Vidian, by the petrosal nerves; the other muscles, by the palatine branches of Meckel's ganglion.

*Actions.* During the *first act* of deglutition, the morsel of food is driven back into the fauces by the pressure of the tongue against the hard palate; the base of the tongue being, at the same time, retracted, and the larynx raised with the pharynx, and carried forwards under it; the epiglottis is pressed over the superior aperture of the larynx, and the morsel glides past it. This constitutes the *second act* of deglutition; then the Palato-glossi muscles, the constrictors of the fauces contract behind it; the soft palate is slightly raised (by the Levator palati), and made tense (by the Tensor palati); and the Palato-pharyngæi contract, and come nearly together, the uvula filling up the slight interval between them. By these means, the food is prevented passing into the upper part of the pharynx or the posterior nares; at the same time, the latter muscles form an inclined plane directed obliquely downwards and backwards, along which the morsel descends into the lower part of the pharynx.

*Surgical Anatomy.* The muscles of the soft palate should be carefully dissected, the relations they bear to the surrounding parts especially examined, and their action attentively studied upon the dead subject, as the surgeon is required to divide one or more of these muscles in the operation of staphyloraphy. Mr. Ferguson has shewn, that in the congenital deficiency, called *cleft palate*, the edges of the fissure are forcibly separated by the action of the Levatores palati and Palato-pharyngæi muscles, producing very considerable impediment to the healing process after the performance of the operation for uniting the margins by adhesion; he has, consequently, recommended the division of these muscles as one of the most important steps in the operation: by these means, the flaps are relaxed, lie perfectly loose and pendulous, and are easily brought and retained in apposition. The Palato-pharyngæi may be divided by cutting across the posterior pillar of the soft palate just below the tonsil, with a pair of blunt-pointed curved scissors; and the anterior pillar may be divided also. To divide the Levator palati, the plan recommended by Mr. Pollock is to be greatly preferred. The flap being put upon the stretch, a double-edged knife is passed through the soft palate, just on the inner side of the hamular process, and above the line of the Levator palati. The handle being now alternately raised and depressed, a sweeping cut is made along the posterior surface of the soft palate, and the knife withdrawn, leaving only a small opening in the mucous membrane on the anterior surface. If this operation is performed on the dead body, and the parts afterwards dissected, the Levator palati will be found completely divided.

## 7. Vertebral Region (Anterior).

| | |
|---|---|
| Rectus Capitis Anticus Major. | Rectus Lateralis. |
| Rectus Capitis Anticus Minor. | Longus Colli. |

The *Rectus Capitis Anticus Major* (fig. 158), broad and thick above, narrow below, appears like a continuation upwards of the Scalenus anticus. It arises by four tendinous slips from the anterior tubercles of the transverse processes of the third, fourth, fifth, and sixth cervical vertebræ, and ascends, converging toward its fellow of the opposite side, to be inserted into the basilar process of the occipital bone.

*Relations.* By its *anterior surface*, with the pharynx, the sympathetic nerve and the sheath enclosing the carotid artery, internal jugular vein, and pneumogastric nerve. By its *posterior surface*, with the Longus colli, the Rectus anticus minor, and the upper cervical vertebræ.

The *Rectus Capitis Anticus Minor* is a short flat muscle, situated immediately beneath the upper part of the preceding. It arises from the anterior surface of the lateral mass of the atlas, and from the root of its transverse process; passing obliquely upwards and inwards, it is inserted into the basilar process immediately behind the preceding muscle.

*Relations.* By its *anterior surface*, with the Rectus anticus major. By its *posterior surface*, with the front of the occipito-atlantal articulation. *Externally*, with the superior cervical ganglion of the sympathetic.

The *Rectus Lateralis* is a short, flat muscle, situated between the transverse process of the atlas, and the jugular process of the occipital bone. It arises from the upper surface of the transverse process of the atlas, and is inserted into the under surface of the jugular process of the occipital bone.

*Relations.* By its *anterior surface*, with the internal jugular vein. By its *posterior surface*, with the vertebral artery.

The *Longus Colli* is a long, flat muscle, situated on the anterior surface of the spine, between the atlas and the third dorsal vertebra, being broad in the middle, narrow and pointed at each extremity. It consists of three portions, a superior oblique, an inferior oblique, and a vertical portion.

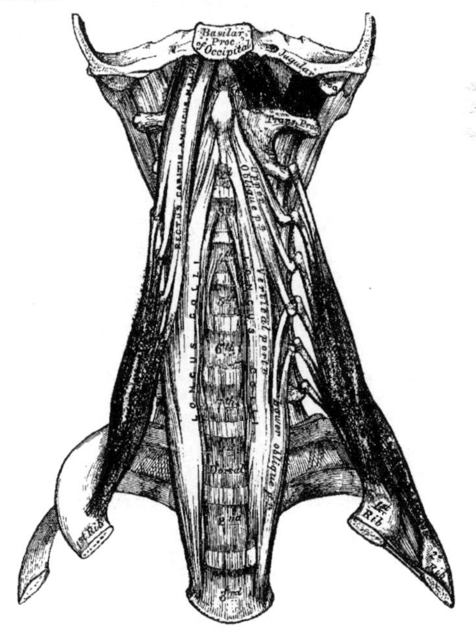

158 – *The pre-vertebral muscles.*

The *superior oblique portion* arises from the anterior tubercles of the transverse processes of the third, fourth, and fifth cervical vertebræ; and, ascending obliquely inwards, is inserted by a narrow tendon into the tubercle on the anterior arch of the atlas.

The *inferior oblique portion*, the smallest part of the muscle, arises from the bodies of the first two or three dorsal vertebræ; and, passing obliquely outwards, is inserted into the transverse processes of the fifth and sixth cervical vertebræ.

The *vertical portion* lies directly on the front of the spine, and is extended between the bodies of the lower three cervical and the upper three dorsal vertebræ below, and the bodies of the second, third, and fourth cervical vertebræ above.

*Relations.* By its *anterior surface*, with the pharynx, the œsophagus, sympathetic nerve, the sheath of the carotid artery, internal jugular vein, and pneumogastric nerve, inferior thyroid artery, and recurrent laryngeal nerve. By its *posterior surface*, with the cervical and dorsal portions of the spine. Its *inner border* is separated from the opposite muscle by a considerable interval below; but they approach each other above.

## 8. Vertebral Region (Lateral).

Scalenus Anticus. Scalenus Medius.
Scalenus Posticus.

The *Scalenus Anticus* is a conical-shaped muscle, situated deeply at the side of the neck, behind the Sterno-mastoid. It arises by a narrow, flat tendon from the tubercle on the inner border and upper surface of the first rib; and, ascending vertically upwards, is inserted into the anterior tubercles of the transverse processes of the third, fourth, fifth, and sixth cervical vertebræ. The lower part of this muscle separates the subclavian artery and vein; the latter being in front, and the former, with the brachial plexus, behind.

*Relations.* It is *covered* by the clavicle, the Subclavius, Sterno-mastoid, and Omo-hyoid muscles, the transversalis colli, and ascending cervical arteries, the subclavian vein, and the phrenic nerve. By its *posterior surface*, with the pleura, the subclavian artery, and brachial plexus of nerves. It is separated from the Longus colli, on the inner side, by the subclavian artery.

The *Scalenus Medius*, the largest and longest of the three Scaleni, arises, by a broad origin, from the upper surface of the first rib, behind the groove for the subclavian artery, as far back as the tubercle; and, ascending along the side of the vertebral column, is inserted, by separate tendinous slips, into the posterior tubercles of the transverse processes of the lower six cervical vertebræ. It is separated from the Scalenus anticus by the subclavian artery below, and the cervical nerves above.

*Relations.* By its *external surface*, with the Sterno-mastoid; it is crossed by the clavicle, and Omo-hyoid muscle. To its *outer side*, is the Levator anguli scapulæ, and the Scalenus posticus muscle.

The *Scalenus posticus*, the smallest of the three Scaleni, arises by a thin tendon from the outer surface of the second rib, behind the attachment of the Serratus magnus, and enlarging as it ascends, is inserted, by two or three separate tendons into the posterior tubercles of the transverse processes of the lower two or three cervical vertebræ. This is the most deeply-placed of the three Scaleni, and is occasionally blended with the Scalenus medius.

*Nerves.* The Rectus capitis anticus major and minor are supplied by the sub-occipital and deep branches of the cervical plexus; the Rectus lateralis, by the sub-occipital; and the Longus colli and Scaleni, by branches from the lower cervical nerves.

*Actions.* The Rectus anticus major and minor are the direct antagonists of those placed at the back of the neck, serving to restore the head to its natural position when drawn backwards by the posterior muscles. These muscles also serve to flex the head, and, from their obliquity, rotate it, so as to turn the face to one or the other side. The Longus colli will flex and slightly rotate the cervical portion of the spine. The Scaleni muscles, taking their fixed point from below, draw down the transverse processes of the cervical vertebræ, flexing the spinal column to one or the other side. If the muscles of both sides act, the spine will be kept erect. When taking their fixed point from above, they elevate the first and second ribs and are, therefore, inspiratory muscles.

# MUSCLES AND FASCIÆ OF THE TRUNK.

The Muscles of the Trunk may be subdivided into four groups:—

1. Muscles of the Back.
2. Muscles of the Abdomen.
3. Muscles of the Thorax.
4. Muscles of the Perinæum.

The Muscles of the Back are very numerous, and may be subdivided into five layers:—

First Layer.

Trapezius.
Latissimus dorsi.

Second Layer.

Levator anguli scapulæ.
Rhomboideus minor.
Rhomboideus major.

Third Layer.

Serratus posticus superior.
Serratus posticus inferior.
Splenius capitis.
Splenius colli.

Fourth Layer.

*Sacral and lumbar regions.*

Erector Spinæ.

*Dorsal region.*

Sacro-lumbalis.
Musculus accessorius ad sacro-lumbalem.
Longissimus dorsi.
Spinalis dorsi.

*Cervical region.*

Cervicalis ascendens.
Transversalis colli.
Trachelo-mastoid.
Complexus.
Biventer cervicis.
Spinalis cervicis.

Fifth Layer.

Semi-spinalis dorsi.
Semi-spinalis colli.
Multifidus spinæ.
Rotatores spinæ.
Supra-spinales.
Inter-spinales.
Extensor coccygis.
Inter-transversales.
Rectus posticus major.
Rectus posticus minor.
Obliquus superior.
Obliquus inferior.

## First Layer.

Trapezius. Latissimus Dorsi.

*Dissection* (fig. 159). The body should be placed in the prone position, with the arms extended over the sides of the table, and the chest and abdomen supported by several blocks, so as to render the muscles tense. An incision should then be made along the middle line of the back, from the occipital protuberance to the coccyx. From the upper end of this, a transverse incision should extend to the mastoid process; and from the lower end, a third incision should be made along the crest of the ilium to about its middle. This large intervening space, for convenience of dissection, should be subdivided by a fourth incision, extending obliquely from the spinous process of the last dorsal vertebra, upwards and outwards, to the acromion process. This incision corresponds with the lower border of the Trapezius muscle. The flaps of integument should then be removed in the direction shewn in figure 159.

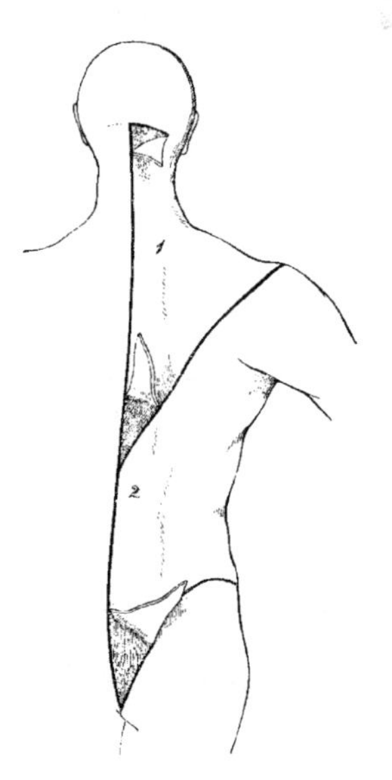

*159 – Dissection of the muscles of the back.*

The *Trapezius* is a broad, flat, triangular muscle, placed immediately beneath the skin, and covering the upper and back part of the neck and shoulders. It arises from the inner third of the superior curved line of the occipital bone; from the ligamentum nuchæ, the spinous process of the seventh cervical, and those of all the dorsal vertebræ; and from the corresponding portion of the supra-spinous ligament. From this origin, the superior fibres proceed downwards and outwards; the inferior ones, upwards and outwards; and the middle fibres, horizontally; and are inserted, the superior ones, into the outer third of the posterior border of the clavicle; the middle fibres, into the upper margin of the acromion process, and into the whole length of the upper border of the spine of the scapula; the inferior fibres converge near the scapula, and are attached to a triangular aponeurosis, which glides over a smooth surface at the inner extremity of the spine, and is inserted into a tubercle in immediate connection with its outer part. The Trapezius is fleshy in the greater part of its extent, but tendinous at its origin and insertion. At its occipital origin, it is connected to the bone by a thin fibrous lamina, firmly adherent to the skin, and wanting the lustrous, shining appearance of aponeurosis. At its origin from the spines of the vertebræ, it is connected by means of a broad semi-elliptical aponeurosis, which occupies the space between the sixth cervical and the third dorsal vertebræ, and forms, with the aponeurosis of the opposite muscle, a tendinous ellipse. The remaining part of the origin is effected by numerous short tendinous fibres. If the Trapezius is dissected on both sides, the two muscles resemble a trapezium, or diamond-shaped quadrangle; two angles, corresponding to the shoulders; a third, to the occipital protuberance; and the fourth, to the spinous process of the last dorsal vertebra.

The clavicular insertion of this muscle varies as to the extent of its attachment; it sometimes advances as far as the middle of the clavicle, and may even become blended with the posterior edge of the Sterno-mastoid, or overlap it. This should be borne in mind in the operation for tying the third part of the subclavian artery.

*Relations.* By its *superficial surface*, with the integument to which it is closely adherent above, but separated below by an aponeurotic lamina. By its *deep surface*, in the neck, with the Complexus, Splenius, Levator anguli scapulæ, and Rhomboideus minor; in the back, with the Rhomboideus major, Supra-spinatus, Infra-spinatus, a small portion of the Serratus posticus superior, the intervertebral aponeurosis which separates it from the Erector spinæ, and with the Latissimus dorsi. The spinal accessory nerve passes beneath the anterior border of this muscle, near the clavicle. The outer margin of its cervical portion forms the posterior boundary of the posterior triangle of the neck, the other boundaries being the Sterno-mastoid in front, and the clavicle below.

The *ligamentum nuchæ* (fig. 160) is a thin band of condensed cellulo-fibrous membrane, placed in the line of union between the two Trapezii in the neck. It extends from the external occipital protuberance to the spinous process of the seventh cervical vertebra, where it is continuous with the supra-spinous ligament. From its anterior surface a fibrous slip is given off to the spinous process of each of the cervical vertebræ, excepting the atlas, so as to form a septum between the muscles on each side of the neck. In man, it is merely the rudiment of an important elastic ligament, which, in some of the lower animals, serves to sustain the weight of the head.

The *Latissimus Dorsi* is a broad flat muscle, which covers the lumbar and lower half of the dorsal regions, and is gradually contracted into a narrow fasciculus at its insertion into the humerus. It arises by an aponeurosis from the spinous processes of the six inferior dorsal, from those of the lumbar and sacral vertebræ, and from the supra-spinous ligament. Over the sacrum, the aponeurosis of this muscle blends with the tendon of the Erector spinæ. It also arises from the external lip of the crest of the ilium, behind the origin of the External oblique, and by fleshy digitations from the three or four lower ribs, being interposed between similar processes of the External oblique muscle. From this extensive origin the fibres pass in different directions, the upper ones horizontally, the middle ones obliquely upwards, and the lower ones vertically upwards, so as to converge and form a thick fasciculus, which crosses the inferior angle of the scapula, and occasionally receives a few fibres from it. The muscle then curves around the lower border of the Teres major, and is twisted upon itself, so that the superior fibres become at first posterior and then inferior, and the vertical fibres at first anterior and then superior. It then terminates in a short quadrilateral tendon, about three inches in length, which, passing in front of the tendon of the Teres major, is inserted into the bottom of the bicipital groove of the humerus, above the insertion of the tendon of the Pectoralis major. The lower border of the tendon of this muscle is united with that of the Teres major, the surfaces of the two being separated by a bursa; another bursa is sometimes interposed between the muscle and the inferior angle of the scapula.

A muscular slip, varying from 3 to 4 inches in length, and from $\frac{1}{4}$ to $\frac{3}{4}$ of an inch broad, occasionally arises from the upper edge of the Latissimus dorsi, about the middle of the posterior fold of the axilla, crosses the axilla in front of the axillary vessels and nerves, to join the under surface of the tendon of the Pectoralis major, the Coraco-brachialis, or the fascia over the Biceps. The position of this abnormal slip, is a point of interest in its relation to the axillary artery, as it crosses the vessel just above the spot usually selected for the application of a ligature, and may mislead the surgeon during the operation. It may be easily recognised by the transverse direction of its fibres. Dr. Struthers found it in 8 out of 105 subjects, occurring 7 times on both sides.

*Relations.* Its *superficial surface* is subcutaneous, excepting at its upper part, where it is covered by the Trapezius. By its *deep surface*, it is in relation with the Erector spinæ, the Serratus posticus inferior, lower Intercostal muscles and ribs, the Serratus magnus, inferior angle of the scapula, Rhomboideus major, Infra-spinatus, and Teres major. Its outer margin is separated below, from the External oblique, by a small triangular interval; and another triangular interval exists between its upper border and the margin of the Trapezius, in which the Intercostal and Rhomboideus major muscles are exposed.

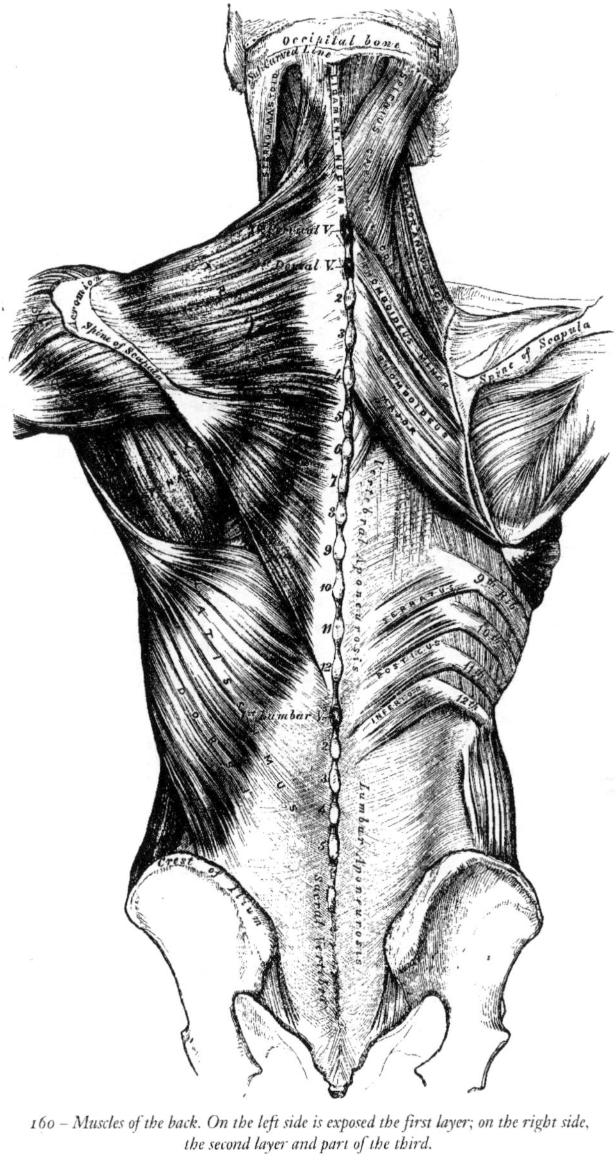

*160 – Muscles of the back. On the left side is exposed the first layer; on the right side, the second layer and part of the third.*

*Nerves.* The Trapezius is supplied by the spinal accessory and cervical plexus; the Latissimus dorsi, by the subscapular nerves.

## Second Layer.

Levator Anguli Scapulæ. Rhomboideus Minor.
Rhomboideus Major.

*Dissection.* The Trapezius must be removed in order to expose the next layer; to effect this, the muscle must be detached from its attachment to the clavicle and spine of the scapula, and turned back towards the spine.

The *Levator Anguli Scapulæ* is situated at the back part and side of the neck. It arises by four tendinous slips from the posterior tubercles of the transverse processes of the three or four upper cervical vertebræ; these becoming fleshy are united so as to form a flat muscle, which, passing downwards and backwards, is inserted into the posterior border of the scapula, between the superior angle and the triangular smooth surface at the root of the spine.

*Relations.* By its *superficial surface*, with the integument, Trapezius, and Sterno-mastoid. By its *deep surface*, with the Splenius colli, Transversalis colli, Cervicalis ascendens, and Serratus posticus superior, and with the transverse cervical and posterior scapular arteries.

The *Rhomboideus Minor* arises from the ligamentum nuchæ, and spinous processes of the seventh cervical and first dorsal vertebræ. Passing downwards and outwards, it is inserted into the margin of the triangular smooth surface at the root of the spine of the scapula. This small muscle is usually separated from the Rhomboideus major by a slight cellular interval.

The *Rhomboideus Major* is situated immediately below the preceding, the adjacent margins of the two being occasionally united. It arises by tendinous fibres from the spinous processes of the four or five upper dorsal vertebræ and the supra-spinous ligament, and is inserted into a narrow, tendinous arch, attached above, to the triangular surface near the spine; below, to the inferior angle, the arch being connected to the border of the scapula by a thin membrane. When the arch extends, as it occasionally does, but a short distance, the muscular fibres are inserted into the scapula itself.

*Relations.* By their *superficial surface*, with the integument, and Trapezius; the Rhomboideus major, with the Latissimus dorsi. By their *deep surface*, with the Serratus posticus superior, posterior scapular artery, part of the Erector spinæ, the Intercostal muscles and ribs.

*Nerves.* These muscles are supplied by branches from the fifth cervical nerve, and additional filaments from the deep branches of the cervical plexus are distributed to the Levator anguli scapulæ.

*Actions.* The movements effected by the preceding muscles are numerous, as may be conceived from their extensive attachment. If the head is fixed, the upper part of the Trapezius will elevate the point of the shoulder, as in supporting weights; when the middle and lower fibres are brought into action, partial rotation of the scapula upon the side of the chest is produced. If the shoulders are fixed, both Trapezii acting together will draw the head directly backwards, or if only one acts, the head is drawn to the corresponding side.

The *Latissimus dorsi*, when it acts upon the humerus, draws it backwards and downwards, and at the same time rotates it inwards. If the arm is fixed, the muscle may act in various ways upon the trunk; thus, it may raise the lower ribs and assist in forcible inspiration, or if both arms are fixed, the two muscles may conspire with the Abdominal and great Pectoral muscles in drawing the whole trunk forwards, as in climbing or walking on crutches.

The *Levator anguli scapulæ* raises the superior angle of the scapula after it has been depressed by the lower fibres of the Trapezius, whilst the Rhomboid muscles carry the inferior angle backwards and upwards, thus producing a slight rotation of the scapula upon the side of the chest. If the shoulder be fixed, the Levator scapulæ may incline the neck to the corresponding side. The Rhomboid muscles acting together with the middle and inferior fibres of the Trapezius, will draw the scapula directly backwards towards the spine.

## Third Layer.

Serratus Posticus Superior. Serratus Posticus Inferior.

Splenius { Splenius Capitis.
Splenius Colli.

*Dissection.* The third layer of muscles is brought into view by the entire removal of the preceding, together with the Latissimus dorsi. To effect this, the Levator anguli scapulæ and Rhomboid muscles should be detached near their insertion, and reflected upwards, thus exposing the Serratus posticus superior; the Latissimus dorsi should then be divided in the middle by a vertical incision carried from its upper to its lower part, and the two halves of the muscle reflected.

The *Serratus Posticus Superior* is a thin, flat muscle, quadrilateral in form, situated at the upper and back part of the thorax. It arises by a thin and broad aponeurosis, from the ligamentum nuchæ and from the spinous processes of the last cervical and two or three upper dorsal vertebræ. Inclining downwards and outwards, it becomes muscular, and is inserted by four fleshy digitations, into the upper borders of the second, third, fourth, and fifth ribs, a little beyond their angles.

*Relations.* By its *superficial surface*, with the Trapezius, Rhomboidei, and Serratus magnus. By its *deep surface*, with the Splenius, upper part of the Erector spinæ, Intercostal muscles and ribs.

The *Serratus Posticus Inferior* is situated opposite the junction of the dorsal and lumbar regions, is of an irregularly quadrilateral form, broader than the preceding, and separated from it by a considerable interval. It arises by a thin aponeurosis from the spinous processes of the last two dorsal and two or three upper lumbar vertebræ, and from the inter-spinous ligaments. Passing obliquely upwards and outwards, it becomes fleshy, and divides into four flat digitations, which are inserted into the lower borders of the four lower ribs, a little beyond their angles.

*Relations.* By its *superficial surface*, it is covered by the Latissimus dorsi, with the aponeurosis of which its own aponeurotic origin is inseparably blended. By its *deep surface*, with the posterior

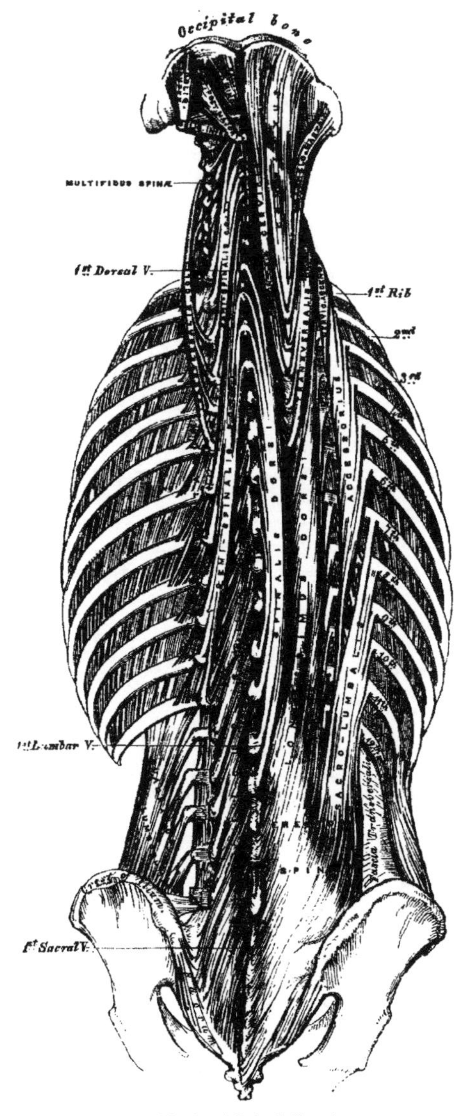


*161 – Muscles of the back. Deep layers.*

aponeurosis of the Transversalis, the Erector spinæ, ribs and Intercostal muscles. Its upper margin is continuous with the vertebral aponeurosis.

The *vertebral aponeurosis* is a thin aponeurotic lamina, extending along the whole length of the back part of the thoracic region, serving to bind down the Erector spinæ, and separating it from those muscles which connect the spine to the upper extremity. It consists of longitudinal and transverse fibres blended together, forming a thin lamella, which is attached in the median line to the spinous processes of the dorsal vertebræ; externally, to the angles of the ribs; and below, to the upper border of the Inferior serratus and tendon of the Latissimus dorsi; above, it passes beneath the Splenius, and blends with the deep fascia of the neck.

The Serratus posticus superior should now be detached from its origin and turned outwards, when the Splenius muscle will be brought into view.

The *Splenius* is situated at the back of the neck and upper part of the dorsal region. At its origin, it is a single muscle, narrow and pointed in form; but it soon becomes broader, and divides into two portions, which have separate insertions. It arises, by tendinous fibres, from the lower half of the ligamentum nuchæ, from the spinous processes of the last cervical and of the six upper dorsal vertebræ, and from the supra-spinous ligament. From this origin, the fleshy fibres proceed obliquely upwards and outwards, forming a broad flat muscle, which divides as it ascends into two portions, the Splenius capitis and Splenius colli.

The *splenius capitis* is inserted into the mastoid process of the temporal bone, and into the rough surface on the occipital bone beneath the superior curved line.

The *splenius colli* is inserted, by tendinous fasciculi, into the posterior tubercles of the transverse processes of the three or four upper cervical vertebræ.

The Splenius is separated from its fellow of the opposite side by a triangular interval, in which is seen the Complexus.

*Relations.* By its *superficial surface*, with the Trapezius, from which it is separated below by the Rhomboidei and the Serratus posticus superior. It is covered at its insertion by the Sterno-mastoid. By its *deep surface*, with the Spinalis dorsi, Longissimus dorsi, Semi-spinalis colli, Complexus, Trachelo-mastoid, and Transversalis colli.

*Nerves.* The Splenius and Superior serratus are supplied from the external posterior branches of the cervical nerves; the Inferior serratus, from the external branches of the dorsal nerves.

*Actions.* The Serrati are respiratory muscles acting in antagonism to each other. The Serratus posticus superior elevates the ribs; it is, therefore, an inspiratory muscle; while the Serratus inferior draws the lower ribs downwards, and is a muscle of expiration. This muscle is also probably a tensor of the vertebral aponeurosis. The Splenii muscles of the two sides, acting together, draw the head directly backwards, assisting the Trapezius and Complexus; acting separately, they draw the head to one or the other side, and slightly rotate it, turning the face to the same side. They also assist in supporting the head in the erect position.

## Fourth Layer.

*Sacral and Lumbar Regions.*
Erector Spinæ.

*Dorsal Region.*
Sacro-lumbalis.
Musculus accessorius ad sacro-lumbalem.
Longissimus dorsi.
Spinalis dorsi.

*Cervical Region.*
Cervicalis ascendens.
Transversalis colli.
Trachelo-mastoid.
Complexus.
Biventer cervicis.
Spinalis cervicis.

*Dissection.*—To expose the muscles of the fourth layer, the Serrati and vertebral aponeurosis should be entirely removed. The Splenius may then be detached by separating its attachments to the spinous processes, and reflecting it outwards.

The *Erector Spinæ* (fig. 161), and its prolongations in the dorsal and cervical regions, fill up the vertebral groove on each side of the spine. It is covered in the lumbar region by the lumbar aponeurosis; in the dorsal region, by the Serrati muscles and the vertebral aponeurosis; and in the cervical region, by a layer of cervical fascia continued beneath the Trapezius. This large muscular and tendinous mass varies in size and structure at different parts of the spine. In the sacral region, the Erector spinæ is narrow and pointed, and its origin chiefly tendinous in structure. In the lumbar region, it becomes enlarged, and forms a large fleshy mass. In the dorsal region, it subdivides into two parts, which gradually diminish in size as they ascend to be inserted into the vertebræ and ribs, and are gradually lost in the cervical region, where a number of special muscles are superadded, which are continued upwards to the head, which they support upon the spine.

The Erector spinæ arises from the sacro-iliac groove, and from the anterior surface of a very broad and thick tendon, which is attached, internally, to the spines of the sacrum, to the spinous processes of the lumbar and three lower dorsal vertebræ, and the supra-spinous ligament; externally, to the back part of the inner lip of the crest of the ilium, and to the series of eminences on the posterior part of the sacrum, representing the transverse processes, where it blends with the great sacro-sciatic ligament. The muscular fibres form a single large fleshy mass, bounded in front by the transverse processes of the lumbar vertebræ, and by the middle lamella of the fascia of the Transversalis muscle. Opposite the last rib, it divides into two parts, the Sacro-lumbalis, and the Longissimus dorsi.

The *Sacro-Lumbalis* (Ilio-Costalis), the external and smaller portion of the Erector spinæ, is inserted, by six or seven flattened tendons, into the angles of the six lower ribs. If this muscle is reflected outwards, it will be seen to be reinforced by a series of muscular slips, which arise from the angles of the ribs; by means of these the Sacro-lumbalis is continued upwards, to be connected with the upper ribs, and with the cervical portion of the spine forming two additional muscles, the Musculus accessorius and the Cervicalis ascendens.

The *musculus accessorius ad sacro-lumbalem* arises by separate flattened tendons, from the angles of the six lower ribs; these become muscular, and are finally inserted, by separate tendons, into the angles of the six upper ribs.

The *cervicalis ascendens* is the continuation of the Accessorius upwards in the neck: it is situated on the inner side of the tendons of the Accessorius arising from the angles of the four or five upper ribs, and is inserted, by a series of slender tendons, into the posterior tubercles of the transverse processes of the fourth, fifth, and sixth cervical vertebræ.

The *Longissimus Dorsi* the inner and larger portion of the Erector spinæ, arises, with the Sacro-lumbalis, from the common origin already mentioned. In the lumbar region, where it is as yet blended with the Sacro-lumbalis, some of the fibres are attached to the posterior surface of the transverse processes of the lumbar vertebræ their whole length, to the tubercles at the back of the articular processes, and to the layer of lumbar fascia connected with the apices of the transverse processes. In the dorsal region, the Longissimus dorsi is inserted, by long thin tendons, into the tips of the transverse processes of all the dorsal vertebræ, and into from seven to eleven ribs between their tubercles and angles. This muscle is continued upwards to the cranium and cervical portion of the spine, by means of two additional fasciculi, the Transversalis colli, and Trachelo-mastoid.

The *transversalis colli*, placed on the inner side of the Longissimus dorsi, arises, by long thin tendons, from the summit of the transverse processes of the third, fourth, fifth, and sixth dorsal vertebræ, and is inserted, by similar tendons, into the posterior tubercles of the transverse processes of the five lower cervical.

The *trachelo-mastoid* lies on the inner side of the preceding, between it and the Complexus muscle. It arises, by four tendons, from the transverse processes of the third, fourth, fifth, and sixth dorsal vertebræ, and by additional separate tendons from the articular processes of the three or four lower cervical; the fibres form a small muscle, which ascends to be inserted into the posterior margin of the mastoid process, beneath the Splenius and Sterno-mastoid muscles. This small muscle is almost always crossed by a tendinous intersection near its insertion into the mastoid process.

The *Spinalis Dorsi* connects the spinous processes of the upper lumbar and the dorsal vertebræ together by a series of muscular and tendinous slips, which are intimately blended with the Longissimus dorsi. It is situated at the inner side of the Longissimus dorsi, arising, by three or four tendons, from the spinous processes of the first two lumbar and the last two dorsal vertebræ: these uniting, form a small muscle, which is inserted, by separate tendons, into the spinous processes of the dorsal vertebræ, the number varying from four to eight. It is intimately united with the Semispinalis dorsi, which lies beneath it.

The *Spinalis Cervicis* is a small muscle, connecting together the spinous processes of the cervical vertebræ, and analogous to the Spinalis dorsi in the dorsal region. It varies considerably in its size, and in its extent of attachment to the vertebræ, not only in different bodies, but on the two sides of the same body. It usually arises by fleshy or tendinous slips, varying from two to four in number, from the spinous processes of the fifth and sixth cervical vertebræ, and occasionally from the first and second dorsal, and is inserted into the spinous process of the axis, and occasionally into the spinous process of the two vertebræ below it. This muscle has been found absent in five cases out of twenty-four.

The *Complexus* is a broad thick muscle, situated at the upper and back part of the neck, beneath the Splenius, and internal to the prolongations from the Longissimus dorsi. It arises, by a series of tendons, about seven in number, from the tips of the transverse processes of the upper three dorsal and seventh cervical, and from the articular processes of the three cervical above this. The tendons uniting form a broad muscle, which passes obliquely upwards and inwards, and is inserted into the innermost depression between the two curved lines of the occipital bone. This muscle, about its middle, is traversed by a transverse tendinous intersection.

The *Biventer Cervicis*, is a small fasciculus, situated on the inner side of the preceding, and in the majority of cases blended with it; it has received its name from having a tendon intervening between two fleshy bellies. It is sometimes described as a separate muscle, arising, by from two to four tendinous slips, from the transverse processes of as many upper dorsal vertebræ, and is inserted, on the inner side of the Complexus, into the superior curved line of the occipital bone.

*Relations.* By their *superficial surface*, with the Trapezius and Splenius. By their *deep surface*, with the Semi-spinalis dorsi and colli and the Recti and Obliqui. The Biventer cervicis is separated from its fellow of the opposite side by the ligamentum nuchæ, and the Complexus from the Semi-spinalis colli by the profunda cervicis artery, the princeps cervicis, a branch of the occipital, and by the posterior cervical plexus of nerves.

*Nerves.* The Erector spinæ and its subdivisions in the dorsal region are supplied by the external posterior branches of the lumbar and dorsal nerves. The Cervicalis ascendens, Transversalis colli, Trachelo-mastoid, and Spinalis cervical by the external posterior branches of the cervical nerves; the Complexus, by the internal posterior branches of the cervical nerves, the sub-occipital and great occipital.

## Fifth Layer.

Semispinalis Dorsi.
Semispinalis Colli.
Multifidus Spinæ.
Rotatores Spinæ.
Supra-spinales.
Inter-spinales.
Extensor Coccygis.
Inter-transversales.
Rectus Capitis Posticus Major.
Rectus Capitis Posticus Minor.
Obliquus Superior.
Obliquus Inferior.

*Dissection.* The muscles of the preceding layer must be removed by dividing and turning aside the Complexus; then detach the Spinalis and Longissimus dorsi from their attachments, and divide the Erector spinæ at its connection below to the sacral and lumbar spines, and turn it outwards. The muscles filling up the interval between the spinous and transverse processes are then exposed.

The *Semispinales muscles* connect the transverse and articular processes to the spinous processes of the vertebræ, extending from the lower part of the dorsal region to the upper part of the cervical.

The *semispinalis dorsi* consists of a thin, narrow, fleshy fasciculus, interposed between tendons of considerable length. It arises by a series of small tendons from the transverse processes of the lower dorsal vertebræ, from the tenth or eleventh to the fifth or sixth; and is inserted, by five or six tendons, into the spinous processes of the upper four dorsal and lower two cervical vertebræ.

The *semispinalis colli*, thicker than the preceding, arises by a series of tendinous and fleshy points from the transverse processes of the upper four dorsal vertebræ and from the articular processes of the cervical vertebræ (lower four); and is inserted into the spinous processes of four cervical vertebræ, from the axis to the fifth cervical. The fasciculus connected with the axis is the largest, and chiefly muscular in structure.

*Relations.* By their *superficial surface*, from below upwards, with the Longissimus dorsi, Spinalis dorsi, Splenius, Complexus, the profunda cervicis and princeps cervicis arteries, and the posterior cervical plexus of nerves. By their *deep surface* with the Multifidus spinæ.

The *Multifidus Spinæ* consists of a number of fleshy and tendinous fasciculi which fill up the groove on either side of the spinous processes of the vertebræ from the sacrum to the axis. In the sacral region, these fasciculi arise from the back of the sacrum, as low as the fourth sacral foramen;

and from the aponeurosis of origin of the Erector spinæ. In the iliac region, from the inner surface of the posterior superior spine, and posterior sacro-iliac ligaments. In the lumbar and cervical regions, they arise from the articular processes; and in the dorsal region from the transverse processes. Each fasciculus, ascending obliquely upwards and inwards, is inserted into the lamina and whole length of the spinous processes of the vertebra above. These fasciculi vary in length; the most superficial, the longest, pass from one vertebra to the third or fourth above; those next in order pass from one vertebra to the second or third above; whilst the deepest connect two contiguous vertebræ.

*Relations.* By its *superficial surface*, with the Longissimus dorsi, Spinalis dorsi, Semispinalis dorsi, and Semispinalis colli. By its *deep surface*, with the lamina and spinous processes of the vertebræ, and with the Rotatores spinæ in the dorsal region.

The *Rotatores Spinæ* are found only in the dorsal region of the spine, beneath the Multifidus spinæ; they are eleven in number on each side. Each muscle, which is small and somewhat quadrilateral in form, arises from the upper and back part of the transverse process, and is inserted into the lower border and outer surface of the lamina of the vertebra above, the fibres extending as far inwards as the root of the spinous process. The first is found between the first and second dorsal; the last, between the eleventh and twelfth. Sometimes, the number of these muscles is diminished by the absence of one or more from the upper or lower end.

The *Supra-Spinales* consist of a series of fleshy bands, which lie on the spinous processes in the cervical region of the spine.

The *Inter-Spinales* are short muscular fasciculi, placed in pairs between the spinous processes of the contiguous vertebræ.

In the *cervical region*, they are most distinct, and consist of six pairs, the first being situated between the axis and third vertebra, and the last between the last cervical and the first dorsal. They are small narrow bundles, attached, above and below, to the apices of the spinous processes.

In the *dorsal region*, they are found between the first and second vertebræ, and occasionally between the second and third; and below, between the eleventh and twelfth.

In the *lumbar region*, there are four pairs of these muscles in the intervals between the five lumbar vertebræ. There is also occasionally one in the interspinous space, between the last dorsal and first lumbar, and between the fifth lumbar and the sacrum.

The *Extensor Coccygis* is a slender muscular fasciculus, occasionally present, which extends over the lower part of the posterior surface of the sacrum and coccyx. It arises by tendinous fibres from the last bone of the sacrum, or first piece of the coccyx, and passes downwards to be inserted into the lower part of the coccyx. It is a rudiment of the Extensor muscle of the caudal vertebræ present in some animals.

The *Inter-Transversales* are small muscles placed between the transverse processes of the vertebræ.

In the *cervical region*, they are most developed, consisting of two rounded muscular and tendinous fasciculi, which pass between the anterior and posterior tubercles of the transverse processes of two contiguous vertebræ, being separated from one another by the anterior branch of a cervical nerve, which lies in the groove between them, and by the vertebral artery and vein. In this region, there are seven pairs of these muscles, the first being between the atlas and axis, and the last between the seventh cervical and first dorsal vertebra.

In the *dorsal region*, they are least developed, consisting chiefly of rounded tendinous cords in the inter-transverse spaces of the upper dorsal vertebræ; but between the transverse processes of the lower three dorsal vertebræ and the first lumbar, they are muscular in structure.

In the *lumbar region*, they are four in number, and consist of a single muscular layer, which occupies the entire interspace between the transverse processes of the lowest lumbar vertebræ, whilst those between the transverse processes of the upper lumbar, are not attached to more than half the breadth of the process.

The *Rectus Capitis Posticus Major*, the larger of the two Recti, arises by a pointed tendinous origin from the spinous process of the axis, and, becoming broader as it ascends, is inserted into the inferior curved line of the occipital bone and the surface of bone immediately below it. As the muscles of the two sides ascend upwards and outwards, they leave between them a triangular space, in which are seen the Recti capitis postici minores muscles.

*Relations.* By its *superficial surface*, with the Complexus, and, at its insertion, with the Superior oblique. By its *deep surface*, with the posterior arch of the atlas, the posterior occipito-atloid ligament, and part of the occipital bone.

The *Rectus Capitis Posticus Minor*, the smallest of the four muscles in this region, is of a triangular shape; it arises by a narrow, pointed tendon from the tubercle on the posterior arch of the atlas, and, becoming broader as it ascends, is inserted into the rough surface beneath the inferior curved line, nearly as far as the foramen magnum, nearer to the middle line than the preceding.

*Relations.* By its *superficial surface*, with the Complexus. By its *deep surface* with the posterior occipito-atloid ligament.

The *Obliquus Inferior*, the largest of the two oblique muscles, arises from the apex of the spinous process of the axis, and passes almost horizontally outwards to be inserted into the apex of the transverse process of the atlas.

*Relations.* By its *superficial surface*, with the Complexus, and is crossed by the posterior branch of the second cervical nerve. By its *deep surface*, with the vertebral artery, and posterior occipito-atloid ligament.

The *Obliquus Superior*, narrow below, wide and expanded above, arises by tendinous fibres from the upper part of the transverse process of the atlas, joining with the insertion of the preceding, and, passing obliquely upwards and inwards is inserted into the occipital bone, between the two curved lines, external to the Complexus. Between the two oblique muscles and the Rectus posticus major, a triangular interval exists, in which is seen the vertebral artery, and the posterior branch of the sub-occipital nerve.

*Relations.* By its *superficial surface*, with the Complexus and Trachelo-mastoid. By its *deep surface*, with the posterior occipito-atloid ligament.

*Nerves.* The Semi-spinalis dorsi and Rotatores spinæ are supplied by the internal posterior branches of the dorsal nerves. The Semispinalis colli, Supra-spinales, and Inter-spinales, by the internal posterior branches of the cervical nerves. The Inter-transversales, by the internal posterior branches of the cervical dorsal, and lumbar nerves. And the Multifidus spinæ, by the same, with the addition of the internal posterior branches of the sacral nerves. The Recti and Obliqui muscles are all supplied by the sub-occipital and great occipital nerves.

*Actions.* The Erector spinæ, comprising the Sacro-lumbalis, with its accessory muscles, the Longissimus dorsi and Spinalis dorsi, serves, as its name implies, to maintain the spine in the erect posture; it also serves to bend the trunk backwards when it is required to counterbalance the influence of any weight at the front of the body, as, for instance, when a heavy weight is suspended from the neck, or when there is any great abdominal development, as in pregnant women or in abdominal dropsy; the peculiar gait under such circumstances depends upon the spine being drawn backwards, by the conterbalancing action of the Erector spinæ muscles. The continuation of these muscles upwards to the neck and head, steady and preserve the upright position of these several parts. If the Sacro-lumbalis and Longissimus dorsi of one side act, they serve to draw down the chest and spine to the corresponding side. The Musculus accessorius, taking its fixed point from the cervical vertebræ, elevates those ribs to which it is attached. The Multifidus spinæ act successively upon the different segments of the spine; thus, the lateral parts of the sacrum furnish a fixed point from which the fasciculi of this muscle act upon the lumbar region; these then become the fixed points for the fasciculi moving the dorsal region, and so on throughout the entire length of the spine; it is by the successive contraction and relaxation of the separate fasciculi of this and other muscles, that the spine preserves the erect posture without the fatigue that would necessarily have existed had this movement been accomplished by the action of a single muscle. The Multifidus spinæ, besides preserving the erect position of the spine, serves to rotate it, so that the front of the trunk is turned to the side opposite to that from which the muscle acts, this muscle being assisted in its action by the Obliquus externus abdominis. The Complexi, the analogues of the Multifidus spinæ in the neck, draw the head directly backwards; if one muscle acts, it draws the head to one side, and rotates it so that the face is turned to the opposite side. The Rectus capitis posticus minor and the Superior oblique draw the head backwards; and the latter, from the obliquity in the direction of its fibres, may turn the face to the opposite side. The Rectus capitis posticus major and the Obliquus inferior, rotate the atlas, and, with it, the cranium round the odontoid process, and turn the face to the same side.

## Muscles of the Abdomen.

The muscles in this region are, the

| | |
|---|---|
| Obliquus Externus. | Rectus. |
| Obliquus Internus. | Pyramidalis. |
| Transversalis. | Quadratus Lumborum. |

*Dissection* (fig. 162). To dissect the abdominal muscles, a vertical incision should be made from the ensiform cartilage to the pubes; a second oblique incision should extend from the umbilicus upwards and outwards to the outer surface of the chest, as high as the lower border of the fifth or sixth rib; and a third, commencing midway between the umbilicus and pubes, should pass transversely outwards to the anterior superior iliac spine, and along the crest of the ilium as far as its posterior third. The three flaps included between these incisions should then be reflected from within outwards, in the line of direction of the muscular fibres. If necessary, the abdominal muscles may be made tense by inflating the peritoneal cavity through the umbilicus.

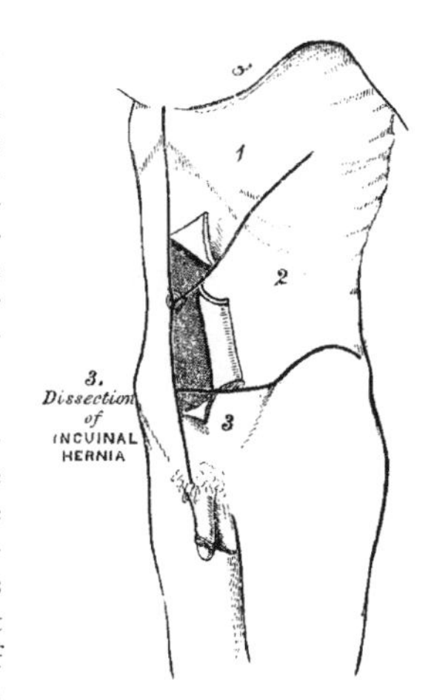


162 – *Dissection of abdomen.*

The *External Oblique Muscle* (fig. 163), so called from the direction of its fibres, is situated on the side and fore part of the abdomen; being the largest and the most superficial of the three flat muscles in this region. It is broad, thin, irregularly quadrilateral in form, its muscular portion occupying the side, its aponeurosis the anterior wall of that cavity. It arises, by eight fleshy digitations, from the external surface and lower borders of the eight inferior ribs; these digitations are arranged in an oblique line running downwards and backwards; the upper ones being attached close to the cartilages of the corresponding ribs; the lowest, to the apex of the cartilage of the last rib; the intermediate ones, to the ribs at some distance from their cartilages. The five superior serrations increase in size from above downwards, and are received between corresponding processes of the Serratus magnus; the three lower ones diminish in size from above downwards, receiving between them corresponding processes from the Latissimus dorsi. From these attachments, the fleshy fibres proceed in various directions. Those

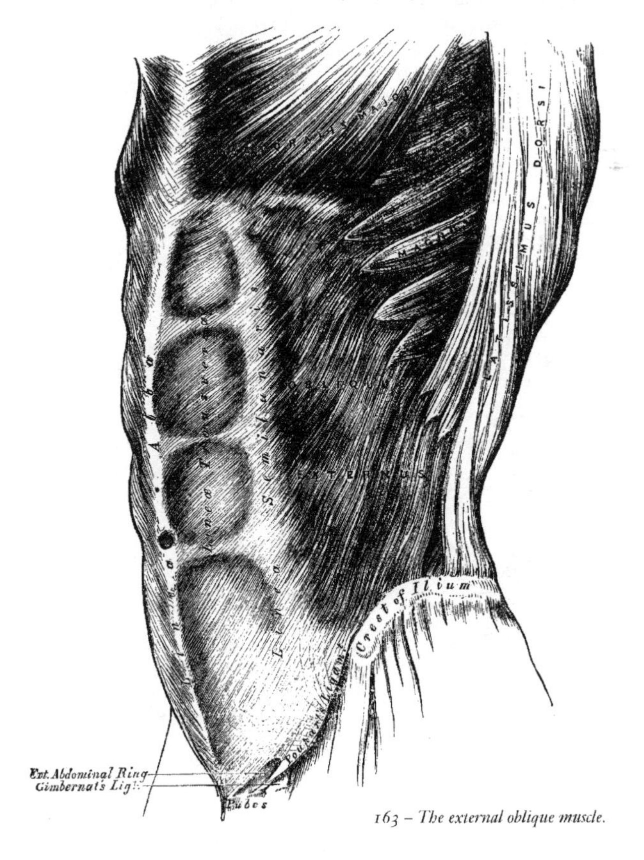


163 – *The external oblique muscle.*

from the lowest ribs pass nearly vertically downwards, to be inserted into the anterior half of the outer lip of the crest of the ilium; the middle and upper fibres, directed downwards and forwards, terminate in tendinous fibres, which spread out into a broad aponeurosis. This aponeurosis, joined with that of the opposite muscle along the median line, covers the whole of the front of the abdomen: above, it is connected with the lower border of the Pectoralis major; below, its fibres are closely aggregated together, and extend obliquely across from the anterior superior spine of the ilium to the spine of the os pubis and the pectineal line. In the median line, it interlaces with the aponeurosis of the opposite muscle, forming the linea alba, and extends from the ensiform cartilage to the symphysis pubis.

That portion of the aponeurosis which extends between the anterior superior spine of the ilium and the spine of the os pubis, is a broad band, folded inwards, and continuous below with the fascia lata; it is called *Poupart's ligament.* The portion which is reflected from Poupart's ligament backwards and inwards into the pectineal line, is called *Gimbernat's ligament.* From the point of attachment of the latter to the pectineal line, a few fibres pass upwards and inwards beneath the inner pillar of the ring, to the linea alba. They diverge as they ascend, and form a thin, triangular, fibrous band, which is called the *triangular ligament.*

In the aponeurosis of the External oblique, immediately above the crest of the os pubis, is a triangular opening, the *external abdominal ring*, formed by a separation of the fibres of the aponeurosis in this situation; it serves for the transmission of the spermatic cord in the male, and the round ligament in the female. This opening is directed obliquely upwards and outwards, and corresponds with the course of the fibres of the aponeurosis. It is bounded, below, by the crest of the os pubis; above, by some curved fibres, which pass across the aponeurosis at the upper angle of the ring, so as to increase its strength; and, on either side, by the margins of the aponeurosis, which are called the *pillars of the ring.* Of these, the external, which is, at the same time, inferior, from the obliquity of its direction, is inserted into the spine of the os pubis. The internal, or superior pillar, being attached to the front of the symphysis pubis, interlaces with the corresponding fibres of the opposite muscle, that of the right being superficial. To the margins of the pillars of the external abdominal ring is attached an exceedingly thin and delicate fascia, which is prolonged down over the outer surface of the cord and testis. This has received the name of *inter-columnar fascia*, from its attachment to the pillars of the ring. It is also called the *external spermatic fascia*, from being the most external of the fasciæ which cover the spermatic cord.

*Relations.* By its *external surface*, with the superficial fascia, superficial epigastric and circumflex iliac vessels, and some cutaneous nerves. By its *internal surface*, with the Internal oblique, the lower part of the eight inferior ribs, and Intercostal muscles, the cremaster, the spermatic cord in the male, and round ligament in the female. Its *posterior border* is occasionally overlapped by the Latissimus dorsi; sometimes an interval exists between the two muscles, in which is seen a portion of the Internal oblique.

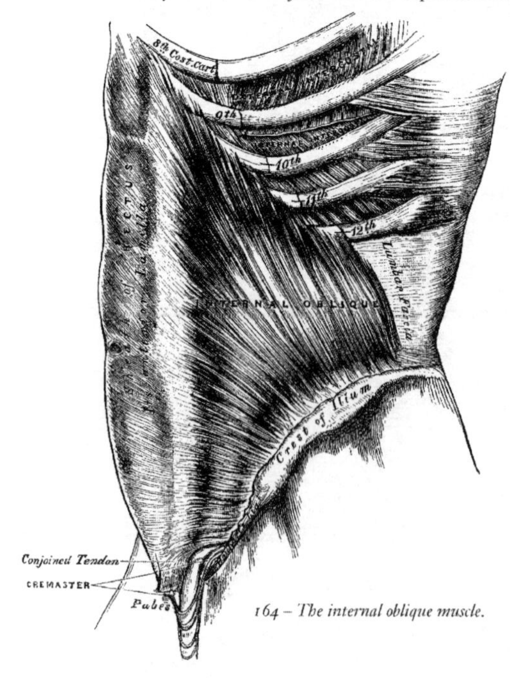


*164 – The internal oblique muscle.*

*Dissection.* The External oblique should now be detached by dividing it across, just in front of its attachment to the ribs, as far as its posterior border, and by separating it below from the crest of the ilium as far as the spine; the muscle should then be carefully separated from the Internal oblique, which lies beneath, and turned towards the opposite side.

The *Internal Oblique Muscle* (fig. 164), thinner and smaller than the preceding, beneath which it lies, is of an irregularly quadrilateral form, and situated at the side and fore part of the abdomen. It arises, by fleshy fibres, from the outer half of Poupart's ligament, being attached to the groove on its upper surface; from the anterior two-thirds of the middle lip of the crest of the ilium, and from the lumbar fascia. From this origin, the fibres diverge in different directions. Those from Poupart's ligament, few in number and paler in colour than the rest, arch downwards and inwards across the spermatic cord, to be inserted, conjointly with those of the Transversalis, into the crest of the os pubis and pectineal line, to the extent of half an inch, forming the conjoined tendon of the Internal oblique and Transversalis; those from the anterior superior iliac spine are horizontal in their direction; whilst those which arise from the fore part of the crest of the ilium pass obliquely upwards and inwards, and terminate in an aponeurosis, which is continued forwards to the linea alba; the most posterior fibres ascend almost vertically upwards, to be inserted into the lower borders of the cartilages of the four lower ribs, being continuous with the Internal Intercostal muscles.

The conjoined tendon of the Internal oblique and Transversalis is inserted into the crest of the os pubis and pectineal line immediately behind the external abdominal ring, serving to protect what would otherwise be a weak point in the abdomen. Sometimes this tendon is insufficient to resist the pressure from within, and is carried forward in front of the protrusion through the external ring, forming one of the coverings of direct inguinal hernia.

The aponeurosis of the Internal oblique is continued forward to the middle line of the abdomen, where it joins with the aponeurosis of the opposite muscle at the linea alba, and extends from the margin of the thorax to the pubes. At the outer margin of the Rectus muscle, this aponeurosis for the upper three-fourths of its extent, divides into two lamellæ, which pass, one in front and the other behind it, enclosing it in a kind of sheath, and reuniting on its inner border at the linea alba: the anterior layer is blended with the aponeurosis of the External oblique muscle; the posterior layer with that of the Transversalis. Along the lower fourth, the aponeurosis passes altogether in front of the Rectus without any separation.

*Relations.* By its *external surface*, with the External oblique, Latissimus dorsi spermatic cord, and external ring. By its *internal surface*, with the Transversalis muscle, fascia transversalis, internal ring, and spermatic cord. Its lower border forms the upper boundary of the spermatic canal.

*Dissection.* The Internal oblique should now be detached in order to expose the Transversalis beneath. This may be effected by dividing the muscle, above, at its attachment to the ribs: below, at its connexion with Poupart's ligament and the crest of the ilium and behind, by a vertical incision extending from the last rib to the crest of the ilium. The muscle should previously be made tense by drawing upon it with the fingers of the left hand, and if its division is carefully effected, the cellular interval between it and the Transversalis, as well as the direction of the fibres of the

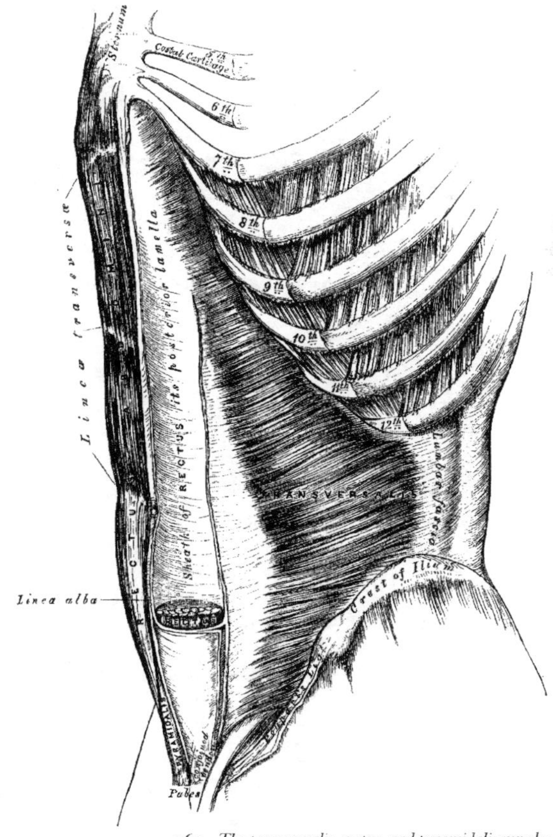


165 – *The transversalis, rectus, and pyramidalis muscles.*

latter muscle, will afford a clear guide to their separation; along the crest of the ilium the circumflex iliac vessels are interposed between them, and form an important guide in separating them. The muscle should then be thrown forwards towards the linea alba.

The *Transversalis muscle* (fig.165), so called from the direction of its fibres is the most internal flat muscle of the abdomen, being placed immediately beneath the Internal Oblique. It arises by fleshy fibres from the outer third of Poupart's ligament, from the inner lip of the crest of the ilium, its anterior two thirds, from the inner surface of the cartilages of the six lower ribs, interdigitating with the Diaphragm, and by a broad aponeurosis from the spinous and transverse processes of the lumbar vertebræ. The lower fibres curve downwards, and are inserted together with those of the Internal oblique, into the crest of the os pubis and pectineal line, forming what was before mentioned as the conjoined tendon of these muscles. Throughout the rest of its extent the fibres pass horizontally inwards and near the outer margin of the Rectus, terminate in an aponeurosis, which is inserted into the linea alba; its upper three-fourths passing behind the Rectus muscle, blending with the posterior lamella of the Internal oblique; its lower fourth passing in front of the Rectus.

*Relations.* By its *external surface*, with the Internal oblique, the inner surface of the lower ribs, and Internal intercostal muscles. Its *inner surface* is lined by the fascia transversalis, which separates it from the peritoneum. Its lower border forms the upper boundary of the spermatic canal.

*Lumbar Fascia* (fig.166). The vertebral aponeurosis of the Transversalis divides into three layers, an anterior, very thin, which is attached to the front part of the apices of the transverse processes of the lumbar vertebræ, and, above, to the lower margin of the last rib, forming the ligamentum arcuatum externum; a middle layer, much stronger, which is attached to the apices of the transverse processes; and a posterior layer, attached to the apices of the spinous processes. Between the anterior and middle layers is situated the Quadratus lumborum, between the middle and posterior, the Erector spinæ. The posterior lamella of this aponeurosis receives the attachment of the Internal oblique; it is also blended with the aponeurosis of the Serratus posticus inferior and with that of the Latissimus dorsi, forming the lumbar fascia.

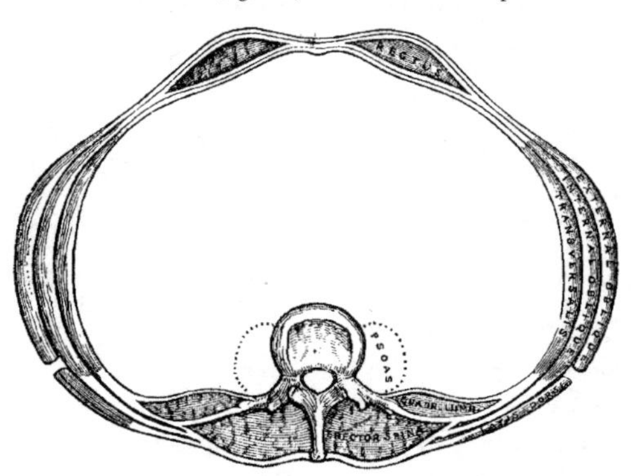

*166 – A transverse section of the abdomen in the lumbar region.*

*Dissection.* To expose the Rectus muscle, its sheath should be opened by a vertical incision extending from the margin of the thorax to the pubes, the two portions should then be reflected from the surface of the muscle, which is easily effected, excepting at the lineæ transversæ, where so close an adhesion exists, that the greatest care is requisite in separating them. The outer edge of the muscle should now be raised, when the posterior layer of the sheath will be seen. By dividing the muscle in the centre, and turning its lower part downwards, the point where the posterior wall of the sheath terminates in a thin curved margin will be seen.

The *Rectus Abdominis* is a long, flat muscle, which extends along the whole length of the front of the abdomen, being separated from its fellow of the opposite side by the linea alba. It is much broader above than below, and arises by two tendons, the external or larger being attached to the crest of the os pubis; the internal, smaller portion, interlacing with its fellow of the opposite side, and being connected with the ligaments covering the symphysis pubis. The fibres ascend vertically upwards, and the muscle becoming broader and thinner at its upper part, is inserted by three portions of unequal size into the cartilages of the fifth, sixth, and seventh ribs. Some fibres are occasionally connected with the costo-xiphoid ligaments, and side of the ensiform cartilage.

The Rectus muscle is traversed by a series of tendinous intersections, which vary from two to five in number, and have received the name lineæ transversæ. One of these is usually situated opposite the umbilicus, and two above that point; of the latter, one corresponds to the ensiform cartilage, and the other, to the interval between the ensiform cartilage and the umbilicus; there is occasionally one below the umbilicus. These intersections pass transversely or obliquely across the muscle in a zigzag course; they rarely extend completely through its substance, sometimes pass only half way across it, and are intimately adherent to the sheath in which the muscle is enclosed.

The Rectus is enclosed in a sheath (fig. 166) formed by the aponeuroses of the Oblique and Transversalis muscles, which are arranged in the following manner. When the aponeurosis of the Internal oblique arrives at the margin of the Rectus, it divides into two lamellæ, one of which passes in front of the Rectus, blending with the aponeurosis of the External oblique; the other, behind it, blending with the aponeurosis of the Transversalis; and these, joining again at its inner border, are inserted into the linea alba. This arrangement of the fasciæ exists along the upper three-fourths of the muscle; at the commencement of the lower fourth the posterior wall of the sheath terminates in a thin curved margin, the concavity of which looks downwards towards the pubes; the aponeuroses of all three muscles passing in front of the Rectus without any separation. The Rectus muscle in the situation where its sheath is deficient, is separated from the peritoneum by the transversalis fascia.

The *Pyramidalis* is a small muscle, triangular in form, placed at the lower part of the abdomen, in front of the Rectus, and contained in the same sheath with that muscle. It arises by tendinous fibres from the front of the os pubis and the anterior public ligament; the fleshy portion of the muscle passes upwards, diminishing in size as it ascends, and terminates by a pointed extremity, which is inserted into the linea alba, midway between the umbilicus and the os pubis. This muscle is sometimes found wanting on one or both sides; the lower end of the Rectus then becomes proportionally increased in size. Occasionally it has been found double on one side, or the muscles of the two sides are of unequal size. Sometimes its length exceeds that stated above.

The *Quadratus Lumborum* is situated in the lumbar region, it is irregularly quadrilateral in shape, broader below than above, and consists of two portions. One portion arises by aponeurotic fibres from the ilio-lumbar ligament, and the adjacent portion of the crest of the ilium for about two inches, and is inserted into the lower border of the last rib, about half its length, and by four small tendons, into the apices of the transverse processes of the third, fourth, and fifth lumbar vertebræ. The other portion of the muscle, situated in front of the preceding, arises from the upper borders of the transverse processes of the third, fourth, and fifth lumbar vertebræ and is inserted into the lower margin of the last rib. The Quadratus lumborum is contained in a sheath formed by the anterior and middle lamellæ of the vertebral aponeurosis of the Transversalis.

*Nerves.* The abdominal muscles are supplied by the lower intercostal, ilio-hypogastric, and ilio-inguinal nerves. The Quadratus lumborum, receives filaments from the anterior branches of the lumbar nerves.

In the description of the abdominal muscles, mention has frequently been made of the linea alba, lineæ semilunares, lineæ transversæ; when the dissection of these muscles is completed, these structures should be examined.

The *linea alba* is a tendinous raphe or cord seen along the middle line of the abdomen, extending from the ensiform cartilage to the pubes. It is placed between the inner borders of the Recti muscles, and formed by the blending of the aponeuroses of the Oblique and Transversalis muscles. It is narrow below, corresponding to the narrow interval existing between the Recti, but broader above, as these muscles diverge from one another in their ascent, becoming of considerable breadth after great distension of the abdomen from pregnancy or ascites. It presents numerous apertures for the passage of vessels and nerves; the largest of these is the umbilicus, which in the fœtus transmits the umbilical vessels, but in the adult is obliterated, the cicatrix being stronger than the neighbouring parts; hence the occurrence of umbilical hernia in the adult above the umbilicus, whilst in the fœtus it occurs at the umbilicus. The linea alba is in relation, in front, with the integument to which it is adherent, especially at the umbilicus; behind, it is separated from the peritoneum by the transversalis fascia; and below, by the urachus, and the bladder, when that organ is distended.

The *lineæ semilunares* are two curved tendinous lines, placed one on each side of the linea alba. Each corresponds with the outer border of the Rectus muscle, extends from the cartilage of the eighth rib to the pubes, and is formed by the aponeurosis of the Internal oblique at its point of division to enclose the Rectus.

The *lineæ transversæ* are three or four narrow transverse lines which intersect the Rectus muscle as already mentioned, they connect the lineæ semilunares with the linea alba.

*Actions.* The abdominal muscles perform a three-fold action.

When the pelvis and thorax are fixed, they can compress the abdominal viscera, by constricting the cavity of the abdomen, in which action they are materially assisted by the descent of the diaphragm. By these means, the fœtus is expelled from the uterus, the fæces from the rectum, the urine from the bladder, and the ingesta from the stomach in vomiting.

If the spine is fixed, these muscles can compress the lower part of the thorax materially assisting in expiration. If the spine is not fixed, the thorax is bent directly forward, if the muscles of both sides act, or to either side if they act alternately, rotation of the trunk at the same time taking place to the opposite side.

If the thorax is fixed, these muscles, acting together, draw the pelvis upwards as in climbing; or, acting singly, the pelvis is drawn upwards, and the verteberal column rotated to one or the other side. The Recti muscles, acting from below, depress the thorax, and consequently flex the vertebral column; when acting from above, they flex the pelvis upon the vertebral column. The Pyramidales are tensors of the linea alba.

## Muscles and Fasciæ of the Thorax.

The muscles exclusively connected with the bones in this region are few in number. They are the

Intercostales Externi. Infra-Costales.
Intercostales Interni. Triangularis Sterni.
Levatores Costarum.

*Intercostal Fasciæ.* A thin but firm layer of fascia covers the outer surface of the External intercostal and the inner surface of the Internal intercostal muscles; and a third layer, more delicate, is interposed between the two planes of muscular fibres. These are the intercostal fasciæ; they are best marked in those situations where the muscular fibres are deficient, as between the External intercostal muscles and sternum, in front; and between the Internal intercostals and spine, behind.

The *Intercostal Muscles* are two thin planes of muscular and tendinous structure, placed one over the other, filling up the intercostal spaces, and being directed obliquely between the margins of the adjacent ribs. They have received the name 'external' and 'internal,' from the position they bear to one another.

The *External Intercostals* are eleven in number on each side, being attached to the adjacent margins of each pair of ribs, and extending from the tubercles of the ribs, behind, to the commencement of the cartilages of the ribs, in front, where they terminate in a thin membranous aponeurosis, which is continued forwards to the sternum. They arise from the outer lip of the groove on the lower border of each rib, and are inserted into the upper border of the rib below. In the two lowest spaces they extend to the end of the ribs. Their fibres are directed obliquely downwards and forwards, in a similar direction with those of the External oblique muscle. They are thicker than the Internal intercostals.

*Relations.* By their *outer surface*, with the muscles which immediately invest the chest, viz., the Pectoralis major and minor, Serratus magnus, Rhomboideus major, Serratus posticus superior and inferior, Scalenus posticus, Sacro-lumbalis and Longissimus dorsi, Cervicalis ascendens, Transversalis colli, Levatores costarum, and the Obliquus externus abdominis. By their *internal surface*, with a thin layer of fascia, which separates them from the intercostal vessels and nerve, the Internal intercostal muscles, and, behind, from the pleura.

The *Internal Intercostals*, also eleven in number on each side, are placed on the inner surface of the preceding, commencing anteriorly at the sternum, in the interspaces between the cartilages of the true ribs, and from the anterior extremities of the cartilages of the false ribs; and extend backwards as far as the angles of the ribs, where they are continued to the vertebral column by a thin aponeurosis. They arise from the inner lip of the groove on the lower border of each rib, as well as from the corresponding costal cartilage, and are inserted into the upper border of the rib below. Their fibres are directed obliquely downwards and backwards, decussating with the fibres of the preceding.

*Relations.* By their *external surface*, with the External intercostals, and the intercostal vessels and nerves. By their *internal surface*, with the pleura costalis, Triangularis sterni, and Diaphragm.

The Intercostal muscles consist of muscular and tendinous fibres, the latter being longer and more numerous than the former; hence these spaces present very considerable strength, to which their crossing materially contributes.

The *Infra-Costales* consist of muscular and aponeurotic fasciculi, which vary in number and length; they arise from the inner surface of one rib, and are inserted into the inner surface of the first, second, or third rib below. Their direction is most usually oblique, like the Internal intercostals. They are most frequent between the lower ribs.

The *Triangularis Sterni* is a thin plane of muscular and tendinous fibres, situated upon the inner wall of the front of the chest. It arises from the lower part of the side of the sternum, from the inner surface of the ensiform cartilage, and from the sternal ends of the costal cartilages of the three or four lower true ribs. Its fibres diverge upwards and outwards, to be inserted by fleshy digitations into the lower border and inner surfaces of the costal cartilages of the second, third, fourth, and fifth ribs. The lowest fibres of this muscle are horizontal in their direction, and are continuous with those of the Transversalis; those which succeed are oblique, whilst the superior fibres are almost vertical. This muscle varies much in its attachment, not only in different bodies, but on opposite sides of the same body.

*Relations. In front* with the sternum, ensiform cartilage, the costal cartilages, the Internal intercostal muscles, and internal mammary vessels. *Behind*, with the pleura, pericardium, and anterior mediastinum.

The *Levatores Costarum*, twelve in number on each side, are small tendinous and fleshy bundles, which arise from the extremities of the transverse processes of the dorsal vertebræ, and passing obliquely downwards and outwards, are inserted into the upper rough surface of the rib below them, between the tubercle and the angle. That for the first rib arises from the transverse process of the last cervical vertebra, and that for the last from the eleventh dorsal. The Inferior levatores divide into two fasciculi, one of which is inserted as above described; the other fasciculus passes down to the second rib below its origin; thus, each of the lower ribs receives fibres from the transverse processes of two vertebræ.

*Nerves.* The muscles of this group are supplied by the intercostal nerves.

*Actions.* The Intercostals are the chief agents in the movement of the ribs in ordinary respiration. The External intercostals raise the ribs, especially their fore part, and so increase the capacity of the chest from before backwards; at the same time they evert their lower borders, and so enlarge the thoracic cavity transversely. The Internal intercostals, at the side of the thorax, depress the ribs, and invert their lower borders, and so diminish the thoracic cavity; but at the fore part of the chest these muscles assist the External intercostals in raising the cartilages. The Levatores Costarum assist the External intercostals in raising the ribs. The Triangularis sterni draws down the costal cartilages; it is therefore an expiratory muscle.

## Diaphragmatic Region.

### Diaphragm

The *Diaphragm* (*Διαφράσσω, to separate two parts*) (fig. 167) is a thin musculo-fibrous septum, placed obliquely at the junction of the upper with the lower two-thirds of the trunk, and separating the thorax from the abdomen, forming the floor of the former cavity and the roof of the latter. It is elliptical, its longest diameter being from side to side, somewhat fan-shaped, the broad elliptical portion being horizontal, the narrow part, which represents the handle, being vertical, and joint at right angles with the former. It is from this circumstance that some anatomists describe it as consisting of two portions, the upper or great muscle of the Diaphragm, and the lower or lesser muscle. It arises from the whole of the internal circumference of the thorax, being attached, in front, by fleshy fibres to the ensiform cartilage; on each side, to the inner surface of the cartilages and bony portions of the six or seven inferior ribs, interdigitating with the Transversalis; and behind, to two aponeurotic arches, named the ligamentum arcuatum externum and internum; and to the lumbar vertebræ. The fibres from these sources vary in length; those arising from the ensiform appendix are very short and occasionally aponeurotic; but those from the ligamenta arcuata, and more especially those from the ribs at the side of the chest, are the longest, describe well marked curves as they ascend, and finally converge, to be inserted into the circumference of the central tendon. Between the sides of the muscular slip from the ensiform appendix and the cartilages of the adjoining ribs, the fibres of the Diaphragm are deficient, the interval being filled by areolar tissue, covered on the thoracic side by the pleuræ; on the abdominal, by the peritoneum. This is, consequently, a weak point, and a portion of the contents of the abdomen may protrude into the chest, forming phrenic or diaphragmatic hernia, or a collection of pus in the mediastinum may descend through it, so as to point at the epigastrium.

The *ligamentum arcuatum internum* is a tendinous arch, thrown across the upper part of the Psoas magnus muscle, on each side of the spine. It is connected by one end, to the outer side of the body of the first, and occasionally the second lumbar vertebra, being continuous with the outer side of the tendon of the corresponding crus; and, by the other end, to the front of the transverse processes the second lumbar vertebra.

The *ligamentum arcuatum externum* is the thickened upper margin of the anterior lamella of the transversalis fascia; it arches across the upper part of the Quadratus lumborum, being attached, by

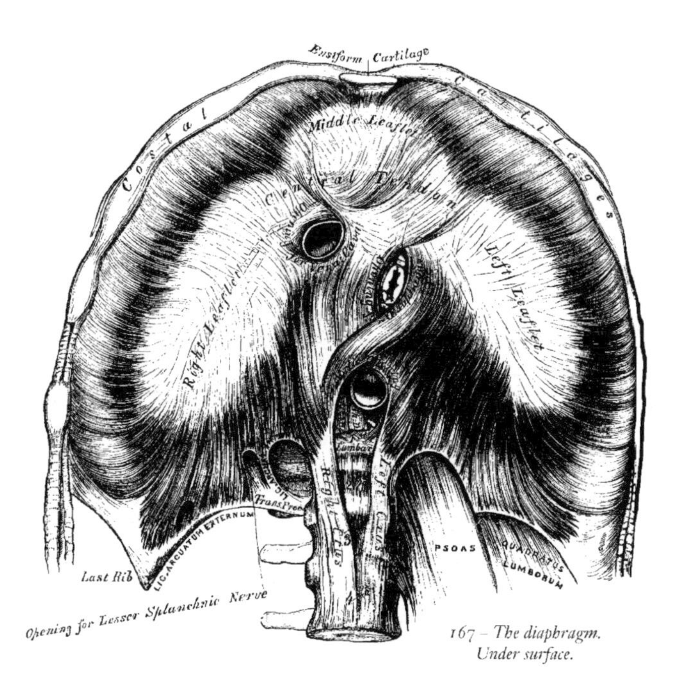


167 – *The diaphragm. Under surface.*

one extremity, to the front of the transverse process of the second lumbar vertebra; and, by the other, to the apex and lower margin of the last rib.

To the spine, the Diaphragm is connected by two crura, which are situated on the bodies of the lumbar vertebræ, one on each side of the aorta. The crura, at their origin, are tendinous in structure; the right crus, larger and longer than the left, arising from the anterior surface of the bodies and intervertebral substances of the second, third, and fourth lumbar vertebræ; the left, from the second and third; both blending with the anterior common ligament of the spine. A tendinous arch is thrown across the front of the vertebral column, from the tendon of one crus to that of the other, beneath which passes the aorta, vena azygos major, and thoracic duct. The tendons terminate in two large fleshy bellies, which, with the tendinous portions above alluded to, are called the *crura*, or *pillars of the diaphragm*. The outer fasciculi of the two crura are directed upwards and outwards to the central tendon; but the inner fasciculi decussate in front of the aorta, and then diverge, so as to surround the œsophagus before ending in the central tendon. The most anterior and larger of these fasciculi is formed by the right crus.

The *Central* or *Cordiform Tendon* of the Diaphragm is a thin tendinous aponeurosis, situated at the centre of the vault of this muscle, immediately beneath the pericardium, with which its circumference is blended. It is shaped somewhat like a trefoil leaf, consisting of three divisions, or leaflets, separated from one another by slight indentations. The right leaflet is the largest; the middle one, directed towards the ensiform cartilage, the next in size; and the left, the smallest. In structure, it is composed of several planes of fibres, which intersect one another at various angles, and unite into straight or curved bundles, an arrangement which affords additional strength to the tendon.

The *Openings* connected with the Diaphragm, are three large and several smaller apertures. The former are the aortic, œsophageal, and the opening for the vena cava.

The *aortic opening* is the lowest and the most posterior of the three large apertures connected with this muscle. It is situated in the middle line, immediately in front of the bodies of the vertebræ; and is, therefore, *behind* the Diaphragm, not in it. It is an osseo-aponeurotic aperture, formed by a tendinous arch thrown across the front of the bodies of the vertebræ, from the crus on one side to that on the other, and transmits the aorta, vena azygos major, thoracic duct, and occasionally the left sympathetic nerve.

The *œsophageal opening*, elliptical in from, muscular in structure, and formed by the two crura, is placed above, and, at the same time, anterior, and a little to the left of the preceding. It transmits the œsophagus and pneumogastric nerves. The anterior margin of this aperture is occasionally tendinous, being formed by the margin of the central tendon.

The *opening for the vena cava* is the highest; it is quadrilateral in form, tendinous in structure, and placed at the junction of the right and middle leaflets of the central tendon, its margins being bounded by four bundles of tendinous fibres, which meet at right angles.

The *right crus* transmits the sympathetic and the greater and lesser splanchnic nerves of the right side; the *left crus*, the greater and lesser splanchnic nerves of the left side, and the vena azygos minor.

The *Serous Membranes* in relation with the Diaphragm, are four in number: three lining its upper or thoracic surface; one its abdominal. The three serous membrances on its upper surface are the pleura on either side, and the serous layer of the pericardium, which covers the middle portion of the tendinous centre. The serous membrane covering its under surface, is a portion of the general peritoneal membrane of the abdominal cavity.

The Diaphragm is arched, being convex towards the chest, and concave to the abdomen. The *right portion* forms a complete arch from before backwards, being accurately moulded over the convex surface of the liver, and having resting upon it the concave base of the right lung. The *left portion* is arched from before backwards in a similar manner; but the arch is narrower in front, being encroached upon by the pericardium, and lower than the right, at its summit, by about three quarters of an inch. It supports the base of the left lung, and covers, the great end of the stomach, the spleen, and left kidney. The *central portion*, which supports the heart, is higher, in front at the sternum, and behind at the vertebræ, than the lateral portions; but deeper, this is reversed.

The height of the Diaphragm is constantly varying during respiration, being carried upwards or downwards from the average level; its height also varies according to the degree of distension of the stomach and intestines, and the size of the liver. After a forced expiration, the right arch is on a level, in front, with the fourth costal cartilage; at the side, with the fifth, sixth, and seventh ribs; and behind, with the eighth rib: the left arch being usually from one to two ribs breadth below the level of the right one. In a forced inspiration, it descends from one to two inches; its slope would then be represented by a line drawn from the ensiform cartilage towards the tenth rib.

*Nerves.* The Diaphragm is supplied by the phrenic nerves.

*Actions.* The action of the Diaphragm modifies considerably the size of the chest, and the position of the thoracic and abdominal viscera. *During a forced inspiration*, the cavity of the thorax is enlarged in the vertical direction from two to three inches, partly from the ascent of the walls of the chest, partly from the descent of the Diaphragm. The chest, consequently, encroaches upon the abdomen; the lungs are expanded, and lowered, in relation with the ribs, nearly two inches; the heart being drawn down about an inch and a half; the descent of the latter organ taking place indirectly through the medium of its connection with the lungs, as well as directly by means of the central tendon to which the pericardium is attached. The abdominal viscera are also pushed down (the liver, to the extent of nearly three inches), so that these organs are no longer protected by the ribs. *During expiration*, when the Diaphragm is passive, it is pushed up by the action of the abdominal muscles; the cavity of the abdomen, (with the organs contained in it,) encroaches upon the chest, by which the lungs and heart are compressed upwards, and the vertical diameter of the thoracic cavity diminished. The Diaphragm is passive when raised or lowered, by the abdominal organs, independently of respiration, in proportion as they are large or small, full or empty hence the oppression felt in the chest after a full meal, or from flatulent distension of the stomach and intestines.

In all expulsive acts, the Diaphragm is called into action, to give additional power to each expulsive effort. Thus, before sneezing, coughing, laughing, and crying; before vomiting; previous to the expulsion of the urine and fæces, or of the fœtus from the womb, a deep inspiration takes place.*

---

* For a detailed description of the general relations of the Diaphragm, and its action, refer to Dr. Sibson's '*Medical Anatomy*.'

# MUSCLES AND FASCIÆ OF THE UPPER EXTREMITY.

The Muscles of the Upper Extremity are divisible into groups, corresponding with the different regions of the limb.

*Anterior Thoracic Region.*

Pectoralis major.
Pectoralis minor.
Subclavius.

*Lateral Thoracic Region.*

Serratus magnus.

*Acromial Region.*

Deltoid.

*Anterior Scapular Region.*

Subscapularis.

*Posterior Scapular Region.*

Supra-spinatus.
Infra-spinatus.
Teres minor.
Teres major.

*Anterior Humeral Region.*

Coraco-brachialis.
Biceps.
Brachialis anticus.

*Posterior Humeral Region.*

Triceps.
Sub-anconeus.

*Anterior Brachial Region.*

Superficial Layer.
- Pronator radii teres.
- Flexor carpi radialis.
- Palmaris longus.
- Flexor carpi ulnaris.
- Flexor sublimis digitorum.

Deep Layer.
- Flexor profundus digitorum.
- Flexor longus pollicis.
- Pronator quadratus.

*Radial Region.*

Supinator longus.
Extensor carpi radialis longior.
Extensor carpi radialis brevior.

*Posterior Brachial Region.*

Superficial Layer.
- Extensor communis digitorum.
- Extensor minimi digiti.
- Extensor carpi ulnaris.
- Anconeus.

Deep Layer.
- Supinator brevis.
- Extensor ossis metacarpi pollicis.
- Extensor primi internodii pollicis.
- Extensor secundi internodii pollicis.
- Extensor indicis.

---

## Muscles of the Hand..

*Radial Region.*

Abductor pollicis.
Flexor ossis metacarpi pollicis (opponens).
Flexor brevis pollicis.
Adductor pollicis.

*Ulnar Region.*

Palmaris brevis.
Abductor minimi digiti.
Flexor brevis minimi digiti.
Flexor ossis metacarpi minimi digiti.

*Palmar Region.*

Lumbricales.
Interossei palmares.
Interossei dorsales.

*Dissection of Pectoral Region and Axilla* (fig. 168). The arm being drawn away from the side nearly at right angles with the trunk, and rotated outwards, a vertical incision should be made through the integument in the median line of the chest, from the upper to the lower part of the sternum; a second incision should be carried along the lower border of the Pectoral muscle, from the ensiform cartilage to the outer side of the axilla; a third, from the sternum along the clavicle, as far as its centre; and a fourth, from the middle of the clavicle obliquely downwards, along the interspace between the Pectoral and Deltoid muscles, as low as the fold of the arm-pit. The flap of integument may then be dissected off in the direction indicated in the figure, but not entirely removed, as it should be replaced on completing the dissection. If a transverse incision is now made from the lower end of the sternum to the side of the chest, as far as the posterior fold of the arm-pit, and the integument reflected outwards, the axillary space will be more completely exposed.

## Fasciæ of the Thorax.

The *superficial fascia* of the thoracic region is a loose cellulo-fibrous layer, continuous with the superficial fascia of the neck and upper extremity above, and of the abdomen below; opposite the mamma, it subdivides into two layers, one of which passes in front, the other behind this gland; and from both of these layers numerous septa pass into its substance, supporting in various lobes: from the anterior layer, fibrous processes pass forward to the integument and nipple, enclosing in their areolæ masses of fat. These processes were called by Sir A. Cooper, the *ligamenta suspensoria*, from the support they afford to the gland in this situation. On removing the superficial fascia, the *deep fascia* of the thoracic region is exposed: it is a thin aponeurotic lamina, covering the surface of the great Pectoral muscle, and sending numerous prolongations between its fasciculi: it is attached in the middle line, to the front of the sternum; and, above, to the clavicle: it is very thin over the upper part of the muscle, somewhat thicker in the interval between the Pectoralis major and Latissimus dorsi where it closes in the axillary space and divides at the margin of the latter muscle into two layers, one of which passes in front, and the other behind it; these proceed as far as the spinous processes of the dorsal vertebræ, to which they are attached. At the lower part of the thoracic region, this fascia is well developed, and is continuous with the fibrous sheath of the Recti muscles.

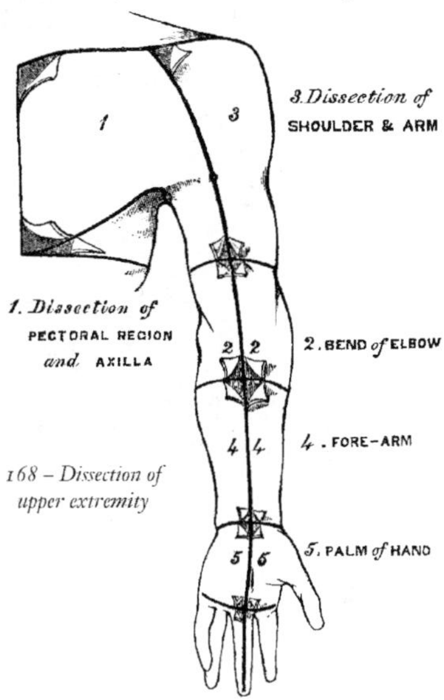


168 – Dissection of upper extremity

## Anterior Thoracic Region.

Pectoralis Major. Pectoralis Minor.
Subclavius.

The *Pectoralis Major* (fig. 169) is a broad, thick, triangular muscle, situated at the upper and fore part of the chest, in front of the axilla. It arises from the sternal half of the clavicle, its anterior surface; from one half the breadth of the front of the sternum, as low down as the attachment of the cartilage of the sixth or seventh rib; its origin consisting of aponeurotic fibres, which intersect with those of the opposite muscle: it also arises from the cartilages of all the true ribs and from the aponeurosis of the External oblique muscle of the abdomen. The fibres from this extensive origin converge towards its insertion, giving to the muscle a radiated appearance. Those fibres which arise from the clavicle, pass obliquely downwards and outwards, and are usually separated from the rest by a cellular interval; those from the lower part of the sternum and the cartilages of the lower true ribs, pass upwards and outwards; whilst the middle fibres pass horizontally. As these three sets of fibres converge, they are so disposed that the upper overlap the middle, and the middle the lower portion, the fibres of the lower portion being folded backwards upon themselves; so that those fibres which are lowest in front, become highest at their point of insertion. They all terminate in a flat tendon, about two inches broad, which is inserted into the anterior bicipital ridge of the humerus. This tendon consists of two laminæ, placed one in front of the other, and usually blended together below. The anterior, the thicker, receives the clavicular and upper half of the sternal portion of the muscle; the posterior lamina receiving the attachment of the lower half of the sternal portion. From this arrangement it results, that the fibres of the upper and middle portions of the muscle are inserted into the lower part of the bicipital ridge; those of the lower portion, into the upper part. The tendon, at its insertion, is connected with that of the Deltoid; it sends up an expansion over the bicipital groove to the head of the humerus; another backwards, which lines the groove; and a third to the fascia of the arm.

*Relations.* By its *anterior surface*, with the Platysma, the mammary gland, the superficial fascia, and integument. By its *posterior surface*: its *thoracic portion*, with the sternum, the ribs and costal cartilages, the Subclavius, Pectoralis minor, Serratus magnus, and the Intercostals; its *axillary portion*

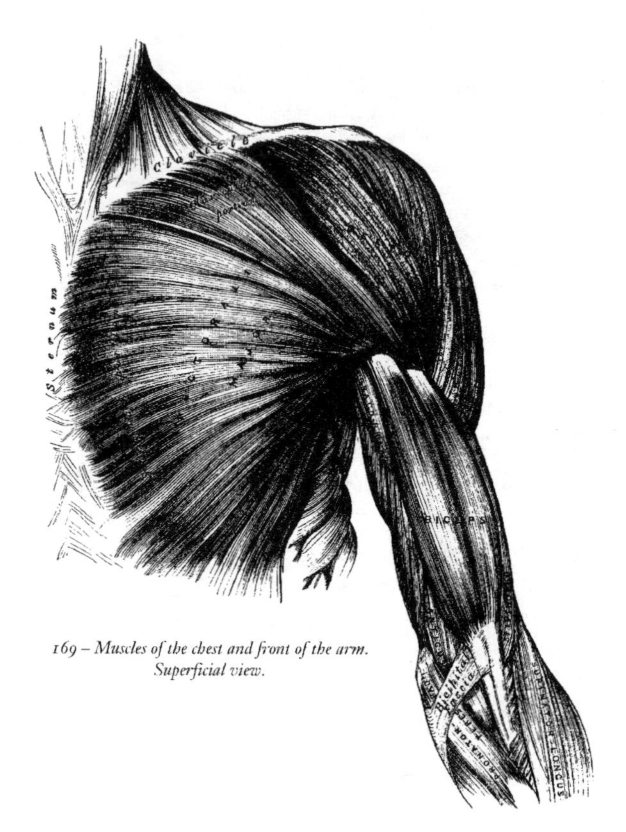


169 – *Muscles of the chest and front of the arm. Superficial view.*

forms the anterior wall of the axillary space, and covers the axillary vessels and nerves. Its *upper border* lies parallel with the Deltoid, from which it is separated by the cephalic vein and descending branch of the thoracico-acromialis artery. Its *lower border* forms the anterior margin of the axilla, being at first separated from the Latissimus dorsi by a considerable interval; but both muscles gradually converge towards the outer part of this space.

*Peculiarities*. In muscular subjects, the sternal origins of the two Pectoral muscles are separated only by a narrow interval; but this interval is enlarged where these muscles are ill developed. Very rarely, the whole of the sternal portion is deficient. Occasionally, one or two additional muscular slips arise from the aponeurosis of the External oblique, and become united to the lower margin of the Pectoralis major. A slender muscular slip is occasionally found lying parallel with the outer margin of the sternum, overlapping the origin of the pectoral muscle. It is attached, by one end, to the upper part of the sternum, near the origin of the sterno-mastoid; and, by the other, to the anterior wall of the sheath of the rectus abdominis. It has received the name 'rectus sternalis.'

*Dissection*. The Pectoralis major should now be detached by dividing the muscle along its attachment to the clavicle, and by making a vertical incision through its substance a little external to its line of attachment to the sternum and costal cartilages. The muscle should then be reflected outwards, and its tendon carefully examined. The Pectoralis minor is now exposed, and immediately above it, in the interval between its upper border and the clavicle, is a strong fascia, the costo-coracoid membrane.

The *costo-coracoid membrane* protects the axillary vessels and nerves, it is very thick and dense externally, where it is attached to the coracoid process, and is continuous with the fascia of the arm; more internally, it is connected with the lower border of the clavicle, as far as the inner extremity of the first rib: traced downwards, it passes behind the Pectoralis minor, surrounding, in a more or less complete sheath, the axillary vessels and nerves; and above, it sends a prolongation behind the Subclavius, which is attached to the lower border of the clavicle, and so encloses the muscle in a kind of sheath. The costo-coracoid membrane is pierced by the cephalic vein, the acromial-thoracic artery and vein, superior thoracic artery, and anterior thoracic nerve.

The *Pectoralis Minor* (fig. 170) is a thin, flat, triangular muscle, situated at the upper part of the thorax, beneath the Pectoralis major. It arises, by three tendinous digitations, from the upper margin and outer surface of the third, fourth, and fifth ribs, near their cartilages, and from the aponeurosis covering the Intercostal muscles: the fibres pass upwards and outwards, and converge to form a flat tendon, which is inserted into the anterior border of the coracoid process of the scapula.

*Relations.* By its *anterior surface*, with the Pectoralis major, and the superior thoracic vessels and nerves. By its *posterior surface*, with the ribs, Intercostal muscles, Serratus magnus, the axillary space, and the axillary vessels and nerves. Its upper border is separated from the clavicle by a triangular interval, broad internally, narrow externally, bounded in front by the costo-coracoid membrane, and internally by the ribs. In this space are seen the axillary vessels and nerves.

The costo-coracoid membrane should now be removed, when the Subclavius muscle will be seen.

The *Subclavius* is a long, thin, spindle-shaped muscle, placed in the interval between the clavicle and the first rib. It arises by a short, thick tendon from the cartilage of the first rib, in front of the rhomboid ligament; the fleshy fibres proceed obliquely outwards to be inserted into a deep groove on the under surface of the middle third of the clavicle.

*Relations.* By its *upper surface*, with the clavicle. By its *under surface*, it is separated from the first rib by the axillary vessels and nerves. Its *anterior surface* is separated from the Pectoralis major by a strong aponeurosis, which with the clavicle, forms an osteo-fibrous sheath in which the muscle is enclosed.

If the costal attachment of the Pectoralis minor is divided across, and the muscle reflected outwards, the axillary vessels and nerves are brought fully into view, and should be examined.

*Nerves.* The Pectoral muscles are supplied by the anterior thoracic nerves; the Subclavius, by a filament from the cord formed by the union of the fifth and sixth cervical nerves.

*Actions.* If the arm has been raised by the Deltoid, the Pectoralis major will, conjointly with the Latissimus dorsi and Teres major, depress it to the side of the chest; and, if acting singly, it will draw the arm across the front of the chest. The Pectoralis minor depresses the point of the shoulder, drawing the scapula downwards and inwards to the thorax. The Subclavius depresses the shoulder, drawing the clavicle downwards and forwards. When the arms are fixed, all three

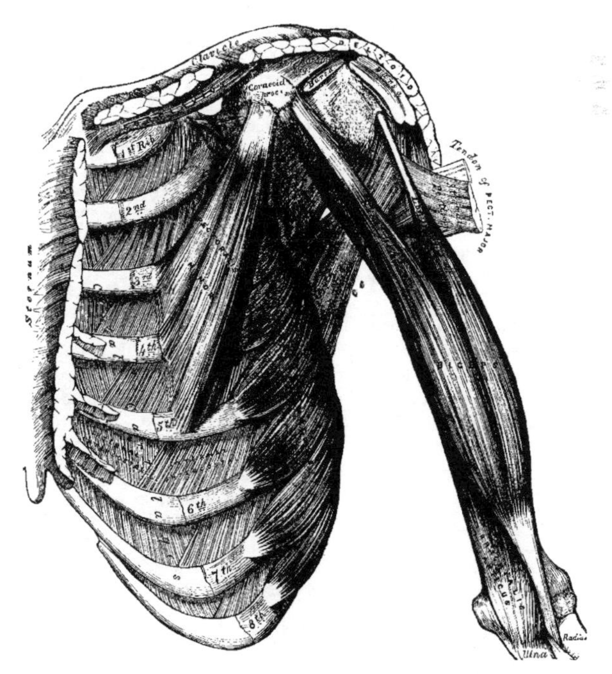


170 – *Muscles of the chest and front of the arm, with the boundaries of the axilla.*

muscles act upon the ribs, drawing them upwards and expanding the chest, thus becoming very important agents in forced inspiration. Asthmatic patients always assume this attitude, fixing the shoulders, so that all these muscles may be brought into action to assist in dilating the cavity of the chest.

## Lateral Thoracic Region.

### Serratus Magnus

The *Serratus Magnus* is a broad, thin, and irregularly quadrilateral muscle, situated at the upper part and side of the chest. It arises by nine fleshy digitations from the outer surface and upper border of the eight upper ribs (the second rib having two), and from the aponeurosis covering the upper intercostal spaces, and is inserted into the whole length of the inner margin of the posterior border of the scapula. This muscle has been divided into three portions, a superior, middle, and inferior, on account of the difference in the direction, and in the extent of attachment of each part. The upper portion, separated from the rest by a cellular interval, is a narrow, but thick fasciculus, which arises by two digitations from the first and second ribs, and from the aponeurotic arch between them; its fibres proceed upwards, outwards and backwards, to be inserted into the triangular smooth surface on the inner side of the superior angle of the scapula. The middle portion of the muscle arises by three digitations from the second, third and fourth ribs, it forms a thin and broad muscular layer, which proceeds horizontally backwards, to be inserted into the posterior border of the scapula, between the superior and inferior angles. The lower portion arises from the fifth, sixth, seventh and eighth ribs, by four digitations, in the intervals between which are received corresponding processes of the External oblique; the fibres pass upwards, outwards, and backwards, to be inserted into the inner surface of the inferior angle of the scapula, by an attachment partly muscular, partly tendinous.

*Relations.* This muscle is covered, in front, by the Pectoral muscles; behind, by the Subscapularis; above, by the axillary vessels and nerves. Its *deep surface* rests upon the ribs and Intercostal muscles.

*Nerves.* The Serratus magnus is supplied by the posterior thoracic nerve.

*Actions.* The Serratus magnus is the most important external inspiratory muscle. When the shoulders are fixed, it elevates the ribs, and so dilates the cavity of the chest, assisting the Pectoral and Subclavius muscles. This muscle, especially its middle and lower segments, draws the base and inferior angle of the scapula forwards, and so raises the point of the shoulder by causing a rotation of the bone on the side of the chest; assisting the Trapezius muscle in supporting weights upon the shoulder, the thorax being at the same time fixed by preventing the escape of the included air.

*Dissection.* After completing the dissection of the axilla, if the muscles of the back have been dissected, the upper extremity should be separated from the trunk. Saw through the clavicle at its centre, and then cut through the muscles which connect the scapula and arm with the trunk, viz., the Pectoralis minor, in front, Serratus magnus, at the side, and behind, the Levator anguli scapulæ, the Rhomboids, Trapezius, and Latissimus dorsi. These muscles should be cleaned and traced to their respective insertions. An incision should then be made through the integument, commencing at the outer third of the clavicle, and extending along the margin of this bone, the acromion process, and spine of the scapula; the integument should be dissected from above downwards and outwards, when the fascia covering the Deltoid is exposed.

The *superficial fascia* of the upper extremity, is a thin cellulo-fibrous lamina, containing between its layers the superficial veins and lymphatics, and the cutaneous nerves. It is most distinct in front of the elbow, and contains between its laminæ in this situation the large superficial cutaneous veins and nerves; in the hand it is hardly demonstrable, the integument being closely adherent to the deep fascia by dense fibrous bands. Small subcutaneous bursæ are found in this fascia, over the acromion, the olecranon, and the knuckles. The deep fascia of the upper extremity comprises the aponeurosis of the shoulder, arm, and forearm, the anterior and posterior annular ligaments of the carpus, and the palmar fascia. These will be considered in the description of the muscles of these several regions.

## Acromial Region.

### Deltoid

The *deep fascia* covering the Deltoid (deltoid aponeurosis), is a thick and strong fibrous layer, which covers the outer surface of the muscle, and sends down numerous prolongations between its fasciculi; it is continuous, internally, with the fascia covering the great Pectoral muscle; behind, with that covering the Infra-spinatus and back of the arm: above, it is attached to the clavicle, the acromion, and spine of the scapula.

The *Deltoid* is a large, thick, triangular muscle, which forms the convexity of the shoulder, and has received its name from its resemblance to the Greek letter Δ reversed. It surrounds the shoulder-joint in the greater part of its extent, covering it on its outer side, and in front and behind. It arises from the outer third of the anterior border and upper surface of the clavicle; from the outer margin and upper surface of the acromion process; and from the whole length of the lower border of the spine of the scapula. From this extensive origin, the fibres converge towards their insertion, the middle passing vertically, the anterior obliquely backwards, the posterior obliquely forwards; they unite to form a thick tendon, which is inserted into a rough prominence on the middle of the outer side of the shaft of the humerus. This muscle is remarkably coarse in texture, and intersected by three or four tendinous laminæ; these are attached, at intervals, to the clavicle and acromion, extend into the substance of the muscle, and give origin to a number of fleshy fibres. The largest of these laminæ extends from the summit of the acromion.

*Relations.* By its *superficial surface*, with the Platysma, supra-acromial nerves, the superficial fascia, and integument. Its *deep surface* is separated from the head of the humerus by a large sacculated synovial bursa, and covers the coracoid process, coraco-acromial ligament, Pectoralis minor, Coraco-brachialis, both heads of the Biceps, tendon of the Pectoralis major, Teres minor, scapular, and external heads of the Triceps, the circumflex vessels and nerve, and the humerus. Its *anterior border* is separated from the Pectoralis major by a cellular interspace, which lodges the cephalic vein and descending branch of the thoracic-acromial artery. Its *posterior border* rests on the Infra-spinatus and Triceps muscles.

*Nerves.* The Deltoid is supplied by the circumflex nerve.

*Actions.* The Deltoid raises the arm directly from the side, so as to bring it at right angles with the trunk. Its anterior fibres, assisted by the Pectoralis major, draw the arm forwards; and its posterior fibres, aided by the Teres major and Latissimus dorsi, draw it backwards.

*Dissection.* Divide the Deltoid across, near its upper part, by an incision carried along the margin of the clavicle the acromion process, and spine of the scapula, and reflect it downwards; the bursa will be seen on its under surface, as well as the circumflex vessels and nerves. The insertion of the muscle should be carefully examined.

## Anterior Scapular Region.

### Subscapularis

The *subscapular aponeurosis* is a thin membrane, attached to the entire circumference of the subscapular fossa, and affording attachment by its inner surface to some of the fibres of the Subscapularis muscle: when this is removed, the Subscapularis muscle is exposed.

The *Subscapularis* is a large triangular muscle, which fills up the subscapular fossa arising from its internal two-thirds, with the exception of a narrow margin along the posterior border, and the inner side of the superior and inferior angles, which afford attachment to the Serratus magnus. Some fibres arise from tendinous laminæ, which intersect the muscle, and are attached to ridges on the bone; and others from an aponeurosis, which separates the muscle from the Teres major and the long head of the Triceps. The fibres pass outwards, and, gradually converging, terminate in a tendon, which is inserted into the lesser tuberosity of the humerus. Those fibres which arise from the axillary border of the scapula, are inserted into the neck of the humerus to the extent of an inch below the tuberosity. The tendon of the muscle is in close contact with the capsular ligament of the shoulder-joint, and glides over a large bursa, which separates it from the base of the coracoid process. This bursa communicates with the cavity of the joint by an aperture in the capsular ligament.

*Relations.* By its *anterior surface*, with the Serratus magnus, Coraco-brachialis, and Biceps, and the axillary vessels and nerves. By its *posterior surface*, with the scapula, the subscapular vessels and nerves, and the capsular ligament of the shoulder-joint. Its *lower border* is contiguous with the Teres major and Latissimus dorsi.

*Nerves.* It is supplied by the subscapular nerves.

*Actions.* The Subscapularis rotates the head of the humerus inwards; when the arm is raised, it draws the humerus downwards. It is a powerful defence to the front of the shoulder-joint, preventing displacement of the head of the bone forwards.

## Posterior Scapular Region.

| Supra-spinatus. | Teres Minor. |
| Infra-spinatus. | Teres Major. |

*Dissection.* To expose these muscles, and to examine their mode of insertion into the humerus, detach the Deltoid and Trapezius from their attachment to the spine of the scapula and acromion process. Remove the clavicle by dividing the ligaments connecting it with the coracoid process, and separate it at its articulation with the scapula: divide the acromion process near its root with a saw, and, the fragments being removed, the tendons of the posterior Scapular muscles will be fully exposed, and can be examined. A block should be placed beneath the shoulder-joint, so as to make the muscles tense.

The *supra-spinous aponeurosis* is a thick and dense membranous layer, which completes the osteo-fibrous case in which the Supra-spinatus muscle is contained; affording attachment, by its inner surface, to some of the fibres of this muscle. It is thick internally, but thinner externally under the coraco-acromion ligament. When this fascia is removed, the Supra-spinatus muscle is exposed.

The *Supra-spinatus* muscle occupies the whole of the supra-spinous fossa, arising from its internal two-thirds, and from a strong fascia which covers its surface. The muscular fibres converge to a tendon, which passes across the capsular ligament of the shoulder-joint, to which it is intimately adherent, and is inserted into the highest of the three facets on the great tuberosity of the humerus.

*Relations.* By its *upper surface*, with the Trapezius, the clavicle, the acromion, the coraco-acromial ligament, and the Deltoid. By its *under surface*, with the scapula, the supra-scapular vessels and nerve, and upper part of the shoulder-joint.

The *infra-spinous aponeurosis* is a dense fibrous membrane, covering-in the Infra-spinatus muscle, and attached to the circumference of the infra-spinous fossa; it affords attachment, by its inner surface, to some fibres of this muscle, is continuous externally with the fascia of the arm, and gives off from its under surface intermuscular septa, which separate the Infra-spinatus from the Teres minor, and the latter from the Teres major.

The *Infra-spinatus* is a thick triangular muscle, which occupies the chief part of the infra-spinous fossa, arising by fleshy fibres, from its internal two-thirds; and by tendinous fibres, from the ridges on its surface: it also arises from a strong fascia which covers it externally, and separates it from the Teres major and minor. The fibres converge to a tendon, which glides over the concave border of the spine of the scapula, and, passing across the capsular ligament of the shoulder-joint, is inserted into the middle facet on the great tuberosity of the humerus. The tendon of this muscle is

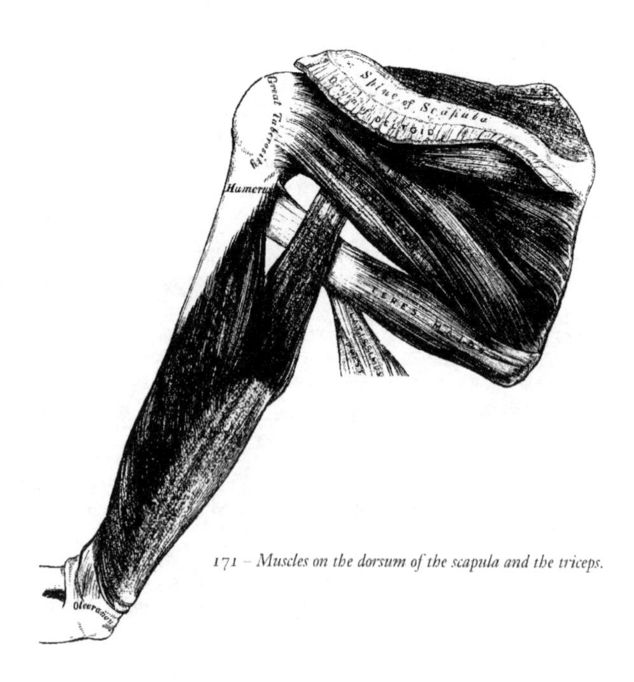


171 – *Muscles on the dorsum of the scapula and the triceps.*

occasionally separated from the spine of the scapula by a synovial bursa, which communicates with the synovial membrane of the shoulder-joint.

*Relations.* By its *posterior surface*, with the Deltoid, the Trapezius, Latissimus dorsi, and the integument. By its *anterior surface*, with the scapula, from which it is separated by the superior and dorsalis scapulæ vessels, and with the capsular ligament of the shoulder-joint. Its *lower border* is in contact with the Teres minor, and occasionally united with it, and with the Teres major.

The *Teres Minor* is a narrow elongated muscle, which lies along the inferior border of the scapula. It arises from the dorsal surface of the axillary border of the scapula for the upper two-thirds of its extent, and from two aponeurotic laminæ, one of which separates this muscle, from the Infra-spinatus, the other from the Teres major; its fibres pass obliquely upwards and outwards, and terminate in a tendon, which is inserted into the lowest of the three facets on the great tuberosity of the humerus, and, by fleshy fibres, into the humerus immediately below it. The tendon of this muscle passes across the capsular ligament of the shoulder-joint.

*Relations.* By its *posterior surface*, with the Deltoid, Latissimus dorsi, and integument. By its *anterior surface*, with the scapula, the dorsal branch of the subscapular artery, the long head of the Triceps, and the shoulder-joint. By its *upper border*, with the Infra-spinatus. By its *lower border*, with the Teres major, from which it is separated anteriorly by the long head of the Triceps.

The *Teres Major* is a broad and somewhat flattened muscle, which arises from the dorsal aspect of the inferior angle of the scapula, and from the fibrous septa interposed between it and the Teres minor and Infra-spinatus; the fibres are directed upwards and outwards, and terminate in a flat tendon, about two inches in length, which is inserted into the posterior bicipital ridge of the humerus. The tendon of this muscle, at its insertion into the humerus, lies behind that of the Latissimus dorsi, from which it is separated by a synovial bursa.

*Relations.* By its *posterior surface*, with the integument, from which it is separated, internally, by the Latissimus dorsi; and externally, by the long head of the Triceps. By its *anterior surface*, with the Subscapularis, Latissimus dorsi, Coraco-brachialis, short head of the Biceps, the axillary vessels, and brachial plexus of nerves. Its *upper border* is at first in relation with the Teres minor, from which it is afterwards separated by the long head of the Triceps. Its *lower border* forms, in conjunction with the Latissimus dorsi, part of the posterior boundary of the axilla.

*Nerves.* The Supra- and Infra-spinati muscles are supplied by the supra-scapular nerve; the Teres minor, by the circumflex; and the Teres major, by the subscapular.

*Actions.* The Supra-spinatus assists the Deltoid in raising the arm from the side; its action must, however, be very feeble, from the very disadvantageous manner in which the force is applied. The Infra-spinatus and Teres minor rotate the head of the humerus outwards; when the arm is raised, they assist in retaining it in that position, and carrying it backwards. One of the most important uses of these three muscles, is the great protection they afford to the shoulder-joint, the Supra-spinatus supporting it above, and preventing displacement of the head of the humerus upwards, whilst the Infra-spinatus and Teres minor protect it behind, and prevent dislocation backwards. The Teres major assists the Latissimus dorsi in drawing the humerus downwards and backwards when previously raised, and rotating it inwards; when the arm is fixed, it may assist the Pectoral and Latissimus dorsi muscles in drawing the trunk forwards.

## Anterior Humeral Region.

Coraco-Brachialis. Biceps. Brachialis Anticus.

*Dissection.* The arm being placed on the table, with the front surface uppermost, make a vertical incision through the integument along the middle line, from the middle of the interval between the folds of the axilla, to about two inches below the elbow-joint, where it should be joined by a transverse incision, extending from the inner to the outer side of the fore-arm; the two flaps being reflected on either side, the fascia should be examined.

The *deep fascia* of the arm, continuous with that covering the shoulder and front of the great Pectoral muscle, is attached, above, to the clavicle, acromion, and spine of the scapula; it forms a thin, loose, membranous sheath investing the muscles of the arm, sending down septa between them, and composed of fibres disposed in a circular or spiral direction, and these being connected together by vertical fibres. It differs in thickness at different parts, being thin over the Biceps, but thicker where it covers the Triceps, and over the condyles of the humerus; and is strengthened by fibrous aponeuroses, derived from the Pectoralis major and Latissimus dorsi, on the inner side; and from the Deltoid, externally. On either side it gives off a strong intermuscular septum, which is attached to the condyloid ridge and condyle of the humerus. These septa serve to separate the muscles of the anterior from those of the posterior brachial region. The external inter-muscular septum extends from the lower part of the anterior bicipital ridge, along the external condyloid ridge, to the outer condyle; it is blended with the tendon of the Deltoid; gives attachment to the Triceps behind, to the Brachialis anticus, Supinator longus, and Extensor carpi radialis longior, in front; and is perforated by the musculo-spiral nerve, and superior profunda artery. The internal intermuscular septum, thicker than the preceding, extends from the lower part of the posterior bicipital ridge below the Teres major, along the internal condyloid ridge to the inner condyle; it is blended with the tendon of the Coraco-brachialis, and affords attachment to the Triceps behind, and the Brachialis anticus in front. It is perforated by the ulnar nerve, and the inferior profunda and anastomotic arteries. At the elbow, the deep fascia takes attachment to all the prominent points round this joint, and is continuous with the fascia of the fore-arm. On the removal of this fascia, the muscles of the anterior humeral region are exposed.

The *Coraco-Brachialis*, the smallest of the three muscles in this region, is situated at the upper and inner part of the arm. It arises by fleshy fibres from the apex of the coracoid process, in common with the short head of the Biceps, and from the intermuscular septum between the two muscles; the fibres pass downwards, backwards, and a little outwards, to be inserted by means of a flat tendon into a rough ridge at the middle of the inner side of the shaft of the humerus. It is perforated by the musculo-cutaneous nerve. The inner border of the muscle forms a guide to the performance of the operation of tying the brachial artery in the upper part of its course.

*Relations.* By its *anterior surface*, with the Deltoid and Pectoralis major above, at its insertion it is crossed by the brachial vessels and median nerve. By its *posterior surface*, with the tendons of the Subscapularis, Latissimus dorsi, and Teres major, the short head of the Triceps, the humerus, and the anterior circumflex vessels. By its *inner border*, with the brachial artery, and the median and musculo-cutaneous nerves. By its *outer border*, with the short head of the Biceps and Brachialis anticus.

The *Biceps* is a long fusiform muscle, situated along the anterior aspect of the arm its entire length, and divided above into two portions or heads, from which circumstance it has received its name. The short head arises by a thick flattened tendon from the apex of the coracoid process, in common with the Coraco-brachialis. The long head, arises from the upper margin of the glenoid cavity, by a long rounded tendon, which is continuous with the glenoid ligament. This tendon arches over the head of the humerus, being enclosed in a special sheath of the synovial membrane of the shoulder joint; it then pierces the capsular ligament at its attachment to the humerus, and descends in the bicipital groove in which it is retained by a fibrous prolongation from the tendon of the Pectoralis major. The fibres from this tendon form a rounded belly, and, about the middle of the arm, join with the short portion of the muscle. The belly of the muscle, narrow and somewhat flattened, terminates above the elbow in a flattened tendon, which is inserted into the back part of the tuberosity of the radius, a synovial bursa being interposed between the tendon and the front of the tuberosity. The tendon of the muscle is thin and broad; as it approaches the radius it becomes narrow and twisted upon itself, being applied by a flat surface to the back part of the tuberosity, and opposite the bend of the elbow gives off, from its inner side, a broad aponeurosis, which passes obliquely downwards and inwards across the brachial artery, and is continuous with the fascia of the fore-arm. The inner border of this muscle forms a guide to the performance of the operation of tying the brachial artery in the middle of the arm.*

*Relations.* Its *anterior surface* is overlapped above by the Pectoralis major and Deltoid; in the rest of its extent it is covered by the superficial and deep fasciæ and the integument. Its *posterior surface* rests on the shoulder-joint and humerus, from which it is separated by the Subscapularis, Teres major, Latissimus dorsi, Brachialis anticus, and the musculo-cutaneous nerve. Its *inner border* is in relation with the Coraco-brachialis, the brachial vessels, and median nerve. By its *outer border*, with the Deltoid and Supinator longus.

The *Brachialis Anticus* is a broad muscle, which covers the elbow joint and the lower half of the front of the humerus. It is somewhat compressed from before backward, and is broader in the middle than at either extremity. It arises from the lower half of the outer and inner surfaces of the

* A third head to the Biceps is occasionally found (Theile says as often as once in eight or nine subjects), arising at the upper and inner part of the Brachialis anticus with the fibres of which it is continuous, and is inserted into the bicipital fascia, and inner side of the tendon of the Biceps. In most cases this additional slip passes behind the brachial artery in its course down the arm. Occasionally the third head consists of two slips, which pass down, one in front, the other behind the artery, concealing this vessel in the lower half of the arm.

shaft of the humerus, commencing above at the insertion of the Deltoid, which it embraces by two angular processes, and extending, below, to within an inch of the margin of the articular surface, and being limited on each side by the external and internal borders. It also arises from the intermuscular septa on each side, but more extensively from the inner than the outer. Its fibres converge to a thick tendon, which is inserted into a rough depression on the under surface of the coronoid process of the ulna, being received into an interval between two fleshy slips of the Flexor digitorum profundus.

*Relations.* By its *anterior surface*, with the Biceps, the brachial vessels, musculo-cutaneous, and median nerves. By its *posterior surface*, with the humerus and front of the elbow joint. By its *inner border*, with the Triceps, ulnar nerve, and Pronator radii teres, from which it is separated by the inter-muscular septum. By its *outer border*, with the musculo-spiral nerve, radial recurrent artery, the Supinator longus, and Extensor carpi radialis longior.

*Nerves.* The muscles of this group are supplied by the musculo-cutaneous nerve. The Brachialis anticus receives an additional filament from the musculo-spiral.

*Actions.* The Coraco-brachialis draws the humerus forwards and inwards, and at the same time assists in elevating it towards the scapula. The Biceps and Brachialis anticus are flexors of the fore-arm; the former muscle is also a supinator, and serves to render tense the fascia of the fore-arm by means of the broad aponeurosis given off from its tendon. When the fore-arm is fixed, the Biceps and Brachialis anticus flex the arm upon the fore-arm, as is seen in efforts of climbing. The Brachialis anticus forms an important defence to the elbow joint.

## Posterior Humeral Region.

Triceps. Subanconeus.

The *Triceps* is situated on the back of the arm, extending the entire length of the posterior surface of the humerus. It is of large size, and divided above into three parts; hence the name of the muscle. These three portions have been named, the middle, or long head, the external, and the internal head.

The *middle* or *long head* arises, by a flattened tendon, from a rough triangular depression, immediately below the glenoid cavity, being blended at its upper part with the capsular and glenoid ligaments; the muscular fibres pass downwards between the two other portions of the muscle, and join with them in the common tendon of insertion.

The *external head* arises from the posterior surface of the shaft of the humerus, between the insertion of the Teres minor and the upper part of the musculo-spiral groove, from the external border of the humerus and external intermuscular septum: the fibres from this origin converge towards the common tendon of insertion.

The *internal head* arises from the posterior surface of the shaft of the humerus, below the groove for the musculo-spiral nerve, commencing above, narrow and pointed, below the insertion of the Teres major, and extending, to within an inch of the trochlear surface; it also arises from the internal border and internal intermuscular septum. The fibres of this portion of the muscle are directed, some downwards to the olecranon, whilst others converge to the common tendon of insertion.

The *common tendon* of the Triceps commences about the middle of the back part of the muscle: it consists of two aponeurotic laminæ, one of which is subcutaneous, and covers the posterior surface of the muscle for the lower half of its extent: the other is more deeply seated in the substance of the muscle; after receiving the attachment of the muscular fibres, they join together above the elbow, and are inserted into the back part of the upper surface of the olecranon process, a small bursa, occasionally multilocular, being interposed between the tendon and the front of this surface.

The long head of the Triceps descends between the Teres minor and Teres major, dividing the triangular space between these two muscles and the humerus into two smaller spaces, one triangular, the other quadrangular (fig. 171). The triangular space transmits the dorsalis scapulæ vessels; it is bounded by the Teres minor above, the Teres major below, and the scapular head of the Triceps, externally: the quadrangular space transmits the posterior circumflex vessels and nerve; it is bounded by the Teres minor above, the Teres major below, the scapular head of the Triceps internally, and the humerus externally.

*Relations.* Its *posterior surface* is overlapped by the Deltoid above, superficial in the rest of its extent. By its *anterior surface*, with the humerus, musculo-spiral nerve, superior profunda vessels, and back part of the elbow-joint. Its *middle* or *long head* is in relation, behind, with the Deltoid and Teres minor; in front, with the Subscapularis, Latissimus dorsi, and Teres major.

The *Subanconeus* is a small muscle, distinct from the Triceps, and analogous to the Subcrureus in the lower limb. It may be exposed by removing the Triceps from the lower part of the humerus. It consists of one or two slender fasciculi, which arise from the humerus, immediately above the olecranon fossa, and are inserted into the posterior ligament of the elbow-joint.

*Nerves.* The Triceps and Subanconeus are supplied by the musculo-spiral nerve.

*Actions.* The Triceps is the great Extensor muscle of the fore-arm; when the fore-arm is flexed, serving to draw it into a right line with the arm. It is the direct antagonist of the Biceps and Brachialis anticus. When the arm is extended, the long head of the muscle may assist the Teres major and Latissimus dorsi in drawing the humerus backwards. The long head of the Triceps protects the under part of the shoulder-joint, and prevents displacement of the head of the humerus downwards and backwards.

## Muscles of the Fore-arm.

*Dissection.* To dissect the fore-arm, place the limb in the position indicated in fig. 168; make a vertical incision along the middle line from the elbow to the wrist, and connect each extremity with a transverse incision; the flaps of integument being removed, the fascia of the fore-arm is exposed.

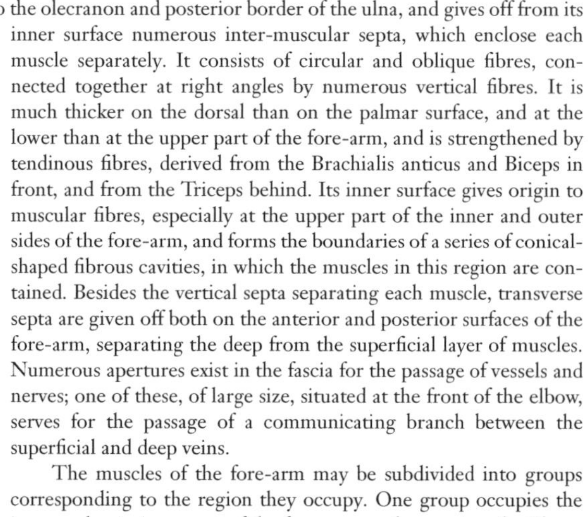

The *deep fascia* of the fore-arm, continuous above with that enclosing the arm, is a dense highly glistening aponeurotic investment, which forms a general sheath enclosing the muscles in this region; it is attached behind to the olecranon and posterior border of the ulna, and gives off from its inner surface numerous inter-muscular septa, which enclose each muscle separately. It consists of circular and oblique fibres, connected together at right angles by numerous vertical fibres. It is much thicker on the dorsal than on the palmar surface, and at the lower than at the upper part of the fore-arm, and is strengthened by tendinous fibres, derived from the Brachialis anticus and Biceps in front, and from the Triceps behind. Its inner surface gives origin to muscular fibres, especially at the upper part of the inner and outer sides of the fore-arm, and forms the boundaries of a series of conical-shaped fibrous cavities, in which the muscles in this region are contained. Besides the vertical septa separating each muscle, transverse septa are given off both on the anterior and posterior surfaces of the fore-arm, separating the deep from the superficial layer of muscles. Numerous apertures exist in the fascia for the passage of vessels and nerves; one of these, of large size, situated at the front of the elbow, serves for the passage of a communicating branch between the superficial and deep veins.

The muscles of the fore-arm may be subdivided into groups corresponding to the region they occupy. One group occupies the inner and anterior aspect of the fore-arm, and comprises the Flexor and Pronator muscles. Another group occupies the outer side of the fore-arm; and a third, its posterior aspect. The two latter groups include all the Extensor and Supinator muscles.

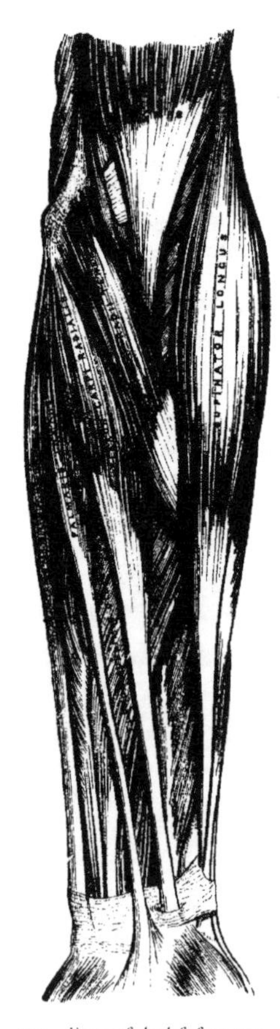


172 – *Front of the left fore-arm. Superficial muscles.*

### Anterior Brachial Region.

*Superficial Layer*

Pronator radii teres.
Flexor carpi radialis.
Palmaris longus.
Flexor carpi ulnaris.
Flexor sublimis digitorum.

These muscles take origin from the internal condyle by a common tendon.

The *Pronator Radii Teres* arises by two heads. One, the largest and most superficial, from the humerus, immediately above the internal condyle, and from the tendon common to the origin of the other muscles; also from the fascia of the fore-arm, and inter-muscular septum between it and the Flexor carpi radialis. The other head is a thin fasciculus, which arises from the inner side of the coronoid process of the ulna, joining the preceding at an acute angle. Between the two heads passes the median nerve. The muscle passes obliquely across the fore-arm from the inner to the outer side, and terminates in a flat tendon, which turns over the outer margin of the radius, and is inserted into a rough ridge at the middle of the outer surface of the shaft of that bone.

*Relations.* By its *anterior surface*, with the deep fascia, the Supinator longus, and the radial vessels and nerve. By its *posterior surface*, with the Brachialis anticus, Flexor sublimis digitorum, the median nerve, and ulnar artery. Its *outer border* forms the inner boundary of a triangular space, in which is placed the brachial artery, median nerve, and tendon of the Biceps muscle. Its *inner border* is in contact with the Flexor carpi radialis.

The *Flexor Carpi Radialis* lies, on the inner side of the preceding muscle. It arises from the internal condyle by the common tendon, from the fascia of the fore-arm, and from the inter-muscular septa between it and the Pronator teres, on the inside; the Palmaris longus, externally; and the Flexor sublimis digitorum, beneath. Slender and aponeurotic in structure at its commencement, it increases in size, and terminates in a tendon which forms the lower two-thirds of its structure. This tendon passes through a canal on the outer side of the annular ligament, runs through a groove in the os trapezium, converted into a canal by a fibrous sheath, lined by a synovial membrane, and is inserted into the base of the metacarpal bone of the index-finger. The radial artery lies between the tendon of this muscle and the Supinator longus, and may easily be secured in this situation.

*Relations.* By its *superficial surface*, with the deep fascia and the integument. By its *deep surface*, with the Flexor sublimis digitorum, Flexor longus pollicis, and wrist joint. By its *outer border*, with the Pronator radii teres, and the radial vessels. By its *inner border*, with the Palmaris longus above, the median nerve below.

The *Palmaris Longus* is a slender fusiform muscle, lying on the inner side of the preceding. It arises from the inner condyle of the humerus by the common tendon, from the deep fascia, and inter-muscular septa, between it and the adjacent muscles. It terminates in a slender flattened tendon, which is inserted into the annular ligament, expanding to end in the palmar fascia.

*Variations.* This muscle is often absent; when present, it presents many varieties. Its fleshy belly is sometimes very long, or it may occupy the middle of the muscle, which is tendinous at either extremity; or it may be muscular at its lower extremity, its upper part being tendinous. Occasionally there is a second Palmaris longus placed on the inner side of the preceding, terminating, below, partly in the annular ligament or fascia, and partly in the small muscles of the little finger.

*Relations.* By its *anterior surface*, with the deep fascia. By its *posterior surface*, with the Flexor digitorum sublimis. *Internally*, with the Flexor carpi ulnaris. *Externally*, with the Flexor carpi radialis.

The *Flexor Carpi Ulnaris* lies along the ulnar side of the fore-arm. It arises by two heads, separated by a tendinous arch, beneath which passes the ulnar nerve, and posterior ulnar recurrent artery. One head arises from the inner condyle of the humerus, by the common tendon; the other, from the inner margin of the olecranon, by an aponeurosis from the upper two-thirds of the posterior border of the ulna, and from the inter-muscular septum between it and the Flexor sublimis digitorum. The fibres terminate in a tendon, which occupies the anterior part of the lower half of the muscle, and is inserted into the pisiform bone, some fibres being prolonged to the annular ligament and base of the metacarpal bone of the little finger. The ulnar artery lies on the outer side of the tendon of this muscle, in the lower two-thirds of the fore-arm; the tendon forming a guide to the operation of including this vessel in a ligature in this situation.

*Relations.* By its *anterior surface*, with the deep fascia, with which it is intimately connected for a considerable extent. By its *posterior surface*, with the Flexor sublimis, the Flexor profundus, the Pronator quadratus, and the ulnar vessels and nerve. By its *outer* or *radial border*, with the Palmaris longus, above; with the ulnar vessels and nerve, below.

The *Flexor Digitorum Sublimis* is placed beneath the preceding muscles; these therefore require to be removed before its attachment is brought into view. It is the largest of the muscles of the superficial layer, and arises by three heads. One from the internal condyle of the humerus by the common tendon, from the internal lateral ligament of the elbow joint, and from the inter-muscular septum common to it and the preceding muscles. The second head arises from the inner side of the coronoid process of the ulna, above the ulnar origin of the Pronator radii teres. The third head arises from the oblique line of the radius, extending from the tubercle to the insertion of the

Pronator radii teres. The fibres pass vertically downwards, forming a broad and thick muscle, which divides into four tendons about the middle of the fore-arm; as these tendons pass beneath the annular ligament into the palm of the hand, they are arranged in pairs, the anterior pair corresponding to the middle and ring fingers; the posterior pair to the index and little fingers. The tendons diverge from one another as they pass onwards, and are finally inserted into the lateral margins of the second phalanges, about their centre. Opposite the base of the first phalanges, each tendon divides, so as to leave a fissured interval, between which passes one of the tendons of the Flexor profundus, and they both enter an osseo-aponeurotic canal, formed by a strong fibrous band which arches across them, and is attached on each side to the margins of the phalanges. The two portions into which the tendon of the Flexor sublimis divides, so as to admit of the passage of the deep flexor, expand somewhat, and form a grooved channel, into which the accompanying deep flexor tendon is received; the two divisions then unite, and finally subdivide a second time to be inserted into the fore part and sides of the second phalanges. The tendons, while contained in the fibro-osseous canals, are connected to the phalanges by slender tendinous filaments, called *vincula accessoria tendinum*. A synovial sheath invests the tendons as they pass beneath the annular ligament; a prolongation from which surrounds each tendon as it passes along the phalanges.

*Relations.* In the fore-arm, By its *anterior surface*, with the deep fascia and all the preceding superficial muscles. By its *posterior surface*, with the Flexor profundus digitorum, Flexor longus pollicis, the ulnar vessels and nerve, and the median nerve. In the hand, its tendons are in relation, in front, with the palmar fascia, superficial palmar arch, and the branches of the median nerve. Behind with the tendons of the deep Flexor and the Lumbricales.

## Anterior Brachial Region.

*Deep Layer*

Flexor Profundus Digitorum. Flexor Longus Pollicis.
Pronator Quadratus.

*Dissection.* Divide each of the superficial muscles at its centre, and turn either end aside, the deep layer of muscles, together with the median nerve and ulnar vessels, will then be exposed.

The *Flexor Profundus Digitorum* (*perforans*) is situated on the ulnar side of the fore-arm, immediately beneath the superficial Flexors. It arises from the upper two-thirds of the anterior and inner surfaces of the shaft of the ulna, embracing above, the insertion of the Brachialis anticus, and extending, below, to within a short distance of the Pronator quadratus. It also arises from a depression on the inner side of the coronoid process, by an aponeurosis from the upper two-thirds of the posterior border of the ulna, and from the ulnar half of the interosseous membrane. The fibres form a fleshy belly of considerable size, which divides into four tendons, these pass beneath the annular ligament beneath the tendons of the Flexor sublimis. Opposite the first phalanges, the tendons pass between the two slips of the tendons of the Flexor sublimis, and are finally inserted into the base of the last phalanges. The tendon of the index finger is distinct; the rest are connected together by cellular tissue and tendinous slips, as far as the palm of the hand.

Four small muscles, the Lumbricales, are connected with the tendons of the Flexor profundus in the palm. They will be described with the muscles in that region.

*Relations.* By its *anterior surface*, in the fore-arm, with the Flexor sublimis digitorum, the Flexor carpi ulnaris, the ulnar vessels and nerve, and the median nerve; and in the hand, with the tendons of the superficial Flexor. By its *posterior surface*, in the fore-arm, with the ulna, the interosseous membrane, the Pronator quadratus; and in the hand, with the Interossei, Adductor pollicis, and deep palmar arch. By its *ulnar border*, with the Flexor carpi ulnaris. By its *radial border*, with the Flexor longus pollicis, the anterior interosseous vessels and nerve being interposed.

The *Flexor Longus Pollicis* is situated on the radial side of the fore-arm, lying on the same plane as the preceding. It arises from the upper two-thirds of the grooved anterior surface of the shaft of the radius; commencing, above, immediately below the tuberosity and oblique line, and extending, below, to within a short distance of the Pronator quadratus. It also arises from the adjacent part of the interosseous membrane, and occasionally by a fleshy slip from the inner side of the base of the coronoid process. The fibres pass downwards and terminate in a flattened tendon, which passes beneath the annular ligament, is then lodged in the inter-space between the two heads of the Flexor brevis pollicis, and entering a tendino-osseous canal, similar to those for the other flexor tendons, is inserted into the base of the last phalanx of the thumb.

*Relations.* By its *anterior surface*, with the Flexor sublimis digitorum, Flexor carpi radialis, Supinator longus, and radial vessels. By its *posterior surface*, with the radius, interosseous membrane, and Pronator quadratus. By its *ulnar border*, with the Flexor profundus digitorum, from which it is separated by the anterior interosseous vessels and nerve.

The *Pronator Quadratus* is a small, flat muscle, quadrilateral in form, extending transversely across the front of the radius and ulna, above their carpal extremities. It arises from the oblique line on the lower fourth of the anterior surface of the shaft of the ulna, and the surface of bone immediately below it; from the internal body of the ulna; and from a strong aponeurosis which covers the inner third of the muscle. The fibres pass horizontally outwards, to be inserted into the lower fourth of the anterior surface and external border of the shaft of the radius.

*Relations.* By its *anterior surface*, with the Flexor profundus digitorum, the Flexor longus pollicis, Flexor carpi radialis, and the radial vessels. By its *posterior surface*, with the radius, ulna, and interosseous membrane.

*Nerves.* All the muscles of the superficial layer are supplied by the median nerve, excepting the Flexor carpi ulnaris, which is supplied by the ulnar. Of the deep layer, the Flexor profundus digitorum is supplied conjointly by the ulnar and anterior interosseous nerves, the Flexor longus pollicis and Pronator quadratus by the anterior interosseous nerve.

*Actions.* These muscles act upon the fore-arm, the wrist, and hand. Those acting on the fore-arm, are the Pronator radii teres and Pronator quadratus, which rotate the radius upon the ulna, rendering the hand prone; when pronation has been fully effected, the Pronator radii teres assists the other muscles in flexing the fore-arm. The flexors of the wrist are the Flexor carpi ulnaris and radialis; and the flexors of the phalanges are the Flexor sublimis and Profundus digitorum; the former flexing the second phalanges, and the latter the last. The Flexor longus pollicis flexes the last phalanx of the thumb. The three latter muscles, after flexing the phalanges, by continuing their action, act upon the wrist, assisting the ordinary flexors of this joint; and all assist in flexing the fore-arm upon the arm. The Palmaris longus is a tensor of the palmar fascia; when this action has been fully effected, it flexes the hand upon the fore-arm.

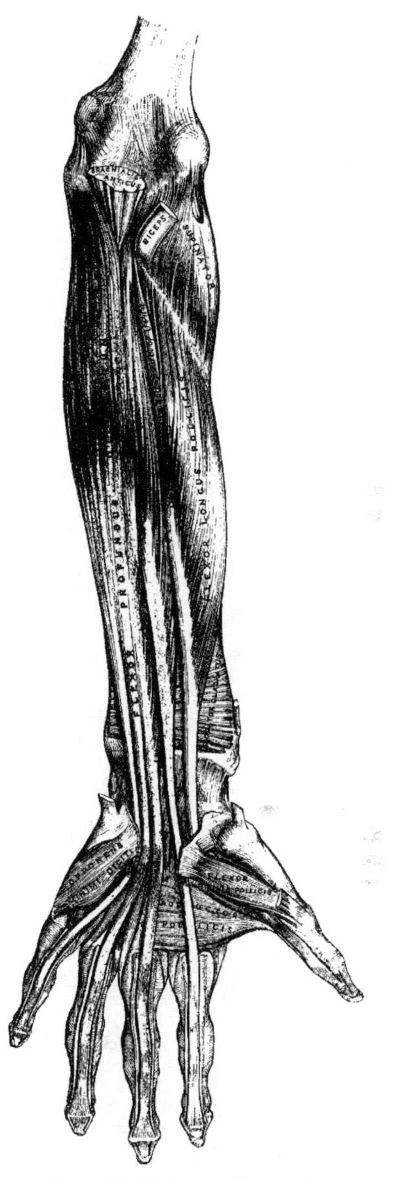


*173 – Front of the left fore-arm. Deep muscles.*

## Radial Region.

Supinator Longus. Extensor Carpi Radialis Longior.
Extensor Carpi Radialis Brevior.

*Dissection.* Divide the integument in the same manner as in the dissection of the anterior brachial region; and after having examined the cutaneous vessels and nerves and deep fascia, they should be removed, when the muscles of this region will be exposed. The removal of the fascia will be considerably facilitated by detaching it from below upwards. Great care should be taken to avoid cutting across the tendons of the muscles of the thumb.

The *Supinator Longus* is the most superficial muscle on the radial side of the fore-arm, fleshy for the upper two-thirds of its extent, tendinous below. It arises from the upper two-thirds of the external condyloid ridge of the humerus, and from the external intermuscular septum, being limited above by the musculo-spiral groove. The fibres terminate above the middle of the fore-arm in a flat tendon which is inserted into the base of the styloid process of the radius.

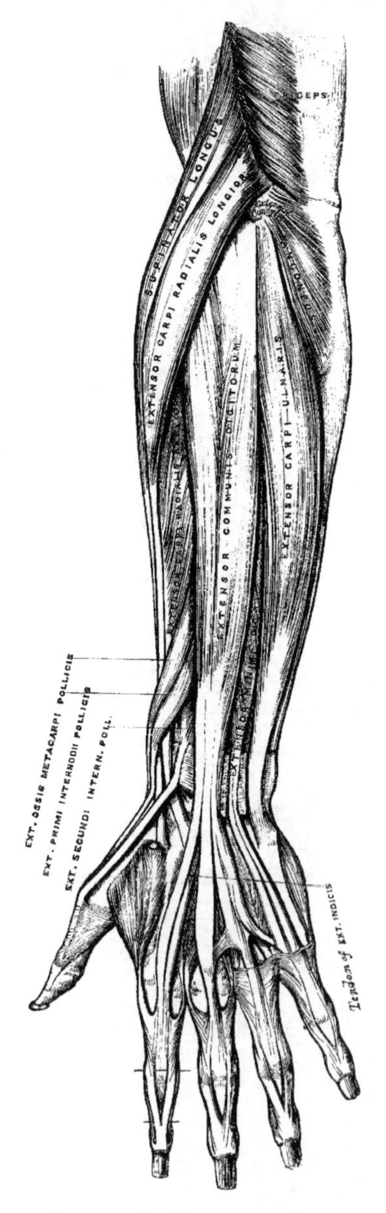


174 – *Posterior surface of fore-arm. Superficial muscles.*

*Relations.* By its *superficial surface*, with the integument and fascia for the greater part of its extent; near its insertion it is crossed by the Extensor ossis metacarpi pollicis and the Extensor primi internodii pollicis. By its *deep surface*, with the humerus, the Extensor carpi radialis longior and brevior, the insertion of the Pronator radii teres, and the Supinator brevis. By its *inner border*, above the elbow, with the Brachialis anticus, the musculo-spiral nerve, and radial recurrent artery; and in the fore-arm, with the radial vessels and nerve.

The *Extensor Carpi Radialis Longior* is placed partly beneath the preceding muscle. It arises from the lower third of the external condyloid ridge of the humerus, and from the external intermuscular septum. The fibres terminate at the upper third of the fore-arm in a flat tendon, which runs along the outer border of the radius, beneath the extensor tendons of the thumb; it then passes through a groove common to it and the Extensor carpi radialis brevior, immediately behind the styloid process; and is inserted into the base of the metacarpal bone of the index finger, its radial side.

*Relations.* By its *superficial surface*, with the Supinator longus, and fascia of the fore-arm. Its *outer side* is crossed obliquely by the extensor tendons of the thumb. By its *deep surface*, with the elbow-joint, the Extensor carpi radialis brevior, and back part of the wrist.

The *Extensor Carpi Radialis Brevior* is shorter, as its name implies, and thicker than the preceding muscle, beneath which it is placed. It arises from the external condyle of the humerus by a tendon common to it and the three following muscles; from the external lateral ligament of the elbow-joint; from a strong aponeurosis which covers its surface; and from the intermuscular septa between it and the adjacent muscles. The fibres terminate about the middle of the fore-arm in a flat tendon, which is closely connected with that of the preceding muscle, accompanies it to the wrist, lying in the same groove on the posterior surface of the radius; passes beneath the annular ligament, and, diverging somewhat from its fellow, is inserted into the base of the metacarpal bone of the middle finger, its radial side.

The tendons of the two preceding muscles pass through the same compartment of the annular ligament, are lubricated by a single synovial membrane, but separated from each other by a small vertical ridge of bone, as they lie in the groove at the back of the radius.

*Relations.* By its *superficial surface*, with the Extensor carpi radialis longior, and crossed by the Extensor muscles of the thumb. By its *deep surface*, with the Supinator brevis, tendon of the Pronator radii teres, radius, and wrist-joint. By its *ulnar border*, with the Extensor communis digitorum.

## Posterior Brachial Region.

*Superficial Layer*

| | |
|---|---|
| Extensor Communis Digitorum. | Extensor Carpi Ulnaris. |
| Extensor Minimi Digiti. | Anconeus. |

The *Extensor Communis Digitorum* is situated at the back part of the fore-arm. It arises from the external condyle of the humerus by the common tendon, from the deep fascia, and the intermuscular septa between it and the adjacent muscles. Just below the middle of the fore-arm it divides into three tendons, which pass together with the Extensor indicis, through a separate compartment of the annular ligament, lubricated by a synovial membrane. The tendons then diverge, the innermost one dividing into two; and all, after passing across the back of the hand, are inserted into the second and third phalanges of the fingers in the following manner: each tendon, opposite its corresponding metacarpo-phalangeal articulation, becomes narrow and thickened, gives off a thin fasciculus upon each side of the joint, and spreads out into a broad aponeurosis, which covers the whole of the dorsal surface of the first phalanx; being reinforced, in this situation, by the tendons of the Interossei and Lumbricales. Opposite the first phalangeal joint, this aponeurosis divides into three slips, a middle, and two lateral; the former is inserted into the base of the second phalanx; and the two lateral, which are continued onwards along the sides of the second phalanx, unite by their contiguous margins, and are inserted into the upper surface of the last phalanx. The tendons of the middle, ring, and little fingers are connected together, as they cross the hand, by small oblique tendinous slips. The tendons of the index and little fingers also receive, before their division, the special extensor tendons belonging to them.

*Relations.* By its *superficial surface*, with the fascia of the fore-arm and hand, the posterior annular ligament, and integument. By its *deep surface*, with the Supinator brevis, the Extensor muscles of the thumb and index finger, posterior interosseous vessels and nerve, the wrist-joint, carpus, metacarpus, and phalanges. By its *radial border*, with the Extensor carpi radialis brevior. By its *ulnar border*, with the Extensor minimi digiti, and Extensor carpi ulnaris.

The *Extensor Minimi Digiti* is a slender muscle, placed on the inner side of the Extensor communis, with which it is generally connected. It arises from the common tendon by a thin tendinous slip; and from the intermuscular septa between it and the adjacent muscles. Its tendon runs through a separate compartment in the annular ligament behind the inferior radio-ulnar joint, subdivides into two as it crosses the hand, and, at the metacarpo-phalangeal articulation, unites with the tendon derived from the long Extensor. The common tendon then spreads into a broad aponeurosis, which is inserted into the second and third phalanges of the little finger in a similar manner to the common extensor tendons of the other fingers.

The *Extensor Carpi Ulnaris* is the most superficial muscle on the ulnar side of the fore-arm. It arises from the external condyle of the humerus, by the common tendon; from the middle third of the posterior border of the ulna below the Anconeus, and from the fascia of the fore-arm. This muscle terminates in a tendon, which runs through a groove behind the styloid process of the ulna, passes through a separate compartment in the annular ligament, and is inserted into the base of the metacarpal bone of the little finger.

*Relations.* By its *superficial surface*, with the fascia of the fore-arm. By its *deep surface*, with the ulna, and the muscles of the deep layer.

The *Anconeus* is a small triangular muscle, placed behind and below the elbow-joint, and appears to be a continuation of the external portion of the Triceps. It arises by a separate tendon from the back part of the outer condyle of the humerus; and is inserted into the side of the olecranon, and upper third of the posterior surface of the shaft of the ulna; its fibres diverge from their origin, the upper ones being directed transversely, the lower obliquely inwards.

*Relations.* By its *superficial surface*, with a strong fascia derived from the Triceps. By its *deep surface*, with the elbow-joint, the orbicular ligament, the ulna, and a small portion of the Supinator brevis.

## Posterior Brachial Region.

*Deep Layer*

Supinator Brevis. Extensor Primi Internodii Pollicis.
Extensor Ossis Metacarpi Pollicis. Extensor Secundi Internodii Pollicis.
Extensor Indicis.

The *Supinator Brevis* is a broad muscle, of a hollow cylindrical form, curved round the upper third of the radius. It arises from the external condyle of the humerus, from the external lateral ligament of the elbow-joint, and the orbicular ligament of the radius, from an oblique ridge on the ulna, extending down from the posterior extremity of the lesser sigmoid cavity, and from the triangular depression in front of it; it also arises from a tendinous expansion which covers its surface. The muscle surrounds the upper part of the radius; the upper fibres forming a sling-like fasciculus, which encircles the neck of the radius above the the tuberosity, to be attached to the back part of its inner surface; the middle fibres are attached to the outer edge of the bicipital tuberosity; the lower fibres to the oblique line, as low down as the insertion of the Pronator radii teres. This muscle is pierced by the posterior interosseous nerve.

*Relations.* By its *superficial surface*, with the superficial Extensor and Supinator muscles, and the radial vessels and nerve. By its *deep surface*, with the elbow-joint, the interosseous membrane, and the radius.

The *Extensor Ossis Metacarpi Pollicis* is the most external and the largest of the deep Extensor muscles, lying immediately below the Supinator brevis, with which it is sometimes united. It arises from the posterior surface of the shaft of the ulna below the origin of the Anconeus, from the interosseous ligament, and from the middle third of the posterior surface of the shaft of the radius. Passing obliquely downwards and outwards, it terminates in a tendon which runs through a groove on the outer side of the styloid process of the radius, accompanied by the tendon of the Extensor primi internodii pollicis, and is inserted into the base of the metacarpal bone of the thumb.

*Relations.* By its *superficial surface*, with the Extensor communis digitorum, Extensor minimi digiti, and fascia of the fore-arm; being crossed by the branches of the posterior interosseous artery and nerve. By its *deep surface*, with the ulna, interosseous membrane, radius, the tendons of the Extensor carpi radialis longior and brevior; and, at the outer side of the wrist, with the radial vessels. By its *upper border*, with the Supinator brevis. By its *lower border*, with the Extensor primi internodii pollicis.

The *Extensor Primi Internodii Pollicis*, the smallest muscle of this group, lies on the inner side of the preceding. It arises from the posterior surface of the shaft of the radius, below the Extensor ossis metacarpi, and from the interosseous membrane. Its direction is similar to that of the Extensor ossis metacarpi, its tendon passing through the same groove on the outer side of the styloid process, to be inserted into the base of the first phalanx of the thumb.

*Relations.* The same as those of the Extensor ossis metacarpi pollicis.

The *Extensor Secondi Internodii Pollicis* is much larger than the preceding muscle, the origin of which it partly covers in. It arises from the posterior surface of the shaft of the ulna, below the origin of the Extensor ossis metacarpi pollicis, and from the interosseous membrane. It terminates in a tendon which passes through a separate compartment in the annular ligament, lying in a narrow oblique groove at the back part of the lower end of the radius. It then crosses obliquely the Extensor tendons of the carpus, being separated from the other Extensor tendons of the thumb by a triangular interval, in which the radial artery is found; and is finally inserted into the base of the last phalanx of the thumb.

*Relations.* By its *superficial surface*, with the same parts as the Extensor ossis metacarpi pollicis. By its *deep surface*, with the ulna, interosseous membrane, radius, the wrist, the radial vessels, and metacarpal bone of the thumb.

The *Extensor Indicis* is a narrow elongated muscle, placed on the inner side of, and parallel with, the preceding. It arises from the posterior surface of the shaft of the ulna, below the origin of the Extensor secundi internodii pollicis, and from the interosseous membrane. Its tendon passes with the Extensor communis digitorum through the same canal in the annular ligament, and subsequently joins that tendon of the Extensor communis which belongs to the index-finger, opposite the lower end of the corresponding metacarpal bone. It is finally inserted into the second and third phalanges of the index-finger, in the manner already described.

*Relations.* They are similar to those of the preceding muscles.

*Nerves.* The Supinator longus, Extensor carpi radialis longior, and Anconeus, are supplied by branches from the musculo-spiral nerve. The remaining muscles of the radial and posterior brachial regions, by the posterior interosseous nerve.

*Actions.* The muscles of the radial and posterior brachial regions, which comprise all the Extensor and Supinator muscles, act upon the fore-arm, wrist, and hand; they are the direct antagonists of the Pronator and Flexor muscles. The Anconeus assists the Triceps in extending the fore-arm. The Supinator longus and brevis are the supinators of the fore-arm and hand; the former muscle more especially acting as a supinator when the limb is pronated. When supination has been produced, the Supinator longus, if still continuing to act, flexes the fore-arm. The Extensor carpi radialis longior and brevior, and Extensor carpi ulnaris muscles, are the Extensors of the wrist; continuing their action, they serve to extend the fore-arm upon the arm; they are the direct antagonists of the Flexor carpi radialis and ulnaris. The common Extensor of the fingers, the Extensors of the thumb, and the Extensors of the index and little fingers, serve to extend the phalanges into which they are inserted; and are the direct antagonists of the Flexors. By continuing their action, they assist in extending the fore-arm. The Extensors of the thumb may assist in supinating the fore-arm, when this part of the hand has been drawn inwards towards the palm, on account of the oblique direction of the tendons of these muscles.

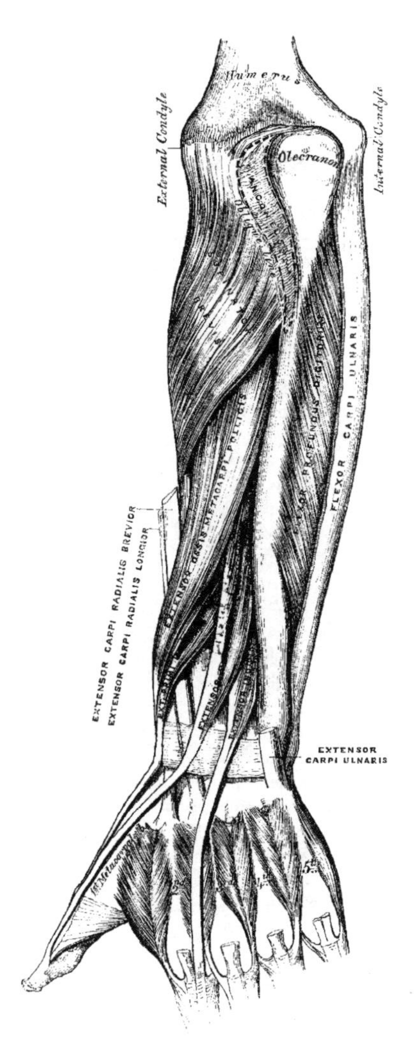


175 – *Posterior surface of the fore-arm. Deep muscles.*

## Muscles and Fasciæ of the Hand.

*Dissection* (fig. 168). Make a transverse incision across the front of the wrist, and a second across the heads of the metacarpal bones, connect the two by a vertical incision in the middle line, and continue it through the centre of the middle finger. The anterior and posterior annular ligaments, and the palmar fascia, should first be dissected.

The *Anterior Annular Ligament* is a strong fibrous band, which arches over the carpus, converting the deep groove on the front of these bones into a canal, beneath which pass the flexor tendons of the fingers. It is attached, internally, to the pisiform bone, and unciform process of the unciform; and externally, to the tuberosity of the scaphoid, and ridge on the trapezium. It is continuous, above, with the deep fascia of the fore-arm, and below, with the palmar fascia. It is crossed by the tendon of the Palmaris longus, by the ulnar vessels and nerve, and the cutaneous branches of the median and ulnar nerves. It has inserted into its upper and inner part, the tendon of the Flexor carpi ulnaris; and has, arising from it below, the small muscles of the thumb and little finger. It is pierced by the tendon of the Flexor carpi radialis; and, beneath it, pass the tendons of the

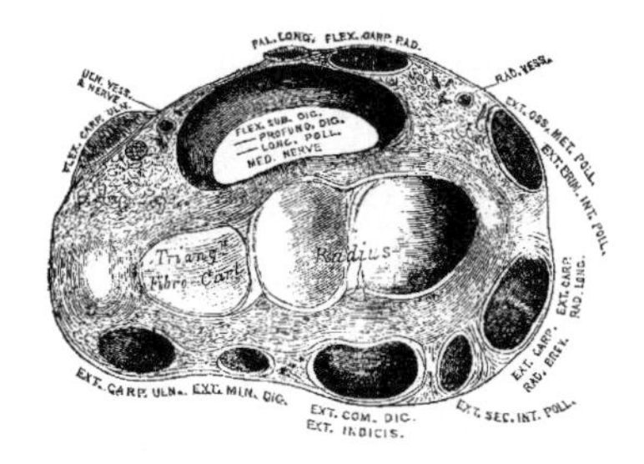


176 – *Transverse section through the wrist, shewing the posterior annular ligament, and the canals for the passage of the extensor tendons.*

Flexor sublimis and profundus digitorum, the Flexor longus pollicis, and the median nerve. There are two synovial membranes beneath this ligament; one of large size, enclosing the tendons of the Flexor sublimis and profundus; and a separate one for the tendon of the Flexor longus pollicis; the latter is also large and very extensive, reaching from above the wrist to the extremity of the last phalanx of the thumb.

The *Posterior Annular Ligament* is a strong fibrous band, extending transversely across the back of the wrist, and continuous with the fascia of the fore-arm. It forms a sheath for the extensor tendons in their passage to the fingers, being attached, internally, to the ulna, the cuneiform and pisiform bones, and palmar fascia; externally, to the margin of the radius; and in its passage across the wrist, to the elevated ridges on the posterior surface of the radius. It presents six compartments for the passage of tendons, each of which is lined by a separate synovial membrane. These are, from without inwards—1. On the outer side of the styloid process for the tendons of the Extensor ossis metacarpi, and Extensor primi internodii pollicis. 2. Behind the styloid process, for the tendons of the Extensor carpi radialis longior and brevior. 3. Opposite the outer side of the posterior surface of the radius, for the tendon of the Extensor secundi internodii pollicis. 4. To the inner side of the latter, for the tendons of the Extensor communis digitorum, and Extensor indicis. 5. For the Extensor minimi digiti, opposite the interval between the radius and ulna. 6. For the tendon of the Extensor carpi ulnaris, grooving the back of the ulna. The synovial membranes lining these sheaths are usually very extensive, reaching from above the annular ligament down upon the tendons, almost to their insertion.

The *palmar fascia* forms a common sheath which invests the muscles of the hand. It consists of a central and two lateral portions.

The *central portion* occupies the middle of the palm, is triangular in shape, of great strength and thickness, and binds down the tendons in this situation. It is narrow above, being attached to the lower margin of the annular ligament, and receives the expanded tendon of the Palmaris longus muscle. Below, it is broad and expanded and opposite the heads of the metacarpal bones divides into four slips, for the four fingers. Each slip subdivides into two processes, which enclose the tendons of the Flexor muscles, and are attached to the sides of the first phalanx, and to the glenoid ligament; by this arrangement, four arches are formed, under which the Flexor tendons pass. The intervals left in the fascia between the four fibrous slips transmit the digital vessels and nerves, and the tendons of the Lumbricales. At the point of division of the palmar fascia into the slips above mentioned, numerous strong transverse fibres bind the separate processes together. The palmar fascia is intimately adherent to the integument by numerous fibrous bands, and gives origin by its inner margin to the Palmaris brevis; it covers the superficial palmar arch, the tendons of the flexor muscles, and the branches of the median and ulnar nerves; and on each side it gives off a vertical septum, which is continuous with the interosseous aponeurosis, and separates the lateral from the middle palmar group of muscles.

The *lateral portions* of the palmar fascia are thin fibrous layers, which cover, on the radial side, the muscles of the ball of the thumb; and, on the ulnar side, the muscles of the little finger; they are continuous with the dorsal fascia, and in the palm, with the middle portion of the palmar fascia.

## Muscles of the Hand.

The muscles of the hand are subdivided into three groups.—1. Those of the thumb, which occupy the radial side. 2. Those of the little finger, which occupy the ulnar side. 3. Those in the middle of the palm and between the interosseous spaces.

## Radial Region.

*Muscles of the Thumb*

Abductor Pollicis.
Opponens Pollicis (Flexor Ossis Metacarpi).
Flexor Brevis Pollicis.
Adductor Pollicis.

The *Abductor Pollicis* is a thin, flat muscle, placed immediately beneath the integument. It arises from the ridge of the os trapezium and annular ligament; and passing outwards and downwards, is inserted by a thin, flat tendon into the radial side of the base of the first phalanx of the thumb.

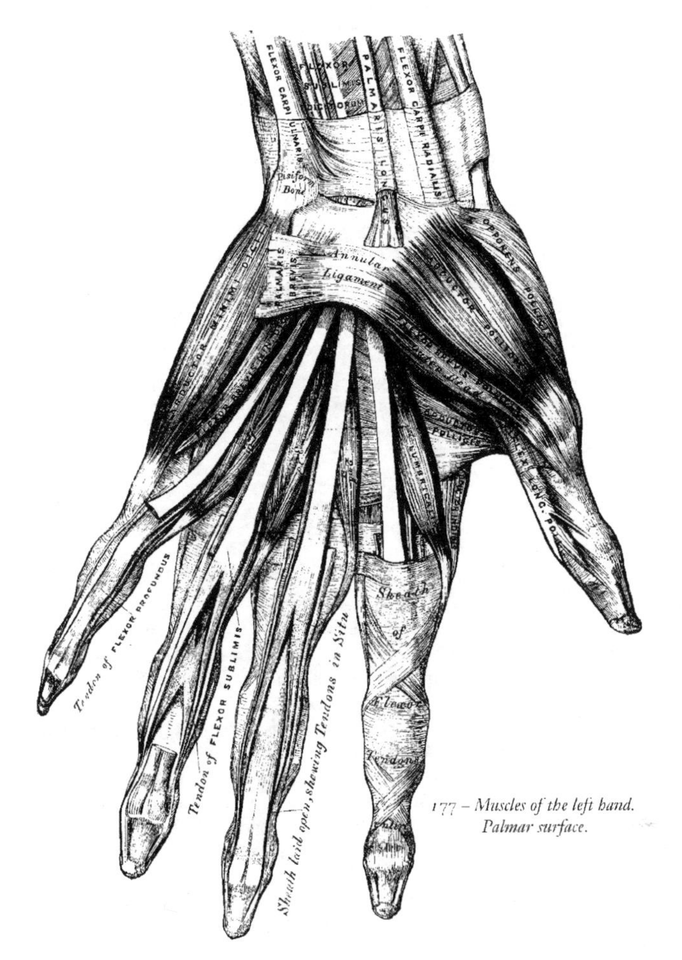


177 – *Muscles of the left hand. Palmar surface.*

*Relations.* By its *superficial surface*, with the palmar fascia. By its *deep surface*, with the Opponens pollicis, from which it is separated by a thin aponeurosis. Its *inner border* is separated from the Flexor brevis pollicis by a narrow cellular interval.

The *Opponens Pollicis* is a small triangular muscle, placed beneath the preceding. It arises from the palmar surface of the trapezium and annular ligament; passing downwards and outwards, it is inserted into the whole length of the metacarpal bone of the thumb on its radial side.

*Relations.* By its *superficial surface*, with the Abductor pollicis. By its *deep surface*, with the trapezio-metacarpal articulation. By its *inner border*, with the Flexor brevis pollicis.

The *Flexor Brevis Pollicis* is much larger than either of the two preceding muscles, beneath which it is placed. It consists of two portions, in the interval between which lies the tendon of the Flexor longus pollicis. The anterior and more superficial portion arises from the trapezium and outer two-thirds of the annular ligament. The deeper portion from the trapezoides, os magnum, base of the third metacarpal bone, and sheath of the tendon of the Flexor carpi radialis. The fleshy fibres unite to form a single muscle; this divides into two portions, which are inserted one on either side of the base of the first phalanx of the thumb, the outer portion being joined with the Abductor, and the inner with the Adductor. A sesamoid bone is developed in each tendon as it passes across the metacarpo-phalangeal joint.

*Relations.* By its *superficial surface*, with the palmar fascia. By its *deep surface*, with the Adductor pollicis, and tendon of the Flexor carpi radialis. By its *external surface*, with the Opponens pollicis. By its *internal surface*, with the tendon of the Flexor longus pollicis.

The *Adductor Pollicis* (fig. 173) is the most deeply seated of this group of muscles. It is of a triangular form, arising, by its broad base, from the whole length of the metacarpal bone of the middle finger on its palmar surface: the fibres, proceeding outwards, converge, to be inserted with the innermost tendon of the Flexor brevis pollicis, into the ulnar side of the base of the first phalanx of the thumb, and into the internal sesamoid bone.

*Relations.* By its *superficial surface*, with the Flexor brevis pollicis, the tendons of the Flexor profundus and Lumbricales. Its *deep surface* covers the first two interosseous spaces, from which it is separated by a strong aponeurosis.

*Nerves.* The Abductor, Opponens, and outer head of the Flexor brevis pollicis are supplied by the median nerve; the inner head of the Flexor brevis, and the Adductor pollicis, by the ulnar nerve.

*Actions.* The actions of the muscles of the thumb are almost sufficiently indicated by their names. This segment of the hand is provided with three Extensors, an Extensor of the metacarpal bone, an Extensor of the first, and an Extensor of the second phalanx; these occupy the dorsal surface of the fore-arm and hand. There are, also, three Flexors on the palmar surface, a Flexor of the metacarpal bone, the Flexor ossis metacarpi (Opponens pollicis), the Flexor brevis pollicis, and the Flexor longus pollicis; there is also an Abductor and an Adductor. These muscles give to the thumb that extensive range of motion which it possesses in an eminent degree.

## Ulnar Region.

*Muscles of the Little Finger*

| | |
|---|---|
| Palmaris Brevis. | Flexor Brevis Minimi Digiti. |
| Abductor Minimi Digiti. | Opponens Minimi Digiti. |

The *Palmaris Brevis* is a thin quadrilateral muscle, placed beneath the integument on the ulnar side of the hand. It arises by tendinous fasciculi, from the annular ligament and palmar fascia; the fleshy fibres pass horizontally inwards to be inserted into the skin on the inner border of the palm of the hand.

*Relations.* By its *superficial surface*, with the integument to which it is intimately adherent, especially by its inner extremity. By its *deep surface*, with the inner portion of the palmar fascia, which separates it from the ulnar vessels and nerve, and from the muscles of the ulnar side of the hand.

The *Abductor Minimi Digiti* is situated on the ulnar border of the palm of the hand. It arises from the pisiform bone, and from an expansion of the tendon of the Flexor carpi ulnaris; and terminates in a flat tendon, which is inserted into the ulnar side of the base of the first phalanx of the little finger.

*Relations.* By its *superficial surface*, with the inner portion of the palmar fascia, and the Palmaris brevis. By its *deep surface*, with the Flexor ossis metacarpi. By its *inner border*, with the Flexor brevis minimi digiti.

The *Flexor Brevis Minimi Digiti* lies on the same plane as the preceding muscle, on its radial side. It arises from the tip of the unciform process of the unciform bone, and anterior surface of the annular ligament, and is inserted into the base of the first phalanx of the little finger, with the preceding. It is separated from the Abductor at its origin, by the deep branches of the ulnar artery and nerve. This muscle is sometimes wanting; the Abductor is then, usually, of large size.

*Relations.* By its *superficial surface*, with the internal portion of the palmar fascia, and the Palmaris brevis. By its *deep surface*, with the Flexor ossis metacarpi.

The *Opponens Minimi Digiti* (fig. 173), is of a triangular form, and placed immediately beneath the preceding muscles. It arises from the unciform process of the unciform bone, and contiguous portion of the annular ligament; its fibres pass downwards and inwards, to be inserted into the whole length of the metacarpal bone of the little finger, along its ulnar margin.

*Relations.* By its *superficial surface*, with the Flexor brevis, and Abductor minimi digiti. By its *deep surface*, with the interossei muscles in the fifth metacarpal space, the metacarpal bone, and the Flexor tendons of the little finger.

*Nerves.* All the muscles of this group are supplied by the ulnar nerve.

*Actions.* The actions of the muscles of the little finger are expressed in their names. The Palmaris brevis corrugates the skin on the inner side of the palm of the hand.

## Middle Palmar Region.

Lumbricales. Interossei Palmares.
Interossei Dorsales.

The *Lumbricales* are four small fleshy fasciculi, accessories to the deep Flexor muscle. They arise by fleshy fibres from the tendons of the deep Flexor: the first and second, from the radial side and palmar surface of the tendons of the index and middle fingers; the third, from the contiguous sides of the tendons of the middle and ring fingers; and the fourth, from the contiguous sides of the tendons of the ring and little fingers. They pass forwards to the radial side of the corresponding fingers, and opposite the metacarpo-phalangeal articulations; each tendon terminates in a broad aponeurosis, which is inserted into the tendinous expansion from the Extensor communis digitorum, which covers the dorsal aspect of each finger.

The *Interossei Muscles* are so named from occupying the intervals between the metacarpal bones. They are divided into two sets, a dorsal and palmar; the former are four in number, one in each metacarpal space; the latter, three in number, lie upon the metacarpal bones.

The *Dorsal Interossei* are four in number, larger than the palmar, and occupy the intervals between the metacarpal bones. They are bipenniform muscles, arising by two heads from the adjacent sides of the metacarpal bones, but more extensively from that side of the metacarpal bone, which corresponds to the side of the finger in which the muscle is inserted. They are inserted into the base of the first phalanges, and into the aponeurosis of the common Extensor tendon. Between the double origin of each of these muscles is a narrow triangular interval, through which passes a perforating branch from the deep palmar arch.

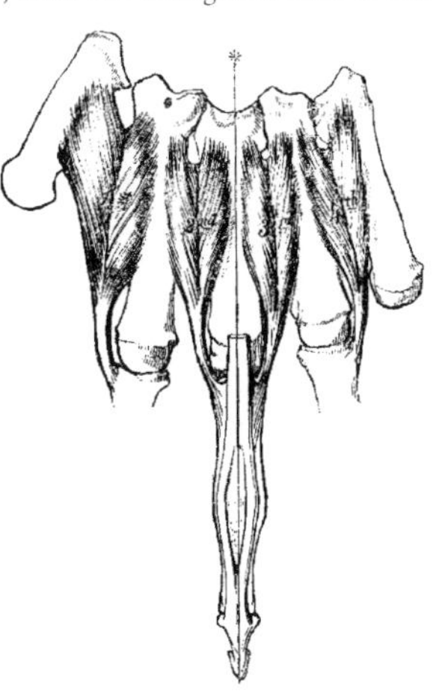

178 – *The dorsal interossei of left hand.*

The *First Dorsal Interosseous* muscle, or Abductor indicis, is larger than the others, and lies in the interval between the thumb and index-finger. It is flat, triangular in form, and arises by two heads, separated by a fibrous arch, for the passage of the radial artery into the deep part of the palm of the hand. The outer head arises from the upper half of the ulnar border of the first metacarpal bone; the inner head, from the entire length of the radial border of the second metacarpal bone; the tendon is inserted into the radial side of the index-finger. The second and third are inserted into the middle finger, the former into its radial, the latter into its ulnar side. The fourth is inserted in the radial side of the ring-finger.

The *Palmar Interossei*, three in number, are smaller than the Dorsal, and placed upon the palmar surface of the metacarpal bones, rather than between them. They arise from the entire length of the metacarpal bone of one finger, and are inserted into the side of the base of the first phalanx and aponeurotic expansion of the common Extensor tendon of the same finger.

The first arises from the ulnar side of the second metacarpal bone, and is inserted into the same side of the index-finger. The second arises from the radial side of the fourth metacarpal bone, and is inserted into the same side of the ring-finger. The third arises from the radial side of the fifth metacarpal bone, and is inserted into the same side of the little finger. From this account it may be seen, that each finger is provided with two Interossei muscles, with the exception of the little finger.

*Nerves.* The two outer Lumbricales are supplied by the median nerve; the rest of the muscles of this group, by the ulnar.

*Actions.* The Dorsal interossei muscles abduct the fingers from an imaginary line drawn longitudinally through the centre of the middle finger; and the Palmar interossei adduct

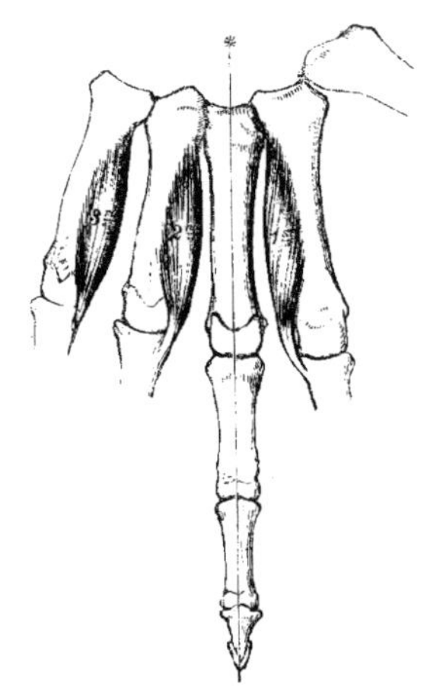

179 – *The palmar interossei of left hand.*

the fingers towards the same line. They usually assist the Extensor muscles; but when the fingers are slightly bent, assist in flexing the fingers.

## Surgical Anatomy.

The Student, having completed the dissection of the muscles of the upper extremity, should consider the effects likely to be produced by the action of the various muscles in fracture of the bones; the causes of displacement are thus easily recognised, and a suitable treatment in each case may be readily adopted.

In considering the actions of the various muscles upon fractures of the upper extremity, the most common forms of injury have been selected, both for illustration and description.

Fracture of the *clavicle* is an exceedingly common accident, and is usually caused by indirect violence, as a fall upon the shoulder; it occasionally, however, occurs from direct force. Its most usual situation is just external to the centre of the bone, but it may occur at the sternal or acromial ends.

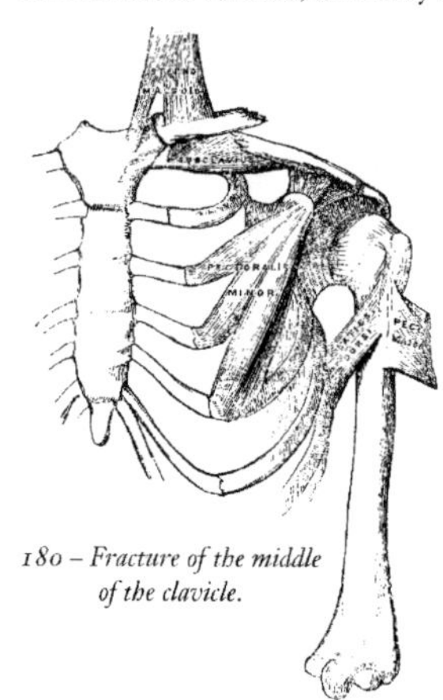

*180 – Fracture of the middle of the clavicle.*

Fracture of the *middle of the clavicle* (fig. 180) is always attended with considerable displacement, the outer fragment being drawn downwards, forwards, and inwards; the inner fragment slightly upwards. The outer fragment is drawn down by the weight of the arm, and the action of the Deltoid, and forwards and inwards by the Pectoralis minor and Subclavius muscles; the inner fragment is slightly raised by the Sterno-cleido mastoid, but only to a very limited extent, as the attachment of the costo-clavicular ligament and Pectoralis major below and in front would prevent any very great displacement upwards. The causes of displacement having been ascertained, it is easy to apply the appropriate treatment. The outer fragment is to be drawn outwards, and, together with the scapula, raised upwards to a level with the inner fragment, and retained in that position.

In fracture of the *acromial end of the clavicle* between the conoid and trapezoid ligaments, only slight displacement occurs, as these ligaments, from their oblique insertion, serve to hold both portions of the bone in apposition. Fracture, also, of the *sternal end*, internal to the costo-clavicular ligament, is attended with only slight displacement, this ligament serving to retain the fragments in close apposition.

Fracture of the *acromion process* usually arises from violence applied to the upper and outer part of the shoulder; it is generally known by the rotundity of the shoulder being lost, from the Deltoid drawing downwards and forwards the fractured portion; and the displacement may easily be discovered by tracing the margin of the clavicle outwards, when the fragment will be found resting on the front and upper part of the head of the humerus. In order to relax the anterior and outer fibres of the Deltoid (the opposing muscle), the arm should be drawn forwards across the chest, and the elbow well raised, so that the head of the bone may press upwards the acromion process, and retain it in its position.

Fracture of the *coracoid process* is an extremely rare accident, and is usually caused by a sharp blow on the point of the shoulder. Displacement is here produced by the combined actions of the Pectoralis minor, short head of the Biceps, and Coraco-brachialis, the former muscle drawing the fragment inwards, the latter directly downwards, the amount of displacement being limited by the connection of this process to the acromion by means of the coraco-acromion ligament. In order to relax these muscles, and replace the fragments in close apposition, the fore-arm should be flexed so as to relax the Biceps, and the arm drawn forwards and inwards across the chest so as to relax the Coraco-brachialis; the humerus should then be pushed upwards against the coraco-acromial ligament, and the arm retained in this position.

Fracture of the *anatomical neck of the humerus* within the capsular ligament is a rare accident, attended with very slight displacement, an impaired condition of the motions of the joint, and crepitus.

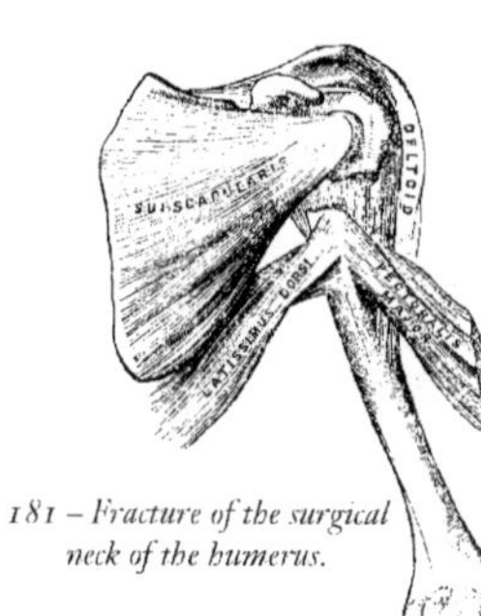

*181 – Fracture of the surgical neck of the humerus.*

Fracture of the *surgical neck* (fig. 181) is very common, is attended with considerable displacement, and its appearances correspond somewhat with those of dislocation of the head of the humerus into the axilla. The upper fragment is slightly elevated under the coraco-acromion ligament by the muscles attached to the greater and lesser tuberosities; the lower fragment is drawn inwards by the Pectoralis major, Latissimus dorsi, and Teres major; and the humerus is thrown obliquely outwards from the side by the Deltoid, and occasionally elevated so as to project beneath and in front of the coracoid process. By fixing the shoulder, and drawing the arm outwards and downwards, the deformity is at once reduced. To counteract the action of the opposing muscles, and to keep the fragments in position, the arm should be drawn from the side, and pasteboard splints applied on its four sides; a large conical-shaped pad should be placed in the axilla with the base turned upwards, and the elbow approximated to the side, and retained there by a broad roller passed round the chest; the fore-arm should then be flexed, and the hand supported in a sling, care being taken not to raise the elbow, otherwise the lower fragment may be displaced upwards.

In fracture of the *shaft of the humerus* below the insertion of the Pectoralis major, Latissimus dorsi, and Teres major, and above the insertion of the Deltoid, there is also considerable deformity, the upper fragment being drawn inwards by the first-mentioned muscles, and the lower fragment drawn upwards and outwards by the Deltoid, producing shortening of the limb, and a considerable prominence at the seat of fracture, from the fractured ends of the bone riding over one another, especially if the fracture takes place in an oblique direction. The fragment may be readily brought into apposition by extension from the elbow, and retained in that position by adopting the same means as in the preceding injury.

In fracture of the *shaft of the humerus* immediately below the insertion of the Deltoid, the amount of deformity depends greatly upon the direction of the fracture. If the fracture occurs in a transverse direction, only slight displacement occurs, the upper fragment being drawn a little forwards: but in oblique fracture, the combined actions of the Biceps and Brachialis anticus muscles in front, and the Triceps behind, draw upwards the lower fragment, causing it to glide over the upper fragment, either backwards or forwards, according to the direction of the fracture. Simple extension reduces the deformity, and the application of splints on the four sides of the arm retain the fragments in apposition Care should be taken not to raise the elbow; but the fore-arm and hand may be supported in a sling.

Fracture of the *humerus* (fig. 182) immediately above the condyles deserves very attentive consideration, as the general appearances correspond somewhat with those produced by separation of the epiphysis of the humerus, and with those of dislocation of the radius and ulna backwards. If the direction of the fracture is oblique from above, downwards, and forwards, the lower fragment is drawn upwards and backwards by the Brachialis anticus and Biceps in front, and the Triceps behind. This injury may be diagnosed from dislocation, by the increased mobility in fracture, the existence of crepitus, and the deformity being remedied by extension, by the discontinuance of which it is again reproduced. The age of the patient is of importance in distinguishing this form of injury from separation of the epiphysis. If fracture occurs in the opposite direction to that shewn in the accompanying figure, the lower fragment is drawn upwards and forwards, causing a considerable prominence in front; and the upper fragment projects backwards beneath the tendon of the Triceps muscle.

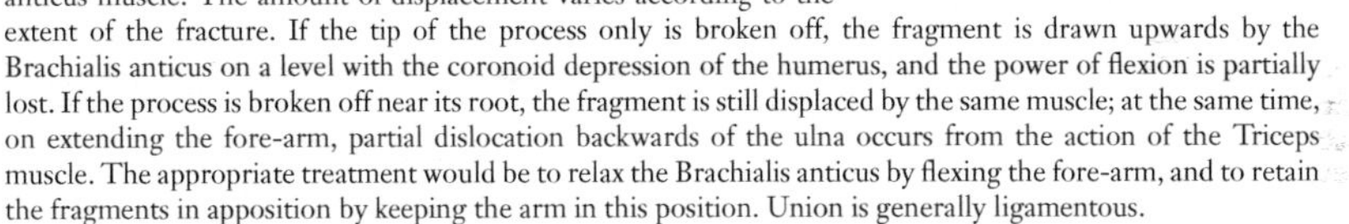

*182 – Fracture of the humerus above the condyles.*

Fracture of the *coronoid process of the ulna* is an accident of rare occurence, and is usually caused by violent action of the Brachialis anticus muscle. The amount of displacement varies according to the extent of the fracture. If the tip of the process only is broken off, the fragment is drawn upwards by the Brachialis anticus on a level with the coronoid depression of the humerus, and the power of flexion is partially lost. If the process is broken off near its root, the fragment is still displaced by the same muscle; at the same time, on extending the fore-arm, partial dislocation backwards of the ulna occurs from the action of the Triceps muscle. The appropriate treatment would be to relax the Brachialis anticus by flexing the fore-arm, and to retain the fragments in apposition by keeping the arm in this position. Union is generally ligamentous.

Fracture of the *olecranon process* (fig. 183) is a more frequent accident, and is caused either by violent action of the Triceps muscle, or by a fall or blow upon the point of the elbow. The detached fragment is displaced upwards, by the action of the Triceps muscle, from half an inch to two inches; the prominence of the elbow is consequently lost, and a deep hollow is felt at the back part of the joint, which is much increased on flexing the limb. The patient at the same time loses the power of extending the fore-arm. The treatment consists in relaxing the Triceps by extending the fore-arm, and retaining it in this position by means of a long straight splint applied to the front of the arm; the fragments are thus brought into closer apposition, and may be further approximated by drawing down the upper fragment. Union is generally ligamentous.

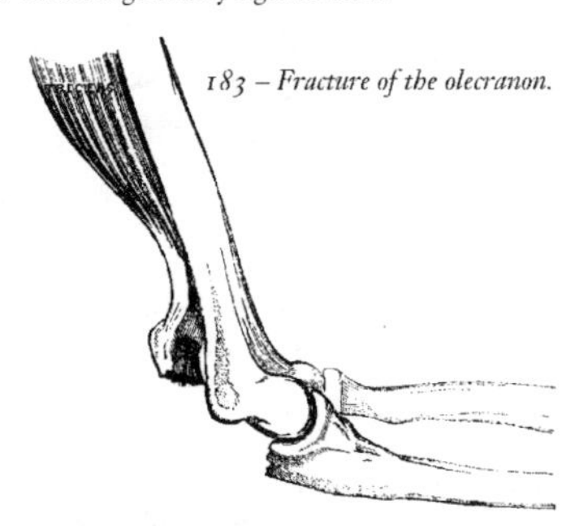

*183 – Fracture of the olecranon.*

Fracture of the *neck of the radius* is an exceedingly rare accident, and is generally caused by direct violence. Its diagnosis is somewhat obscure, on account of the slight deformity visible from the large number of muscles which surround it; but the movements of pronation and supination are entirely lost. The upper fragment is drawn outwards by the Supinator brevis, its extent of displacement being limited by the attachment of the orbicular ligament. The lower fragment is drawn forwards and slightly upwards by the Biceps, and inwards by the Pronator radii teres, its displacement forwards and upwards being counteracted in some degree by the Supinator brevis. The treatment essentially consists in relaxing the Biceps, Supinator brevis, and Pronator radii teres muscles, by flexing the fore-arm, and placing it in a position midway between pronation and supination, extension having been previously made so as to bring the parts in apposition.

Fracture of the *radius* (fig. 184) is more common than fracture of the ulna, on account of the connection of the former with the wrist. Fracture of the shaft of the radius near its centre may occur from direct violence, but more frequently from a fall forwards, the entire weight of the body being received on the wrist and hand. The upper fragment is drawn upwards by the Biceps, and inwards by the Pronator radii teres, holding a position midway between pronation and supination, and a degree of fulness in the upper half of the fore-arm is thus produced; the lower fragment is drawn downwards and inwards towards the ulna by the Pronator quadratus, and thrown into a state of pronation by the same muscle; at the same time, the Supinator longus, by elevating the styloid process, into which it is inserted, will serve to depress still more the upper end of the lower fragment towards the ulna. In order to relax the opposing muscles the fore-arm should be bent, and the limb placed in a position midway between pronation and supination; the fracture is then easily reduced by extension from the wrist and elbow: well padded splints should then be applied on both sides of the fore-arm from the elbow to the wrist; the hand being allowed to fall, will, by its own weight, counteract the action of the Pronator quadratus and Supinator longus, and elevate the lower fragment to the level of the upper one.

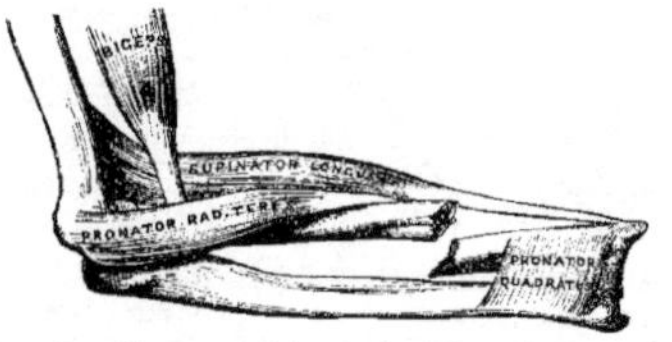


*184 – Fracture of the shaft of the radius.*

Fracture of the *shaft of the ulna* is not a common accident; it is usually caused by direct violence. Its more protected position on the inner side of the limb, the greater strength of its shaft, and its indirect connection with the wrist, render it less liable to injury than the radius. It usually occurs a little below the centre, which is the weakest part of the bone. The upper fragment retains its usual position; but the lower fragment is drawn outwards towards the radius by the Pronator quadratus, producing a well marked depression at the seat of fracture, and some fulness on the dorsal and palmar surfaces of the fore-arm. The fracture is easily reduced by extension from the wrist and fore-arm. The fore-arm should be flexed, and placed in a position midway between pronation and supination, and well padded splints applied from the elbow to the ends of the fingers.

Fracture of the *shafts of the radius and ulna together* is not a common accident; it may arise from a direct blow, or from indirect violence. The lower fragments are drawn upwards, sometimes forwards, sometimes backwards, according to the direction of the fracture, by the combined actions of the Flexor and Extensor muscles, producing a degree of fulness on the dorsal or palmar surface of the fore-arm; at the same time the two fragments are drawn into contact by the Pronator quadratus, the radius in a state of pronation: the upper fragment of the radius is drawn upwards and inwards by the Biceps and Pronator radii teres to a higher level than the ulna; the upper portion of the ulna is slightly elevated by the Brachialis anticus. The fracture may be reduced by extension from the wrist and elbow, and the fore-arm should be placed in the same position as in fracture of the ulna.

In the treatment of all cases of fracture of the bones of the fore-arm, the greatest care is requisite to prevent the ends of the bones from being drawn inwards towards the interosseous space: if this is not carefully attended to, the radius and ulna may become anchylosed, and the movements of pronation and supination entirely lost. To obviate this, the splints applied to the limb should be well padded, so as to press the muscles down into their normal situation in the interosseous space, and so prevent the approximation of the fragments.

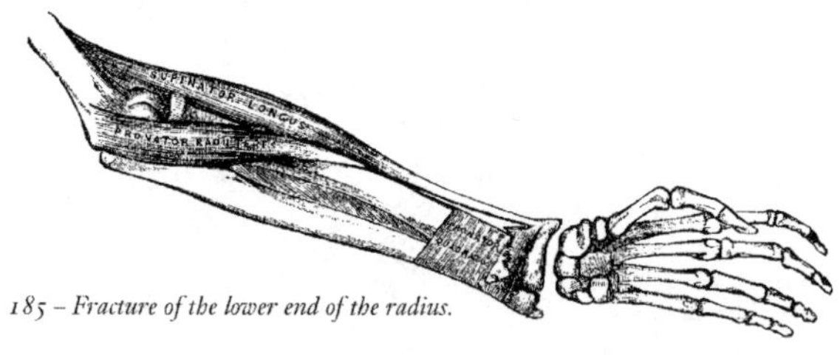


*185 – Fracture of the lower end of the radius.*

Fracture of the *lower end of the radius* (fig. 185) is usually called *Colles fracture*, from the name of the eminent Dublin surgeon who first accurately described it. It is generally produced from the patient falling from a height, and alighting upon the hand, which received the entire weight of the body. This fracture usually takes place from half an inch to an inch above the articular surface if it occurs in the adult; but in the child, before the age of sixteen, it is more frequently a separation of the epiphysis from the apophysis. The displacement which is produced is very considerable, and bears some resemblance to dislocation of the carpus backwards, from which it should be carefully distinguished. The lower fragment is drawn upwards and backwards behind the upper fragment by the combined actions of the Supinator longus and the flexors and the extensors of the thumb and carpus, producing a well marked prominence on the back of the wrist, with a deep depression behind. The upper fragment projects forwards, often lacerating the substance of the Pronator quadratus, and is drawn by this muscle into close contact with the lower end of the ulna, causing a projection on the anterior surface of the fore-arm, immediately above the carpus, from the flexor tendons being thrust forwards. This fracture may be distinguished from dislocation by the deformity being removed on making sufficient extension, when crepitus may be occasionally detected; at the same time, on extension being discontinued, the parts immediately resume their deformed appearance. The age of the patient will also assist in determining whether the injury is fracture or separation of the epiphysis. The treatment consists in flexing the fore-arm, and making powerful extension from the wrist and elbow, depressing at the same time the radial side of the hand, and retaining the parts in this position by well padded *pistol-shaped* splints.

# MUSCLES AND FASCIÆ OF THE LOWER EXTREMITY.

The Muscles of the Lower Extremity are subdivided into groups, corresponding with the different regions of the limb.

*Iliac region.*

Psoas magnus.
Psoas parvus.
Iliacus.

THIGH.

*Anterior femoral region.*

Tensor vaginæ femoris.
Sartorius.
Rectus.
Vastus externus.
Vastus internus.
Cruræus.
Subcruræus.

*Internal femoral region.*

Gracilis.
Pectineus.
Adductor longus.
Adductor brevis.
Adductor magnus.

HIP.

*Gluteal region.*

Gluteus maximus.
Gluteus medius.
Gluteus minimus.
Pyriformis.
Gemellus superior.
Obturator internus.
Gemellus inferior.
Obturator externus.
Quadratus femoris.

*Posterior femoral region.*

Biceps.
Semi-tendinosus.
Semi-membranosus.

LEG.

*Anterior tibio-fibular region.*

Tibialis anticus.
Extensor longus digitorum.
Extensor proprius pollicis.
Peroneus tertius.

*Posterior tibio-fibular region.*
*Superficial layer.*

Gastrocnemius.
Plantaris.
Soleus.

*Deep layer.*

Popliteus.
Flexor longus pollicis.
Flexor longus digitorum.
Tibialis posticus.

*Fibular region.*

Peroneus longus.
Peroneus brevis.

FOOT.

*Dorsal region.*

Extensor brevis digitorum.
Interossei dorsales.

*Plantar region.*

*First layer.*

Abductor pollicis.
Flexor brevis digitorum.
Abductor minimi digiti.

*Second layer.*

Musculus accessorius.
Lumbricales.

*Third layer.*

Flexor brevis pollicis.
Adductor pollicis.
Flexor brevis minimi digiti.
Transversus pedis.

*Fourth layer.*

Interossei plantares.

## ILIAC REGION.

Psoas Magnus. Psoas Parvus. Iliacus.

*Dissection.* No detailed description is required for the dissection of these muscles. They are exposed after the removal of the viscera from the abdomen, covered by the peritoneum and a thin layer of fascia, the fascia iliaca.

The *iliac fascia* is the aponeurotic layer which lines the back part of the abdominal cavity, and encloses the Psoas and Iliacus muscles throughout their whole extent. It is thin above; and becomes gradually thicker below, as it approaches the femoral arch.

The *portion investing the Psoas*, is attached, above, to the ligamentum arcuatum internum; internally, to the sacrum; and by a series of arched processes to the inter-vertebral substances, and prominent margins of the bodies of the vertebræ; the intervals left opposite the constricted portions of the bodies, transmitting the lumbar arteries and sympathetic filaments of nerves. Externally, it is continuous with the fascia lumborum.

The *portion investing the Iliacus* is connected, externally, to the whole length of the inner border of the crest of the ilium. Internally, to the brim of the true pelvis, where it is continuous with the periosteum, and receives the tendon of insertion of the Psoas parvus. External to the femoral vessels, this fascia is intimately connected with Poupart's ligament, and is continuous with the fascia transversalis; but, corresponding to the point where the femoral vessels pass down into the thigh, it is prolonged down behind them, forming the posterior wall of the femoral sheath. Below this point, the iliac fascia surrounds the Psoas and Iliacus muscles to their termination, and becomes continuous with the iliac portion of the fascia lata. Internal to the femoral vessels the iliac fascia is connected to the iliopectineal line, and is continuous with the pubic portion of the fascia lata. The iliac vessels lie in front of the iliac fascia, but all the branches of the lumbar plexus, behind it; it is separated from the peritoneum by a quantity of loose areolar tissue. In abscess accompanying caries of the lower part of the spine, the matter makes its way to the femoral arch, distending the sheath of the Psoas; and when it accumulates in considerable quantity, this muscle becomes absorbed, and the nervous cords contained in it are dissected out, and lie exposed in the cavity of the abscess; the femoral vessels, however, remain intact, and the peritoneum seldom becomes implicated notwithstanding the extreme thinness of the membrane.

Remove this fascia, and the muscles of the iliac region will be exposed.

The *Psoas Magnus* (fig. 187) is a long fusiform muscle, placed on the side of the lumbar region of the spine and margin of the pelvis. It arises from the sides of the bodies, from the corresponding inter-vertebral substances, and from the front of the bases of the transverse processes of the last dorsal and all the lumbar vertebræ. The muscle is connected to the bodies of the vertebræ by five slips; each slip is attached to the upper and lower margins of two vertebræ, and to the intervertebral substance between them; the slips themselves being connected by tendinous arches extending across the constricted part of the bodies, beneath which pass the lumbar arteries and sympathetic nervous filaments. These tendinous arches also give origin to muscular fibres and protect the blood-vessels and nerves from pressure during the action of the muscle. The first slip is attached to the contiguous margins of the last dorsal and first lumbar vertebræ; the last, to the contiguous margins of the fourth and fifth lumbar, and inter-vertebral substance. From these points, the muscle passes down across the brim of the pelvis, and diminishing gradually in size, passes beneath Poupart's ligament, and terminates in a tendon, which, after receiving the fibres of the Iliacus, is inserted into the lesser trochanter of the femur.

*Relations.* In the lumbar region. By its *anterior surface*, which is placed behind the peritoneum, with the ligamentum arcuatum internum, the kidney, Psoas parvus, renal vessels, ureter, spermatic vessels, genito-crural nerve, the colon, and along its pelvic border, with the common and external iliac artery and vein. By its *posterior surface*, with the transverse processes of the lumbar vertebræ and the Quadratus lumborum, from which it is separated by the anterior lamella of the aponeurosis of the Transversalis; the anterior crural nerve is at first situated in the substance of the muscle, and emerges from its outer border at its lower part. The lumbar plexus is situated in the posterior part of the substance of the muscle. By its *inner side*, with the bodies of the lumbar vertebræ, the lumbar arteries, the sympathetic ganglia, and its communicating branches with the spinal nerves; the lumbar glands, with the vena cava on the right, and the aorta on the left side. In the thigh it is in relation, in front, with the fascia lata; behind, with the capsular ligament of the hip, from which it is separated by a synovial bursa, which sometimes communicates with the cavity of the joint through an opening of variable size. By its *inner border*, with the Pectineus and the femoral artery, which slightly overlaps it. By its *outer border*, with the crural nerve and Iliacus muscle.

The *Psoas Parvus* is a long slender muscle, placed in front of the preceding. It arises from the sides of the bodies of the last dorsal and first lumbar vertebræ, and from the inter-vertebral substance between them. It forms a small flat muscular bundle, which terminates in a long, flat tendon, which is inserted into the ilio-pectineal eminence, being continuous, by its outer border, with the iliac fascia. This muscle is present, according to M. Theile, in one out of every twenty subjects examined.

*Relations.* It is covered by the peritoneum, and at its origin by the ligamentum arcuatum internum; it rests on the Psoas magnus.

The *Iliacus* is a flat radiated muscle, which fills up the whole of the internal iliac fossa. It arises from the iliac fossa, and inner margin of the crest of the ilium; behind, from the ilio-lumbar ligament, and base of the sacrum; in front, from the anterior superior and anterior inferior spinous processes of the ilium, the notch between them, and by a few fibres from the capsule of the hip-joint. The fibres converge to be inserted into the outer side of the tendon of the Psoas, some of them being prolonged into the oblique line which extends from the lesser trochanter to the linea aspera.

*Relations. Within the pelvis:* by its *anterior surface*, with the iliac fascia, which separates the muscle from the peritoneum, and with the external cutaneous nerve; on the right side, with the cæcum; on the left side, with the sigmoid flexure of the colon. By its *posterior surface*, with the iliac fossa. By its *inner border*, with the Psoas magnus, and anterior crural nerve. In the thigh, it is in relation, by its *anterior surface*, with the fascia lata, Rectus and Sartorius; behind, with the capsule of the hip-joint, a synovial bursa common to it, and the Psoas magnus being interposed.

*Nerves.* The Psoæ muscles are supplied by the anterior branches of the lumbar nerves. The Iliacus from the anterior crural.

*Actions.* The Psoas and Iliacus muscles, acting from above, flex the thigh upon the pelvis, and, at the same time, rotate the femur outwards, from the obliquity of their insertion into the inner and back part of that bone. Acting from below, the femur being fixed, the muscles of both sides bend the lumbar portion of the spine and pelvis forwards. They also serve to maintain the erect position, by supporting the spine and pelvis upon the femur, and assist in raising the trunk when the body is in the recumbent posture.

The *Psoas parvus* is a tensor of the iliac fascia.

## Anterior Femoral Region.

Tensor Vaginæ Femoris.
Sartorius.
Rectus.
Vastus Externus.
Vastus Internus.
Cruræus.
Subcruræus.

*Dissection.* To expose the muscles and fasciæ in this region, an incision should be made along Poupart's ligament, from the spine of the ilium to the pubes, from the centre of this, a vertical incision must be carried along the middle line of the thigh to below the knee-joint, and connected with a transverse incision, carried from the inner to the outer side of the leg. The flaps of integument having been removed, the superficial and deep fasciæ should be examined. The more advanced student would commence the study of this region by an examination of the anatomy of femoral hernia, and Scarpa's triangle, the incisions for the dissection of which are marked out in the accompanying figure.

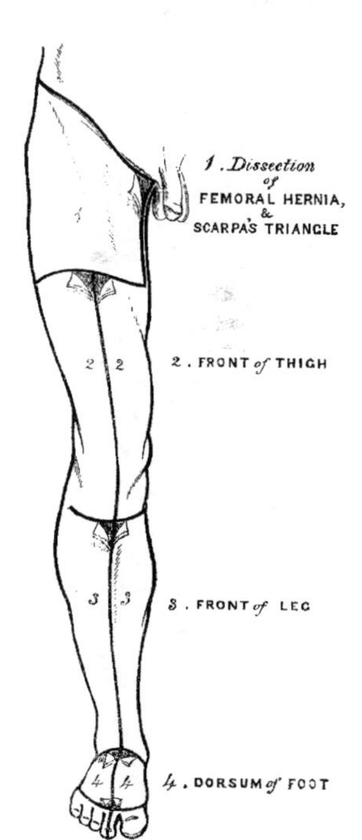


186 – *Dissection of lower extremity. Front view.*

## Fasciæ of the Thigh.

The *superficial fascia*, forms a continuous layer over the whole of the lower extremity, consisting of areolar tissue, containing in its meshes much adipose matter, and capable of being separated into two or more layers, between which are found the superficial vessels and nerves. It varies in thickness in different parts of the limb; in the sole of the foot it is so thin, as to be scarcely demonstrable, the integument being closely adherent to the deep fascia beneath, but in the groin it is thicker, and the two layers are separated from one another by the superficial inguinal glands, the internal saphenous vein, and several smaller vessels. Of these two layers, the most superficial is continuous above with the superficial fascia of the abdomen, the deep layer becoming blended with the fascia lata, a little below Poupart's ligament. The deep layer of superficial fascia is intimately adherent to the margins of the saphenous opening in the fascia lata, and pierced in this situation by numerous small blood and lymphatic vessels, hence the name *cribriform fascia*, which has been applied to it. Subcutaneous bursæ are found in the superficial fascia over the patella, point of the heel, and phalangeal articulations of the toes.

The *deep fascia* of the thigh is exposed on the removal of the superficial fascia, and is named, from its great extent, the fascia lata; it forms a uniform investment for the whole of this region of the limb, but varies in thickness in different parts; thus, it is thickest in the upper and outer side of the thigh, where it receives a fibrous expansion from the Gluteus maximus muscle, and the Tensor vaginæ femoris is inserted between its layers, it is very thin behind, and at the upper and inner side, where it covers the Adductor muscles, and again becomes stronger around the knee, receiving fibrous expansions from the tendon of the Biceps externally, and from the Sartorius, Gracilis, Semitendinosus, and Quadriceps extensor cruris in front. The fascia lata is attached, above, to Poupart's ligament, and crest of the ilium; behind, to the margin of the sacrum and coccyx; internally, to the pubic arch and pectineal line; and below, to all the prominent points around the knee-joint, the condyles of the femur, tuberosities of the tibia, and head of the fibula. That portion which invests the Gluteus medius (the Gluteal aponeurosis) is very thick and strong, and gives origin, by its inner surface, to some of the fibres of that muscle; at the upper border of the Gluteus maximus, it divides into two layers; the most superficial, very thin, covers the surface of the Gluteus maximus, and is continuous below with the fascia lata: the deep layer is thick above, and blends with the great sacro-sciatic ligament, thin below, where it separates the Gluteus maximus from the deeper muscles. From the inner surface of the fascia lata, are given off two strong intermuscular septa, which are attached to the whole length of the linea aspera; the external and stronger one, which extends from the insertion of the Gluteus maximus to the outer condyle, separates the Vastas externus in front from the short head of the Biceps behind, and gives partial origin to these muscles; the inner one, the thinner of the two, separates the Vastus internus from the Adductor muscles. Besides these, there are numerous smaller septa, separating the individual muscles, and enclosing each in a distinct sheath. At the upper and inner part of the thigh, a little below Poupart's ligament, a large oval-shaped aperture is observed: it transmits the internal saphenous vein, and other smaller vessels, and is termed the *saphenous opening*. In order more correctly to consider the mode of formation of this aperture, the fascia lata is described as consisting, in this part of the thigh, of two portions, an iliac portion, and a pubic portion.

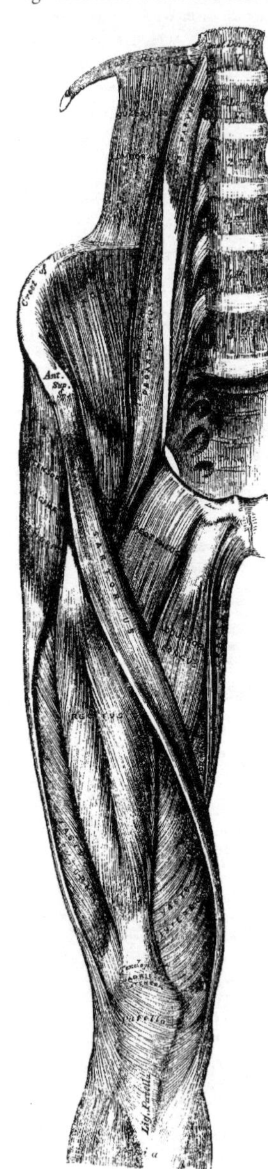

187 – *Muscles of the iliac and anterior femoral regions.*

The *iliac portion* is all that part of the fascia lata placed on the outer side of the saphenous opening. It is attached, externally, to the crest of the ilium, and its anterior superior spine, to the whole length of Poupart's ligament, as far internally as the spine of the pubes, and to the pectineal line in conjunction with Gimbernat's ligament. From the spine of the pubes, it is reflected downwards and outwards, forming an arched margin, the superior cornu, or outer boundary of the saphenous opening; this margin overlies, and is adherent to the anterior layer of the sheath of the femoral vessels; to its edge is attached the cribriform fascia, and, below, it is continuous with the pubic portion of the fascia lata.

The *pubic portion* is situated at the inner side of the saphenous opening; at the lower margin of this aperture it is continuous with the iliac portion; traced upwards, it is seen to cover the surface of the Pectineus muscle, and passing behind the sheath of the femoral vessels, to which it is closely united, is continuous with the sheath of the Psoas and Iliacus muscles, and is finally lost in the fibrous capsule of the hip-joint. This fascia is attached above to the pectineal line in front of the insertion of the aponeurosis of the External oblique, and internally to the margin of the pubic arch. From this description it may be observed, that the iliac portion of the fascia lata passes in front of the femoral vessels, the pubic portion behind them, an apparent aperture consequently exists between the two, through which the internal saphenous joins the femoral vein.

The fascia should now be removed from the surface of the muscles. This may be effected by pinching it up between the forceps, dividing it, and separating it from each muscle in the course of its fibres.

The *Tensor Vaginæ Femoris* is a short flat muscle, situated at the upper and outer side of the thigh. It arises from the anterior part of the outer lip of the crest of the ilium, and from the outer surface of the anterior superior spinous process, between the Gluteus medius, and Sartorius. The muscle passes obliquely downwards, and a little backwards, to be inserted into the fascia lata, about one-fourth down the outer side of the thigh.

*Relations.* By its *superficial surface*, with the fascia lata and the integument. By its *deep surface*, with the Gluteus medius, Rectus femoris, Vastus externus, and the ascending branches of the external circumflex artery. By its *anterior border*, with the Sartorius, from which it is separated below by a triangular space, in which is seen the Rectus femoris. By its *posterior border*, with the Gluteus medias.

The *Sartorius*, the longest muscle in the body, is a flat, narrow, riband-like muscle, which arises by tendinous fibres from the anterior superior spinous process of the ilium and upper half of the notch below it; it passes obliquely across the upper and anterior part of the thigh, from the outer to the inner side of the limb, then descends vertically, as far as the inner side of the knee, passing behind the inner condyle of the femur, and terminates in a tendon, which curving obliquely forwards, expands into a broad aponeurosis, which is inserted into the upper part of the inner surface of the shaft of the tibia, nearly as far forwards as the crest. This expansion covers the insertion of the tendons of the Gracilis and Semitendinosus, with which it is partially united, a synovial bursa being interposed between them. An offset is derived from this aponeurosis, which blends with the fibrous capsule of the knee-joint, and another, given off from its lower border, blends with the fascia on the inner side of the leg. The relations of this muscle to the femoral artery should be carefully examined, as its inner border forms the chief guide in the operation of including this vessel in a ligature. In the upper third of the thigh, it forms, with the Adductor longus, the side of a triangular space, Scarpa's triangle, the base of which, turned upwards, is formed by Poupart's ligament; the femoral artery passes perpendicularly through the centre of this space from its base to its apex. In the middle third of the thigh, the femoral artery lies first along the inner border, and then beneath the Sartorius.

*Relations.* By its *superficial surface*, with the fascia lata and integument. By its *deep surface*, with the Iliacus, Psoas, Rectus, Vastus internus, anterior crural nerve, sheath of the femoral vessels, Adductor longus, Adductor magnus, Gracilis, long saphenous nerve, and internal lateral ligament of the knee-joint.

The *Quadriceps extensor* includes the four remaining muscles on the front of the thigh. It is the great Extensor muscle of the leg, forming a large fleshy mass, which covers the front and sides of the femur, being united below into a single tendon, attached to the tibia, and above subdividing into separate portions, which have received separate names. Of these, one occupying the middle of the thigh, connected above with the ilium, is called the *Rectus femoris*, from its straight course. The other divisions lie in immediate connection with the shaft of the femur, which they cover from the condyles to the trochanters. The portion on the outer side of the femur is termed the *Vastus externus*; that covering the inner side, the *Vastus internus*; and that covering the front of the femur, the *Cruræus*. The two latter portions are, however, so intimately blended, as to form but one muscle.

The *Rectus femoris* is situated in the middle of the anterior region of the thigh; it is fusiform in shape, and its fibres are arranged in a bipenniform manner. It arises by two tendons; one, the straight tendon, from the anterior inferior spinous process of the ilium; the other is flattened, and curves outwards, to be attached to a groove above the brim of the acetabulum; this is the reflected tendon of the Rectus, it unites with the straight tendon at an acute angle, and then spreads into an aponeurosis, from which the muscular fibres arise. The muscle terminates in a broad and thick aponeurosis, which occupies the lower two-thirds of its posterior surface, and, gradually becoming narrowed into a flattened tendon, is inserted into the patella in common with the Vasti and Cruræus.

*Relations.* By its *superficial surface*, with the anterior fibres of the Gluteus medius, the Tensor vaginæ femoris, Sartorius, and the Psoas and Iliacus; by its lower three-fourths, with the fascia lata.

By its *posterior surface*, with the hip-joint, the external circumflex vessels, and the Cruræus and Vasti muscles.

The three remaining muscles have been described collectively by some anatomists, separate from the Rectus; under the name of the *Triceps extensor cruris*; in order to expose them, divide the Sartorius and Rectus muscles across the middle, and turn them aside, when they will be fully brought into view.

The *Vastus externus* is the largest part of the Quadriceps extensor. It arises by a broad aponeurosis, which is attached to the anterior border of the great trochanter, to a horizontal ridge on its outer surface, to a rough line, leading from the trochanter major to the linea aspera, and to the whole length of the outer lip of the linea aspera; this aponeurosis covers the upper three-fourths of the muscle, and from its inner surface, many fibres arise. A few additional fibres arise from the tendon of the Gluteus maximus, and from the external intermuscular septum between the Vastus externus, and short head of the Biceps. These fibres form a large fleshy mass, which is attached to a strong aponeurosis, placed on the under surface of the muscle at its lowest part; this becomes contracted and thickened into a flat tendon, which is inserted into the outer part of the upper border of the patella, blending with the great extensor tendon.

*Relations.* By its *superficial surface*, with the Rectus, the Tensor vaginæ femoris, the fascia lata, and the Gluteus maximus, from which it is separated by a synovial bursa. By its *deep surface*, with the Cruræus, some large branches of the external circumflex artery and anterior crural nerve being interposed.

The *Vastus internus* and *Cruræus*, are so inseparably connected together, as to form but one muscle. It is the smallest portion of the Quadriceps extensor. The anterior portion of it, which is covered by the Rectus, being called the Cruræus; the internal portion, which lies immediately beneath the fascia lata, is called the Vastus Internus. It arises by an aponeurosis, which is attached to the lower part of the line that extends from the inner side of the neck of the femur to the linea aspera, from the whole length of the inner lip of the linea aspera, and internal intermuscular septum. It also arises from nearly the whole of the internal, anterior, and external surfaces of the shaft of the femur, limited, above, by the line between the two trochanters, and extending, below, to within the lower fourth of the bone. From these different origins, the fibres converge to a broad aponeurosis, which covers the anterior surface of the middle portion of the muscle (the Cruræus), and the deep surface of the inner division of the muscle (the Vastus internus), becoming joined and gradually narrowing, it is inserted into the patella, blending with the other portions of the Quadriceps extensor.

*Relations.* By their *superficial surface*, with the Psoas and Iliacus, the Rectus, Sartorius, Pectineus, Adductors, and fascia lata, femoral vessels, and saphenous nerve. By its *deep surface*, with the femur, subcruræus, and synovial membrane of the knee-joint.

The student will observe the striking analogy that exists between the Quadriceps extensor, and the Triceps brachialis in the upper extremity. So close is this similarity, that M. Cruvelhier has described it under the name of the *Triceps femoralis*. Like the Triceps brachialis, it consists of three distinct divisions or heads; a middle or long head, analogous to the long head of the Triceps, attached to the ilium, and of two other portions which have respectively received the names of the external and internal heads of the muscle. These, it will be noticed, are strictly analogous to the outer and inner heads of the Triceps brachialis.

The *tendons* of the different portions of the Quadriceps extensor unite at the lower part of the thigh, so as to form a single strong tendon, which is inserted into the upper part of the patella. More properly speaking, the patella may be regarded as a sesamoid bone, developed in the tendon of the Quadriceps; and the ligamentum patellæ, which is continued from the lower part of the patella to the tuberosity of the tibia, as the proper tendon of insertion of this muscle. A synovial bursa is interposed between the tendon and the upper part of the tuberosity of the tibia. From the tendons corresponding to the Vasti, a fibrous prolongation is derived, which is attached below to the upper extremities of the tibia and fibula. It serves to protect the knee-joint, which is strengthened on its outer side by the fascia lata.

The *Subcruræus* is a small muscle, usually distinct from the superficial muscle, which arises from the anterior surface of the lower part of the shaft of the femur, and is inserted into the upper part of the synovial pouch that extends upwards from the knee-joint behind the patella. This fasciculus is occasionally united with the Cruræus. It sometimes consists of two separate muscular bundles.

*Nerves.* The Tensor vaginæ femoris is supplied by the superior gluteal nerve; the other muscles of this region, by branches from the anterior crural.

*Actions.* The Tensor vaginæ femoris is a tensor of the fascia lata; continuing its action, the oblique direction of its fibres enables it to rotate the thigh inwards. In the erect posture, acting from below, it will serve to steady the pelvis upon the head of the femur. The Sartorius flexes the leg upon the thigh, and, continuing to act, the thigh upon the pelvis, at the same time drawing the limb inwards, so as to cross one leg over the other. Taking its fixed point from the leg, it flexes the pelvis upon the thigh, and, if one muscle acts, assists in rotating it. The Quadriceps extensor extends the leg upon the thigh. Taking its fixed point from the leg, as in standing, this muscle will act upon the femur, supporting it perpendicularly upon the head of the tibia, thus maintaining the entire weight of the body. The Rectus muscle assists the Psoas and Iliacus, in supporting the pelvis and trunk upon the femur, or in bending it forwards.

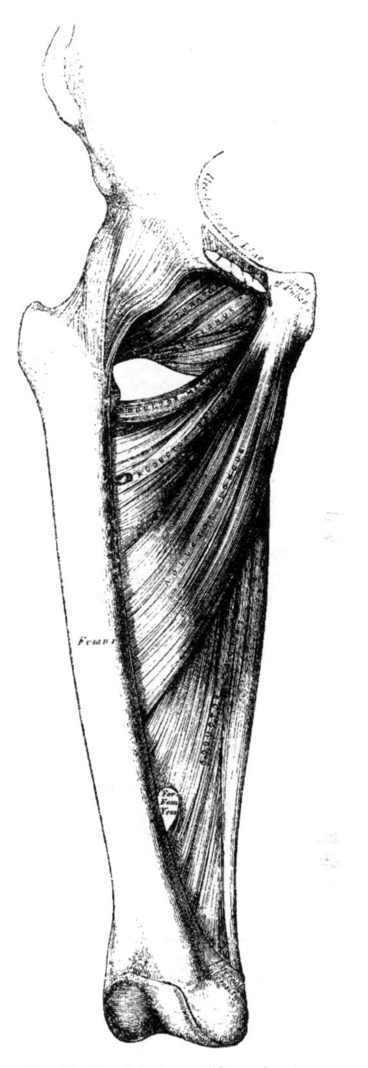
188 – *Muscles of the internal femoral region.*

## Internal Femoral Region.

Gracilis.
Pectineus.
Adductor Longus.
Adductor Brevis.
Adductor Magnus.

*Dissection.* These muscles are at once exposed by removing the fascia from the fore part and inner side of the thigh. The limb should be abducted, so as to render the muscles tense, and easier of dissection.

The *Gracilis* is the most superficial muscle on the inner side of the thigh. It is thin and flattened, broad above, narrow and tapering below. It arises by a thin aponeurosis between two and three inches in breadth, from the inner margin of the ramus of the pubes and ischium. The fibres pass vertically downwards, and terminate in a rounded tendon which passes behind the internal condyle of the femur; curving round the inner tuberosity of the tibia, it becomes flattened and is inserted into the upper part of the inner surface of the shaft of the tibia, below the tuberosity. The tendon of this muscle is situated immediately above that of the Semi-tendinosus, and beneath the aponeurosis of the Sartorius, with which it is in part blended. As it passes across the internal lateral ligament of the knee-joint, it is separated from it by a synovial bursa common to it and the Semi-tendinosus muscle.

*Relations.* By its *superficial surface* with the fascia lata and the Sartorius below; the internal saphenous vein crosses it obliquely near its lower part, lying superficial to the fascia lata. By its *deep surface*, with the three Adductors, and the internal lateral ligament of the knee-joint.

The *Pectineus* is a flat quadrangular muscle, situated at the anterior part of the upper and inner aspect of the thigh. It arises from the linea ilio-pectinea, from the surface of bone in front of it, between the pectineal eminence and spine of the pubes, and from a tendinous prolongation of Gimbernat's ligament, which is attached to the crest of the pubes, and is continuous with the fascia covering the outer surface of the muscle; the fibres pass downwards, backwards, and outwards, to be inserted into a rough line leading from the trochanter minor to the linea aspera.

*Relations.* By its *anterior surface*, with the pubic portion of the fascia lata, which separates it from the femoral vessels and internal saphenous vein. By its *posterior surface*, with the hip-joint, the Adductor brevis and Obturator externus muscles, the obturator vessels and nerve being interposed.

By its *outer border*, with the Psoas, a cellular interval separating them, upon which lies the femoral artery. By its *inner border*, with the margin of the Adductor longus.

The *Adductor Longus*, the most superficial of the three Adductors, is a flat triangular muscle, lying on the same plane as the Pectineus, with which it is often blended above. It arises, by a flat narrow tendon, from the front of the pubes, at the angle of junction of the crest with the symphysis; it soon expands into a broad fleshy belly, which, passing downwards, backwards, and outwards, is inserted, by an aponeurosis, into the middle third of the linea aspera, between the Vastus internus and the Adductor magnus.

*Relations.* By its *anterior surface*, with the fascia lata, and, near its insertion, with the femoral artery and vein. By its *posterior surface*, with the Adductor brevis and magnus, the anterior branches of the obturator vessels and nerve, and with the profunda artery and vein near its insertion. By its *outer border*, with the Pectineus. By its *inner border*, with the Gracilis.

The Pectineus and Adductor longus should now be divided near their origin, and turned downwards, when the Adductor brevis and Obturator externus will be exposed.

The *Adductor Brevis* is situated immediately beneath the two preceding muscles. It is somewhat triangular in form, and arises by a narrow origin from the outer surface of the descending ramus of the pubes, between the Gracilis and Obturator externus. Its fibres, passing backwards, outwards, and downwards, are inserted, by an aponeurosis, into the upper part of the linea aspera, immediately behind the Pectineus and upper part of the Adductor longus.

*Relations.* By its *anterior surface*, with the Pectineus, Adductor longus, and anterior branches of the obturator vessels and nerve. By its *posterior surface*, with the Adductor magnus, and posterior branches of the obturator vessels and nerve. By its *outer border*, with the Obturator externus, and conjoined tendon of the Psoas and Iliacus. By its *inner border*, with the Gracilis and Adductor magnus. This muscle is pierced, near its insertion, by the middle perforating branch of the profunda artery.

The Adductor brevis should now be cut away near its origin, and turned outwards, when the entire extent of the Adductor magnus will be exposed.

The *Adductor Magnus* is a large triangular muscle, forming a septum between the muscles on the inner, and those on the back of the thigh. It arises from a small part of the descending ramus of the pubes, from the ascending ramus of the ischium, and from the outer margin and under surface of the tuberosity of the ischium. Those fibres which arise from the ramus of the pubes, are very short, horizontal in direction, and are inserted into the rough line leading from the great trochanter to the linea aspera, internal to the Gluteus maximus; those from the ramus of the ischium are directed downwards and outwards with different degrees of obliquity, to be inserted, by means of a broad aponeurosis, into the whole length of the linea aspera and upper part of its internal bifurcation below. The internal portion of the muscle, consisting principally of those fibres which arise from the tuberosity of the ischium, forms a thick fleshy mass consisting of coarse bundles which descend almost vertically, and terminate about the lower third of the thigh in a rounded tendon, which is inserted into the tubercle above the inner condyle of the femur, being connected by a fibrous expansion to the line leading upwards from the tubercle to the linea aspera. Between the two portions of the muscle, an angular interval is left, tendinous in front, fleshy behind, for the passage of the femoral vessels into the popliteal space. The external portion of the muscle is pierced by four apertures: the three superior, for the three superior perforating arteries; the fourth, for the passage of the profunda. This muscle gives off an aponeurosis, which passes in front of the femoral vessels, and joins with the Vastus internus.

*Relations.* By its *anterior surface*, with the Pectineus, Adductor brevis, Adductor longus and the femoral vessels. By its *posterior surface*, with the great sciatic nerve, the Gluteus maximus, Biceps, Semi-tendinosus, and Semi-membranosus. By its *superior* or *shortest border*, it lies parallel with the Quadrates femoris. By its *internal* or *longest border*, with the Gracilis, Sartorius, and fascia lata. By its *external* or *attached border*, it is inserted into the femur behind the Adductor brevis and Adductor longus, which separate it, in front, from the Vastas internus; and in front of the Gluteus maximus and short head of the Biceps, which separate it from the Vastus externus.

*Nerves.* All the muscles of this group are supplied by the obturator nerve. The Pectineus receives additional branches from the accessory obturator and anterior crural; and the Adductor magnus an additional branch from the great sciatic.

*Actions.* The Pectineus and three Adductors adduct the thigh powerfully; they are especially used in horse-exercise, the flanks of the horse being firmly grasped between the knees by the action of these muscles. From their oblique insertion into the linea aspera, they rotate the thigh outwards, assisting the external Rotators, and when the limb has been abducted, they draw it inwards, carrying the thigh across that of the opposite side. The Pectineus and Adductor brevis and longus assist the Psoas and Iliacus in flexing the thigh upon the pelvis. In progression, also, all these muscles assist in drawing forwards the hinder limb. The Gracilis assists the Sartorius in flexing the leg and drawing it inwards; it is also an Adductor of the thigh. If the lower extremities are fixed, these muscles may take their fixed point from below and act upon the pelvis, serving to maintain the body in the erect posture; or, if their action is continued, to flex the pelvis forwards upon the femur.

## Gluteal Region.

| | |
|---|---|
| Gluteus Maximus. | Gemellus Superior. |
| Gluteus Medius. | Obturator Internus. |
| Gluteus Minimus. | Gemellus Inferior. |
| Pyriformis. | Obturator Externus. |
| Quadratus Femoris. | |

*Dissection*, (fig. 189). The subject should be turned on its face, a block placed beneath the pelvis to make the buttocks tense, and the limbs allowed to hang over the end of the table, the foot inverted, and the limb abducted. An incision should be made through the integument along the back part of the crest of the ilium and margin of the sacrum to the tip of the coccyx, from which point a second incision should be carried obliquely downwards and outwards to the outer side of the thigh, four inches below the great trochanter. The portion of integument included between these incisions, together with the superficial fascia, should be removed in the direction shewn in the figure, when the Gluteus maximus and the dense fascia covering the Gluteus medius will be exposed.

The *Gluteus Maximus*, the most superficial muscle in the gluteal region, is a very broad and thick fleshy mass, of a quadrilateral shape, which forms the prominence of the nates. Its large size is one of the most characteristic points in the muscular system in man, connected as it is with the power he has of maintaining the trunk in the erect posture. In structure it is remarkably coarse, being made up of muscular fasciculi lying parallel with one another, and collected together into large bundles, separated by deep cellular intervals. It arises from the superior curved line of the ilium, and the portion of bone, including the crest immediately behind it; from the posterior surface of the last piece of the sacrum, the side of the coccyx, and posterior surface of the great sacro-sciatic and posterior sacro-iliac ligaments. The fibres are directed obliquely downwards and outwards; those forming the upper and larger portion of the muscle (after converging somewhat) terminate in a thick tendinous lamina, which passes across the great trochanter, and is inserted into the fascia lata covering the outer side of the thigh, the lower portion of the muscle being inserted into the rough line leading from the great trochanter to the linea aspera, between the Vastus externus and Adductor magnus.

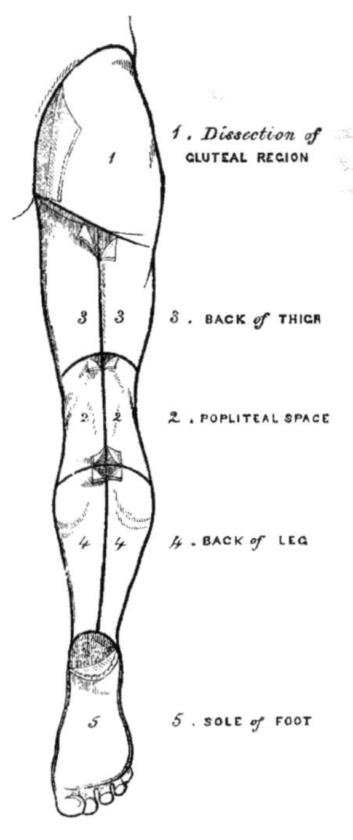

189 – *Dissection of lower extremity. Posterior view.*

Three *synovial bursæ* are usually found separating the under surface of this muscle from the eminences which it covers. One of these, of large size, and generally multilocular, separates it from the great trochanter. A second, often wanting, is situated on the tuberosity of the ischium. A third, between the tendon of this muscle and the Vastus externus.

*Relations.* By its *superficial surface*, with a thin fascia, which separates it from cellular membrane, fat, and the integument. By its *deep surface*, from above downwards, with the ilium, sacrum, coccyx, and great sacro-sciatic ligament, part of the Gluteus medius, Pyriformis, Gemelli, Obturator internus, Quadratus femoris, the tuberosity of the ischium, great trochanter, the origin of the Biceps, Semi-tendinosus,

Semi-membranosus, and Adductor magnus muscles, the gluteal vessels and nerve are seen issuing from the pelvis above the Pyriformis muscle, the ischiatic and internal pudic vessels and nerves, and the nerve to the Obturator internus muscle below it. Its *upper border* is thin, and connected with the Gluteus medius by the fascia lata. Its *lower border*, free and prominent, forms the fold of the nates, and is directed towards the perineum.

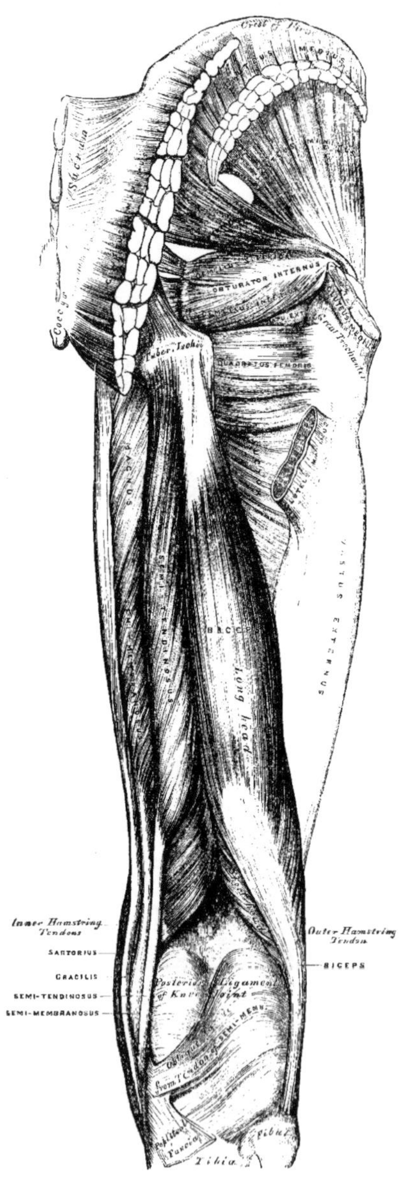


190 – *Muscles of the hip and thigh.*

*Dissection.* The Gluteus maximus should now be divided near its origin by a vertical incision carried from its upper to its lower border: a cellular interval will be exposed, separating it from the Gluteus medius and External rotator muscles beneath. The upper portion of the muscle should be altogether detached, and the lower portion turned outwards; the loose areolar tissue filling up the interspace between the trochanter major and tuberosity of the ischium being removed, the parts already enumerated as exposed by the removal of this muscle will be seen.

The *Gluteus Medius* is a broad, thick, radiated muscle, situated on the outer surface of the pelvis. Its posterior third is covered by the Gluteus maximus; its anterior two-thirds are covered by the fascia lata, which separates it from the integument. It arises from the outer surface of the ilium, between the superior and middle curved lines, and from the outer lip of that portion of the crest which is between them; it also arises from the dense fascia covering its anterior part. The fibres converge to a strong flattened tendon, which is inserted into the oblique line which traverses the outer surface of the great trochanter. A synovial bursa separates the tendon of the muscle from the surface of the trochanter in front of its insertion.

*Relations.* By its *superficial surface*, with the Gluteus maximus behind, the Tensor vaginæ femoris, and deep fascia in front. By its *deep surface*, with the Gluteus minimus and the gluteal vessels and nerve. Its *anterior border* is blended with the Gluteus minimus. Its *posterior border* lies parallel with the Pyriformis the gluteal vessels intervening.

This muscle should now be divided near its insertion and turned upwards, when the Gluteus minimus will be exposed.

The *Gluteus Minimus*, the smallest of the three glutei, is placed immediately beneath the preceding. It is fan-shaped, arising from the outer surface of the ilium, between the middle and inferior curved lines, and behind, from the margin of the great sacro-sciatic notch; the fibres converge to the deep surface of a radiated aponeurosis, which, terminating in a tendon, is inserted into an impression on the anterior border of the great trochanter. A synovial bursa is interposed between the tendon and the great trochanter.

*Relations.* By its *superficial surface*, with the Gluteus medius, and the gluteal vessels and nerve. By its *deep surface*, with the ilium, the reflected tendon of the Rectus femoris, and capsular ligament of the hip-joint. Its *anterior margin* is blended with the Gluteus medius. Its *posterior margin* is often joined with the tendon of the Pyriformis.

The *Pyriformis* is a flat muscle, pyramidal in shape, lying almost parallel with the lower margin of the Gluteus minimus. It is situated partly within the pelvis at its posterior part, and partly at the back of the hip-joint. It arises from the front of the sacrum by three fleshy digitations, attached to the portions of bone interposed between the second, third, and fourth anterior sacral foramina, and also from the grooves leading from the foramina: a few fibres also arise from the margin of the great sacro-sciatic foramen, and from the anterior surface of the great sacro-sciatic ligament. The muscle passes out of the pelvis through the great sacro-sciatic foramen, the upper part of which it fills, and is inserted, by a rounded tendon, into the upper border of the great trochanter, being generally blended with the tendon of the Obturator internus.

*Relations.* By its *anterior surface, within the pelvis*, with the Rectum (especially on the left side), the sacral plexus of nerves, and the internal iliac vessels; *external to the pelvis*, with the os innominatum and capsular ligament of the hip-joint. By its *posterior surface, within the pelvis*, with the sacrum; and *external to it*, with the Gluteus maximus. By its *upper border*, with the Gluteus medius, from which it is separated by the gluteal vessels and nerves. By its *lower border*, with the Gemellus superior; the ischiatic vessels and nerves, the internal pudic vessels and nerve, and the nerve to the Obturator internus, passing from the pelvis in the interval between them.

*Dissection.* The next muscle, as well as the origin of the Pyriformis, can only be seen when the pelvis is divided, and the viscera contained in this cavity removed.

The *Obturator Internus*, like the preceding muscle, is situated partly within the cavity of the pelvis, partly at the back of the hip-joint. It arises from the inner surface of the anterior and external wall of the pelvis, being attached to the margin of bone around the inner side of the obturator foramen; viz., from the descending ramus of the pubes, and the ascending ramus of the ischium; and laterally, from the inner surface of the body of the ischium, between the margin of the obturator foramen in front, the great sacro-sciatic notch behind, and the brim of the true pelvis above. It also arises from the inner surface of the obturator membrane and from the tendinous arch which completes the canal for the passage of the obturator vessels and nerve. The fibres are directed backwards and downwards, and terminate in four or five tendinous bands, which are found on its deep surface; these bands are reflected at a right angle over the inner surface of the tuberosity of the ischium, which is covered with cartilage, grooved for their reception, and lined with a synovial bursa. The muscle leaves the pelvis by the lesser sacro-sciatic notch; and the tendinous bands unite into a single flattened tendon, which passes horizontally outwards, and, after receiving the attachment of the Gemelli, is inserted into the upper border of the great trochanter in front of the Pyriformis. A synovial bursa, narrow and elongated in form, is usually found between the tendon of this muscle and the capsular ligament of the hip. It occasionally communicates with that between the tendon and the tuberosity of the ischium, the two forming a single sac.

In order to display the peculiar appearances presented by the tendon of this muscle, it should be divided near its insertion and reflected outwards.

*Relations. Within the pelvis*, this muscle is in relation, by its *anterior surface*, with the obturator membrane and inner surface of the anterior wall of the pelvis; by its *posterior surface*, with the pelvic and obturator fasciæ, which separate it from the Levator ani; and it is crossed by the internal pudic vessels and nerve. This surface forms the outer boundary of the ischio-rectal fossa. *External to the pelvis*, it is covered by the great sciatic nerve and Gluteus maximus, and rests on the back part of the hip-joint.

The *Gemelli* are two small muscular fasciculi, accessories to the tendon of the Obturator internus, which is received into a groove between them. They have received the names *superior* and *inferior* from the position they occupy.

The *Gemellus Superior*, the smaller of the two, arises from the outer surface of the spine of the ischium, and passing horizontally outwards, becomes blended with the upper part of the tendon of the Obturator internus, and is inserted with it into the upper border of the great trochanter. This muscle is sometimes wanting.

*Relations.* By its *superficial surface*, with the Gluteus maximus and the ischiatic vessels and nerves. By its *deep surface*, with the capsule of the hip joint. By its *upper border*, with the lower margin of the Pyriformis. By its *lower border*, with the tendon of the Obturator internus.

The *Gemellus Inferior* arises from the upper part of the outer border of the tuberosity of the ischium, and, passing horizontally outwards, is blended with the lower part of the tendon of the Obturator internus, and inserted with it into the upper border of the great trochanter.

*Relations.* By its *superficial surface*, with the Gluteus maximus, and the ischiatic vessels and nerves. By its *deep surface*, it covers the capsular ligament of the hip-joint. By its *upper border*, with the tendon of the Obturator internal. By its *lower border*, with the tendon of the Obturator externus and Quadratus femoris.

The *Quadratus Femoris* is a short, flat muscle, quadrilateral in shape (hence its name), situated between the Gemellus inferior and the upper margin of the Adductor magnus. It arises from the outer border of the tuberosity of the ischium, and proceeding horizontally outwards, is inserted into the upper part of the linea quadrati, on the posterior surface of the trochanter major. A synovial bursa is often found between the under surface of this muscle and the lesser trochanter, which it covers.

*Relations.* By its *posterior surface*, with the Gluteus maximus and the sciatic vessels and nerves. By its *anterior surface*, with the tendon of the Obturator externus and trochanter minor. By its *upper border*, with the Gemellus inferior. Its *lower border* is separated from the Adductor magnus by the terminal branches of the internal circumflex vessels.

*Dissection.* In order to expose the next muscle (the Obturator externus), it is necessary to remove the Psoas, Iliacus, Pectineus, and Adductor brevis and longus muscles, from the front and inner side of the thigh; and the Gluteus maximus and Quadratus femoris, from the back part. Its dissection should consequently be postponed until the muscles of the anterior and internal femoral regions have been examined.

The *Obturator Externus* is a flat triangular muscle, which covers the outer surface of the anterior wall of the pelvis. It arises from the margin of bone immediately around the inner side of the obturator foramen, viz., from the body and ramus of the pubes, and the ramus of the ischium; it also arises from the inner two-thirds of the outer surface of the obturator membrane, and from the tendinous arch which completes the canal for the passage of the obturator vessels and nerves. The fibres converging pass outwards and backwards, and terminate in a tendon which runs across the back part of the hip-joint, and is inserted into the digital fossa of the femur.

*Relations.* By its *anterior surface*, with the Psoas, Iliacus, Pectineus, Adductor longus, Adductor brevis, and Gracilis; and more externally, with the neck of the femur and capsule of the hip-joint. By its *posterior surface*, with the obturator membrane and Quadratus femoris.

*Nerves.* The Gluteus maximus is supplied by the inferior gluteal nerve and a branch from the sacral plexus. The Gluteus medius and minimus, by the superior gluteal. The Pyriformis, Gemelli, Obturator internus, and Quadratus femoris, by branches from the sacral plexus. And the Obturator externus, by the obturator nerve.

*Actions.* The Glutei muscles, when they take their fixed point from the pelvis are all abductors of the thigh. The Gluteus maximus and the posterior fibres of the Gluteus medius, rotate the thigh outwards; the anterior fibres of the Gluteus medius and the Gluteus minimus rotate it inwards. The Gluteus maximus serves to extend the femur, and the Gluteus medius and minimus draw it forwards. The Gluteus maximus is also a tensor of the fascia lata. Taking their fixed point from the femur, the Glutei muscles act upon the pelvis, supporting it and the whole trunk upon the head of the femur, which is especially obvious in standing on one leg. In order to gain the erect posture after the effort of stooping, these muscles draw the pelvis backwards, assisted by the Biceps, Semi-tendinosus, and Semimembranosus muscles. The remaining muscles are powerful rotators of the thigh outwards. In the sitting posture, when the thigh is flexed upon the pelvis, their action as rotators ceases, and they become abductors, with the exception of the Obturator externus, which still rotates the femur outwards. When the femur is fixed, the Pyriformis and Obturator muscles serve to draw the pelvis forwards if it has been inclined backwards, and assist in steadying it upon the head of the femur.

## Posterior Femoral Region.

Biceps. Semi-tendinosus. Semi-membranosus.

*Dissection* (fig. 189). Make a vertical incision along the middle of the thigh, from the lower fold of the nates to about three inches below the back of the knee-joint, and there connect it with a transverse incision, carried from the inner to the outer side of the leg. A third incision should then be made transversely at the junction of the middle with the lower third of the thigh. The integument having been removed from the back of the knee, and the boundaries of the popliteal space examined, the removal of the integument from the remaining part of the thigh should be continued, when the fascia and muscles of this region will be exposed.

The *Biceps* is a large muscle, of considerable length, situated on the posterior and outer aspect of the thigh. It arises by two heads. One, the long head, from an impression at the upper and back part of the tuberosity of the ischium, by a tendon common to it and the Semi-tendinosus. The femoral or short head, from the outer lip of the linea aspera, between the Adductor magnus and Vastus externus, extending from two inches below the insertion of the Gluteus maximus, to within two inches of the outer condyle; it also arises from the external intermuscular septum. The fibres of the long head form a fusiform belly, which, passing obliquely downwards and a little outwards, terminate in an aponeurosis which covers the posterior surface of the muscle, and receives the fibres of the short head; this aponeurosis becomes gradually contracted into a tendon, which is inserted into the outer side of the head of the fibula. At its insertion, the tendon divides into two portions, which embrace the external lateral ligament of the knee-joint, a strong prolongation being sent forwards to the outer tuberosity of the tibia, which gives off an expansion to the fascia of the leg. The tendon of this muscle forms the outer ham-string.

*Relations.* By its *superficial surface*, with the Gluteus maximus above, the fascia lata and integument in the rest of its extent. By its *deep surface*, with the Semi-membranosus, Adductor magnus, and Vastus externus, the great sciatic nerve, popliteal artery and vein, and near its insertion, with the external head of the Gastrocnemius, Plantaris, and superior external articular artery.

The *Semitendinosus*, remarkable for the great length of its tendon, is situated at the posterior and inner aspect of the thigh. It arises from the tuberosity of the ischium by a tendon common to it and the long head of the Biceps; it also arises from an aponeurosis which connects the adjacent surfaces of the two muscles to the extent of about three inches after their origin. It forms a fusiform muscle, which, passing downwards and inwards, terminates a little below the middle of the thigh in a long round tendon, which lies along the inner side of the popliteal space; curving around the inner tuberosity of the tibia, it is inserted into the upper part of the inner surface of the shaft of that bone, nearly as far forwards as its anterior border. This tendon lies beneath the expansion of the Sartorius, and below that of the Gracilis, to which it is united. A tendinous intersection is usually observed about the middle of the muscle.

*Relations.* By its *superficial surface*, with the Gluteus maximus and fascia lata. By its *deep surface*, with the Semi-membranosus, Adductor magnus, inner head of the Gastrocnemius, and internal lateral ligament of the knee-joint.

The *Semi-membranosus*, so called from the membranous expansion on its anterior and posterior surfaces, is situated at the back part and inner side of the thigh. It arises by a thick tendon from the upper and outer part of the tuberosity of the ischium, above and to the outer side of the Biceps and Semi-tendinosus, and is inserted into the inner and back part of the inner tuberosity of the tibia, beneath the internal lateral ligament. The tendon of the muscle at its origin expands into an aponeurosis, which covers the upper part of its anterior surface; from this aponeurosis, muscular fibres arise, and converge to another aponeurosis, which covers the lower part of its posterior surface, and this contracts into the tendon of insertion. The tendon of the muscle at its insertion divides into three portions; the middle portion is the fasciculus of insertion into the back part of the inner tuberosity, it sends down an expansion to cover the Popliteus muscle. The internal portion is horizontal, passing forwards beneath the internal lateral ligament, to be inserted into a groove along the inner side of the internal tuberosity. The posterior division passes upwards and backwards, to be inserted into the back part of the outer condyle of the femur, forming the chief part of the posterior ligament of the knee-joint.

The tendons of the two preceding muscles, with those of the Gracilis and Sartorius, form the inner ham-string.

*Relations.* By its *superficial surface*, with the Semi-tendinosus, Biceps, and fascia lata. By its *deep surface*, with the popliteal vessels, Adductor magnus and the inner head of the Gastrocnemius, from which it is separated by a synovial bursa. By its *inner border*, with the Gracilis. By its *outer border*, with the great sciatic nerve, and its internal popliteal branch.

*Nerves.* The muscles of this region are supplied by the great sciatic nerve.

*Actions.* The three ham-string muscles flex the leg upon the thigh. When the knee is semi-flexed, the Biceps, from its oblique direction downwards and outwards, rotates the leg slightly outwards; and the Semi-membranosus, in consequence of its oblique direction, rotates the leg inwards, assisting the Popliteus. Taking their fixed point from below, these muscles serve to support the pelvis upon the head of the femur, and to draw the trunk directly backwards, as is seen in feats of strength, when the body is thrown backwards in the form of an arch.

*Surgical Anatomy*. The tendons of these muscles occasionally require subcutaneous division in some forms of spurious anchylosis of the knee-joint, dependent upon permanent contraction and rigidity of the Flexor muscles, or from stiffening of the ligamentous and other tissues surrounding the joint, the result of disease. This is easily effected by putting the tendon upon the stretch, and inserting a narrow sharp-pointed knife between it and the skin; the cutting edge being then turned towards the tendon, it should be divided taking care that the wound in the skin is not at the same time enlarged. This operation has been attended with considerable success in some cases of stiffened knee from rheumatism, gradual extension being kept up for some time after the operation.

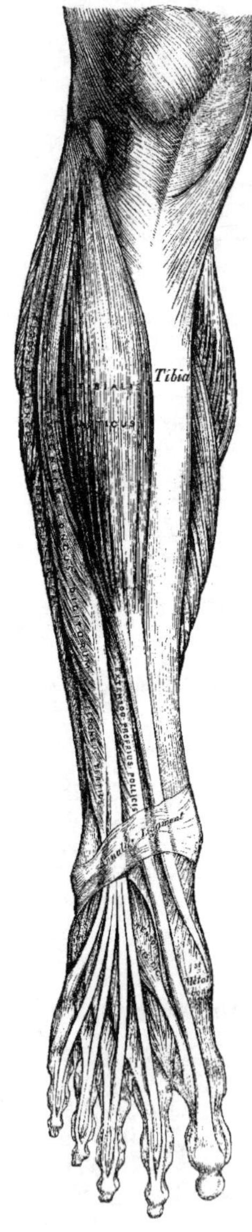


191 – *Muscles of the front of the leg.*

## Muscles and Fasciæ of the Leg.

*Dissection* (fig. 186). The knee should be bent, a block placed beneath it, and the food kept in an extended position; an incision should then be made through the integument in the middle line of the leg to the ankle, and continued along the dorsum of the foot to the toes. A second incision should be made transversely across the ankle, and a third in the same direction across the bases of the toes: the flaps of integument included between these incisions should be removed, and the deep fascia of the leg examined.

The *fascia of the leg* forms a complete investment to the whole of this region of the limb, excepting to the inner surface of the tibia, to which it is unattached. It is continuous above with the fascia lata, receiving an expansion from the tendon of the Biceps on the outer side, and from the tendons of the Sartorius, Gracilis, and Semi-tendinosus on the inner side; in front it blends with the periosteum covering the tibia and fibula; below, it is continuous with the annular ligaments of the ankle. It is thick and dense in the upper and anterior part of the leg, and gives attachment, by its inner surface, to the Tibialis anticus and Extensor longus digitorum muscles; but thinner behind, where it covers the Gastrocnemius and Soleus muscles. Its inner surface gives off, on the outer side of the leg, two strong intermuscular septa, which enclose the Peronæi muscles, and separate them from the muscles on the anterior and posterior tibial regions, and several smaller and more slender processes enclose the individual muscles in each region; at the same time, a broad transverse intermuscular septum intervenes between the superficial and deep muscles in the posterior tibio-fibular region.

The fascia should now be removed by dividing it in the same direction as the integument, excepting opposite the ankle, where it should be left entire. The removal of the fascia should be commenced from below, opposite the tendons, and detached in the line of direction of the muscular fibres.

## Muscles of the Leg.

These may be subdivided into three groups: those on the anterior, those on the posterior, and those on the outer side.

## Anterior Tibio-Fibular Region.

Tibialis Anticus.
Extensor Proprius Pollicis.
Extensor Longus Digitorum.
Peroneus Tertius.

The *Tibialis Anticus* is situated on the outer side of the tibia; it is thick and fleshy at its upper part, tendinous below. It arises from the outer tuberosity and upper two-thirds of the external surface of the shaft of the tibia; from the adjoining part of the interosseous membrane; from the deep fascia of the leg; and from the intermuscular septum between it and the Extensor communis digitorum: the fibres pass vertically downwards, and terminate in a tendon, which is apparent on the anterior surface of the muscle at the lower third of the leg. After passing through the innermost compartment of the anterior annular ligament, it is inserted into the inner and under surface of the internal cuneiform bone, and base of the metatarsal bone of the great toe.

*Relations.* By its *anterior surface*, with the deep fascia, and with the annular ligament. By its *posterior surface*, with the interosseous membrane, tibia, ankle-joint, and inner side of the tarsus. By its *inner surface*, with the tibia. By its *outer surface*, with the Extensor longus digitorum, and Extensor proprius pollicis, the anterior tibial vessels and nerve lying between it and the last mentioned muscle.

The *Extensor Proprius Pollicis* is a thin, elongated, and flattened muscle, situated between the Tibialis anticus and Extensor longus digitorum. It arises from the anterior surface of the fibula for about the middle two-fourths of its extent, its origin being internal to the Extensor longus digitorum; it also arises from the interosseous membrane to a similar extent. The fibres pass downwards, and terminate in a tendon, which occupies the anterior border of the muscle, passes through a distinct compartment in the annular ligament, and is inserted into the base of the last phalanx of the great toe. Opposite the metatarso-phalangeal articulation, the tendon gives off a thin prolongation on each side, which covers its surface.

*Relations.* By its *anterior border*, with the deep fascia, and the anterior annular ligament. By its *posterior border*, with the interosseous membrane, fibula, tibia, ankle-joint, and Extensor brevis digitorum. By its *outer side*, with the Extensor longus digitorum above, the dorsalis pedis vessels and anterior tibial nerve below. By its *inner side*, with the Tibialis anticus, and the anterior tibial vessels above.

The *Extensor Longus Digitorum* is an elongated, flattened, semi-penniform muscle, situated the most externally of all the muscles on the fore-part of the leg. It arises from the outer tuberosity of the tibia; from the upper three-fourths of the anterior surface of the shaft of the fibula; from the interosseous membrane, and deep fascia; and from the intermuscular septa between it and the Tibialis anticus on the inner, and the Peronei on the outer side. The muscle terminates in three tendons, which pass through a canal in the annular ligament, with the Peroneus tertius, run across the dorsum of the foot, and, the innermost tendon having subdivided into two, they are inserted into the second and third phalanges of the four lesser toes. The mode in which the tendons are inserted, is the following: Each tendon opposite the metatarso-phalangeal articulation is joined, on its outer side, by the tendon of the Extensor brevis digitorum (except the fourth), and receives a fibrous expansion from the Interossei and Lumbricales; it then spreads into a broad aponeurosis, which covers the dorsal surface of the first phalanx: this aponeurosis, at the articulation of the first with the second phalanx, divides into three slips, a middle one, which is inserted into the base of the second phalanx; and two lateral slips, which, after uniting on the dorsal surface of the second phalanx, are continued onwards, to be inserted into the base of the third.

*Relations.* By its *anterior surface*, with the deep fascia, and the annular ligament. By its *posterior surface*, with the fibula, interosseous membrane, ankle-joint, and Extensor brevis digitorum. By its *inner side*, with the Tibialis anticus, Extensor proprius pollicis, and anterior tibial vessels and nerve. By its *outer side*, with the Peroneus longus and brevis.

The *Peroneus Tertius* may be considered as part of the Extensor longus digitorum, being almost always intimately united with it. It arises from the lower fourth of the anterior surface of the fibula, on its outer side; from the lower part of the interosseous membrane; and from an intermuscular septum between it and the Peroneus brevis. Its tendon, after passing through the same canal in the annular ligament as the Extensor longus digitorum, is inserted into the base of the metatarsal bone of the little toe, on its dorsal surface. This muscle is sometimes wanting.

*Nerves.* These muscles are supplied by the anterior tibial nerve.

*Actions.* The Tibialis anticus and Peroneus tertius are the direct flexors of the tarsus upon the leg; the former muscle, from the obliquity in the direction of its tendon, raises the inner border of the foot; and the latter, acting with the Peroneus brevis and longus, will draw the outer border of the foot upwards, and the sole outwards. The Extensor longus digitorum and Extensor proprius pollicis extend the phalanges of the toes, and, continuing their action, flex the tarsus upon the leg. Taking their origin from below, in the erect posture, all these muscles serve to fix the bones of the leg in a perpendicular direction, and give increased strength to the ankle-joint.

## Posterior Tibio-Fibular Region.

*Dissection* (fig. 189). Make a vertical incision along the middle line of the back of the leg, from the lower part of the popliteal space to the heel, connecting it below by a transverse incision extending between the two malleoli; the flaps of integument being removed, the fascia and muscles should be examined.

The muscles in this region of the leg are subdivided into two layers, superficial, and deep. The superficial layer constitutes a powerful muscular mass, forming what is called the calf of the leg. Their large size is one of the most characteristic features of the muscular apparatus in man, and bears a direct connection with his ordinary attitude and mode of progression.

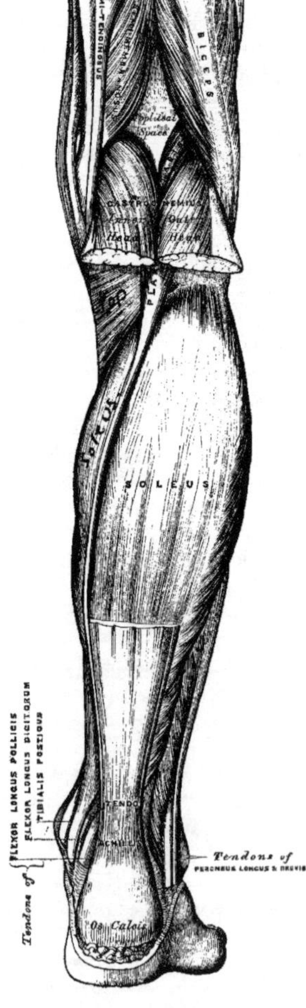


192 – *Muscles of the back of the leg. Superficial layer.*

*Superficial Layer*

Gastrocnemius. Soleus.
Plantaris.

The *Gastrocnemius* is the most superficial muscle, and forms the greater part of the calf. It arises by two heads, which are connected to the condyles of the femur by two strong flat tendons. The inner head, the larger, and a little the most posterior, is attached to a depression at the upper and back part of the inner condyle. The outer head, to the upper and back part of the external condyle, immediately above the origin of the Popliteus. Both heads, also, arise by a few tendinous and fleshy fibres from the ridges which are continued upwards from the condyles to the linea aspera. Each tendon spreads into an aponeurosis, which covers the posterior surface of that portion of the muscle to which it belongs; that covering the inner head being longer and thicker than the outer. From the anterior surface of these tendinous expansions, muscular fibres are given off; the fibres in the median line, which correspond to the accessory portions of the muscle derived from the bifurcations of the linea aspera, unite at an angle upon a median tendinous raphe below. The remaining fibres converge to the posterior surface of an aponeurosis which covers the front of the muscle, and this, gradually contracting, unites with the tendon of the Soleus, and forms with it the tendo Achillis.

*Relations.* By its *superficial surface*, with the fascia of the leg, which separates it from the external saphenous vein and nerve. By its *deep surface*, with the posterior ligament of the knee-joint, the Popliteus, Soleus, Plantaris, popliteal vessels, and internal popliteal nerve. The tendon of the inner head corresponds with the back part of the inner condyle, from which it is separated by a synovial bursa, which, in some cases, communicates with the cavity of the knee-joint. The tendon of the outer head contains a sesamoid fibro-cartilage (rarely osseous), where it plays over the corresponding outer condyle; and one is occasionally found in the tendon of the inner head.

The Gastrocnemius should be divided across, just below its origin, and turned downwards, in order to expose the next muscles.

The *Soleus* is a broad flat muscle, situated immediately beneath the preceding. It has received its name from the fancied resemblance it bears to a sole-fish. It arises by tendinous fibres from the back part of the head, and from the upper half of the posterior surface of the shaft of the fibula; from the oblique line of the tibia, and from the middle third of its internal border; some fibres also arise from a tendinous arch placed between the tibial and fibular origins of the muscle, and beneath which the posterior tibial vessels and nerve pass into the leg. The fibres pass backwards to an aponeurosis which covers the posterior surface of the muscle, and this, gradually becoming thicker and narrower, joins with the tendon of the Gastrocnemius, and forms with it the tendo Achillis.

*Relations.* By its *superficial surface*, with the Gastrocnemius and Plantaris. By its *deep surface*, with the Flexor longus digitorum, Flexor longus pollicis, Tibialis posticus, and posterior tibial vessels and nerve, from which it is separated by the transverse intermuscular septum.

The *tendo Achillis*, the common tendon of the Gastrocnemius and Soleus is the thickest and strongest tendon in the body. It is about six inches in length, and formed by the junction of the aponeuroses of the two preceding muscles. It commences about the middle of the leg, but receives fleshy fibres on its anterior surface, nearly to its lower end. Gradually becoming contracted below, it is inserted into the lower part of the posterior tuberosity of the os calcis, a synovial bursa being interposed between the tendon and the upper part of the tuberosity. The tendon is covered by the fascia and the integument, and is separated from the deep muscles and vessels, by a considerable interval filled up with areolar and adipose tissue. Along its outer side, but superficial to it, is the external saphenous vein.

The *Plantaris* is an extremely diminutive muscle, placed between the Gastrocnemius and Soleus, and remarkable for its long and delicate tendon. It arises from the lower part of the outer bifurcation of the linea aspera, and from the posterior ligament of the knee-joint. It forms a small fusiform belly, about two inches in length, which terminates in a long slender tendon, which crosses obliquely between the two muscles of the calf, and, running along the inner border of the tendo Achillis, is inserted with it into the posterior part of the os calcis. This muscle is occasionally double, and is sometimes wanting. Occasionally, its tendon is lost in the internal annular ligament, or in the fascia of the leg.

*Nerves.* These muscles are supplied by the internal popliteal nerve.

*Actions.* The muscles of the calf possess considerable power, and are constantly called into use in standing, walking, dancing, and leaping; hence the large size they usually present. In walking, these muscles draw powerfully upon the os calcis, raising the heel, and, with it, the entire body, from the ground; the body being thus supported on the raised foot, the opposite limb can be carried forwards. In standing, the Soleus, taking its fixed point from below, steadies the leg upon the foot, and prevents the body from falling forwards, to which there is a constant tendency from the superincumbent weight. The Gastrocnemius, acting from below, serves to flex the femur upon the tibia, assisted by the Popliteus. The Plantaris is the rudiment of a large muscle which exists in some of the lower animals, and serves as a tensor of the plantar fascia.

## Posterior Tibio-Fibular Region.

*Deep Layer*

| | |
|---|---|
| Popliteus. | Flexor Longus Digitorum. |
| Flexor Longus Pollicis. | Tibialis Posticus. |

*Dissection.* Detach the Soleus from its attachment to the fibula and tibia, and turn it downwards, when the deep layer of muscles is exposed, covered by the deep fascia of the leg.

The *deep fascia* of the leg is a broad, transverse, intermuscular septum, interposed between the superficial and deep muscles in the posterior tibio-fibular region. On each side it is connected to the margins of the tibia and fibula. Above, where it covers the Popliteus, it is thick and dense, and receives an expansion from the tendon of the Semi-membranosus; it is thinner in the middle of the leg; but below, where it covers the tendons passing behind the malleoli, it is thickened. It is continued onwards in the interval between the ankle and the heel, where it covers the vessels, and is blended with the internal annular ligament.

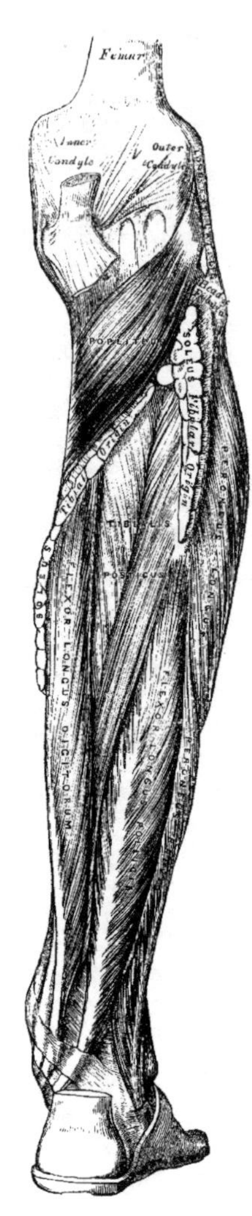

193 – *Muscles of the back of the leg. Deep layers.*

This fascia should now be removed, commencing from below opposite the tendons, and detaching it from the muscles in the direction of their fibres.

The *Popliteus* is a thin, flat, triangular muscle, which forms the floor of the popliteal space, and is covered by a tendinous expansion, derived from the Semi-membranosus muscle. It arises by a strong flat tendon, about an inch in length; from a deep depression on the outer side of the external condyle of the femur; and from the posterior ligament of the knee-joint; and is inserted into the inner two-thirds of the triangular surface above the oblique line on the posterior surface of the shaft of the tibia, and into the tendinous expansion covering the surface of the muscle. The tendon of the muscle is covered by that of the Biceps and the external lateral ligament of the knee-joint; it grooves the outer surface of the external semi-lunar cartilage, and is invested by the synovial membrane of the knee-joint.

*Relations.* By its *superficial surface*, with the fascia above mentioned, which separates it from the Gastrocnemius, Plantaris, popliteal vessels, and internal popliteal nerve. By its *deep surface*, with the superior tibio-fibular articulation, and back of the tibia.

The *Flexor Longus Pollicis* is situated on the fibular side of the leg, and is the most superficial, and largest of the three next muscles. It arises from the lower two-thirds of the internal surface of the shaft of the fibula, with the exception of an inch below; from the lower part of the interosseous membrane; from an intermuscular septum between it and the Peronei, externally; and from the fascia covering the Tibialis posticus. The fibres pass obliquely downwards and backwards, and terminate round a tendon which occupies nearly the whole length of the posterior surface of the muscle. This tendon passes through a groove on the posterior surface of the tibia, external to that for the Tibialis posticus and Flexor longus digitorum; it then passes through another groove on the posterior extremity of the astragalus, and along a third groove, beneath the tubercle of the os calcis, into the sole of the foot, where it runs forwards between the two heads of the Flexor brevis pollicis, and is inserted into the base of the last phalanx of the great toe. The grooves in the astragalus and os calcis which contain the tendon of the muscle, are converted by tendinous fibres into distinct canals, lined by synovial membrane; and as the tendon crosses the sole of the foot, it is connected to the common flexor by a tendinous slip.

*Relations.* By its *superficial surface*, with the Soleus and tendo Achillis, from which it is separated by the deep fascia. By its *deep surface*, with the fibula, Tibialis posticus, the peroneal vessels, the lower part of the interosseus membrane, and the ankle-joint. By its *outer border*, with the Peronei. By its *inner border*, with the Tibialis posticus, and Flexor longus digitorum.

The *Flexor Longus Digitorum* is situated on the tibial side of the leg. At its origin, it is thin and pointed, but gradually increases in size as it descends. It arises from the posterior surface of the shaft of the tibia, immediately below the oblique line, to within three inches of its extremity, internal to the tibial origin of the Tibialis posticus; some fibres also arise from the inter-muscular septum, between it and the Tibialis posticus. The fibres terminate in a tendon, which runs nearly the whole length of the posterior surface of the muscle. This tendon passes, behind the inner malleolus, in a groove, common to it, and the Tibialis posticus, but separated from the latter by a fibrous septum; each tendon being contained in a special sheath lined by a separate synovial membrane. It then passes, obliquely, forwards and outwards, beneath the arch of the os calcis into the sole of the foot, where, crossing beneath the tendon of the Flexor longus pollicis, to which it is connected by a strong tendinous slip, it becomes expanded, is joined by the Flexor accessorius, and finally divides into four tendons, which are inserted into the bases of the last phalanges of the four lesser toes, each tendon passing through a fissure in the tendon of the Flexor brevis digitorum, opposite the middle of the first phalanges.

*Relations.* In the leg. By its *superficial surface*, with the Soleus, and the posterior tibial vessels and nerve, from which it is separated by the deep fascia. By its *deep surface*, with the tibia and Tibialis posticus. *In the foot*, it is covered by the Abductor Pollicis, and Flexor brevis digitorum, and crosses beneath the Flexor longus pollicis.

The *Tibialis Posticus* lies between the two preceding muscles, and is the most deeply seated of all the muscles in the leg. It commences above, by two pointed processes, separated by an angular interval, through which, the anterior tibial vessels pass forwards to the front of the leg. It arises from the posterior surface of the interosseous membrane, its whole length, excepting its lowest part, from the posterior surface of the shaft of the tibia, external to the Flexor longus digitorum between the commencement of the oblique line above, and the middle of the external border of the bone below, and from the upper two-thirds of the inner surface of the shaft of the fibula; some fibres also arise from the deep fascia, and from the intermuscular septa, separating it from the adjacent muscles on each side. This muscle, in the lower fourth of the leg, passes in front of the Flexor longus digitorum, terminates in a tendon, which passes through a groove with it behind the inner malleolus, but enclosed in a separate sheath; it then passes through another sheath, over the internal lateral ligament, and beneath the calcaneo-scaphoid articulation, and is inserted into the tuberosity of the scaphoid, and internal cuneiform bones. The tendon of this muscle, contains a sesamoid bone, near its insertion, and gives off fibrous expansions, one of which, passes, backwards to the os calcis, others outwards to the middle and external cuneiform, and some forwards to the bases of the third and fourth metatarsal bones.

*Relations.* By its *superficial surface*, with the Soleus, and Flexor longus digitorum, the posterior tibial vessels and nerve, and the peroneal vessels, from which it is separated by the deep fascia. By its *deep surface*, with the interosseous ligament, the tibia, fibula, and ankle-joint.

*Nerves.* The Popliteus is supplied by the internal popliteal nerve, the remaining muscles of this group, by the posterior tibial nerve.

*Actions.* The Popliteus assists in flexing the leg upon the thigh; when the leg is flexed, it may rotate the tibia inwards. The Tibialis posticus is a direct extensor of the tarsus upon the leg; acting in conjunction with the Tibialis anticus, it turns the sole of the foot inwards, antagonizing the Peroneus longus which turns it outwards. The Flexor longus digitorum, and Flexor longus pollicis, are the direct Flexors of the phalanges, and, continuing their action, extend the foot upon the leg; they assist the Gastrocnemius and Soleus in extending the foot, as in the act of walking, or in standing on tiptoe. In consequence of the oblique direction of the tendon of the long extensor, the toes would be drawn inwards, were it not for the Flexor accessorius muscle, which is inserted into the outer side of that tendon, and draws it to the middle line of foot, during its action. Taking their fixed point from the foot, these muscles serve to maintain the upright posture, by steadying the tibia and fibula, perpendicularly, upon the ankle-joint. They also serve to raise these bones from the oblique position they assume in the stooping posture.

## Fibular Region.

Peroneus Longus. Peroneus Brevis.

*Dissection.* These muscles are readily exposed, by removing the fascia, covering their surface, from below upwards, in the line of direction of their fibres.

The *Peroneus Longus* is situated at the upper part of the outer side of the leg, and is the more superficial of the two muscles. It arises from the head, and upper two-thirds of the outer surface of the shaft of the fibula, from the deep fascia, and from the intermuscular septa, between it and the muscles on the front, and those on the back of the leg. It terminates in a long tendon, which passes behind the outer malleolus, in a groove, common to it, and the Peroneus brevis, the groove being converted into a canal by a fibrous band, and the tendons, invested by a common synovial membrane; it is then reflected, obliquely forwards, across the outer side of the os calcis, being contained in a separate fibrous sheath, lined by a prolongation of the synovial membrane, from the groove behind the malleolus. Having reached the outer side of the cuboid bone, it runs, in a groove, on its under surface, which is converted into a canal, by the long calcaneo-cuboid ligament, lined by a synovial membrane, and crossing, obliquely, the sole of the foot, is inserted into the outer side of the base of the metatarsal bone of the great toe. The tendon of the muscle has a double reflection, first, behind the external malleolus, secondly, on the outer side of the cuboid bone; in both of these situations, the tendon is thickened, and, in the latter, a sesamoid bone is usually developed in its substance.

*Relations.* By its *superficial surface*, with the fascia and integument. By its *deep surface*, with the fibula, the Peroneus brevis, os calcis, and cuboid bone. By its *anterior border*, an inter-muscular septum intervenes between it and the Extensor longus digitorum. By its *posterior border*, an inter-muscular septum separates it from the Soleus above, and the Flexor longus pollicis below.

The *Peroneus Brevis* lies beneath the Peroneus longus, and is shorter and smaller than it. It arises from the middle third of the external surface of the shaft of the fibula, internal to the Peroneus longus; from the anterior and posterior borders of the bone; and from the intermuscular septa separating it from the adjacent muscles on the front and back part of the leg. The fibres pass vertically downwards, and terminate in a tendon, which runs through the same groove as the preceding muscle, behind the external malleolus, being contained in the same fibrous sheath, and lubricated by the same synovial membrane; it then passes through a separate sheath on the outer side of the os calcis, above that for the tendon of the Peroneus longus, and is finally inserted into the base of the metatarsal bone of the little toe, on its dorsal surface.

*Relations.* By its *superficial surface*, with the Peroneus longus and the fascia of the leg and foot. By its *deep surface*, with the fibula and outer side of the os calcis.

*Nerves.* The Peroneus longus and brevis are supplied by the musculo-cutaneous branch of the external popliteal nerve.

*Actions.* The Peroneus longus and brevis extend the foot upon the leg, in conjunction with the Tibialis posticus, antagonizing the Tibialis anticus and Peroneus tertius, which are flexors of the foot. The Peroneus longus also everts the side of the foot; hence the extreme eversion observed in fracture of the lower end of the fibula, where that bone offers no resistance to the action of this muscle. Taking their fixed point below, they serve to steady the leg upon the foot. This is especially the case in standing upon one leg, when the tendency of the superincumbent weight is to throw the leg inwards; and the Peroneus longus overcomes this by drawing on the outer side of the leg, and thus maintains the perpendicular direction of the limb.

*Surgical Anatomy*. The student should now consider the position of the tendons of the various muscles of the leg, their relation with the ankle-joint and surrounding blood vessels, and especially their action upon the foot, as their rigidity and contraction give rise to one or the other forms of deformity known as *club-foot*. The most simple and common deformity, and one that is rarely if ever congenital, is the *talipes equinus*, the heel being raised by rigidity and contraction of the Gastrocnemius muscle, and the patient walking upon the ball of the foot. In the *talipes varus*, which is the more common congenital form, the heel is raised by the tendo Achillis, the inner border of the foot drawn upwards by the Tibialis anticus, and the anterior two-thirds of the foot twisted inwards by the Tibialis posticus and Flexor longus digitorum, the patient walking upon the outer edge of the foot, and in severe cases upon the dorsum and outer ankle. In the *talipes vulgus* the outer edge of the foot is raised by the Peronei muscles, and the patient walks on the inner ankle. In the *talipes calcaneus* the toes are raised by the extensor muscles, the heel is depressed, and the patient walks upon it. Other varieties of deformity are met with, as the *talipes equino-varus*, *equino-vulgus*, and *calcaneo-vulgus*, whose names sufficiently indicate their nature. Each of these deformities may be successfully relived (after other remedies fail) by division of the opposing tendons and fascia; by this means, the foot regains its proper position, and the tendons heal by the organization of lymph thrown out between the divided ends. The operation is easily performed by putting the contracted tendon upon the stretch, and dividing it by means of a narrow sharp-pointed knife inserted between it and the skin.

## Muscles and Fasciæ of the Foot.

The fibrous bands which bind down the tendons in front of and behind the ankle in their passage to the foot, should now be examined; they are termed the *annular ligaments*, and are three in number, anterior, internal, and external.

The *Anterior Annular Ligament* consists of a superior or vertical portion, which binds down the extensor tendons as they descend on the front of the tibia and fibula; and an inferior or horizontal portion, which retains them in connection with the tarsus: the two portions being connected by a thin intervening layer of fascia. The vertical portion is attached externally to the lower end of the fibula internally to the tibia, and above is continuous with the fascia of the leg; it contains two separate sheaths, one internally, for the tendon of the Tibialis anticus; one externally, for the tendons of the Extensor longus digitorum and Peroneus tertius, the tendon of the Extensor proprius pollicis, and the anterior tibial vessels and nerve pass beneath it. The horizontal portion is attached externally to the upper surface of the os calcis, in front of the depression for the interosseous ligament, and internally to the inner malleolus and plantar fascia: it contains three sheaths; the most internal for the tendon of the Tibialis anticus, the next in order for the tendon of the Extensor proprius pollicis, and the most external for the Extensor communis digitorum and Peroneus tertius: the anterior tibial vessels and nerve lie altogether beneath it. These sheaths are lined by separate synovial membranes.

The *Internal Annular Ligament* is a strong fibrous band, which extends from the inner malleolus above, to the internal margin of the os calcis below, converting a series of bony grooves in this situation into osteo-fibrous canals, for the passage of the tendons of the flexor muscles and vessels into the sole of the foot. It is continuous above with the deep fascia of the leg, below with the planter fascia and the fibres of origin of the Abductor pollicis muscle. The three canals which it forms, transmit from within outwards, first, the tendon of the *Tibialis posticus*; second, the tendon of the Flexor longus digitorum, then the posterior tibial vessels and nerve, which run through a broad space beneath the ligament; lastly, in a canal formed partly by the astragalus, the tendon of the Flexor longus pollicis. Each of these canals is lined by a separate synovial membrane.

The *External Annular Ligament* extends from the extremity of the outer malleolus to the outer surface of the os calcis, it binds down the tendons of the Peronei muscles in their passage beneath the outer ankle. The two tendons are enclosed in one synovial sac.

*Dissection of the Sole of the Foot.* The foot should be placed on a high block with the sole uppermost, and firmly secured in that position. Carry an incision round the heel and along the inner and outer borders of the foot to the great and little toes. This incision should divide the integument and thick layer of granular fat beneath, until the fascia is visible; it should then be removed from the fascia in a direction from behind forwards, as seen in fig. 189.

The *Plantar Fascia*, the densest of all the fibrous membranes, is of great strength, and consists of dense pearly-white glistening fibres, disposed, for the most part longitudinally; it is divided into a central and two lateral portions.

The *central portion*, the thickest, is narrow behind and attached to the inner tuberosity of the os calcis, behind the origin of the Flexor brevis digitorum, and becoming broader and thinner in front, divides opposite the middle of the metatarsal bones into five processes, one for each of the toes. Each of these processes divides opposite the metatarso-phalangeal articulation into two slips, which embrace the sides of the flexor tendons of the toes, and are inserted into the sides of the metatarsal bones, and into the transverse metatarsal ligament, thus forming a series of arches through which the tendons of the short and long flexors pass to the toes. The intervals left between the five processes allow the digital vessels and nerves, and the tendons of the Lumbricales and Interossei muscles to become superficial. At the point of division of the fascia into processes and slips, numerous transverse fibres are superadded, which serve to increase the strength of the fascia at this part, by binding the processes together and connecting them with the integument. The central portion of the plantar fascia is continuous with the lateral portions at each side, and sends upwards into the foot, at their point of junction, two strong vertical intermuscular septa, broader in front than behind, which separate the middle from the external and internal plantar group of muscles; from these again thinner transverse septa are derived, which separate the various layers of muscles in this region. The upper surface of this fascia gives attachment behind to the Flexor brevis digitorum muscle.

The *lateral portions* of the plantar fascia are thinner than the central piece and cover the sides of the foot.

The *outer portion* covers the under surface of the Abductor minimi digiti; it is thick behind, thin in front, and extends from the os calcis forwards to the base of the fifth metatarsal bone, into the outer side of which it is attached; it is continuous internally with the middle portion of the plantar fascia, and externaly with the dorsal fascia.

The *inner portion* is very thin, and covers the Abductor pollicis muscle; it is attached behind to the internal annular ligament, is continuous around the side of the foot with the dorsal fascia, and externally with the middle portion of the plantar fascia.

## Muscles of the Foot.

These are divided into two groups: 1. Those on the dorsum; 2. Those on the plantar surface.

## 1. Dorsal Region.

Extensor Brevis Digitorum

The *Fascia* on the dorsum of the foot is a thin membranous layer, continuous above with the anterior margin of the annular ligament; it becomes gradually lost opposite the heads of the metatarsal bones, and on each side blends with the lateral portions of the plantar fascia: it forms a sheath for the tendons placed on the dorsum of the foot. On the removal of this fascia, the muscles of the dorsal region of the foot are exposed.

The *Extensor Brevis Digitorum* is a broad thin muscle, which arises from the outer side of the os calcis, in front of the groove for the Peroneus brevis; from the astragalo-calcanean ligament; and from the horizontal portion of the anterior annular ligament: passing obliquely across the dorsum of the foot, it terminates in four tendons. The innermost, which is the largest, is inserted into the first phalanx of the great toe; the other three, into the outer sides of the long extensor tendons of the second, third, and fourth toes.

*Relations.* By its *superficial surface*, with the fascia of the foot, the tendons of the Extensor longus digitorum, and Extensor proprius pollicis. By its *deep surface*, with the tarsal and metatarsal bones, and the Dorsal interossei muscles.

*Nerves.* It is supplied by the anterior tibial nerve.

*Actions.* The Extensor brevis digitorum is an accessory to the long Extensor, extending the phalanges of the four inner toes, but acting only on the first phalanx of the great toe. The obliquity of its direction counteracts the oblique movements given to the toes by the long Extensor, so that, both muscles acting together, the toes are evenly extended.

194 – *Muscles of the sole of the foot. First layer.*

## 2. Plantar Region.

The muscles in the plantar region of the foot may be divided into three groups, in a similar manner to those in the hand. Those of the internal plantar region, are connected with the great toe, and correspond with those of the thumb; those of the external plantar region, are connected with the little toe, and correspond with those of the little finger; and those of the middle plantar region, are connected with the tendons intervening between the two former groups. In order to facilitate the dissection of these muscles, it will be found more convenient to divide them into three layers, as they present themselves, in the order in which they are successively exposed.

*First Layer*

Abductor Pollicis. Flexor Brevis Digitorum.
Abductor Minimi Digiti.

*Dissection.* Remove the fascia on the inner and outer sides of the foot, commencing in front over the tendons, and proceeding backwards. The central portion should be divided transversely in the middle of the foot, and the two flaps dissected forwards and backwards.

The *Abductor Pollicis* lies along the inner border of the foot. It arises from the inner tuberosity on the under surface of the os calcis; from the internal annular ligament; from the plantar fascia; and from the intermuscular septum between it and the Flexor brevis digitorum. The fibres terminate in a tendon, which is inserted, together with the innermost tendon of the Flexor brevis pollicis, into the inner side of the base of the first phalanx of the great toe.

*Relations.* By its *superficial surface*, with the plantar fascia. By its *deep surface*, with the Flexor brevis pollicis, the

Flexor accessorius, and the tendons of the Flexor longus digitorum and Flexor longus pollicis, the Tibialis anticus and posticus, the plantar vessels and nerves, and the articulations of the tarsus.

The *Flexor Brevis Digitorum* lies in the middle line of the sole of the foot, immediately beneath the plantar fascia, with which it is firmly united. It arises, by a narrow tendinous process, from the inner tuberosity of the os calcis; from the central part of the plantar fascia; and from the intermuscular septa between it and the adjacent muscles. It passes forwards, and divides into four tendons. Opposite the middle of the first phalanges, each tendon presents a longitudinal slit, to allow of the passage of the corresponding tendon of the Flexor longus digitorum, the two portions forming a groove for its reception, and, after reuniting, divides a second time into two processes, which are inserted into the sides of the second phalanges. The mode of division of the tendons of the Flexor brevis digitorum, and their insertion into the phalanges, is analogous to the Flexor sublimis in the hand.

*Relations.* By its *superficial surface*, with the plantar fascia. By its *deep surface*, with the Flexor accessorius, the Lumbricales, the tendons of the Flexor longus digitorum, and the external plantar vessels and nerve, from which it is separated by a thin layer of fascia. The *outer* and *inner borders* are separated from the adjacent muscles by means of vertical prolongations of the plantar fascia.

The *Abductor Minimi Digiti* lies along the outer border of the foot. It arises, by a very broad origin, from the outer tuberosity of the os calcis, from the under surface of the os calcis in front of both tubercles, from the plantar fascia, and the intermuscular septum between it and the Flexor brevis digitorum. Its tendon, after gliding over a smooth facet on the under surface of the base of the fifth metatarsal bone, is inserted with the short flexor of the little toe into the outer side of the base of the first phalanx of the little toe.

*Relations.* By its *superficial surface*, with the plantar fascia. By its *deep surface*, with the Flexor accessorius, the Flexor brevis minimi digiti, the long plantar ligament, and Peroneus longus. On its *inner side* are the external plantar vessels and nerve, and it is separated from the Flexor brevis digitorum by a vertical septum of fascia.

*Dissection.* The muscles of the superficial layer should be divided at their origin, by inserting the knife beneath each, and cutting obliquely backwards, so as to detach them from the bone; they should then be drawn forwards, in order to expose the second layer, but not separated at their insertion. The two layers are separated by a thin membrane, the deep plantar fascia, on the removal of which is seen the tendon of the Flexor longus digitorum, with its accessory muscle, the Flexor longus pollicis, and the Lumbricales. The long flexor tendons cross each other at an acute angle, the Flexor longus pollicis running along the inner side of the foot, on a plane superior to that of the Flexor longus digitorum, the direction of which is obliquely outwards.

*Second Layer*

Flexor Accessorius.
Lumbricales.

The *Flexor Accessorius* arises by two heads: the inner or larger, which is muscular, being attached to the inner concave surface of the os calcis, and to the calcaneo-scaphoid ligament; the outer head, flat and tendinous, to the under surface of the os calcis, in front of its outer tuberosity, and to the long plantar ligament: the two portions become united at an acute angle, and are inserted into the outer margin and upper and under surfaces of the tendon of the Flexor longus digitorum, forming a kind of groove, in which the tendon is lodged.

*Relations.* By its *superficial surface*, with the muscles of the superficial layer, from which it is separated by the external plantar vessels and nerves. By its *deep surface*, with the os calcis and long calcaneo-cuboid ligament.

The *Lumbricales* are four small muscles, accessory to the tendons of the Flexor longus digitorum: they arise

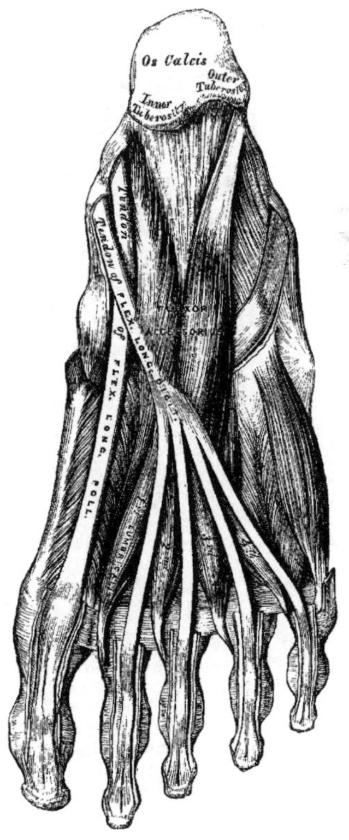


*195 – Muscles of the sole of the foot. Second layer.*

from the tendons of the long flexor, as far back as their angle of division, each arising from two tendons, except the internal one. Each muscle terminates in a tendon, which passes forwards on the inner side of each of the lesser toes, and is inserted into the expansion of the long extensor and base of the second phalanx of the corresponding toe.

*Dissection.* The flexor tendons should be divided at the back part of the foot, and the Flexor accessorius at its origin, and drawn forwards, in order to expose the third layer.

### *Third Layer*

| | |
|---|---|
| Flexor Brevis Pollicis. | Flexor Brevis Minimi Digiti. |
| Adductor Pollicis. | Transversus Pedis. |

The *Flexor Brevis Pollicis* arises, by a pointed tendinous process, from the inner border of the cuboid bone, from the contiguous portion of the external cuneiform, and from the prolongation of the tendon of the Tibialis posticus, which is attached to that bone. The muscle divides, in front, into two portions, which are inserted into the inner and outer sides of the base of the first phalanx of the great toe, a sesamoid bone being developed in each tendon at its insertion. The inner head of this muscle is blended with the Abductor pollicis previous to its insertion; the outer head, with the Adductor pollicis; and the tendon of the Flexor longus pollicis lies in a groove between them.

196 – *Muscles of the sole of the foot. Third layer.*

*Relations.* By its *superficial surface*, with the Abductor pollicis, the tendon of the Flexor longus pollicis and plantar fascia. By its *deep surface*, with the tendon of the Peroneus longus, and metatarsal bone of the great toe. By its *inner border*, with the Abductor pollicis. By its *outer border*, with the Adductor pollicis.

The *Adductor Pollicis* is a large, thick, fleshy mass, passing obliquely across the foot, and occupying the hollow space between the four outer metatarsal bones. It arises from the tarsal extremities of the second, third, and fourth metatarsal bones, and from the sheath of the tendon of the Peroneus longus, and is inserted, together with the outer head of the Flexor brevis pollicis, into the outer side of the base of the first phalanx of the great toe.

The *Flexor Brevis Minimi Digiti* lies on the metatarsal bone of the little toe. It arises from the base of the metatarsal bone of the little toe, and from the sheath of the Peroneus longus; its tendon is inserted into the base of the first phalanx of the little toe, on its outer side.

*Relations.* By its *superficial surface*, with the plantar fascia and tendon of the Abductor minimi digiti. By its *deep surface*, with the fifth metatarsal bone.

The *Transversus Pedis* is a narrow, flat, muscular fasciculus, stretched transversely across the heads of the metatarsal bones, between them and the flexor tendons. It arises from the under surface of the head of the fifth metatarsal bone, and from the transverse ligament of the metatarsus; and is inserted into the outer side of the first phalanx of the great toe; its fibres being blended with the tendon of insertion of the Adductor pollicis.

*Relations.* By its *under surface*, with the tendons of the long and short Flexors and Lumbricales. By its *upper surface*, with the Interossei.

## The Interossei.

The Interossei muscles in the foot are similar to those in the hand. They are seven in number, and consist of two groups, dorsal, and plantar.

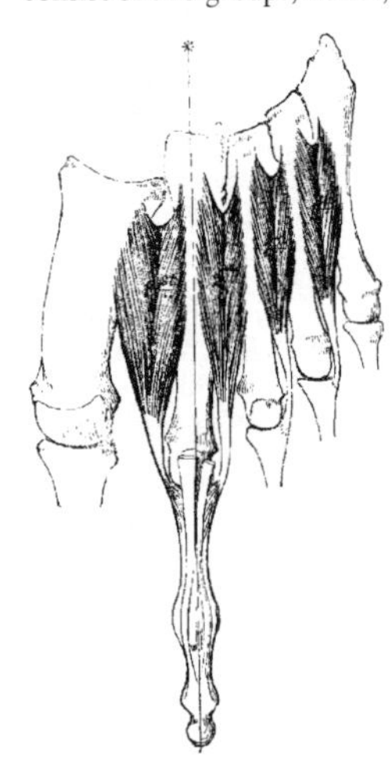

197 – *The dorsal interossei. Left foot.*

The *Dorsal Interossei*, four in number, are situated between the metatarsal bones. They are bipenniform muscles, arising by two heads from the adjacent sides of the metatarsal bones between which they are placed, their tendons being inserted into the bases of the first phalanges, and into the aponeurosis of the common extensor tendon. In the angular interval left between each muscle at its posterior extremity, the perforating arteries pass to the dorsum of the foot; except in the first Interosseous muscle, where the interval allows the passage of the communicating branch of the dorsalis pedis artery. The first Dorsal interosseous muscle is inserted into the inner side of the second toe; the other three are inserted into the outer sides of the second, third, and fourth toes. They are all abductors from an imaginary line or axis drawn through the second toe.

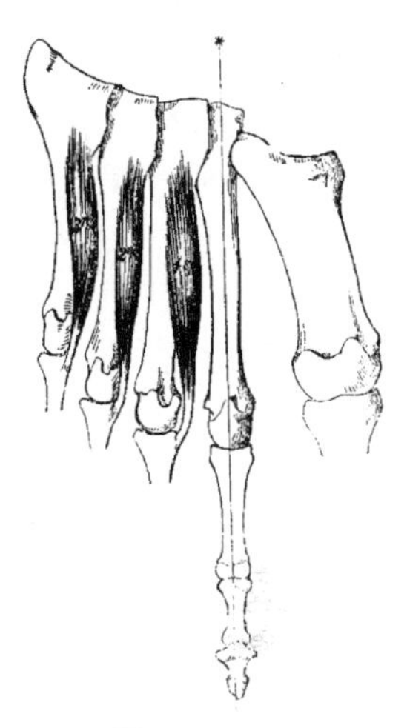

198 – *The plantar interossei. Left foot.*

The *Plantar Interossei*, three in number, lie beneath, rather than between, the metatarsal bones. They are single muscles, and are each connected with but one metatarsal bone. They arise from the base and inner sides of the shaft of the third, fourth, and fifth metatarsal bones, and are inserted into the inner sides of the bases of the first phalanges of the same toes, and into the aponeurosis of the common extensor tendon. These muscles are all adductors towards as imaginary line, extending through the second toe.

*Nerves.* The internal plantar nerve supplies the Abductor pollicis, Flexor brevis digitorum, Flexor brevis pollicis, and the first and second Lumbricales. The external plantar nerve supplies the Abductor minimi digiti, Flexor accessorius, third and fourth Lumbricales, Adductor pollicis, Flexor brevis minimi digiti, Transversus pedis, and all the Interossei.

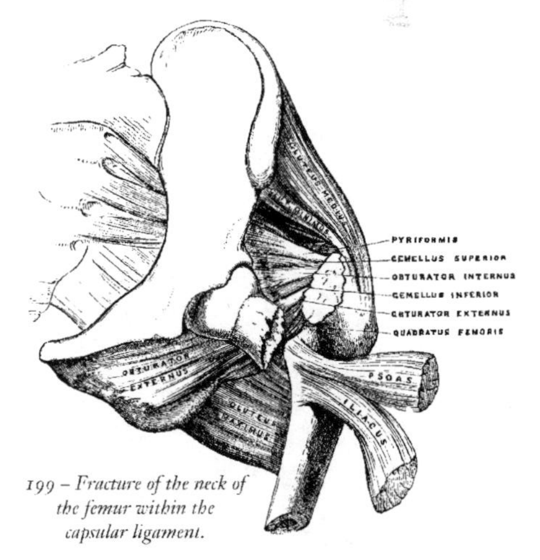


199 – *Fracture of the neck of the femur within the capsular ligament.*

## Surgical Anatomy.

The student should now consider the effects produced by the action of the various muscles in fractures of the bones of the lower extremity. The more common forms of fracture have been especially selected for illustration and description.

Fracture of the *neck of the femur internal to the capsular ligament* (fig.199) is a very common accident, and is most frequently caused by indirect violence, such as slipping off the edge of the kerbstone, the impetus and weight of the body falling upon the neck of the bone. It usually occurs in females, and seldom under fifty years of age. At this period of life, the cancellous tissue of the neck of the bone not unfrequently is atrophied, becoming soft and infiltrated with fatty matter; the compact tissue is partially absorbed: hence the bone is more brittle, and

200 – *Fracture of the femur below the trochanters.*

more liable to fracture. The characteristic marks of this accident are slight shortening of the the limb, and eversion of the foot, neither of which symptoms occur, however, in some cases until some time after the injury. The eversion is caused by the combined action of the external rotator muscles, as well as by the Psoas and Iliacus, Pectineus, Adductors, and Glutei muscles. The shortening and retraction is produced by the action of the Glutei, and by the Rectus femoris in front, and the Biceps, Semi-tendinosus, and Semi-membranosus behind.

Fracture of the *femur just below the trochanters* (fig. 200) is an accident of not unfrequent occurrence, and is attended with great displacement producing considerable deformity. The upper fragment, the portion chiefly displaced, is tilted forwards almost at right angles with the pelvis, by the combined action of the Psoas and Iliacus; and, at the same time, everted and drawn outwards by the external rotator and Glutei muscles, causing a marked prominence at the upper and outer side of the thigh, and much pain from the bruising and laceration of the muscles. The limb is shortened, from the lower fragment being drawn upwards by the Rectus in front, and the Biceps, Semi-membranosus, and Semi-tendinosus behind; and, at the same time, everted, and the upper end thrown outwards, the lower inwards, by the Pectineus and Adductor muscles. This fracture may be reduced in two different methods: either by direct relaxation of all the opposing muscles, to effect which, the limb should be placed on a double inclined plane; or by overcoming the contraction of the muscle by continued extension, which may be effected by means of the long splint.

Oblique fracture of the femur *immediately above the condyles* (fig. 201) is a formidable injury, and attended with considerable displacement. On examination of the limb, the lower fragment may be felt deep in the popliteal space, being drawn backwards by the Gastrocnemius, Soleus, and Plantaris muscles; and upwards by the posterior femoral, and Rectus muscles. The pointed end of the apper fragment is drawn inwards by the Pectineus and Adductor muscles, and tilted forwards by the Psoas and Iliacus, piercing the Rectus muscle, and, occasionally, the integument. Relaxation of these muscles, and direct approximation of the broken fragments is effected by placing the limb on a double inclined plane. The greatest care is requisite in keeping the pointed extremity of the upper fragment in proper apposition; otherwise, after union of the fracture, extension of the limb is partially destroyed, from the Rectus muscle being held down by the fractured end of the bone, and from the patella, when elevated, being drawn upwards against it.

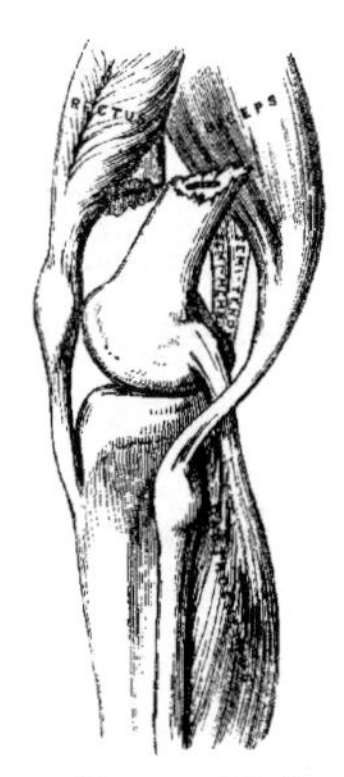

201 – *Fracture of the femur above the condyles.*

202 – *Fracture of the patella.*

Fracture of the *patella* (fig. 202), may be produced by muscular action, or by direct violence. When produced by muscular action, it occurs thus: a person in danger of falling forwards, attempts to recover himself by throwing the body backwards, and the violent action of the Quadriceps extensor upon the patella snaps that bone transversely across. The upper fragment is drawn up the thigh by the Quadriceps extensor, the lower fragment being retained in its position by the ligamentum patellæ; the extent of separation of the two fragments depending upon the degree of laceration of the ligamentous structures around the bone. The patient is totally unable to straighten the limb; the prominence of the patella is lost; and a marked but varying interval can be felt between the fragments. The treatment consists in relaxing the opposing muscles, which may be effected by raising the trunk, and slightly elevating the limb, which should be kept in a straight position. Union is usually ligamentous. In fracture from direct violence, the bone is generally comminuted, or fractured obliquely or perpendicularly.

Oblique fracture of the *shaft of the tibia* (fig. 203), usually occurs at the lower fourth of the bone, this being the narrowest and weakest part, and is usually accompanied with fracture of the fibula. If the fracture has taken place obliquely from above, downwards, and forwards the fragments ride over one another, the lower fragment being drawn backwards and upwards by the powerful action of the muscles of the calf; the pointed extremity of the upper fragment projects forwards immediately beneath the integument, often protruding through it, and rendering the fracture a compound one. If the direction of the fracture is the reverse of that shown in the figure, the pointed extremity of the lower fragment projects forwards, riding upon the lower end of the upper one. By bending the knee, which relaxes the opposing muscles, and making extension from the knee and ankle, the fragments may be brought into apposition. It is often necessary, however, in compound fracture, to remove a portion of the projecting bone with the saw before complete adaptation can be effected.

Fracture of the *fibula, with displacement of the tibia* (fig. 204), commonly known as 'Pott's Fracture,' is one of the most frequent injuries of the ankle-joint. The end of the tibia is displaced from the corresponding surface of the astragalus; the internal lateral ligament is ruptured; and the inner malleolus projects inwards beneath the

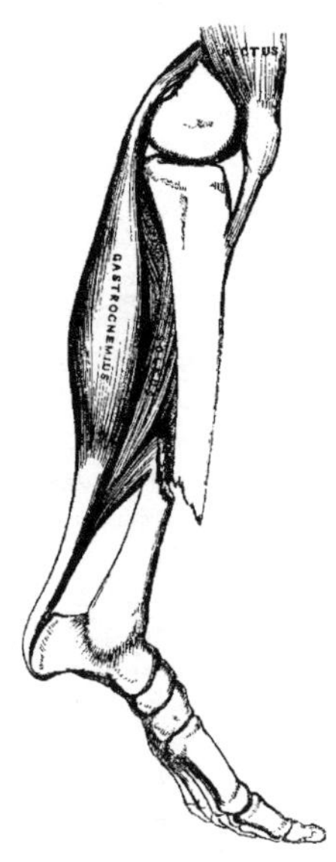


203 – *Oblique fracture of the shaft of the tibia.*

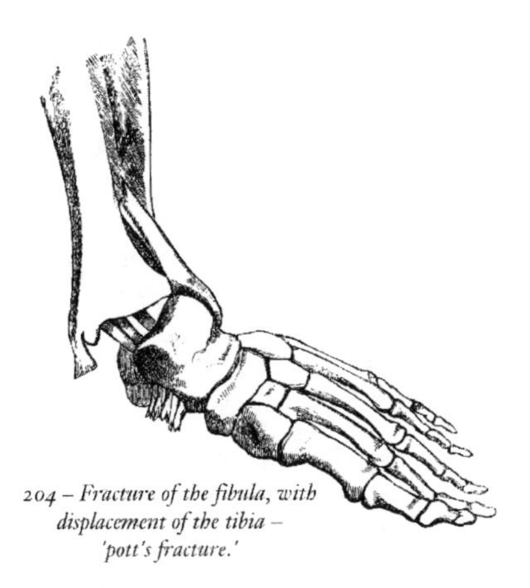

204 – *Fracture of the fibula, with displacement of the tibia – 'pott's fracture.'*

integument, which is tightly stretched over it, and in danger of bursting. The fibula is broken, usually from two to three inches above the ankle, and occasionally that portion of the tibia with which it is more directly connected below; the foot is everted by the action of the Peroneus longus, its inner border resting upon the ground, and, at the same time, the heel is drawn up by the muscles of the calf. This injury may be at once reduced by flexing the leg at right angles with the thigh, which relaxes all the opposing muscles, and by making slight extension from the knee and ankle.

---

For a detailed account of the Minute Anatomy of Muscle, reference should be made to the following sources of information:—Quain's 'Elements of Anatomy.'—Kölliker's 'Handbook of Human Microscopic Anatomy,' before alluded to.—Todd and Bowman's 'Physiological Anatomy.'—To the article, 'Muscle and Muscular Motion,' by W. Bowman, in the Cyclopedia of Anatomy; and 'On the Minute Structure of Voluntary Muscle,' by the same author, in the Phil. Trans. 1840, 1841.

On the Descriptive Anatomy of the Muscles, refer to Cruvelhier's 'Anatomie Descriptive.'—Traite de Myologie et d'Angeiologie, by F. G. Theile, 'Encyclopédie Anatomique,' Paris, 1843; and Henle's 'Handbuch der Systematischen Anatomie,' before alluded to.

# Of the Arteries.

The Arteries are cylindrical tubular vessels, which serve to convey blood from both ventricles of the heart to every part of the body. These vessels were named arteries (*ἀὴρ τηρεῖν, to contain air*), from the belief entertained by the ancients that they contained air. To Galen is due the honour of refuting this opinion; he showed that these vessels, though for the most part empty after death contained blood in the living body.

The pulmonary artery, which arises from the right ventricle of the heart, carries venous blood directly into the lungs, from whence it is returned by the pulmonary veins into the left auricle. This constitutes the lesser or pulmonic circulation. The great artery, which arises from the left ventricle, the aorta, conveys arterial blood to the body generally; from whence it is brought back to the right side of the heart by means of the veins. This constitutes the greater or systemic circulation.

The distribution of the systemic arteries is like a highly ramified tree, the common trunk of which, formed by the aorta, commences at the left ventricle of the heart, the smallest ramifications corresponding to the circumference of the body and the contained organs. The arteries are found in nearly every part of the animal body, with the exception of the hairs, nails, epidermis, cartilages, and cornea; and the larger trunks usually occupy the most protected situations, running, in the limbs, along the flexor side, where they are less exposed to injury.

There is considerable variation in the mode of division of the arteries; occasionally a short trunk subdivides into several branches at the same point, as we observe in the cœliac and thyroid axes; or the vessel may give off several branches in succession, and still continue as the main trunk, as is seen in the arteries of the limbs; but the usual division is dichotomous, as, for instance, the aorta dividing into the two common iliacs; and the common carotid, into the external and internal.

The branches of arteries arise at very variable angles; some, as the superior intercostal arteries, arise from the aorta at an obtuse angle; others, as the lumbar arteries, at a right angle; or, as the spermatic, at an acute angle. An artery from which a branch is given off, is smaller in size, but retains a uniform diameter until a second branch is derived from it. A branch of an artery is smaller than the trunk from which it arises; but if an artery divides into two branches, the combined area of the two vessels is, in nearly every instance, somewhat greater than that of the trunk; and the combined area of all the arterial branches greatly exceeds the area of the aorta; so that the arteries collectively may be regarded as a cone, the apex of which corresponds to the aorta; the base, to the capillary system.

The arteries, in their distribution, communicate freely with one another, forming what is called an *anastomosis* (*ἀνὰ, between; στόμα, mouth*), or inosculation; and this communication is very free between the large, as well as between the smaller branches. The anastomoses between trunks of equal size is found where great freedom and activity of the circulation is requisite, as in the brain; here, the two vertebral arteries unite to form the basilar, and the two internal carotid arteries are connected by a short inter-communicating trunk; it is also found in the abdomen, the intestinal arteries having very free anastomoses between their larger branches. In the limbs, the anastomoses are most frequent and of largest size around the joints; the branches of an artery above, freely inosculating with branches from the vessel below; these anastomoses are of considerable interest to the surgeon, as it is by their enlargement that a collateral circulation is established after the application of a ligature to an artery for the cure of aneurism. The smaller branches of arteries anastomose more frequently than the larger; and between the smallest twigs, these inosculations become so numerous as to constitute a close network that pervades nearly every tissue of the body.

Throughout the body generally, the larger arterial branches pursue a perfectly straight course; but in certain situations they are tortuous: thus, the facial artery in its course over the face, and the labial arteries of the lips, are extremely tortuous in their course, to accommodate themselves to the movements of these parts. The uterine arteries are also tortuous, to accommodate themselves to the increase of size which this organ undergoes during pregnancy. Again, the internal carotid and vertebral arteries, previous to their entering the cavity of the skull, describe a series of curves, which are evidently intended to diminish the velocity of the current of blood, by increasing the extent of surface over which it moves, and adding to the amount of impediment which is produced from friction.

The arteries are dense in structure, of considerable strength, highly elastic, and, when divided, they preserve, although empty, their cylindrical form.

The arteries are composed of three coats,—internal, middle, and external.

The *internal*, the thinnest, consists usually of two layers, an inner or epithelial; and an outer, elastic. The former consists of a single layer of fusiform-shaped epithelial cells with round or oval nuclei. The latter is a transparent, colourless, shining membrane, perforated with small elongated apertures (hence the name, *fenestrated*), and marked with numerous reticulations. This layer is perfectly smooth when the artery is distended; but when empty, presents numerous longitudinal and transverse folds.

In arteries above the size of the capillaries, the elastic layer is very delicate, and the epithelium clearly demonstrable.

In arteries of less than a line in diameter, the internal coat consists of two layers, as above described; but in middle-sized arteries, several lamellæ composed of elastic fibres and connective tissue, are interposed between the epithelial and elastic coats. In the largest arteries, the inner coat is usually much thickened, especially in the aorta; and consists of a homogeneous substance, occasionally striated or fibrillated, traversed by longitudinal elastic networks, which are very fine in the lamellæ immediately beneath the epithelium, but increase in thickness from within outwards. The internal and middle coats are separated, by either a dense elastic reticulated coat, or a true fenestrated membrane.

The *middle coat*, thicker than the preceding, consists of muscular and elastic fibres, and connective tissue, disposed chiefly in the transverse direction. In the largest arteries, this coat is of great thickness, of a yellow colour, and highly elastic; it diminishes in thickness, and becomes redder in colour as the arteries become smaller; becomes very thin, and finally disappears. In small arteries, this coat is purely muscular, consisting of muscular fibre-cells united to form lamellæ which vary in number according to the size of the arteries, the very small arteries having only a single layer, and those not larger than the $\frac{1}{10}$th of a line three or four layers. In arteries of medium size, this coat becomes thicker in proportion with the size of the vessel; its layers of muscular tissue are more numerous, and intermixed with numerous fine elastic fibres which unite to form broad meshed networks. In the larger vessels of this class, as the femoral, superior mesenteric, cœliac, external iliac, brachial, and popliteal arteries, the elastic fibres unite to form lamellæ, which alternate with the layers of muscular fibre. In the largest arteries, the muscular tissue is only slightly developed, and forms about one-third, or one-fourth of the whole substance of the middle coat; this is especially the case in the aorta, and trunk of the pulmonary artery, in which the individual cells of the muscular layer are imperfectly formed; while, in the carotid, axillary, iliac, and subclavian arteries, the muscular tissue of the middle coat is more developed. The elastic lamellæ are well marked, may amount to fifty or sixty in number, and alternate regularly with the layers of muscular fibre. They are most distinct, and arranged with most regularity in the abdominal aorta, innominate artery, and common carotid.

The *external*, or areolar and elastic coat, the thickest of the three, consists of connective tissue, and elastic fibres. It is very thin in the largest arteries; but in those of medium size, and in small arteries, is as thick, or thicker, than the middle coat. In small arteries, this coat consists of connective tissue, and fine elastic fibres. In arteries just above the capillaries, the elastic fibres are wanting; the connective tissue composing the coat becoming, the nearer it approaches the capillaries, more homogeneous, being gradually reduced to a thin membranous envelope which finally disappears. In arteries of medium size, this coat is composed of two distinct layers, an inner composed of elastic tissue; the outer, composed of connective tissue, containing elastic networks irregularly connected with each other. The inner elastic layer is very distinct in the carotid, femoral, brachial, profunda, mesenteric, and cœliac arteries, the elastic fibres being often arranged in lamellæ.

Some arteries have extremely thin coats in proportion to their size; this is especially the case in those situated in the cavity of the cranium and spinal canal, the difference depending upon the greater thinness of the external and middle coats.

The arteries, in their distribution throughout the body, are included in a thin areolo-fibrous investment, which forms what is called their sheath. In the limbs, this is usually formed by a prolongation of the deep fascia; in the upper part of the thigh, it consists of a continuation downwards of the transversalis and iliac fasciæ of the abdomen; in the neck, of a prolongation of the deep cervical fascia. The included vessel is loosely connected with its sheath by a delicate areolar tissue;

and the sheath usually encloses the accompanying veins, and sometimes a nerve. Some arteries, as those in the cranium, are not included in sheaths.

All the larger arteries are supplied with blood-vessels like the other organs of the body; they are called *vasa vasorum*. These nutrient vessels arise from a branch of the artery or from a neighbouring vessel, at some considerable distance from the point at which they are distributed; they ramify in the loose areolar tissue connecting the artery with its sheath, and are distributed to the external and middle coats, and, according to Arnold and others, supply the internal coat. Minute veins serve to return the blood from these vessels; they empty themselves into the venæ comites in connection with the artery. Arteries are also provided with nerves; they are derived chiefly from the sympathetic, but partly from the cerebro-spinal system. They form intricate plexuses upon the surface of the larger trunks, the smaller branches being usually accompanied by single filaments; their exact mode of distribution is unknown. According to Kölliker, the majority of the arteries of the brain and spinal cord, those of the choroid, of the placenta, as well as many arteries of muscles, glands, and membranes, are unprovided with them.

The smaller arterial branches (excepting those of the cavernous structures of the sexual organs, and in the uterine placenta), terminate in a network of vessels which pervade nearly every tissue of the body. These vessels, from their minute size, are termed capillaries (*capillus*, 'a hair'). They are interposed between the smallest branches of the arteries and the commencing veins, constituting a network, the branches of which maintain the same diameter throughout, the meshes of the network being more uniform in shape and size than those formed by the anastomoses of the small arteries and veins.

The *diameter* of the capillaries varies in the different tissues of the body, their usual size being about $\frac{1}{3000}$th part of an inch. The smallest are those of the brain, and the mucous membrane of the intestines; the largest, those of the skin, and the marrow of bones.

The *form* of the capillary net varies in the different tissues, being modifications chiefly of rounded or elongated meshes. The *rounded form of mesh* is most common, and prevails where there is a dense network, as in the lungs, in most glands and mucous membranes, and in the cutis; the meshes being more or less angular, sometimes nearly quadrangular, or polygonal; more frequently, irregular. *Elongated meshes* are observed in the bundles of fibres and tubes composing muscles and nerves, the meshes being usually of a parallelogram form, the long axis of the mesh running parallel with the long axis of the nerve or fibre. Sometimes, the capillaries have a *looped* arrangement, a single vessel projecting from the common network, and returning after forming one or more loops, as in the papillæ of the tongue and skin.

The number of the capillaries, and the size of the meshes, determines the degree of vascularity of a part. The closest network, and the smallest interspaces, are found in the lungs and in the choroid coat of the eye. In the liver and lung, the interspaces are smaller than the capillary vessels themselves. In the kidney, in the conjunctiva, and in the cutis, the interspaces are from three to four times as large as the capillaries which form them; and from eight to ten times as large as the capillaries of the brain in their long diameter, and from four to six times as large in their transverse diameter. In the cellular coat of the arteries, the width of the meshes is ten times that of the capillary vessels. As a general rule, the more active the function of an organ is, the closer is its capillary net, and the larger its supply of blood. The network being very narrow in all growing parts, in the glands, and in the mucous membranes; wider in bones and ligaments, which are comparatively inactive; and nearly altogether absent in tendons and cartilages, in which very little organic change occurs after their formation.

*Structure.* The walls of the capillaries consist of a fine, transparent, homogeneous membrane, in which are embedded, at intervals, minute oval corpuscles, probably the remains of the nuclei of the cells from which the vessel was originally formed. The largest capillaries have a trace of an epithelial lining, and a few filaments circularly dispersed.

In the description of the arteries, we shall first consider the efferent trunk of the systemic circulation, the aorta, and its branches; and then the efferent trunk of the pulmonic circulation, the pulmonary artery.

## The Aorta.

The *aorta* (ἀορτή); *arteria magna*) is the main trunk of a series of vessels, which, arising from the heart, conveys the red oxygenated blood to every part of the body for its nutrition. This vessel commences at the upper part of the left ventricle, and after ascending for a short distance, arches backwards to the left side, over the root of the left lung, descends within the thorax on the left side of the vertebral column, passes through the aortic opening in the Diaphragm, and entering the abdominal cavity, terminates, considerably diminished in size, opposite the fourth lumbar vertebra, where it divides into the right and left common iliac arteries. Hence its subdivision into the arch of the aorta, the thoracic aorta and the abdominal aorta, from the direction or position peculiar to each part.

## Arch of the Aorta.

*Dissection.* In order to examine the arch of the aorta, the thorax should be opened, by dividing the cartilages of the ribs on each side of the sternum, and raising this bone from below upwards, and then sawing through the sternum on a level with its articulation with the clavicle. By this means, the relations of the large vessels to the upper border of the sternum and root of the neck are kept in view.

The arch of the aorta extends from the origin of the vessel at the upper part of the left ventricle, to the lower border of the body of the third dorsal vertebra. At its commencement, it ascends behind the sternum, obliquely upwards and forwards towards the right side, and opposite the upper border of the second costal cartilage of the right side, passes transversely from right to left, and from before backwards to the left side of the second dorsal vertebra; it then descends upon the left side of the body of the third dorsal vertebra, at the lower border of which it becomes the thoracic aorta. Hence this portion of the vessel is divided into an ascending, a transverse, and a descending portion. The artery in its course describes a curve, the convexity of which is directed upwards and to the right side, the concavity in the opposite direction.

## Ascending Part of the Arch.

The ascending portion of the arch of the aorta is about two inches in length. It commences at the upper part of the left ventricle, in front of the left auriculo-ventricular orifice, and opposite the

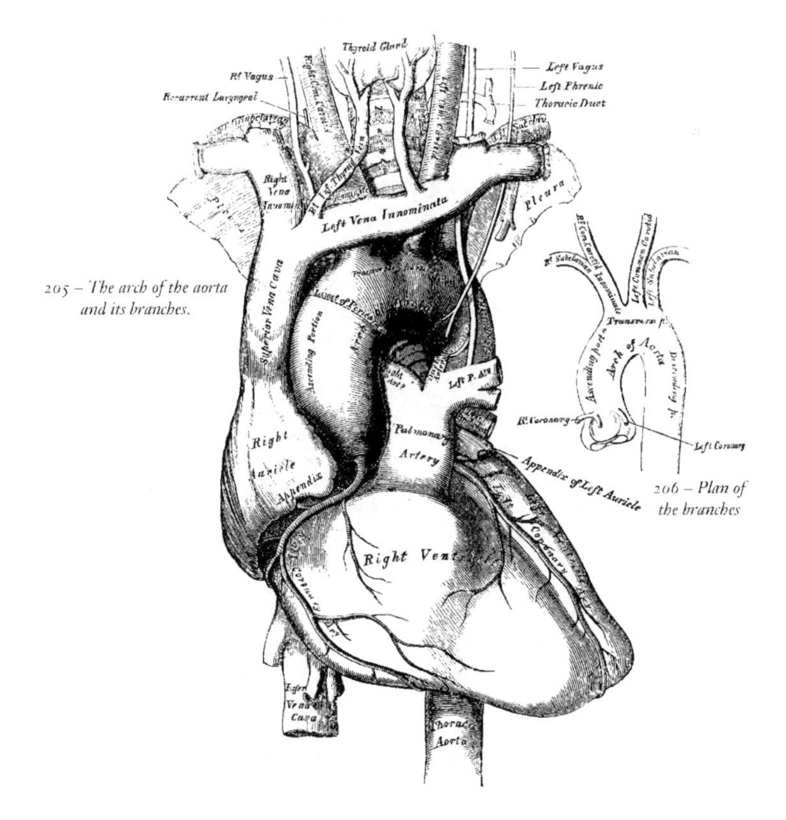


205 – *The arch of the aorta and its branches.*

206 – *Plan of the branches*

middle of the sternum on a line with its junction to the third costal cartilage; it passes obliquely upwards in the direction of the heart's axis, to the right side, as high as the upper border of the second costal cartilage, describing a slight curve in its course, and being situated, when distended, about a quarter of an inch behind the posterior surface of the sternum. A little above its commencement, it is somewhat enlarged, and presents three small dilatations, called the *sinuses of the aorta* (sinuses of Valsalva) opposite to which are attached the three semi-lunar valves, which serve the purpose of preventing any regurgitation of blood into the cavity of the ventricle. A section of the aorta opposite this part has a somewhat triangular figure; but below the attachment of the valves it is circular. This portion of the arch is contained in the cavity of the pericardium, and together with the pulmonary artery, is invested in a tube of serous membrane, continued on to them from the surface of the heart.

*Relations.* The ascending part of the arch is covered at its commencement by the trunk of the pulmonary artery and the right auricular appendage, and, higher up, is separated from the sternum by the pericardium, some loose areolar tissue, and the remains of the thymus gland; *behind*, it rests upon the right pulmonary vessels and root of the right lung. On the *right side*, it is in relation with the superior vena cava and right auricle; on the *left side*, with the pulmonary artery.

## Plan of the Relations of the ascending part of the Arch.

*In front.*
Pulmonary artery.
Right auricular appendage.
Pericardium.
Remains of thymus gland.

*Right side.*
Superior cava.
Right auricle.

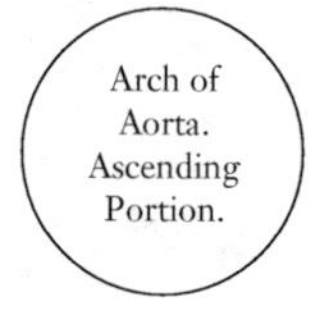


*Left side.*
Pulmonary artery.

*Behind.*
Right pulmonary vessels.
Root of right lung.

## Transverse Part of the Arch.

The second or transverse portion of the arch commences at the upper border of the second costo-sternal articulation of the right side in front, and passes from right to left, and from before backwards, to the left side of the second dorsal vertebra behind. Its upper border is usually about an inch below the upper margin of the sternum.

*Relations.* Its *anterior surface* is covered by the left pleura and lung, and crossed towards the left side by the left pneumogastric and phrenic nerves, and cardiac branches of the sympathetic. Its *posterior surface* lies on the trachea, just above its bifurcation, on the great cardiac plexus, the œsophagus, thoracic duct, and left recurrent laryngeal nerve. Its *upper border* is in relation with the left innominate vein; and from its upper part are given off the innominate, left carotid, and left subclavian arteries. By its *lower border*, with the bifurcation of the pulmonary artery, and the remains of the ductus arteriosus, which is connected with the left division of that vessel; the left recurrent laryngeal nerve winds round it from before backwards, whilst the left bronchus passes below it.

## Plan of the Relations of the Transverse Part of the Arch.

*Above.*
Left innominate vein.
Arteria innominata.
Left carotid.
Left subclavian.

*In front.*
Left pleura and lung.
Left pneumogastric nerve.
Left phrenic nerve.
Cardiac nerves.

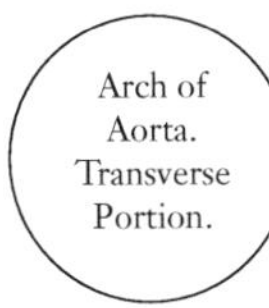


*Behind.*
Trachea.
Cardiac plexus.
Œsophagus.
Thoracic duct.
Left recurrent nerve.

*Below.*
Bifurcation of pulmonary artery.
Remains of ductus arteriosus.
Left recurrent nerve.
Left bronchus.

## Descending Part of the Arch.

The descending portion of the arch has a straight direction, inclining downwards on the left side of the body of the third dorsal vertebra, at the lower border of which it becomes the thoracic aorta.

*Relations.* Its *anterior surface* is covered by the pleura and root of the left lung; *behind*, it lies on the left side of the body of the third dorsal vertebra. On its *right side* lies the œsophagus and thoracic duct; on its *left side* it is covered by the pleura.

## Plan of the Relations of the Descending Part of the Arch.

*In front.*
Pleura.
Root of left lung.

*Right side.*
Œsophagus.
Thoracic duct.

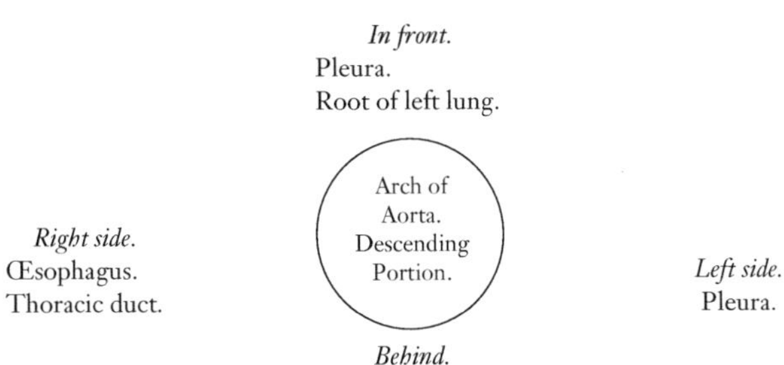


*Left side.*
Pleura.

*Behind.*
Left side of body of third dorsal vertebra.

The ascending, transverse, and descending portions of the arch vary in position according to the movements of respiration, being lowered, together with the trachea, bronchi, and pulmonary vessels, during inspiration by the descent of the Diaphragm, and elevated during expiration, when the Diaphragm ascends. These movements are greater in the ascending than the transverse, and in the latter than the descending part.

*Peculiarities.* The height to which the aorta rises in the chest is usually about an inch below the upper border of the sternum; but it may ascend nearly to the top of that bone Occasionally it is found an inch and a half, more rarely, three inches below this point.

*Direction.* Sometimes the aorta arches over the root of the right instead of the left lung as in birds, and passes down on the right side of the spine. In such cases, all the viscera of the thoracic and abdominal cavities are transposed. Less frequently, the aorta, after arching over the root of the right lung, is directed to its usual position on the left side of the spine, this peculiarity not being accompanied by any transposition of the viscera.

*Conformation.* The aorta occasionally divides into an ascending and a descending trunk, as in some quadrupeds, the former being directed vertically upwards, subdivides into three branches, to supply the head and upper extremities. Sometimes the aorta subdivides soon after its origin into two branches, which soon reunite. In one of these cases, the œsophagus and trachea were found to pass through the interval left by their division; this is the normal condition of the vessel in the reptilia.

*Surgical Anatomy.* Of all the vessels of the arterial system, the aorta, and more especially its arch, is most frequently the seat of disease; hence it is important to consider some of the consequences that may ensue from aneurism of this part.

It will be remembered, that the ascending part of the arch is contained in the pericardium, just behind the sternum, its commencement being crossed by the pulmonary artery and right auricular appendage, having the root of the right lung behind, the vena cava on the right side, and the pulmonary artery and left auricle on the left side.

Aneurism of the ascending aorta, in the situation of the aortic sinuses, in the great majority of cases, affects the right coronary sinus; this is mainly owing to the regurgitation of blood upon the sinuses, taking place chiefly on the right anterior aspect of the vessel. As the aneurismal sac enlarges, it may compress any or all of the structures in immediate proximity with it, but chiefly projects towards the right anterior side; and, consequently, interferes mainly with those structures that have a corresponding relation with the vessel. In the majority of cases, it bursts in the cavity of the pericardium, the patient suddenly drops down dead, and, upon a post-mortem examination, the pericardial bag is found full of blood: or it may compress the right auricle, or the pulmonary artery, and adjoining part of the right ventricle, and open into one or the other of these parts, or it may compress the superior cava.

Aneurism of the ascending aorta, originating above the sinuses, most frequently implicates the right anterior wall of the vessel; this is probably mainly owing to the blood being impelled against this part. The direction of the aneurism is also chiefly towards the right of the median line. If it attains a large size and projects forwards, it may absorb the sternum and the cartilages of the ribs, usually on the right side, and appear as a pulsating tumour on the front of the chest, just below the manubrium; or it may burst into the pericardium, may compress or even open into the right lung, the trachea, bronchi, or œsophagus.

Regarding the transverse part of the arch, the student is reminded that the vessel lies on the trachea, the œsophagus, and thoracic duct; that the recurrent laryngeal nerve winds around it; and that from its upper part are given off three large trunks, which supply the head, neck, and upper extremities. Now an aneurismal tumour taking origin from the posterior part or right aspect of the vessel, its most usual site, may press upon the trachea, impede the breathing, or produce cough, hæmoptysis, or stridulous breathing, or it may ultimately burst into that tube, producing fatal hæmorrhage. Again, its pressure on the laryngeal nerves may give rise to symptoms which so accurately resemble those of laryngitis, that the operation of tracheotomy has in some cases been resorted to from the supposition that disease existed in the larynx; or it may press upon the thoracic duct, and destroy life by inanition; or it may involve the œsophagus, producing dysphagia; or may burst into this tube, when fatal hæmorrhage will occur. Again, the innominate artery, or the left carotid, or subclavian, may be so obstructed by clots, as to produce a weakness, or even a disappearance, of the pulse in one or the other wrist; or the tumour may present itself at or above the manubrium, generally either in the median line, or to the right of the sternum.

Aneurism affecting the descending part of the arch is usually directed backwards and to the left side, causing absorption of the vertebræ and corresponding ribs; or it may press upon the trachea, left bronchus, œsophagus, and the right and left lungs, generally the latter: when rupture of the sac occurs, this usually takes place into the left pleural cavity; less frequently into the left bronchus, the right pleura, or into the substance of the lungs or trachea. In this form of aneurism, pain is almost a constant and characteristic symptom, existing either in the back or chest, and usually radiating from the spine around the left side. This symptom depends upon the aneurismal sac compressing the intercostal nerves against the bone.

## Branches of the Arch of the Aorta (Figs. 205, 206).

The branches given off from the arch of the aorta are five in number. Two of small size from the ascending portion, the right and left coronary; and three of large size from the transverse portion, the innominate artery, the left carotid, and the left subclavian.

*Peculiarities. Position of the Branches.* The branches, instead of arising from the highest part of the arch (their usual position), may be moved more to the right, arising from the commencement of the transverse or upper part of the ascending portion; or the distance from one another at their origin may be increased or diminished, the most frequent change in this respect being the approximation of the left carotid, towards the innominate artery.

*The Number* of the primary branches may be reduced to two: the left carotid arising from the innominate artery; or (more rarely), the carotid and subclavian arteries of the left side arising from a left innominate artery. But the number may be increased to four, from the right carotid and subclavian arteries arising directly from the aorta, the innominate being absent. In most of these latter cases, the right subclavian arose from the left end of the arch; in other cases, it was the second or third branch given off instead of the first. Lastly, the number of trunks from the arch may be increased to five or six; in these instances, the external and internal carotids arose separately from the arch, the common carotid being absent on one or both sides.

*Number usual, Arrangement different.* When the aorta arches over to the right side, the three branches have an arrangement the reverse of what is usual, the innominate supplying the left side; and the carotid and subclavian (which arise separately) the right side. In other cases, where the aorta takes its usual course, the two carotids may be joined in a common trunk, and the subclavians arise separately from the arch, the right subclavian generally arising from the left end of the arch.

*Secondary Branches* sometimes arise from the arch; most commonly it is the left vertebral, which usually takes origin between the left carotid, and left subclavian, or beyond them. Sometimes, a thyroid branch is derived from the arch, or the right internal mammary, or left vertebral, or, more rarely, both vertebrals.

## The Coronary Arteries.

The coronary arteries supply the heart; they are two in number, right and left, arising near the commencement of the aorta immediately above the free margin of the semi-lunar valves.

The *Right Coronary Artery*, about the size of a crow's quill, arises from the aorta immediately above the free margin of the right semi-lunar valve, between the pulmonary artery, and the appendix of the right auricle. It passes forwards to the right side in the groove between the right auricle and ventricle, and curving around the right border of the heart, runs along its posterior surface as far as the posterior inter-ventricular groove, where it divides into two branches, one of which continues onwards in the groove between the left auricle and ventricle, and anastomoses with the left coronary; the other descends along the posterior inter-ventricular furrow, supplying branches to both ventricles, and to the septum; anastomosing at the apex of the heart with the descending branch of the left coronary.

This vessel sends a large branch along the thin margin of the right ventricle to the apex, and numerous small branches to the right auricle and ventricle, and commencement of the pulmonary artery.

The *Left Coronary*, smaller than the former, arises immediately above the free edge of the left semi-lunar valve, a little higher than the right; it passes forwards between the pulmonary artery and the left appendix auriculæ, and descends obliquely towards the anterior inter-ventricular groove, where it divides into two branches. Of these, one passes transversely outwards in the left auriculo-ventricular groove, and winds around the left border of the heart to its posterior surface, where it anastomoses with the superior branch of the right coronary; the other descends along the anterior inter-ventricular groove to the apex of the heart, where it anastomoses with the descending branch of the right coronary. The left coronary supplies the left auricle and its appendix, both ventricles, and numerous small branches to the pulmonary artery, and commencement of the aorta.

*Peculiarities.* These vessels occasionally arise by a common trunk, or their number may be increased to three; the additional branch being of small size. More rarely, there are two additional branches.

## Arteria Innominata.

The innominate artery is the largest branch given off from the arch of the aorta. It arises from the commencement of the transverse portion in front of the left carotid, and, ascending obliquely to the upper border of the right sterno-clavicular articulation, divides into the right carotid and subclavian arteries. This vessel varies from an inch and a half to two inches in length.

*Relations. In front*, it is separated from the first bone of the sternum by the Sterno-hyoid and Sterno-thyroid muscles, the remains of the thymus gland, and by the left innominate and right inferior thyroid veins which cross its root. *Behind*, it lies upon the trachea which it crosses obliquely. On the *right side* is the right vena innominata, right pneumogastric nerve, and the pleura; and on the *left side*, the remains of the thymus gland, and origin of the left carotid artery.

## Plan of the Relations of the Innominate Artery.

*In front.*

Sternum.
Sterno-hyoid and Sterno-thyroid.
Remains of thymus gland.
Left innominate and right inferior thyroid veins.

| *Right side.* | | *Left side.* |
|---|---|---|
| Right vena innominata. | Innominate Artery. | Remains of thymus. |
| Right pneumogastric nerve. | | Left carotid. |
| Pleura. | | |

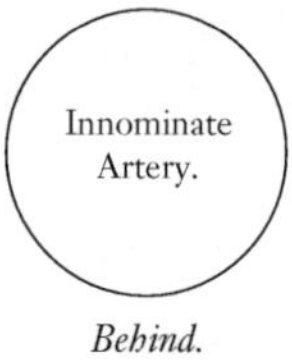

*Behind.*
Trachea.

*Peculiarities in point of division.* When the bifurcation of the innominate artery varies from the point above mentioned, it sometimes ascends a considerable distance above the sternal end of the clavicle; less frequently it divides below it. In the former class of cases, its length may exceed two inches; and, in the latter, be reduced to an inch or less. These are points of considerable interest for the surgeon to remember in connection with the operation of including this vessel in a ligature.

*Branches.* The arteria innominata occasionally supplies a thyroid branch (middle thyroid artery), which ascends along the front of the trachea to the thyroid gland; and sometimes, a thymic or bronchial branch. The left carotid is frequently joined with the innominate artery at its origin. Sometimes, there is no innominate artery, the right subclavian arising directly from the arch of the aorta.

*Position.* When the aorta arches over to the right side, the innominate is directed to the left side of the neck, instead of the right.

*Surgical Anatomy.* Although the operation of tying the innominate artery, has been performed by several surgeons, for aneurism of the right subclavian extending inwards as far as the Scalenus, in no instance has it been attended with success. An important fact has, however, been established; viz., that the circulation in the parts supplied by the artery, can be supported after the operation; a fact which cannot but encourage surgeons to have recourse to it whenever the urgency of the case may require it, notwithstanding that it must be regarded as peculiarly hazardous.

The failure of the operation in those cases where it has been performed, has depended on subsequent repeated secondary hæmorrhage, or on inflammation of the adjoining pleural sac and lung. The main obstacles to its performance are, as the student will perceive from his dissection of this vessel, its deep situation behind and beneath the sternum, and the number of important structures which surround it in every part.

In order to apply a ligature to this vessel, the patient is placed upon his back, with the shoulders raised, and the head bent a little backwards, so as to draw out the artery from behind the sternum into the neck. An incision two inches long is then made along the anterior border of the Sterno-mastoid muscle, terminating at the sternal end of the clavicle. From this point, a second incision is to be carried about the same length along the upper border of the clavicle. The skin is to be dissected back, and the Platysma being exposed, must be divided on a director: the sternal end of the Sterno-mastoid is now brought into view, and a director being passed beneath it, and close to its under surface, so as to avoid any small vessels, it must be divided transversely throughout the greater part of its attachment. Pressing aside any loose cellular tissue or vessels that may now appear, the Sterno-hyoid and Sterno-thyroid muscles will be exposed, and must be divided, a director being previously passed beneath them. The inferior thyroid veins now come into view, and must be carefully drawn either upwards or downwards, by means of a blunt hook. On no account should these vessels be divided, as it would add much to the difficulty of the operation, and endanger its ultimate success. After tearing through a strong fibro-cellular lamina, the right carotid is brought into view, and being traced downward, the arteria innominata is arrived at. The left vena innominata should now be depressed, the right vena innominata, the internal jugular vein, and pneumogastric nerve drawn to the right side; and a curved aneurism needle may then be passed around the vessel, close to its surface, and in a direction from below upwards and inwards; care being taken to avoid the right pleural sac, the trachea, and cardiac nerves. The ligature should be applied to the artery as high as possible, in order to allow room between it and the aorta for the formation of a coagulum.

It has been seen that the failure of this operation depends either upon repeated secondary hæmorrhage, or inflammation of the pleural sac and lung. The importance of avoiding the thyroid plexus of veins during the primary steps of the operation, and the pleural sac whilst including the vessel in the ligature, should be most carefully attended to.

## Common Carotid Arteries.

The common carotid arteries, although occupying a nearly similar position in the neck, differ in position, and, consequently, in their relations at their origin. The right carotid arises from the arteria innominata, behind the right sterno-clavicular articulation; the left from the highest part of the arch of the aorta. The left carotid is, consequently, longer and placed more deeply in the thorax. It will, therefore, be more convenient to describe first the course and relations of that portion of the left carotid which intervenes between the arch of the aorta and the left sterno-clavicular articulation (see fig. 205).

The left carotid within the thorax passes obliquely outwards from the arch of the aorta to the root of the neck. In *front*, it is separated from the first piece of the sternum by the Sterno-hyoid and Sterno-thyroid muscles, the left innominate vein, and the remains of the thymus gland; *behind*, it lies on the trachea, œsophagus, and thoracic duct. *Internally*, it is in relation with the arteria innominate; *externally*, with the left pneumogastric nerve, and left subclavian artery.

### Plan of the Relations of the Left Common Carotid.

### Thoracic Portion.

*In front.*

Sternum.
Sterno-hyoid and Sterno-thyroid muscles.
Left innominate vein.
Remains of thymus gland.

*Internally.*
Arteria innominata.

Left Common. Carotid. Thoracic Portion.

*Externally.*
Left pneumogastric nerve.
Left Subclavian artery.

*Behind.*
Trachea.
Œsophagus.
Thoracic duct.

In the neck, the two common carotids resemble each other so closely, that one description will apply to both. Starting from each side of the neck, each vessel passes obliquely upwards, from behind the sterno-clavicular articulation, to a level with the upper border of the thyroid cartilage, where it divides into the external and internal carotid; these names being derived, the former from its distribution to the external parts of the head and face, the latter from its distribution to the internal parts of the cranium. The course of the vessel is indicated by a line drawn from the sternal end of the clavicle below, to a point midway between the angle of the jaw and the mastoid process above.

At the lower part of the neck the two common carotid arteries are separated from each other by a very small interval, which corresponds to the trachea; but at the upper part, the thyroid body, the larynx and pharynx project forwards between these vessels, and give the appearance of their being placed further back in this situation. The common carotid artery is contained in a sheath, derived from the deep cervical fascia, which also encloses the internal jugular vein and pneumogastric nerve, the vein lying on the outer side of the artery, and the nerve between the artery and vein, on a plane posterior to both. On opening the sheath, these three structures are seen to be separated from one another, each being enclosed in a separate fibrous investment.

*Relations.* At the lower part of the neck the common carotid artery is very deeply seated, being covered by the superficial fascia, Platysma, and deep fascia, the Sterno-mastoid, Sterno-hyoid, and Sterno-thyroid muscles, and by the Omo-hyoid opposite the cricoid cartilage; but in the upper part of its course, near its termination, it is more superficial, being covered merely by the integument, the superficial fascia, Platysma, and deep fascia, and inner margin of the Sterno-mastoid, and is contained in a triangular space, bounded behind by the Sterno-mastoid, above by the posterior belly of the Digastric, and below by the anterior belly of the Omo-hyoid. This part of the artery is crossed obliquely from within outwards by the sterno-mastoid artery; it is also crossed by the facial, lingual and superior thyroid veins, which terminate in the internal jugular, and, descending on its sheath in front, is seen the descendens noni nerve, this filament being joined with branches from the cervical nerves, which cross the vessel from without inwards. Sometimes the descendens noni is contained within the sheath. The middle thyroid vein crosses it about its centre, and the anterior jugular vein below. *Behind*, the artery lies in front of the cervical portion of the spine, resting first on the Longus colli muscle, then on the Rectus anticus major, from which it is separated by the sympathetic nerve. The recurrent laryngeal nerve and inferior thyroid artery cross behind the vessel at its lower part. *Internally*, it is in relation with the trachea and thyroid gland, the inferior thyroid artery and recurrent laryngeal nerve being interposed; higher up, with the larynx and pharynx. On its *outer side* are placed the internal jugular vein and pneumogastric nerve.

At the lower part of the neck, the internal jugular vein on the right side recedes from the artery, but on the left side it approaches it, and often crosses its lower part. This is an important fact to bear in mind during the performance of any operation on the lower part of the left common carotid artery.

## Plan of the Relations of the Common Carotid Artery.

*In front.*

Integument and fascia.
Platysma.
Sterno-mastoid.
Sterno-hyoid.
Sterno-thyroid.

Omo-hyoid.
Descendens noni nerve.
Sterno-mastoid artery.
Thyroid, lingual, and facial veins.
Anterior jugular vein.

*Externally.*

Internal jugular vein.
Pneumogastric nerve.

Common Carotid.

*Internally.*

Trachea.
Thyroid gland.
Recurrent laryngeal nerve.
Inferior thyroid artery.
Larynx.
Pharynx.

*Behind.*

Longus colli.
Rectus anticus major.

Sympathetic nerve.
Inferior thyroid artery.

Recurrent laryngeal nerve.

*Peculiarities as to Origin.* The *right common carotid* may arise above or below its usual point (the upper border of the sterno-clavicular articulation). This variation occurs in one out of about eight cases and a half, and is more frequently above than below the point stated; or its origin may be transferred to the arch of the aorta, or it may arise in conjunction with the left carotid. The *left common carotid* varies more frequently in its origin than the right. In the majority of cases it arises with the innominate artery, as where the innominate artery was absent, the two carotids arose usually by a single trunk. This vessel has a tendency towards the right side of the arch, occasionally being the first branch given off from the transverse portion. It rarely joins with the left subclavian, except in cases of transposition of the arch.

*Point of Division.* The most important peculiarities of this vessel, in a surgical point of view, relate to its place of division in the neck. In the majority of cases, this occurs higher than usual, the artery dividing into two branches opposite the hyoid bone, or even higher; more rarely, it occurs below its usual place, opposite the middle of the larynx, or the lower border of the cricoid cartilage; and one case is related by Morgagni, where this vessel, only an inch and a half in length, divided at the root of the neck. Very rarely, the common carotid ascends in the neck without any subdivision, the internal carotid being wanting; and in two cases the common carotid has been found to be absent, the external and internal carotids arising directly from the arch of the aorta. This peculiarity existed on both sides in one subject, on one side in another.

*Occasional Branches.* The common carotid usually gives off no branches, but it occasionally gives origin to the superior thyroid, or a laryngeal branch, the inferior thyroid, or, more rarely, the vertebral artery.

*Surgical Anatomy.* The operation of tying the common carotid artery may be necessary in a wound of that vessel or its branches, in an aneurism, or in a case of pulsating tumour of the orbit or skull. If the wound involves the trunk of the common carotid, it will be necessary to tie the artery above and below the wounded part. If, however, one of the branches of that vessel is wounded, or has an aneurismal tumour connected with it, a ligature may be applied to any part of it, excepting its origin and termination. When the case is such as to allow of a choice being made, the lower part of the carotid should never be selected as the spot upon which a ligature should be placed, for not only is the artery in this situation placed very deeply in the neck, but it is covered by three layers of muscles, and on the left side the jugular vein, in the great majority of cases, passes obliquely in

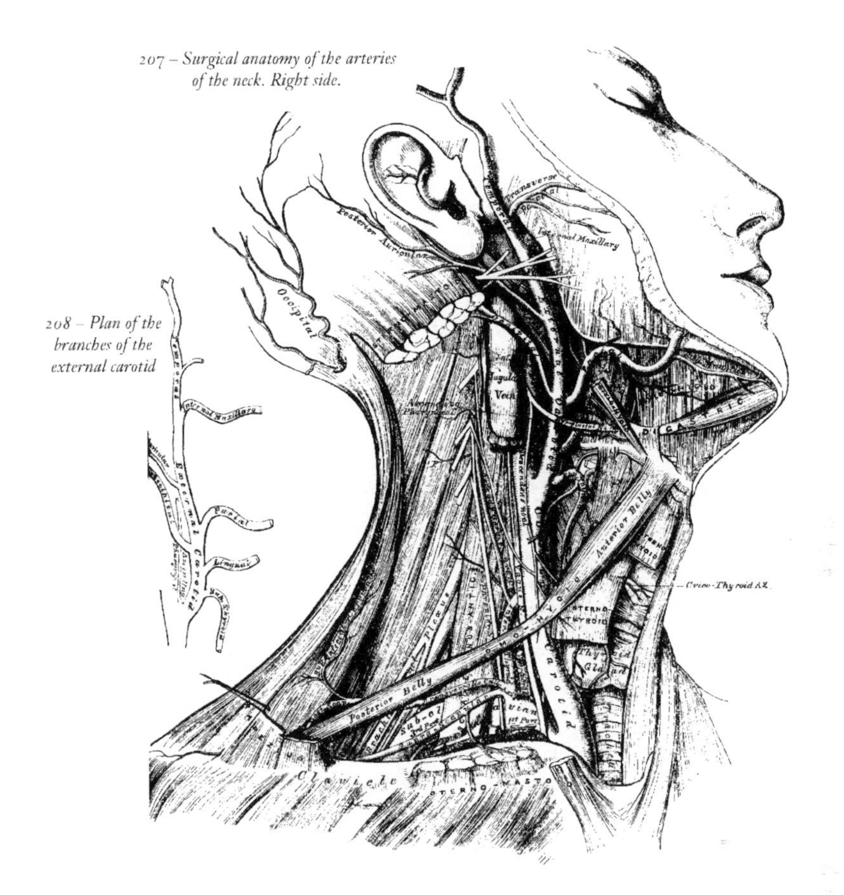


*207 – Surgical anatomy of the arteries of the neck. Right side.*

*208 – Plan of the branches of the external carotid*

front of it. Neither should the upper end be selected, for here the superior thyroid, lingual, and facial veins would give rise to very considerable difficulty in the application of a ligature. The point most favourable for the operation is opposite the lower part of the larynx, and here a ligature may be applied on the vessel, either above or below the point where it is crossed by the Omo-hyoid muscle. In the former situation the artery is most accessible, and it may be tied there in cases of wounds, or aneurism of any of the large branches of the carotid; whilst in cases of aneurism of the upper part of the carotid, that part of the vessel may be selected which is below the Omo-hyoid. It occasionally happens that the carotid artery bifurcates below its usual position: if the artery be exposed at its point of bifurcation, both divisions of the vessel should be tied near their origin, in preference to tying the trunk of the artery near its termination; and if, in consequence of the entire absence of the common carotid, or from its early division, two arteries, the external and internal carotids, are met with, the ligature should be placed on that vessel which is found on compression to be connected with the disease.

In this operation, the direction of the vessel and the inner margin of the Sterno-mastoid are the chief guides to its performance.

*To tie the Common Carotid, above the Omo-hyoid.* The patient should be placed on his back with the head thrown back; an incision is to be made, three inches long, in the direction of the anterior border of the Sterno-mastoid, from a little below the angle of the jaw to a level with the cricoid cartilage: after dividing the integument, superficial fascia, and Platysma, the deep fascia must be cut through on a director, so as to avoid wounding numerous small veins that are usually found beneath. The head may now be brought forwards so as to relax the parts somewhat, and the margins of the wound must be held asunder by copper spatulæ. The descendens noni nerve is now exposed, and must be avoided, and the sheath of the vessel having been raised by forceps, is to be opened over the artery to a small extent. The internal jugular vein will now present itself alternately distended and relaxed; this should be compressed both above and below, and drawn outwards, in order to facilitate the operation. The aneurism needle is now passed from the outside, care being taken to keep the needle in close contact with the artery, and thus avoid the risk of injuring the jugular vein, or including the vagus nerve. Before the ligature is secured, it should be ascertained that nothing but the artery is included in it.

*To tie the Common Carotid, below the Omo-hyoid.* The patient should be placed in the same situation as above mentioned. An incision about three inches in length is to be made, parallel with the inner edge of the

Sterno-mastoid, commencing on a level with the cricoid cartilage. The inner border of the Sterno-mastoid having been exposed, the sterno-mastoid artery and a large vein, the middle thyroid will be seen, and must be carefully avoided; the Sterno-mastoid is to be drawn outwards, and the Sterno-hyoid and thyroid muscles inwards. The deep fascia must now be divided below the Omo-hyoid muscle, and the sheath having been exposed, must be opened, care being taken to avoid the descendens noni, which here runs on the inner or tracheal side. The jugular vein and vagus nerve being then pressed to the outer side, the needle must be passed round the artery from without inwards, great care being taken to avoid the inferior thyroid artery, the recurrent laryngeal, and sympathetic nerves which lie behind it.

*Collateral Circulation.* After ligature of the common carotid, the collateral circulation can be perfectly established, by the free communication which exists between the carotid arteries of opposite sides both without and within the cranium—and by enlargement of the branches of the subclavian artery on the side corresponding to that on which the vessel has been tied, the chief communication outside the skull taking place between the superior and inferior thyroid arteries, and the profunda cervicis, and arteria princeps cervicis of the occipital; the vertebral taking the place of the internal carotid within the cranium.

## External Carotid Artery.

The external carotid artery (fig. 207), arises opposite the upper border of the thyroid cartilage, and taking a slightly curved course, ascends upwards and forwards, and then inclines backwards, to the space between the neck of the condyle of the lower jaw, and the external meatus, where it divides into the temporal and internal maxillary arteries. It rapidly diminishes in size as it ascends the neck, owing to the number and large size of the branches given off from it. In the child, it is somewhat smaller than the internal carotid; but in the adult, the two vessels are of nearly equal size. At its commencement, this artery is more superficial, and placed nearer the middle line than the internal carotid, and is contained in the triangular space bounded by the Sterno-mastoid behind, the Omo-hyoid below, and the posterior belly of the Digastric and Stylo-hyoid above; it is covered by the skin, Platysma, deep fascia, and anterior margin of the Sterno-mastoid, crossed by the hypoglossal nerve, and by the lingual and facial veins; it is afterwards crossed by the Digastric and Stylo-hyoid muscles, and higher up passes deeply into the substance of the parotid gland, where it lies beneath the facial nerve, and the junction of the temporal and internal maxillary veins.

*Internally* is the hyoid bone, the wall of the pharynx, and the ramus of the jaw, from which it is separated by a portion of the parotid gland.

*Behind* it, near its origin, is the superior laryngeal nerve; and, higher up, it is separated from the internal carotid by the Stylo-glossus and Stylo-pharyngeus muscles, the glosso-pharyngeal nerve, and part of the parotid gland.

### Plan of the Relations of the External Carotid.

*In front.*
Integument, superficial fascia.
Platysma and deep fascia.
Hypoglossal nerve.
Lingual and facial veins.
Digastric and Stylo-hyoid muscles.
Facial nerve and parotid gland.
Temporal and maxillary veins.

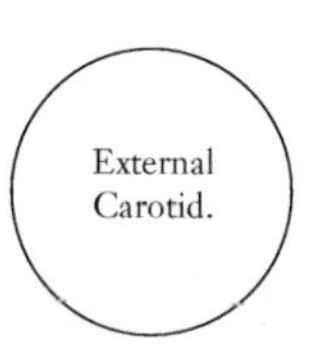


*Behind.*
Superior Laryngeal nerve.
Stylo-glossus.
Stylo-pharyngeus.
Glosso-pharyngeal nerve.
Parotid gland.

*Internally.*
Hyoid bone.
Pharynx.
Parotid gland.
Ramus of jaw.

*Surgical Anatomy.* The application of a ligature to the external carotid may be required in cases of wounds of this vessel, or of its branches when these cannot be tied; this, however, is an operation very rarely performed, ligature of the common carotid being preferable, on account of the number of branches given off from the external. To tie this vessel near its origin, below the point where it is crossed by the Digastric, an incision about three inches in length should be made along the margin of the Sterno-mastoid, from the angle of the jaw to the cricoid cartilage, as in the operation for tying the common carotid. To tie the vessel above the Digastric, between it and the parotid gland, an incision should be made from the lobe of the ear to the great cornu of the os-hyoides, dividing successively the skin, Platysma, and fascia. By separating the posterior belly of the Digastric and Stylo-hyoid muscles which are seen at the lower part of the wound, from the parotid gland, the vessel will be exposed, and a ligature may be applied to it.

*Branches.* The external carotid artery gives off eight branches, which, for convenience of description, may be divided into four sets. (See fig. 208, Plan of the Branches.)

| *Anterior.* | *Posterior.* | *Ascending.* | *Terminal.* |
|---|---|---|---|
| Superior thyroid. | Occipital. | Ascending | Temporal. |
| Lingual. | Posterior auricular. | pharyngeal. | Internal maxillary. |
| Facial. | | | |

The student is here reminded that many variations are met with in the number, origin, and course of these branches in different subjects; but the above arrangement is that which is found in the great majority of cases.

The Superior Thyroid Artery (figs. 207 and 212), is the first branch given off from the external carotid, being derived from that vessel just below the greater cornu of the hyoid bone. At its commencement, it is quite superficial, being covered by the integument, fascia, and Platysma, and is contained in the triangular space bounded by the Sterno-mastoid, Digastric, and Omo-hyoid muscles. After ascending upwards and inwards for a short distance, it curves downwards and forwards in an arched and tortuous manner to the upper part of the thyroid gland, passing beneath the Omo-hyoid, Sterno-hyoid, and Sterno-thyroid muscles; and distributes numerous branches to its anterior surface, anastomosing with its fellow of the opposite side, and with the inferior thyroid arteries. Besides the arteries distributed to the muscles and substance of the gland, its branches are the following.

| | |
|---|---|
| Hyoid. | Superior laryngeal. |
| Superficial descending branch. | Crico-thyroid. |

The *hyoid* is a small branch which runs along the lower border of the os hyoides, beneath the Thyro-hyoid muscle; after supplying the muscles connected to that bone, it forms an arch, by anastomosing with the vessel of the opposite side.

The *superficial descending branch* runs downwards and outwards across the sheath of the common carotid artery, and supplies the Sterno-mastoid and neighbouring muscles and integument. It is of importance that the situation of this vessel be remembered, in the operation for tying the common carotid artery.

The *superior laryngeal*, larger than either of the preceding, accompanies the superior laryngeal nerve, beneath the Thyro-hyoid muscle; it pierces the thyro-hyoid membrane, and supplies the muscles, mucous membrane, and glands of the larynx and epiglottis, anastomosing with the branch from the opposite side.

The *crico-thyroid* (inferior laryngeal) is a small branch which runs transversely across the crico-thyroid membrane, communicating with the artery of the opposite side. The position of this vessel should be remembered, as it may prove the source of troublesome hæmorrhage during the operation of laryngotomy.

*Surgical Anatomy.* The superior thyroid, or some of its branches, are occasionally divided in cases of cut throat, giving rise to considerable hæmorrhage. In such cases, the artery should be secured, the wound being enlarged for that purpose, if necessary. The operation may be easily performed, the position of the artery being very superficial, and the only structures of importance covering it, being a few small veins. The operation of tying the superior thyroid artery, in bronchocele, has been performed in numerous instances with partial or temporary success. When, however, the collateral circulation between this vessel with the artery of the opposite side, and with the inferior thyroid is completely re-established, the tumour usually regains its former size.

The Lingual Artery (fig. 212) arises from the external carotid between the superior thyroid and facial; it runs obliquely upwards and inwards to the great cornu of the hyoid bone, then passes horizontally forwards parallel with the great cornu, and, ascending perpendicularly to the under surface of the tongue, turns forwards on its under surface as far as the tip of that organ, under the name of the ranine artery.

*Relations.* Its first, or oblique portion, is superficial, being contained in the triangular intermuscular space already described, resting upon the Middle constrictor of the pharynx, and covered by the Platysma and fascia of the neck. Its second, or horizontal portion, also lies upon the Middle constrictor, being covered at first by the tendon of the Digastric, and the Stylo-hyoid muscle, and afterwards by the Hyo-glossus, the latter muscle separating it from the hypoglossal nerve. Its third, or ascending portion, lies between the Hyo-glossus and Genio-hyo-glossus muscles. The fourth, or terminal part, under the name of the ranine, runs along the under surface of the tongue to its tip, it is very superficial, being covered only by the mucous membrane, and rests on the Lingualis on the outer side of the Genio-hyo-glossus. The hypoglossal nerve lies nearly parallel with the lingual artery, separated from it, in the second part of its course, by the Hyo-glossus muscle.

The branches of the lingual artery are, the

| | |
|---|---|
| Hyoid. | Sublingual. |
| Dorsalis Linguæ. | Ranine. |

The *hyoid* branch runs along the upper border of the hyoid bone, supplying the muscles attached to it, and anastomosing with its fellow of the opposite side.

The *dorsalis linguæ* (fig. 212) arises from the lingual artery beneath the Hyo-glossus muscle; ascending to the dorsum of the tongue, it supplies its mucous membrane, the tonsil, soft palate, and epiglottis; anastomosing with its fellow from the opposite side.

The *sublingual*, a branch of bifurcation of the lingual artery, arises at the anterior margin of the Hyo-glossus muscle, and, running forwards and outwards beneath the Mylo-hyoid to the sublingual gland, supplies its substance, giving branches to the Mylo-hyoid and neighbouring muscles, the mucous membrane of the mouth and gums.

The *ranine* may be regarded as the continuation of the lingual artery; it runs along the under surface of the tongue, resting on the Lingualis, and covered by the mucous membrane of the mouth; it lies on the outer side of the Genio-hyo-glossus, and is covered by the Hyo-glossus and Stylo-glossus, accompanied by the gustatory nerve. On arriving at the tip of the tongue, it anastomoses with the artery of the opposite side. These vessels in the mouth are placed one on each side of the frænum.

*Surgical Anatomy.* The lingual artery may be divided near its origin in cases of cut throat, a complication that not unfrequently happens in this class of wounds, or severe hæmorrhage which cannot be restrained by ordinary means, may ensue from a wound, or deep ulcer of the tongue. In the former case, the primary wound may be enlarged if necessary, and the bleeding vessel at once secured. In the latter case, it has been suggested that the lingual artery should be tied near its origin. If the student, however, will observe the depth at which this vessel is placed from the surface, the number of important parts which surround it on every side, and its occasional irregularity of origin, the great difficulty of such an operation will be apparent; under such circumstances, it is more advisable that the external or common carotid should be tied.

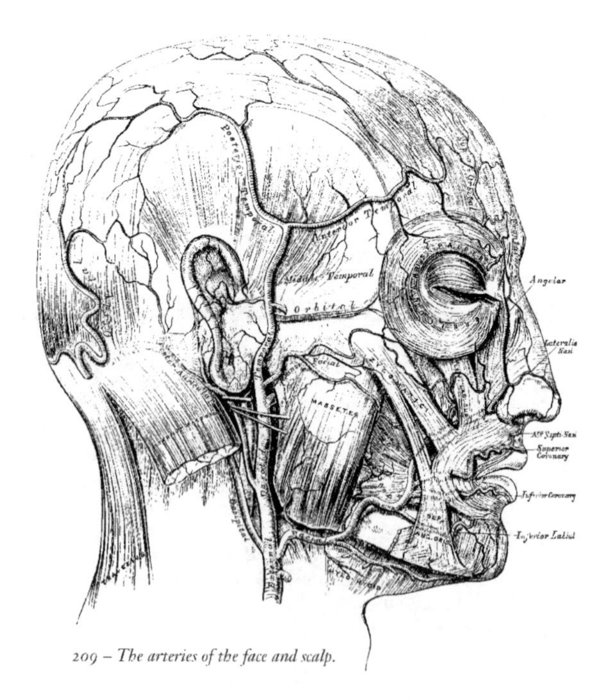


209 – *The arteries of the face and scalp.*

Troublesome hæmorrhage may occur in the division of the frænum in children, if the ranine artery, which lies on each side of it, is cut through. The student should remember that the operation is always to be performed with a pair of blunt-pointed scissors, which should be so held as to divide the part in the direction downwards and backwards; the ranine artery and veins are then avoided.

The FACIAL ARTERY (fig. 209), arises a little above the lingual, and ascends obliquely forwards and upwards, beneath the body of the lower jaw, to the submaxillary gland, in which it is imbedded; this may be called the cervical part of the artery. It then curves upwards over the body of the jaw at the anterior inferior angle of the Masseter muscle, ascends forwards and upwards across the cheek to the angle of the mouth, passes up along the side of the nose, and terminates at the inner canthus of the eye, under the name of the angular artery. This vessel both in the neck, and on the face, is remarkably tortuous; in the former situation, to accommodate itself to the movements of the pharynx in deglutition; and in the latter, to the movements of the jaw, and the lips and cheeks.

*Relations. In the neck*, its origin is superficial, being covered by the integument, Platysma, and fascia; it then passes beneath the Digastric and Stylo-hyoid muscles, and the submaxillary gland. *On the face*, where passing over the body of the lower jaw, it is comparatively superficial, being covered by the Platysma. In this situation, its pulsation may be distinctly felt, and compression of the vessel effectually made against the bone. In its course over the face, it is covered by the integument, the fat of the cheek, and, near the angle of the mouth, by the Platysma and Zygomatic muscles. It rests on the Buccinator, the Levator anguli oris, and the Levator labii superioris alæque nasi. It is accompanied by the facial vein throughout its entire course; the vein is not tortuous like the artery, and, on the face, is separated from that vessel by a considerable interval. The branches of the facial nerve cross this vessel, and the infra-orbital nerve lies beneath it.

The branches of this vessel may be divided into two sets, those given off below the jaw (cervical); and those on the face (facial).

| *Cervical Branches.* | *Facial Branches.* |
|---|---|
| Inferior or Ascending Palatine. | Muscular. |
| Tonsillar. | Inferior Labial. |
| Submaxillary. | Inferior Coronary. |
| Submental. | Superior Coronary. |
| | Lateralis Nasi. |
| | Angular. |

The *inferior* or *ascending palatine* (fig. 212) passes up between the Stylo-glossus and Stylopharyngeus to the outer side of the pharynx. After supplying these muscles, the tonsil, and Eustachian tube, it divides, near the Levator palati, into two branches; one follows the course of the Tensor palati, supplies the soft palate and the palatine glands; the other passes to the tonsil, which it supplies, anastomosing with the tonsillar artery. These vessels inosculate with the posterior palatine branch of the internal maxillary artery.

The *tonsillar* branch (fig. 212) passes up along the side of the pharynx, and, perforating the Superior constrictor, ramifies in the substance of the tonsil and root of the tongue.

The *submaxillary* consist of three or four large branches, which supply the submaxillary gland, some being prolonged to the neighbouring muscles, lymphatic glands, and integument.

The *submental*, the largest of the cervical branches, is given off from the facial artery, just as that vessel quits the submaxillary gland; it runs forwards upon the Mylo-hyoid muscle, just below the body of the jaw, and beneath the Digastric; after supplying the muscles attached to the jaw, and anastomosing with the sublingual artery, it arrives at the symphysis of the chin, where it divides into a superficial and a deep branch; the former turns round the chin, and, passing between the integument and Depressor labii inferioris, supplies both, and anastomoses with the inferior labial. The deep branch passes between the latter muscle and the bone, supplies the lip, and anastomoses with the inferior labial and mental arteries.

The *muscular* branches are distributed to the internal Pterygoid, Masseter, and Buccinator.

The *inferior labial* passes beneath the Depressor anguli oris, to supply the muscles and integument of the lower lip, anastomosing with the inferior coronary and submental branches of the facial, and with the mental branch of the inferior dental artery.

The *inferior coronary* is derived from the facial artery near the angle of the mouth; it passes upwards and inwards beneath the Depressor anguli oris, and, penetrating the Orbicularis muscle, runs in a tortuous course along the edge of the lower lip between this muscle and the mucous membrane, inosculating with the artery of the opposite side. This artery supplies the labial glands, the mucous membrane, and muscles of the lower lip; and anastomoses with the inferior labial, and mental branch of the inferior dental artery.

The *superior coronary* is larger, and more tortuous in its course than the preceding. It follows the same course along the edge of the upper lip, lying between the mucous membrane and the Orbicularis, and anastomoses with the artery of the opposite side. It supplies the textures of the upper lip, and gives off in its course two or three vessels which ascend to the nose. One, named the artery of the septum, ramifies on the septum of the nares as far as the point of the nose; another supplies the ala of the nose.

The *lateralis nasi* is derived from the facial, as that vessel is ascending along the side of the nose; it supplies the ala and dorsum of the nose, anastomosing with its fellow, the nasal branch of the ophthalmic, the artery of the septum, and the infra-orbital.

The *angular artery* is the termination of the trunk of the facial; it ascends to the inner angle of the orbit, accompanied by a large vein, the angular; it distributes some branches on the cheek which anastomose with the infra-orbital, and, after supplying the lachrymal sac, and Orbicularis muscle, terminates by anastomosing with the nasal branch of the ophthalmic artery.

The anastomoses of the facial artery are very numerous, not only with the vessel of the opposite side, but with other vessels from different sources; viz., with the sub-lingual branch of the lingual, with the mental branch of the inferior dental as it emerges from the dental foramen, with the ascending pharyngeal and posterior palatine, and with the ophthalmic, a branch of the internal carotid; it also inosculates with the transverse facial, and with the infra-orbital.

*Peculiarities*. The facial artery not unfrequently arises by a common trunk with the lingual. This vessel also is subject to some variations in its size, and in the extent to which it supplies the face. It occasionally terminates as the submental, and not unfrequently supplies the face only as high as the angle of the mouth or nose. The deficiency is then supplied by enlargement of one of the neighbouring arteries.

*Surgical Anatomy*. The passage of the facial artery over the body of the jaw would appear to afford a favourable position for the application of pressure in cases of hæmorrhage from the lips, the result either of an accidental wound, or from an operation; but its application is useless, on account of the free communication of this vessel with its fellow, and with numerous branches from different sources. In a wound involving the lip, it is better to seize the part between the fingers, and evert it, when the bleeding vessel may be at once secured with a tenaculum. In order to prevent hæmorrhage in cases of excision, or in the removal of diseased growths from the part, the lip should be compressed on each side between the finger and thumb, whilst the surgeon excises the diseased part. In order to stop hæmorrhage where the lip has been divided in an operation, it is necessary in uniting the edges of the wound, to pass the sutures through the cut edges, almost as deep as its mucous surface; by these means, not only are the cut surfaces more neatly adapted to each other, but the possibility of hæmorrhage is prevented by including in the suture the divided artery. If the suture is, on the contrary, passed through merely the cutaneous portion of the wound, hæmorrhage occurs into the cavity of the mouth. The student should, lastly, observe the relation of the angular artery to the lachrymal sac, and it will be seen that, as the vessel passes up along the inner margin of the orbit it ascends on its nasal side. In operating for fistula lachrymalis, the sac should always be opened on its outer side, in order that this vessel may be avoided.

The Occipital Artery arises from the posterior part of the external carotid, opposite the facial, near the lower margin of the Digastric muscle. At its origin, it is covered by the posterior belly of the Digastric and Stylo-hyoid muscles, and part of the parotid gland, the hypoglossal nerve winding around it from behind forwards; higher up, it passes across the internal carotid artery, the internal jugular vein, and the pneumogastric and spinal accessory nerves; it then ascends to the interval between the transverse process of the atlas, and the mastoid process of the temporal bone, passes horizontally backwards, grooving the surface of the latter bone, being covered by the Sterno-mastoid, Splenius, Digastric, and Trachelo-mastoid muscles, resting upon the Complexus, Superior oblique, and Rectus posticus major muscles; it then ascends vertically upwards, piercing the cranial attachment of the Trapezius, and passes in a tortuous course over the occiput, as high as the vertex, where it divides into numerous branches.

The branches given off from this vessel are,

| | |
|---|---|
| Muscular. | Inferior meningeal. |
| Auricular. | Arteria princeps cervicis. |

The *muscular branches* supply the Digastric, Stylo-hyoid, Sterno-mastoid, Splenius, and Trachelo-mastoid muscles. The branch distributed to the Sterno-mastoid is of large size.

The *auricular branch* supplies the back part of the concha.

The *meningeal branch* ascends with the internal jugular vein, and enters the skull through the foramen lacerum posterius, to supply the dura mater in the posterior fossa.

The *arteria princeps cervicis* (fig. 212), is a large branch which descends along the back part of the neck, and divides into a superficial and deep branch. The former runs beneath the Splenius, giving off branches which perforate that muscle to supply the Trapezius, anastomosing with the superficial cervical artery; the latter passes beneath the Complexus, between it and the Semi-spinalis colli, and anastomoses with the vertebral, and deep cervical branch of the superior intercostal. The anastomosis between these vessels serves mainly to establish the collateral circulation after ligature of the carotid or subclavian artery.

The cranial branches of the occipital artery are distributed upon the occiput; they are very tortuous, and lie between the integument and Occipito-frontalis, anastomosing with the artery of the opposite side, the posterior auricular, and temporal arteries. They supply the back part of the Occipito-frontalis muscle, the integument, pericranium, and one or two branches occasionally pass through the perietal or mastoid foramina, to supply the dura mater.

The Posterior Auricular Artery (fig. 209) is a small vessel, which arises from the external carotid, above the Digastric and Stylo-hyoid muscles, opposite the apex of the styloid process. It ascends, under cover of the parotid gland, to the groove between the cartilage of the ear and the mastoid process, immediately above which it divides into two branches, an anterior, which passes forwards to anastomose with the posterior division of the temporal; and a posterior, which communicates with the occipital. Just before arriving at the mastoid process, this artery is crossed by the portio dura, and has beneath it the spinal accessory nerve.

Besides several small branches to the Digastric, Stylo-hyoid, and Sterno-mastoid muscles, and to the parotid gland, this vessel gives off two branches.

Stylo-mastoid. Auricular.

The *stylo-mastoid branch* enters the stylo-mastoid foramen, and supplies the tympanum, mastoid cells, and semi-circular canals. In the young subject, a branch from this vessel forms, with the tympanic branch from the internal maxillary, a vascular circle, which surrounds the auditory meatus, and from which delicate vessels ramify on the membrana tympani.

The *auricular branch* is distributed to the back part of the cartilage of the ear, upon which it minutely ramifies, some branches curving round its margin, others perforating the fibro-cartilage, to supply its anterior surface.

The Ascending Pharyngeal Artery (fig. 212), the smallest branch of the external carotid, is a long slender vessel, deeply seated in the neck, beneath the other branches of the external carotid and Stylo-pharyngeus muscle. It arises from the back part of the external carotid, near the commencement of that vessel, and passes up to the under surface of the base of the skull, ascending the neck between the internal carotid and the side of the pharynx, and lying on the Rectus capitis anticus major. Its branches may be subdivided into three sets: 1. Those directed outwards to supply muscles and nerves. 2. Those directed inwards to the pharynx. 3. Meningeal branches.

The *external branches* are numerous small vessels, which supply the Recti antici muscles, the sympathetic, hypoglossal and pneumogastric nerves, and the lymphatic glands of the neck, anastomosing with the ascending cervical artery.

The *pharyngeal branches* are three or four in number. Two of these desecend to supply the middle and inferior Constrictors and the Stylo-pharyngeus, ramifying in their substance and in the mucous membrane lining them. The largest of the pharyngeal branches passes inwards, running upon the Superior constrictor, and sends ramifications to the soft palate, Eustachian tube, and tonsil, which take the place of the ascending palatine branch of the facial artery, when that vessel is of small size.

The *meningeal branches* consist of several small vessels, which pass through foramina in the base of the skull, to supply the dura mater. One, the posterior meningeal, enters the cranium through the foramen lacerum posterius with the internal jugular vein. A second passes through the foramen lacerum basis cranii; and occasionally a third through the anterior condyloid foramen. They are all distributed to the dura mater.

The Temporal Artery (fig. 209), the smaller of the two terminal branches of the external carotid, appears, from its direction, to be the continuation of that vessel. It commences in the substance of the parotid gland, in the interspace between the neck of the condyle of the lower jaw and the external meatus; crossing over the root of the zygoma, immediately beneath the integument, it divides about two inches above the zygomatic arch into two branches, an anterior and a posterior.

The *anterior temporal* inclines forwards over the forehead, supplying the muscles, integument, and pericranium in this region, and anastomoses with the supra-orbital and frontal arteries, its branches being directed from before backwards.

The *posterior temporal*, larger than the anterior, curves upwards and backwards along the side of the head, lying above the temporal fascia, and inosculates with its fellow of the opposite side, and with the posterior auricular and occipital arteries.

The temporal artery, as it crosses the zygoma, is covered by the Attrahens aurem muscle, and by a dense fascia given off from the parotid gland; it is also usually crossed by one or two veins, and accompanied by branches of the facial and temporo-auricular nerves. Besides some twigs to the parotid gland, the articulation of the jaw, and to the Masseter muscle, its branches are the

Transverse facial. Middle temporal.
Anterior auricular.

The *transverse facial* is given off from the temporal before that vessel quits the parotid gland; running forwards through its substance, it passes transversely across the face, between Stenon's duct and the lower border of the zygoma, and divides on the side of the face into numerous branches, which supply the parotid gland, the Masseter muscle, and the integument, anastomosing with the facial and infra-orbital arteries. This vessel rests on the Masseter, and is accompanied by one or two branches of the facial nerve.

The *middle temporal artery* arises immediately above the zygomatic arch, and perforating the temporal fascia, supplies the Temporal muscle, anastomosing with the deep temporal branches of the internal maxillary. It occasionally gives of an orbital branch, which runs along the upper border of the zygoma, between the two layers of the temporal fascia, to the outer angle of the orbit; it supplies the Orbicularis, and anastomoses with the lachrymal and palpebral branches of the ophthalmic artery.

The *anterior auricular branches* are distributed to the anterior portion of the pinna, the lobule, and part of the external meatus, anastomosing with branches of the posterior auricular.

*Surgical Anatomy*. It occasionally happens that the surgeon is called upon to perform the operation of arteriotomy upon this vessel in cases of inflammation of the eye or brain. Under these circumstances, the anterior branch is the one usually selected. If the student will consider the relations of the trunk of this vessel as it crosses the zygomatic arch, with the surrounding structures, he will observe that it is covered by a thick and dense fascia, crossed by one or two veins, and accompanied by branches of the facial and temporo-auricular nerves. Bleeding should not be performed in this situation, as much difficulty may arise from the dense fascia covering this vessel preventing a free flow of blood, and considerable pressure is requisite afterwards to repress it. Again, a varicose aneurism may be formed by the accidental opening of one of the veins covering it; or severe neuralgic pain may arise from the operation implicating one of the nervous filaments which accompany the artery.

The anterior branch is, on the contrary, subcutaneous, is a large vessel, and as readily compressed as any other portion of the artery; it should consequently always be selected for the operation.

The Internal Maxillary, the larger of the two terminal branches of the external carotid, passes inwards, at right angles from that vessel, to the inner side of the neck of the condyle of the lower jaw, to supply the deep structures of the face. At its origin, it is imbedded in the substance of the parotid gland, being on a level with the lower extremity of the lobe of the ear.

In the first part of its course (maxillary portion), the artery passes horizontally forwards and inwards, between the ramus of the jaw, and the internal lateral ligament. The artery here lies parallel with the auriculo-temporal nerve; it crosses the inferior dental nerve, and lies beneath the narrow portion of the External pterygoid muscle.

In the second part of its course (pterygoid portion), it ascends obliquely forwards and upwards upon the outer surface of the External pterygoid muscle, being covered by the ramus of the lower jaw, and lower part of the Temporal muscle.

In the third part of its course (spheno-maxillary portion), it approaches the superior maxillary bone, and enters the spheno-maxillary fossa, in the interval between the processes of origin of the External pterygoid, where it lies in relation with Meckel's ganglion, and gives off its terminal branches.

*Peculiarities*. Occasionally, this artery passes between the two Pterygoid muscles. The vessel in this case passes forwards to the interval between the processes of origin of the External pterygoid, in order to reach the maxillary bone. Sometimes, the vessel escapes from beneath the External pterygoid by perforating the middle of this muscle.

The branches of this vessel may be divided into three groups, corresponding with its three divisions.

## Branches from the Maxillary Portion.

| | |
|---|---|
| Tympanic. | Small meningeal. |
| Middle meningeal. | Inferior dental. |

The *tympanic branch* passes upwards behind the articulation of the lower jaw, enters the tympanum through the fissura Glaseri, supplies the Laxator tympani, and ramifies upon the membrana tympani, anastomosing with the stylo-mastoid and Vidian arteries.

The *middle meningeal* is the largest of the branches which supply the dura mater. It arises from the internal maxillary between the internal lateral ligament, and the neck of the jaw, and ascends vertically upwards to the foramen spinosum in the spinous process of the sphenoid bone. On entering the cranium, it divides into two branches, an anterior, and a posterior. The anterior branch, the larger, crosses the great ala of the sphenoid, and reaches the groove, or canal, in the anterior inferior angle of the parietal bone; it then divides into branches which spread out between the dura mater and internal surface of the cranium, some passing upwards over the parietal bone as far as the vertex, and others backwards to the occipital bone. The posterior branch crosses the squamous portion of the temporal, and on the inner surface of the parietal bone divides into branches which supply the posterior part of the dura mater and cranium. The branches of this vessel are distributed to the dura mater, but chiefly to the bones; they anastomose with the arteries of the opposite side, and with the anterior and posterior meningeal.

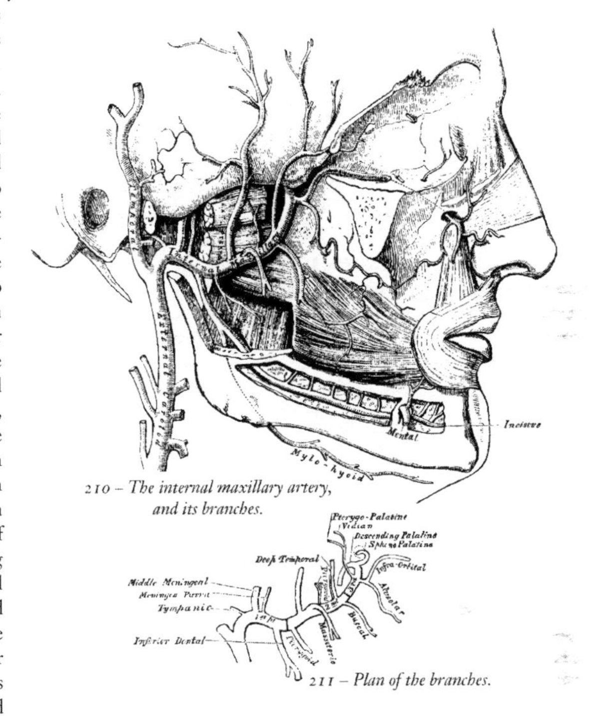


210 – *The internal maxillary artery, and its branches.*

211 – *Plan of the branches.*

The middle meningeal, on entering the cranium, gives off the following collateral branches: 1. Numerous small vessels to the ganglion of the fifth nerve, and to the dura mater in this situation. 2. A branch to the facial nerve, which enters the hiatus Fallopii, supplies the facial nerve, and anastomoses with the stylo-mastoid branch of the posterior auricular artery. 3. Orbital branches, which pass through the sphenoidal fissure, or through separate canals in the great wing of the sphenoid, to anastomose with the lachrymal or other branches of the ophthalmic artery. 4. Temporal branches, which pass through foramina in the great wing of the sphenoid, and anastomose in the temporal fossa with the deep temporal arteries.

The *small meningeal* is sometimes derived from the preceding. It enters the skull through the foramen ovale, and supplies the Casserian ganglion and dura mater. Before entering the cranium, it gives off a branch to the nasal fossa and soft palate.

The *inferior dental* descends with the dental nerve, to the foramen on the inner side of the ramus of the jaw. It runs along the dental canal in the substance of the bone, accompanied by the nerve, and opposite the bicuspid tooth divides into two branches, incisor and mental, the former is continued forwards beneath the incisor teeth as far as the symphysis, where it anastomoses with the artery of the opposite side; the mental branch, escapes with the nerve at the mental foramen, supplies the structures composing the chin, and anastomoses with the submental, inferior labial, and inferior coronary arteries. As the dental artery enters the foramen, it gives off a mylo-hyoid branch, which runs in the mylo-hyoid groove, and ramifies on the under surface of the Mylo-hyoid muscle. The dental and incisor arteries during their course through the substance of the bone, give off a few twigs which are lost in the diploë, and a series of branches which correspond in number to the roots of the teeth; these enter the minute apertures at the extremities of the fangs, and ascend to supply the pulp of the teeth.

## Branches of the Second, or Pterygoid Portion.

| | |
|---|---|
| Deep temporal. | Masseteric. |
| Pterygoid. | Buccal. |

These branches are distributed, as their names imply, to the muscles in the maxillary region.

The *deep temporal branches*, two in number, anterior, and posterior, each occupy that part of the temporal fossa indicated by its name. Ascending between the Temporal muscle and pericranium, they supply that muscle, and anastomose with the other temporal arteries; the anterior branch communicating with the lachrymal through small branches which perforate the malar bone.

The *pterygoid branches*, irregular in their number and origin, supply the Pterygoid muscles.

The *masseteric* is a small branch which passes outwards above the sigmoid notch of the lower jaw, to the deep surface of the Masseter. It supplies that muscle, and anastomoses with the masseteric branches of the facial and transverse facial arteries.

The *buccal* is a small branch which runs obliquely forwards between the Internal pterygoid, and the ramus of the jaw, to the outer surface of the Buccinator, to which it is distributed, anastomosing with branches of the facial artery.

## Branches of the Third, or Spheno-Maxillary Portion.

| | |
|---|---|
| Alveolar. | Vidian. |
| Infra-orbital. | Pterygo-palatine. |
| Posterior or Descending palatine. | Nasal or Spheno-palatine. |

The *alveolar* is given off from the internal maxillary by a common branch with the infra-orbital, and just as the trunk of the vessel is passing into the spheno-maxillary fossa. Descending upon the tuberosity of the superior maxillary bone, it divides into numerous branches; one, the superior dental, larger than the rest, supplies the molar and bicuspid teeth, its branches entering the foramina in the alveolar process; some branches pierce the bone to supply the lining of the antrum, and others are continued forwards on the alveolar process to supply the gums.

The *infra-orbital* appears, from its direction, to be the continuation of the trunk of the internal maxillary. It arises from that vessel by a common trunk with the preceding branch, and runs along the infra-orbital canal with the superior maxillary nerve, emerging upon the face at the infra-orbital foramen, beneath the Levator labii superioris. Whilst contained in the canal, it gives off branches which ascend into the orbit, and supply the Inferior rectus, and Inferior oblique muscles, and the lachrymal gland. Other branches descend through canals in the bone, to supply the mucous membrane of the antrum, and the front teeth of the upper jaw. On the face, it supplies the lachrymal sac, and inner angle of the orbit, anastomosing with the facial artery and nasal branch of the ophthalmic; and other branches descend beneath the elevator of the upper lip, and anastomose with the transverse facial and buccal branches.

The four remaining branches arise from that portion of the internal maxillary which is contained in the spheno-maxillary fossa.

The *descending palatine* passes down along the posterior palatine canal with the posterior palatine branches of Meckel's ganglion, and emerging from the posterior palatine foramen, runs forwards in a groove on the inner side of the alveolar border of the hard palate, to be distributed to the gums, the mucous membrane of the hard palate, and palatine glands. Whilst it is contained in the palatine canal, it gives off branches, which descend in the accessory palatine canals to supply the soft palate, anastomosing with the ascending palatine artery; and anteriorly it terminates in a small vessel, which ascends in the anterior palatine canal, and anastomoses with the artery of the septum, a branch of the spheno-palatine.

The *Vidian branch* passes backwards along the Vidian canal with the Vidian nerve. It is distributed to the upper part of the pharynx and Eustachian tube, sending a small branch into the tympanum.

The *pterygo-palatine* is also a very small branch, which passes backwards through the pterygo-palatine canal with the pharyngeal nerve, and is distributed to the upper part of the pharynx and Eustachian tube.

The *nasal* or *spheno-palatine* passes through the spheno-palatine foramen into the cavity of the nose, at the back part of the superior meatus, and divides into two branches; one internal, the artery of the septum, passes obliquely downwards and forwards along the septum nasi, supplies the mucous membrane, and anastomoses in front with the ascending branch of the descending palatine. The external branches, two or three in number, supply the mucous membrane covering the lateral wall of the nares, the antrum, and the ethmoid and sphenoid cells.

## Surgical Anatomy of the Triangles of the Neck.

The student having considered the relative anatomy of the large arteries of the neck and their branches, and the relations they bear to the veins and nerves, should now examine these structures collectively, as they present themselves in certain regions of the neck, in each of which important operations are being constantly performed.

For this purpose, the Sterno-mastoid, or any other muscles that have been divided in the dissection of these vessels, should be replaced in their normal position; the head should be supported by placing a block at the back of the neck, and the face turned to the side opposite to that which is being examined.

The side of the neck presents a somewhat quadrilateral outline, limited, above, by the lower border of the body of the jaw, and an imaginary line extending from the angle of the jaw to the mastoid process; below, by the prominent upper border of the clavicle; in front, by the median line of the neck; behind, by the anterior margin of the Trapezius muscle. This space is subdivided into two large triangles by the Sterno-mastoid muscle, which passes obliquely across the neck, from the sternum and clavicle, below, to the mastoid process, above. The triangular space in front of this muscle, is called the *anterior triangle;* and that behind it, the *posterior triangle.*

## Anterior Triangular Space.

The anterior triangle is limited, in front, by a line extending from the chin to the sternum; behind, by the anterior margin of the Sterno-mastoid; its base, directed upwards, is formed by the lower border of the body of the jaw, and a line extending from the angle of the jaw to the mastoid process; its apex is formed below by the sternum. This space is covered by the integument, superficial fascia, Platysma, deep fascia, crossed by branches of the facial and superficial cervical nerves; and subdivided into three smaller triangles by the Digastric muscle, above, and the anterior belly of the Omo-hyoid, below. These are named, from below upwards, the inferior carotid triangle, the superior carotid triangle, and the submaxillary triangle.

The *Inferior Carotid Triangle* is limited, in front, by the median line of the neck; behind, by the anterior margin of the Sterno-mastoid; above, by the anterior belly of the Omo-hyoid; and it is covered by the integument, superficial fascia, Platysma, and deep fascia; ramifying between which, is seen the descending branch of the superficial cervical nerve. Beneath these superficial structures, are the Sterno-hyoid and Sterno-thyroid muscles, which, together with the anterior margin of the Sterno-mastoid, conceal the lower part of the common carotid artery. This vessel is enclosed within its sheath, together with the internal jugular vein, and pneumogastric nerve; the vein lying on the outer side of the artery on the right side of the neck, but overlapping it, or passing directly across it on the left side; the nerve lying between the artery and vein, on a plane posterior to both. In front of the sheath are a few filaments descending from the loop of communication between the descendens and communicans noni; behind the sheath is seen the inferior thyroid artery, the recurrent laryngeal and sympathetic nerves; and on its inner side, the trachea, the thyroid gland, much more prominent in the female than in the male, and the lower part of the larynx. In the upper part of this space, the common carotid artery may be tied below the Omo-hyoid muscle.

The *Superior Carotid Triangle* is bounded, behind, by the Sterno-mastoid; below, by the anterior belly of the Omo-hyoid; and above, by the posterior belly of the Digastric muscle. Its floor is formed by parts of the Thyro-hyoid, Hyo-glossus, and the inferior and middle Constrictor muscles of the pharynx; and it is covered by the integument, superficial fascia, Platysma, and deep fascia; ramifying between which, are branches of the facial and superficial cervical nerves. This space contains the upper part of the common carotid artery, which bifurcates opposite the upper border of the thyroid cartilage into the external and internal carotid. These vessels are concealed from view by the anterior margin of the Sterno-mastoid muscle, which overlaps them. The external and internal carotids lie side by side, the external being the most anterior of the two. The following branches of the external carotid are also met with in this space: the superior thyroid, which runs forwards and downwards; the lingual, which passes directly forwards; the facial, forwards and upwards; the occipital is directed backwards; and the ascending pharyngeal runs directly upwards on the inner side of the internal carotid. The veins met with are: the internal jugular, which lies on the outer side of the common and internal carotid vessels; and veins corresponding to the above-mentioned branches of the external carotid, viz., the superior thyroid, the lingual, facial, ascending pharyngeal, and sometimes the occipital; all of which accompany their corresponding arteries, and terminate in the internal jugular. In front of the sheath of the common carotid is the descendens noni, the hypoglossal, from which it is derived, crossing both carotids above, curving round the occipital artery at its origin. Within the sheath, between the artery and vein, and behind both, is the pneumogastric nerve; behind the sheath, the sympathetic. On the outer side of the vessels, the spinal accessory nerve runs for a short distance before it pierces the Sterno-mastoid muscle; and on the inner side of the internal carotid, just below the hyoid bone, may be seen the superior laryngeal nerve; and still more inferiorly, the external laryngeal nerve. The upper part of the larynx and the pharynx, are also found in the front part of this space.

The *Submaxillary Triangle* corresponds to that part of the neck immediately beneath the body of the jaw. It is bounded, above, by the lower border of the body of the jaw, the parotid gland, and mastoid process; behind by the posterior belly of the Digastric and Stylo-hyoid muscles; in front, by the middle line of the neck. The floor of this space is formed by the anterior belly of the Digastric, the Mylo-hyoid, and Hyo-glossus muscles; and it is covered by the integument, superficial fascia, Platysma, and deep fascia; ramifying between which, are branches of the facial and ascending filaments of the superficial cervical nerve. This space contains, in front, the submaxillary gland, imbedded in the surface of which is the facial artery and vein, and its glandular branches; beneath this gland, on the surface of the Mylo-hyoid muscle, is the submental artery, and the mylohyoid artery and nerve. The back part of this space is separated from the front part, by the stylomaxillary ligament; it contains the external carotid artery, ascending deeply in the substance of the parotid gland; this vessel here lies in front of, and superficial to, the internal carotid, being crossed by the facial nerve, and gives off in its course the posterior auricular, temporal, and internal maxillary branches; more deeply is the internal carotid, the internal jugular vein, and the pneumogastric nerve, separated from the external carotid by the Stylo-glossus and Stylo-pharyngeus muscles, and the glosso-pharyngeal nerve.

## Posterior Triangular Space.

The posterior triangular space is bounded, in front, by the Sterno-mastoid muscle; behind, by the anterior margin of the Trapezius; its base corresponds to the upper border of the clavicle; its apex, to the occiput. This space is crossed about an inch above the clavicle, by the posterior belly of the Omo-hyoid, which divides it unequally into two, an upper or occipital, and a lower or subclavian.

The *Occipital*, the larger of the two posterior triangles, is bounded, in front, by the Sterno-mastoid; behind, by the Trapezius; below, by the Omo-hyoid. Its floor is formed from above downwards by the Splenius, Levator anguli scapulæ, and the middle and posterior Scaleni muscles. It is covered by the integument, the Platysma below, the superficial and deep fasciæ; and crossed, above, by the ascending branches of the cervical plexus; the spinal accessory nerve is directed obliquely across the space from the Sterno-mastoid, which it pierces, to the under surface of the Trapezius; below, it is crossed by the descending branches of the same plexus, and the transversalis colli artery and vein. A chain of lymphatic glands is also found running along the posterior border of the Sterno-mastoid, from the mastoid process to the root of the neck.

The *Subclavian*, the smaller of the two posterior triangles, is bounded, above, by the posterior belly of the Omo-hyoid; below, by the clavicle; its base, directed forwards, being formed by the Sterno-mastoid. The size of this space varies according to the extent of attachment of the clavicular portion of the Sterno-mastoid and Trapezius muscles, and also according to the height at which the Omo-hyoid crosses the neck above the clavicle. The height also of this space varies much, according to the position of the arm, being much diminished on raising the limb, on account of the ascent of the clavicle, and increased on drawing the arm downwards, when this bone is consequently depressed. This space is covered by the integument, superficial and deep fascia; and crossed by the descending branches of the cervical plexus. Just above the level of the clavicle, the third portion of the subclavian artery curves outwards and downwards from the outer margin of the Scalenus anticus, across the first rib, to the axilla. Sometimes, this vessel rises as high as an inch and a-half above the clavicle, or to any point intermediate between this and its usual level. Occasionally, it passes in front of the Scalenus anticus, or pierces the fibres of this muscle. The subclavian vein lies beneath the clavicle, and is usually not seen in this space; but it occasionally rises as high up as the artery, and has even been seen to pass with that vessel behind the Scalenus anticus. The brachial plexus of nerves lies above the artery, and in close contact with it. Passing transversely across the clavicular margin of the space, are the supra-scapular vessels; and traversing its upper angle in the same direction, the transverse cervical vessels. The external jugular vein descends vertically downwards behind the posterior border of the Sterno-mastoid, to terminate in the Subclavian; it receives the transverse cervical and supra-scapular veins, which occasionally form a plexus in front of the artery, and a small vein which crosses the clavicle from the cephalic. The small nerve to the Subclavius also crosses this space about its centre.

## Internal Carotid Artery.

The internal carotid artery commences at the bifurcation of the common carotid, opposite the upper border of the thyroid cartilage, and ascends perpendicularly upwards, in front of the transverse processes of the three upper cervical vertebræ, to the carotid foramen in the petrous portion of the temporal bone. After ascending in it for a short distance, it passes forwards and inwards through the carotid canal, and, ascending a little by the side of the sella Turcica, curves upwards by the anterior clinoid process, where it pierces the dura mater, and divides into its terminal branches.

This vessel supplies the anterior part of the brain, the eye, and its appendages. Its size, in the adult, is equal to that of the external carotid. In the child, it is larger than that vessel. It is remarkable for the number of curvatures that it presents in different parts of its course. In its cervical portion it occasionally presents one or two flexures near the base of the skull, whilst through the rest of its extent it describes a double curvature, which resembles the italic letter ∫ placed horizontally ⏝.

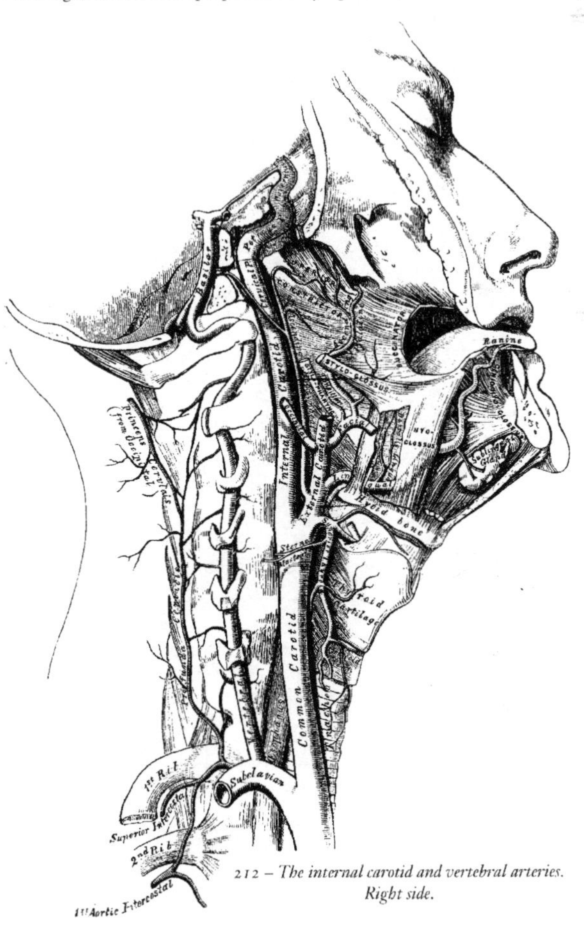


212 – *The internal carotid and vertebral arteries. Right side.*

These curvatures most probably diminish the velocity of the current of blood, by increasing the extent of surface over which it moves, and adding to the amount of impediment produced from friction. In considering the course and relations of this vessel, it may be conveniently divided into four portions, a cervical, petrous, cavernous, and cerebral.

*Cervical Portion.* This portion of the internal carotid at its commencement is very superficial, being contained in the superior carotid triangle, on the same level but behind the external carotid, overlapped by the Sterno-mastoid, and covered by the Platysma, deep fascia, and integument; it then passes beneath the parotid gland, being crossed by the hypoglossal nerve, the Digastric and Stylo-hyoid muscles, and the external carotid and occipital arteries. Higher up, it is separated from the external carotid by the Stylo-glossus and Stylo-pharyngeus muscles, the glosso-pharyngeal nerve, and pharyngeal branch of the vagus. It is in relation *behind*, with the Rectus anticus major, the superior cervical ganglion of the sympathetic, and superior laryngeal nerve; *externally*, with the internal jugular vein and pneumogastric nerve; *internally*, with the pharynx, tonsil, and ascending pharyngeal artery.

*Petrous Portion.* When the internal carotid artery enters the canal in the petrous portion of the temporal bone, it first ascends a short distance, then curves forwards and inwards, and again ascends as it leaves the canal to enter the cavity of the skull. In this canal, the artery lies at first anterior to the tympanum from which it is separated by a thin bony lamella, which is cribriform in the young subject, and often absorbed in old age. It is separated from the bony wall of the carotid canal by a prolongation of dura mater, and is surrounded by filaments of the carotid plexus.

*Cavernous Portion.* The internal carotid artery, in this part of its course, first ascends to the posterior clinoid process, then passes forwards by the side of the body of the sphenoid bone, being situated on the inner wall of the cavernous sinus, in relation, externally, with the sixth nerve, and covered by the lining membrane of the sinus. The third, fourth, and ophthalmic nerves are placed on the outer wall of the sinus, being separated from its cavity by the lining membrane.

*Cerebral Portion.* On the inner side of the anterior clinoid process the internal carotid curves upwards, perforates the dura mater bounding the sinus, and is received into a sheath of the arachnoid. This portion of the artery is on the outer side of the optic nerve; it lies at the inner extremity of the fissure of Sylvius, having the third nerve externally.

*Peculiarities.* The length of the internal carotid varies according to the length of the neck, and also according to the point of bifurcation of the common carotid. Its origin sometimes takes place from the arch of the aorta; in such rare instances, this vessel was placed nearer the middle line of the neck than the external carotid, as far upwards as the larynx, when the latter vessel crossed the internal carotid. The course of the vessel instead of being straight, may be very tortuous. A few instances are recorded in which this vessel was altogether absent: in one of these the common carotid ascended the neck, and gave off the usual branches of the external carotid: the cranial portion of the vessel being replaced by two branches of the internal maxillary, which entered the skull through the foramen rotundum and ovale, and joined to form a single vessel.

*Surgical Anatomy.* The cervical part of the internal carotid is sometimes wounded by a stab or gun-shot wound in the neck, or even occasionally by a stab from within the mouth, as when a person receives a thrust from the end of a parasol, or falls down with a tobacco-pipe in his mouth. In such cases a ligature should be applied to the common carotid. Its relation with the tonsil should be especially remembered, as instances have occurred in which the artery has been wounded, during the operation of scarifying the tonsil, and fatal hæmorrhage has supervened.

The branches given off from the internal carotid are:

| | |
|---|---|
| *From Petrous Portion* | Tympanic. |
| *From Cavernous Portion* | Arteria receptaculi.<br>Anterior meningeal.<br>Ophthalmic. |
| *From Cerebral Portion* | Anterior cerebral.<br>Middle cerebral.<br>Posterior communicating.<br>Anterior choroid. |

The cervical portion of the internal carotid gives off no branches.

The *tympanic* is a small branch which enters the cavity of the tympanum, through a minute foramen in the carotid canal, and anastomoses with the tympanic branch of the internal maxillary, and stylo-mastoid arteries.

The *arteriæ receptaculi* are numerous small vessels, derived from the internal carotid in the cavernous sinus; they supply the pituitary body, the Casserian ganglion, and the walls of the cavernous and inferior petrosal sinuses. One of these branches, distributed to the dura mater, is called the *anterior meningeal*; it anastomoses with the middle meningeal.

The Ophthalmic Artery arises from the internal carotid, just as that vessel is emerging from the cavernous sinus, on the inner side of the anterior clinoid process, and enters the orbit through the optic foramen, below, and on the outer side, of the optic nerve. It then crosses above, and to the inner side of this nerve, to the inner wall of the orbit, and, passing horizontally forwards, beneath the lower border of the Superior oblique muscle, passes to the inner angle of the eye, where it divides into two terminal branches, the frontal, and nasal.

*Branches.* The branches of this vessel may be divided into an orbital group, which are distributed to the orbit and surrounding parts; and an ocular group, which supply the muscles and globe of the eye.

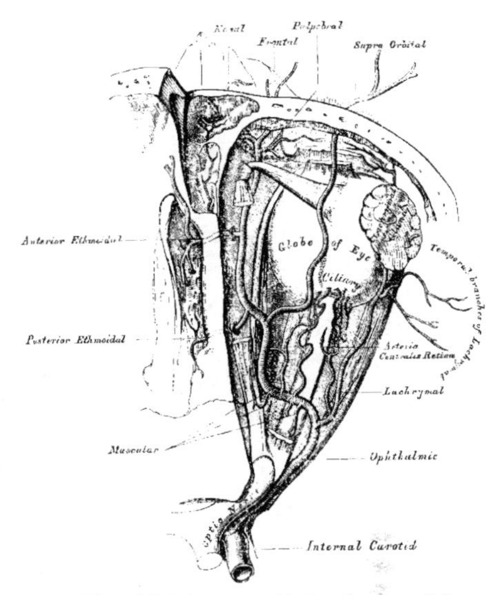


213 – *The ophthalmic artery and its branches, the roof of the orbit having been removed.*

| *Orbital Group.* | *Ocular Group.* |
|---|---|
| Lachrymal. | Muscular. |
| Supra-orbital. | Anterior ciliary. |
| Posterior ethmoidal. | Short ciliary. |
| Anterior ethmoidal. | Long ciliary. |
| Palpebral. | Arteria centralis retinæ. |
| Frontal. | |
| Nasal. | |

The *lachrymal* is the first, and one of the largest branches, derived from the ophthalmic, arising close to the optic foramen, and not unfrequently from that vessel before entering the orbit. It accompanies the lachrymal nerve along the upper border of the External rectus muscle, and is distributed to the lachrymal gland. Its terminal branches, escaping from the gland, are distributed to the upper eyelid and conjunctiva, anastomosing with the palpebral arteries. The lachrymal artery gives off one or two malar branches; one of which passes through a foramen in the malar bone to reach the temporal fossa and anastomoses with the deep temporal arteries. The other appears on the cheek, and anastomoses with the transverse facial. A branch is also sent backwards, through the sphenoidal fissure, to the dura mater, which anastomoses with a branch of the middle meningeal artery.

*Peculiarities.* The lachrymal artery is sometimes derived from one of the anterior branches of the middle meningeal artery.

The *supra-orbital artery*, the largest branch of the ophthalmic, arises from that vessel above the optic nerve. Ascending so as to rise above all the muscles of the orbit, it passes forwards, with the frontal nerve, between the periosteum and Levator palpebræ; and, passing through the supra-orbital foramen, divides into a superficial and deep branch, which supply the muscles and integument of the forehead and pericranium, anastomosing with the temporal, angular branch of the facial, and the artery of the opposite side. This artery in the orbit supplies the Superior rectus and the Levator palpebræ, sends a branch inwards, across the pulley of the Superior oblique muscle, to supply the parts at the inner canthus; and at the supra-orbital foramen, frequently transmits a branch to the diploë.

The *ethmoidal branches* are two in number; posterior and anterior. The former, which is the smaller, passes through the posterior ethmoidal foramen, supplies the posterior ethmoidal cells, and, entering the cranium, gives off a meningeal branch, which supplies the adjacent dura mater, and nasal branches which descend into the nose through apertures in the cribriform plate, anastomosing with branches of the spheno-palatine. The anterior ethmoidal artery accompanies the nasal nerve

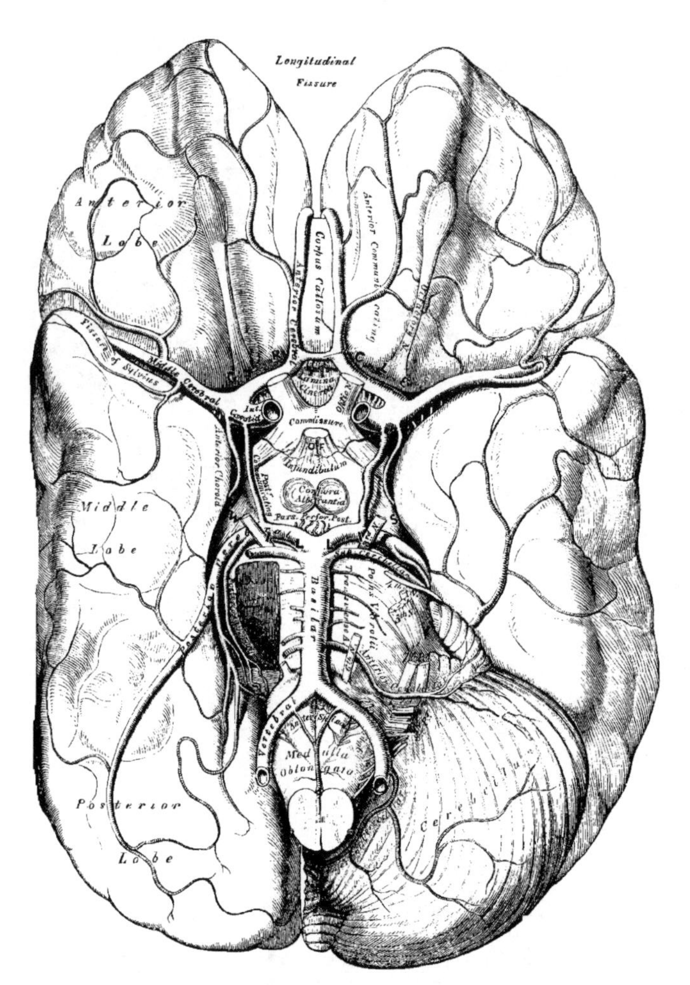


*214 – The arteries of the base of the brain. The right half of the cerebellum and pons have been removed.*

through the anterior ethmoidal foramen, supplies the anterior ethmoidal cells, and frontal sinuses, and, entering the cranium, divides into a meningeal branch, which supplies the adjacent dura mater, and a nasal branch which descends into the nose, through an aperture in the cribriform plate.

The *palpebral arteries,* two in number, superior and inferior, arise from the ophthalmic, opposite the pulley of the Superior oblique muscle; they encircle the eyelids near their free margin, forming a superior and an inferior arch, which lie between the Orbicularis muscle and tarsal cartilages; the superior palpebral inosculating at the outer angle of the orbit with the orbital branch of the temporal artery; the inferior palpebral anastomosing with the orbital branch of the infra-orbital artery, at the inner side of the lid. From this anastomosis, a branch passes to the nasal duct, ramifying, in its mucous membrane, as far as the inferior meatus.

The *frontal artery*, one of the terminal branches of the ophthalmic, passes from the orbit at its inner angle, and, ascending on the forehead, supplies the muscles, integument, and pericranium, anastomosing with the supra-orbital artery.

The *nasal artery*, the other terminal branch of the ophthalmic, emerges from the orbit above the tendo oculi, and, after giving a branch to the lachrymal sac, divides into two, one of which anastomoses with the angular artery, the other branch, the dorsalis nasi, runs along the dorsum of the nose, supplies its entire surface, and anastomoses with the artery of the opposite side.

The *ciliary arteries* are divisible into three groups, the short, long, and anterior.

The *short ciliary arteries*, from twelve to fifteen in number, arise from the ophthalmic, or some of its branches; they surround the optic nerve as they pass forwards to the posterior part of the eyeball, pierce the sclerotic coat around the entrance of this nerve, and supply the choroid coat and ciliary processes.

The *long ciliary arteries*, two in number, also pierce the posterior part of the sclerotic, and run forwards, along each side of the eyeball, between the sclerotic and choroid, to the ciliary ligament, where they divide into two branches; these form an arterial circle around the circumference of the iris, from which numerous radiating branches pass forwards, in its substance, to its free margin, where they form a second arterial circle around its pupillary margin.

The *anterior ciliary arteries* are derived from the muscular branches; they pierce the sclerotic a short distance from the cornea, and terminate in the great arterial circle of the iris.

The *arteria centralis retinæ*, is one of the smallest branches of the ophthalmic artery. It arises near the optic foramen, pierces the optic nerve obliquely, and runs forwards, in the centre of its substance, to the retina, in which its branches are distributed as far forwards as the ciliary processes. In the human fœtus, a small vessel passes forwards, through the vitreous humour, to the posterior surface of the capsule of the lens.

The *muscular branches*, two in number, superior and inferior, supply the muscles of the eyeball. The superior, the smaller, often wanting, supplies the Levator palpebræ, Superior rectus, and Superior oblique. The inferior, more constant in its existence, passes forwards, between the optic nerve and Inferior rectus, and is distributed to the External and Inferior recti, and Inferior oblique. This vessel gives off most of the anterior ciliary arteries.

The *cerebral branches* of the internal carotid are, the anterior cerebral, the middle cerebral, the posterior communicating, and the anterior choroid.

The *anterior cerebral* arises from the internal carotid, and the inner extremity of the fissure of Sylvius. It passes forwards in the great longitudinal fissure between the two anterior lobes of the brain, being connected, soon after its origin, with the vessel of the opposite side by a short anastomosing trunk, about two lines in length, the anterior communicating. The two anterior cerebral arteries, lying side by side curve round the anterior border of the corpus callosum, and run along its upper surface to its posterior part, where they terminate by anastomosing with the posterior cerebral arteries. They supply the olfactory and optic nerves, the under surface of the anterior lobes, the third ventricle, the anterior perforated space, the corpus callosum, and the inner surface of the hemispheres.

The *anterior communicating artery* is a short branch, about two lines in length but of moderate size, connecting together the two anterior cerebral arteries across the longitudinal fissure. Sometimes this vessel is wanting, the two arteries joining together to form a single trunk, which afterwards subdivides. Or the vessel may be wholly or partially subdivided into two; frequently, it is longer and smaller than usual.

The *middle cerebral artery*, the largest branch of the internal carotid, passes obliquely outwards along the fissure of Sylvius, within which it divides into three branches: an anterior, which supplies the pia mater, investing the surface of the anterior lobe; a posterior, which supplies the middle lobe; and a median branch which supplies the small lobe at the outer extremity of the Sylvian fissure. Near its origin, this vessel gives off numerous small branches, which enter the substantia perforata, to be distributed to the corpus striatum.

The *posterior communicating artery* arises from the back part of the internal carotid, runs directly backwards, and anastomoses with the posterior cerebral, a branch of the basilar. This artery varies considerably in size, being sometimes small, and occasionally so large that the posterior cerebral may be considered as arising from the internal carotid rather than from the basilar. It is frequently larger on one than on the other side.

The *anterior choroid* is a small but constant branch which arises from the back part of the internal carotid, near the posterior communicating artery. Passing backwards and outwards, it enters the descending horn of the lateral ventricle, beneath the edge of the middle lobe of the brain. It is distributed to the hippocampus major, corpus fimbriatum, and choroid plexus.

## ARTERIES OF THE UPPER EXTREMITY.

The artery which supplies the upper extremity, continues as a single trunk from its commencement, as far as the elbow; but different portions of it have received different names, according to the region through which it passes. Thus, that part of the vessel which extends from its origin, as far as the outer border of the first rib, is termed the subclavian; beyond this point to the lower border of the axilla, it is termed the axillary; and from the lower margin of the axillary space to the bend of the elbow, it is termed brachial; here, the single trunk terminates by dividing into two branches, the radial, and ulnar, an arrangement precisely similar to what occurs in the lower limb.

## Subclavian Arteries.

The subclavian artery on the right side arises from the arteria innominata, opposite the right sterno-clavicular articulation; on the left side, it arises from the arch of the aorta. It follows, therefore, that these two vessels must, in the first part of their course, differ in their length, their direction, and in their relation with neighbouring parts.

In order to facilitate the description of these vessels, more especially in a surgical point of view, each subclavian artery has been divided into three parts. The first portion, on the right side, ascends obliquely outwards, from the origin of the vessel to the inner border of the Scalenus anticus. On the left side, it ascends perpendicularly to the inner border of that muscle. The second part passes outwards, behind the Scalenus anticus: and the third part passes from the outer margin of that muscle, beneath the clavicle, to the lower border of the first rib, where it becomes the axillary artery. The first portions of these two vessels, differ so much in their course, and in their relation with neighbouring parts, that they will be described separately. The second and third parts are precisely alike on both sides.

## First Part of the Right Subclavian Artery (Figs. 205, 207).

It arises from the arteria innominata, opposite the right sterno-clavicular articulation, passes upwards and outwards across the root of the neck, and terminates at the inner margin of the Scalenus anticus muscle. In this part of its course, it ascends a little above the clavicle, the extent to which it does so varying in different cases. It is covered, *in front*, by the integument, superficial fascia, Platysma, deep fascia, the clavicular origin of the Sterno-mastoid, the Sterno-hyoid and Sterno-thyroid muscles, and another layer of the deep fascia. It is crossed by the internal jugular and vertebral veins, and by the pneumogastric, the cardiac branches of the sympathetic, and phrenic nerves. *Beneath*, the artery is invested by the pleura, and *behind*, it is separated by a cellular interval from the Longus colli, the transverse process of the seventh cervical vertebra, and the sympathetic; the recurrent laryngeal nerve winding around the lower and back part of this vessel. The subclavian vein lies below the subclavian artery, immediately behind the clavicle.

## Plan of Relations of First Portion of Right Subclavian Artery.

*In front.*

Integument, and superficial fascia.
Platysma and deep fascia.
Clavicular origin of Sterno-mastoid.
Sterno-hyoid and Sterno-thyroid.
Internal jugular and vertebral veins.
Pneumogastric, cardiac and phrenic nerves.

Right Subclavian Artery First portion.

*Beneath.*
Pleura.

*Behind.*

Recurrent laryngeal nerve.
Sympathetic.
Longus colli.
Transverse process of seventh cervical vertebra.

## First Part of the Left Subclavian Artery (Fig. 205).

It arises from the end of the transverse portion of the arch of the aorta, opposite the second dorsal vertebra, and ascends to the inner margin of the first rib, behind the insertion of the Scalenus anticus muscle. This vessel is, therefore, longer than the right, situated more deeply in the cavity of the chest, and directed almost vertically upwards, instead of arching outwards like the vessel of the opposite side.

It is in relation, *in front*, with the pleura, the left lung, the pneumogastric, phrenic, and cardiac nerves, which lie parallel with it, the left carotid artery, left internal jugular and innominate veins, and is covered by the Sterno-thyroid, Sterno-hyoid, and Sterno-mastoid muscles; *behind*, with the œsophagus, thoracic duct, inferior cervical ganglion of the sympathetic, Longus colli, and, vertebral column, To its *inner side* is the œsophagus, trachea, and thoracic duct; to its *outer side*, the pleura.

## Plan of Relations of First Portion of Left Subclavian Artery.

*In front.*

Pleura and left lung.
Pneumogastric, cardiac, and phrenic nerves.
Left carotid artery.
Left internal jugular and innominate veins.
Sterno-thyroid, Sterno-hyoid, and Sterno-mastoid muscles.

| *Inner side.* | Left Subclavian Artery First portion. | *Outer side.* |
|---|---|---|
| Œsophagus.<br>Trachea.<br>Thoracic duct. | | Pleura. |

*Behind.*

Œsophagus and thoracic duct.
Inferior cervical ganglion of sympathetic.
Longus colli and vertebral column.

The relations of the second and third portions of the subclavian arteries are precisely similar on both sides.

The *Second Portion of the Subclavian Artery* lies between the two Scaleni muscles; it is very short, and forms the highest part of the arch described by that vessel.

*Relations.* It is covered, *in front*, by the integument, Platysma, Sterno-mastoid, cervical fascia, and by the phrenic nerve, which is separated from the artery by the Scalenus anticus muscle. *Behind*, it is in relation with the Middle scalenus. *Above*, with the brachial plexus of nerves. *Below*, with the pleura. The subclavian vein lies below and in front of the artery, separated from it by the Scalenus anticus.

## Plan of Relations of Second Portion of Subclavian Artery.

*In front.*

| | |
|---|---|
| Platysma. | Scalenus anticus. |
| Sterno-mastoid. | Phrenic nerve. |
| Cervical fascia. | Subclavian Vein. |

| *Above.* | Subclavian Artery. Second portion. | *Below.* |
|---|---|---|
| Brachial plexus. | | Pleura. |

*Behind.*

Middle scalenus.

The *Third Portion of the Subclavian Artery* passes downwards and outwards from the outer margin of the Scalenus anticus to the lower border of the first rib, where it becomes the axillary artery. This portion of the vessel is the most superficial, and is contained in a triangular space, the base of which is formed in front by the Anterior scalenus, and the two sides by the Omo-hyoid above and the clavicle below.

*Relations.* It is covered, *in front*, by the integument, the superficial fascia, the Platysma, deep fascia; and by the clavicle, the Subclavius muscle, and the supra-scapular artery and vein; the clavicular descending branches of the cervical plexus and the nerve to the Subclavius pass vertically downwards in front of the artery. The external jugular vein crosses it at its inner side, and receives the supra-scapular and transverse cervical veins, which occasionally form a plexus in front of it. The subclavian vein is below the artery, lying close behind the clavicle. *Behind*, it lies on the Middle scalenus muscle. *Above* it, and to its outer side, is the brachial plexus, and Omo-hyoid muscle. *Below*, it rests on the outer surface of the first rib.

## Plan of Relations of Third Portion of Subclavian Artery.

*In front.*

Integument, fasciæ, and Platysma.
The external jugular, supra-scapular, and transverse cervical veins.
Descending branches of cervical plexus.
Subclavius muscle, supra-scapular artery, and clavicle.

| *Above.* | Subclavian Artery. Third portion. | *Below.* |
|---|---|---|
| Brachial plexus.<br>Omo-hyoid. | | First rib. |

*Behind.*
Scalenus medius.

*Peculiarities.* The subclavian arteries vary in their origin, their course, and in the height to which they rise in the neck.

*The origin* of the right subclavian from the innominate takes place, in some cases, above the sterno-clavicular articulation; more frequently in the cavity of the thorax, below that joint. Or the artery may arise as a separate trunk from the arch of the aorta; in such cases it may be either the first, second, third, or even the last branch derived from that vessel: in the majority of cases, it is the first or last, rarely the second or third.

When it is the first branch, it occupies the ordinary position of the innominate artery; when the second or third, it gains its usual position by passing behind the right carotid: and when the last branch, it arises from the left extremity of the arch, at its upper or back part, and passes obliquely towards the right side, behind the œsophagus and right carotid, sometimes between the œsophagus and trachea, to the upper border of the first rib, where it follows its ordinary course. In very rare instances, this vessel arises from the thoracic aorta, as low down as the fourth dorsal vertebra. Occasionally it perforates the Anterior scalenus; more rarely it passes in front of this muscle: sometimes the subclavian vein passes with the artery behind the Scalenus. The artery sometimes ascends as high as an inch and a half above the clavicle, or to any intermediate point between this and the upper border of this bone, the right subclavian usually ascending higher than the left.

The left subclavian is occasionally joined at its origin with the left carotid.

*Surgical Anatomy.* The relations of the subclavian arteries of the two sides having been examined, the student should direct his attention to consider the best position in which compression of the vessel may be effected, or in what situation a ligature may be best applied in cases of aneurism or wounds.

*Compression of the subclavian artery* is required in cases of operations about the shoulder, in the axilla, or at the upper part of the arm; and the student will observe that there is only one situation in which it can be effectually applied, viz., where the artery passes across the outer surface of the first rib. In order to compress the vessel in this situation, the shoulder should be depressed, and the surgeon, grasping the side of the neck, may press with his thumb in the hollow behind the clavicle downwards against the rib; if from any cause the shoulder cannot be sufficiently depressed, pressure may be made from before backwards, so as to compress the artery against the Middle scalenus and transverse process of the seventh cervical vertebra.

*Ligature of the subclavian artery* may be required in cases of wounds of the axillary artery, or in aneurism of that vessel; and the third part of the artery is consequently that which is most favourable for such an operation, on account of its being comparatively superficial, and most remote from the origin of the large branches. In those cases where the clavicle is not displaced, this operation may be performed with comparative facility; but where the clavicle is elevated from the presence of a large aneurismal tumour in the axilla, the artery is placed at a great depth from the surface, which materially increases the difficulty of the operation. Under these circumstances, it becomes a matter of importance to consider the height to which this vessel reaches above the bone. In ordinary cases, its arch is about half an inch above the clavicle, occasionally as high as an inch and a half, and sometimes so low as to be on a level with its upper border. If displacement of the clavicle occurs, these variations will necessarily make the operation more or less difficult, according as the vessel is more or less accessible.

The chief points in the operation of tying the third portion of the subclavian artery are as follows. The patient being placed on a table in the horizontal position, and the shoulder depressed as much as possible, the integument should be drawn downwards upon the clavicle and an incision made through it upon that bone from the anterior border of the Trapezius to the posterior border of the Sterno-mastoid, to which may be added a short vertical incision meeting the centre of the preceding; the Platysma and cervical fascia should be divided upon a director, and if the interval between the Trapezius and Sterno-mastoid muscles be insufficient for the performance of the operation, a portion of one or both may be divided. The external jugular vein will now be seen towards the inner side of the wound; this and the supra-scapular and transverse cervical veins which terminate in it should be held aside, and if divided both ends should be included in a ligature: the supra-scapular artery should be avoided, and the Omo-hyoid muscle must now be looked for, and held aside if necessary. In the space beneath this muscle, careful search must be made for the vessel; the deep fascia having been divided with the finger-nail or silver scalpel, the outer margin of the Scalenus muscle must be felt for, and the finger being guided by it to the first rib, the pulsation of the subclavian artery will be felt as it passes over its surface. The aneurism needle may then be passed around the vessel from before backwards, by which means the vein will be avoided, care being taken not to include a branch of the brachial plexus instead of the artery in the ligature. If the clavicle is so raised by the tumour that the application of the ligature cannot be effected in this situation, the artery may be tied above the first rib, or even behind the Scalenus muscle: the difficulties of the operation in such a case will be materially increased, on account of the greater depth of the artery, and alteration of the surrounding parts.

The second division of the subclavian artery, from being that portion which rises highest in the neck, has been considered favourable for the application of the ligature, where it is difficult to apply it in the third part of its course. There are, however, many objections to the operation in this situation. It is necessary to divide the Scalenus anticus muscle upon which lies the phrenic nerve, and at the inner side of which is situated the internal jugular vein; a wound of either of these structures might lead to the most dangerous consequences. Again, the artery is in contact, below, with the pleura, which must also be avoided; and, lastly, the proximity of so many of its larger branches arising internal to this point, must be a still further objection to the operation. If, however, it has been determined upon to perform the operation in this situation, it should be remembered that it occasionally happens, that the artery passes in front of the Scalenus anticus, or through the fibres of that muscle; or that the vein sometimes passes with the artery behind the Scalenus anticus.

In those cases of aneurism of the axillary or subclavian artery which encroach upon the outer portion of the Scalenus muscle to such an extent that a ligature cannot be applied in that situation, it may be deemed advisable, as a last resource, to tie the first portion of the subclavian artery. On the left side, this operation is quite impracticable; the great depth of the artery from the surface, its intimate relation with the pleura, and its close proximity with so many important veins and nerves, present a series of difficulties which it is impossible to overcome. On the right side, the operation is practicable, and has been performed, though not with success. The main objection to the operation in this situation is the smallness of the interval which usually exists between the commencement of the vessel, and the origin of the nearest branch. This operation may be performed in the following manner. The patient being placed on a table in the horizontal position, with the neck extended, an incision should be made parallel with the inner part of the clavicle, and a second along the inner border of the Sterno-mastoid, meeting it at right angles. The sternal attachment of the Sterno-mastoid may now be divided on a director, and turned outwards; a few small arteries and veins, and occasionally the anterior jugular, must be avoided, and the Sterno-hyoid and thyroid muscles divided in the same manner as the preceding muscle. After tearing through the deep fascia with the finger-nail, the internal jugular vein will be seen crossing the artery; this should be pressed aside, and the artery secured by passing the needle from below upwards, by which the pleura is more effectually avoided. The exact position of the vagus nerve, the recurrent laryngeal, the phrenic and sympathetic nerves, should be remembered, and the ligature should be applied near the origin of the vertebral, in order to afford as much room as possible for the formation of a coagulum between the ligature and the origin of the vessel. It should be remembered, that the right subclavian artery is occasionally deeply placed in the first part of its course, when it arises from the left side of the aortic arch, and passes in such cases behind the œsophagus, or between it and the trachea.

*Collateral Circulation.* After ligature of the third part of the subclavian artery, the collateral circulation is mainly established by three sets of vessels.

'1. A posterior set, consisting of the supra-scapular and posterior scapular branches of the subclavian, which anastomosed with the infra-scapular from the axillary.

'2. An internal set, produced by the connection of the internal mammary on the one hand, with the short and long thoracic arteries, and the infra-scapular, on the other.

'3. A middle or axillary set, which consisted of a number of small vessels derived from branches of the subclavian, above; and passing through the axilla, to terminate either in the main trunk, or some of the branches of the axillary, below. This last set presented most conspicuously the peculiar character of newly-formed, or, rather, dilated arteries,' being excessively tortuous, and forming a complete plexus.

'The chief agent in the restoration of the axillary artery below the tumour, was the infra-scapular artery, which communicated most freely with the internal mammary, supra-scapular, and posterior scapular branches of the subclavian, from all of which it received so great an influx of blood as to dilate it to three times its natural size.'*

## Branches of the Subclavian Artery.

These are four in number. Three arising from the first portion of the vessel, the vertebral, the internal mammary, and the thyroid axis; and one from the second portion, the superior intercostal. The vertebral arises from the upper and back part of the first portion of the artery; the thyroid axis from the front, and the internal mammary from the under part of this vessel. The superior intercostal is given off from the upper and back part of the second portion of the artery. On the left side, the second portion usually gives off no branch, the superior intercostal arising at the inner side of the Scalenus anticus. On both sides of the body, the first three branches arise close together at the inner margin of the Scalenus anticus; in the majority of cases, a free interval of half an inch to an inch existing between the commencement of the artery and the origin of the nearest branch; in a smaller number of cases, an interval of more than an inch existed, never exceeding an inch and three-quarters. In a very few instances, the interval was less than half an inch.

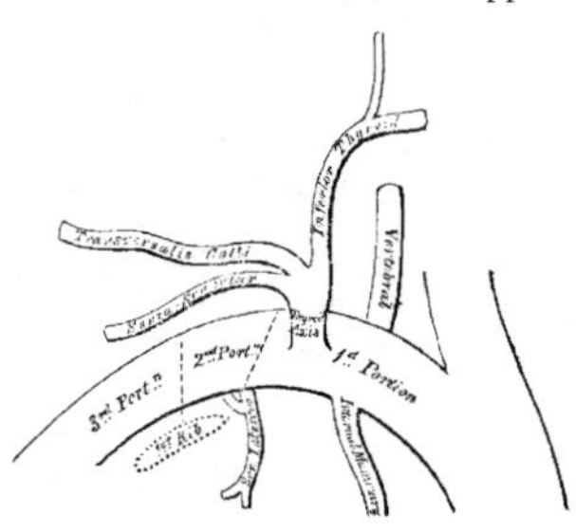

*215 – Plan of the branches of the right subclavian artery.*

The Vertebral Artery (fig. 212) is generally the first and largest branch of the subclavian; it arises from the upper and back part of the first portion of the vessel, and, passing upwards, enters the foramen in the transverse process of the sixth cervical vertebra, and ascends through the foramina in the transverse processes of all the vertebræ above this. Above the upper border of the axis, it inclines outwards and upwards to the foramen in the transverse process of the atlas, through which it passes; it then winds backwards behind its articular process, runs in a deep groove on the surface of the posterior arch of this bone, and, piercing the posterior occipito-atloid ligament and dura mater, enters the skull through the foramen magnum. It then passes in front of the medulla oblongata, and unites with the vessel of the opposite side at the lower border of the pons Varolii, to form the basilar artery.

At its origin, it is situated behind the internal jugular vein, and inferior thyroid artery; and, near the spine, lies between the Longus colli and Scalenus anticus muscles, having the thoracic duct in front of it on the left side. Within the foramina formed by the transverse processes of the vertebræ, it is accompanied by a plexus of nerves from the sympathetic, and lies between the vertebral vein, which is in front, and the cervical nerves, which issue from the intervertebral foramina behind it. Whilst winding round the articular process of the atlas, it is contained in a triangular space formed by the Rectus posticus minor, the Superior and Inferior oblique muscles; and it is covered by the Rectus posticus major and Complexus. And within the skull, as it winds round the medulla oblongata, it is placed between the hypoglossal and anterior root of the sub-occipital nerves.

*Branches.* These may be divided into two sets, those given off in the neck, and those within the cranium.

| *Cervical Branches.* | *Cranial Branches.* |
| --- | --- |
| Lateral spinal. | Posterior meningeal. |
| Muscular. | Anterior spinal. |
| | Posterior spinal. |
| | Inferior cerebellar. |

The *lateral spinal branches* enter the spinal canal through the intervertebral foramina, each dividing into two branches. Of these, one passes along the roots of the nerves, to supply the spinal cord and

* *Guy's Hospital Reports*, vol. i. 1836. Case of axillary aneurism, in which Mr. Aston Key had tied the subclavian artery on the outer edge of the Scalenus muscle, twelve years previously.

its membranes, anastomosing with the other spinal arteries; the other is distributed to the posterior surface of the bodies of the vertebræ.

*Muscular branches* are given off to the deep muscles of the neck, where the vertebral artery curves round the articular process of the atlas. They anastomose with the occipital and deep cervical arteries.

The *posterior meningeal* are one or two small branches given off from the vertebral opposite the foramen magnum. They ramify between the bone and dura mater in the cerebellar fossæ, and supply the falx cerebelli.

The *anterior spinal* is a small branch, larger than the posterior spinal, which arises near the termination of the vertebral, and unites with its fellow of the opposite side in front of the medulla oblongata. The single trunk thus formed descends a short distance on the front of the spinal cord, and joins with a succession of small branches which enter the spinal canal through some of the intervertebral foramina; these branches are derived from the vertebral and ascending cervical, in the neck; from the intercostal, in the dorsal region; and from the lumbar, ilio-lumbar, and lateral sacral arteries in the lower part of the spine. They unite, by means of ascending and descending branches, to form a single anterior median artery, which extends as far as the lower part of the spinal cord. This vessel is placed beneath the pia mater along the anterior median fissure; it supplies that membrane and the substance of the cord, and sends off branches at its lower part, to be distributed to the cauda equina.

The *posterior spinal* arises from the vertebral, at the side of the medulla oblongata; passing backwards to the posterior aspect of the spinal cord, it descends on either side, lying behind the posterior roots of the spinal nerves; and is reinforced by a succession of small branches, which enter the spinal canal through the intervertebral foramina, and by which it is continued to the lower part of the cord, and to the cauda equina. Branches from these vessels form a free anastomosis round the posterior roots of the spinal nerves, and communicate, by means of very tortuous transverse branches, with the vessel of the opposite side. At its commencement, it gives off an ascending branch, which terminates on the side of the fourth ventricle.

The *inferior cerebellar artery*, the largest branch of the vertebral, winds backwards round the upper part of the medulla oblongata, passing between the origin of the spinal accessory and pneumogastric nerves, over the restiform body, to the under surface of the cerebellum, where it divides into two branches; an internal one, which is continued backwards to the notch between the two hemispheres of the cerebellum; and an external one, which supplies the under surface of the cerebellum, as far as its outer border, where it anastomoses with the superior cerebellar. Branches from this artery supply the choroid plexus of the fourth ventricle.

The *Basilar artery*, so named from its position at the base of the skull, is a single trunk, formed by the junction of the two vertebral arteries; it extends from the posterior to the anterior border of the pons Varolii, where it divides into two terminal branches, the posterior cerebral arteries. Its branches are, on each side, the following:

| | |
|---|---|
| Transverse. | Superior cerebellar. |
| Anterior cerebellar. | Posterior cerebral. |

The *transverse* branches supply the pons Varolii and adjacent parts of the brain; one accompanies the auditory nerve into the internal auditory meatus; and another, of larger size, passes along the crus cerebelli, to be distributed to the anterior border of the under surface of the cerebellum. It is called the *anterior (inferior) cerebellar artery*.

The *superior cerebellar arteries* arise near the termination of the basilar. They wind round the crus cerebri, close to the fourth nerve, and, arriving at the upper surface of the cerebellum, divide into branches which supply the pia mater, covering its surface, anastomosing with the inferior cerebellar. It gives several branches to the pineal gland, and also to the velum interpositum.

The *posterior cerebral arteries*, the two terminal branches of the basilar, are larger than the preceding, from which they are separated near their origin by the third nerves. Winding round the crus cerebri, they pass to the under surface of the posterior lobes of the cerebrum, which they supply, anastomosing with the anterior and middle cerebral arteries. Near their origin, they give off numerous branches, which enter the posterior perforated spot, and receive the posterior communicating arteries from the internal carotid. They also give off a branch, the posterior choroid, which supplies the velum interpositum and choroid plexus, entering the interior of the brain, beneath the posterior border of the corpus callosum.

*Circle of Willis.* The remarkable anastomosis which exists between the branches of the internal carotid, and vertebral arteries at the base of the brain, constitutes the circle of Willis. It is formed, in front, by the anterior cerebral and anterior communicating arteries; on each side, by the trunk of the internal carotid, and the posterior communicating; behind, by the posterior cerebral, and point of the basilar. It is by this anastomosis that the cerebral circulation is equalized, and provision made for effectually carrying it on if one or more of the branches are obliterated. The parts of the brain included within this arterial circle are, the lamina cinerea, the commissure of the optic nerves, the infundibulum, the tuber cinereum, the corpora albicantia, and the pars perforata postica.

The Thyroid Axis is a short, thick trunk, which arises from the fore part of the first portion of the subclavian artery, close to the inner side of the Scalenus anticus muscle, and divides, almost immediately after its origin, into three branches, the inferior thyroid, supra-scapular, and transversalis colli.

The Inferior Thyroid Artery passes upwards, in a serpentine course, behind the sheath of the common carotid vessel and sympathetic nerve (the middle cervical ganglion resting upon it), and is distributed to the under surface of the thyroid gland, anastomosing with the superior thyroid, and with the corresponding artery of the opposite side. Its branches are the

| | |
|---|---|
| Laryngeal. | Œsophageal. |
| Tracheal. | Ascending cervical. |

The *laryngeal* branch ascends upon the trachea to the back part of the larynx and supplies the muscles and the mucous membrane of this part.

The *tracheal* branches are distributed upon the trachea, anastomosing below with the bronchial arteries.

The *œsophageal* branches are distributed to the œsophagus.

The *ascending cervical* is a small branch which arises from the inferior thyroid, just where that vessel is passing behind the common carotid artery, and runs up the neck in the interval between the Scalenus anticus, and Rectus anticus major. It gives branches to the muscles of the neck, which communicate with those sent out from the vertebral, and sends one or two through the intervertebral foramina along the cervical nerves, to supply the bodies of the vertebræ, the spinal cord and its membranes.

The Supra-Scapular Artery, smaller than the transversalis colli, passes obliquely from within outwards, across the root of the neck. It at first lies on the lower part of the Scalenus anticus, being covered by the Sterno-mastoid; it then crosses the subclavian artery, and runs outwards behind and parallel with the clavicle and Subclavius muscle, and beneath the posterior belly of the Omo-hyoid, to the superior border of the scapula, where it passes over the transverse ligament of the scapula to the supra-spinous fossa. In this situation it lies close to the bone, and ramifies between it and the Supra-spinatus muscle to which it is mainly distributed, giving off a communicating branch, which crosses the neck of the scapula, to reach the infra-spinous fossa, where it anastomoses with the dorsal branch of the subscapular artery. Besides distributing branches to the Sterno-mastoid, and neighbouring muscles, it gives off a supra-acromial branch, which, piercing the Trapezius muscle, supplies the cutaneous surface of the acromion, anastomosing with the acromial thoracic artery. As the artery passes across the supra-scapular notch, a branch descends into the subscapular fossa, ramifies beneath that muscle, and anastomoses with the posterior and subscapular arteries. It also supplies the shoulder joint.

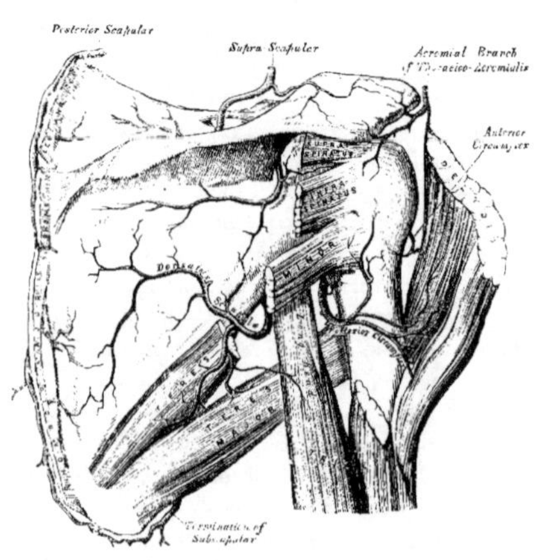


216 – *The scapular and circumflex arteries.*

The Transversalis Colli passes transversely outwards, across the upper part of the subclavian triangle, to the anterior margin of the Trapezius muscle, beneath which it divides into two branches, the superficial cervical, and the posterior scapular. In its passage across the neck, it crosses in front of the Scaleni muscles and the brachial plexus, between the divisions of which it sometimes passes and is covered by the Platysma, Sterno-mastoid, Omo-hyoid, and Trapezius muscles.

The *superficial cervical* ascends beneath

the anterior margin of the Trapezius distributing branches to it, and to the neighbouring muscles and glands in the neck.

The *posterior scapular*, the continuation of the transversalis colli, passes beneath the Levator anguli scapulæ to the superior angle of the scapula, and descends along the posterior border of that bone as far as the inferior angle, where it anastomoses with the subscapular branch of the axillary. In its course it is covered by the Rhomboid muscles, supplying these, the Latissimus dorsi and Trapezius, and anastomosing with the supra-scapular and subscapular arteries, and with the posterior branches of some of the intercostal arteries.

*Peculiarities.* The *superficial cervical* frequently arises as a separate branch from the thyroid axis; and the posterior scapular, from the third, more rarely from the second, part of the subclavian.

The Internal Mammary arises from the under surface of the first portion of the subclavian artery, opposite the thyroid axis. It descends behind the clavicle, to the inner surface of the anterior wall of the chest, resting upon the costal cartilages, a short distance from the margin of the sternum; and, at the interval between the sixth and seventh cartilages, divides into two branches, the musculo-phrenic, and superior epigastric.

At its origin, it is covered by the internal jugular and subclavian veins, and crossed by the phrenic nerve. In the upper part of the thorax, it lies upon the costal cartilages, and internal Intercostal muscles in front, covered by the pleura behind. At the lower part of the thorax, the Triangularis sterni separates this vessel from the pleura. It is accompanied by two veins, which join at the upper part of the thorax into a single trunk.

The branches of the internal mammary are,

Comes nervi phrenici (superior phrenic).
Mediastinal.
Pericardiac.
Sternal.
Anterior intercostal.
Perforating.
Musculo-phrenic.
Superior epigastric.

The *comes nervi phrenici* (*superior phrenic*), is a long slender branch, which accompanies the phrenic nerve, between the pleura and pericardium, to the Diaphragm, to which it is distributed; anastomosing with the other phrenic arteries from the internal mammary, and abdominal aorta.

The *mediastinal branches* are small vessels, which are distributed to the areolar tissue in the anterior mediastinum, and the remains of the thymus gland.

The *pericardiac branches* supply the upper part of the pericardium, the lower part receiving branches from the musculo-phrenic artery. Some *sternal* branches are distributed to the Triangularis sterni, and both surfaces of the sternum.

The *anterior intercostal arteries* supply the five or six upper intercostal spaces. The branch corresponding to each space passes outwards, and soon divides into two, which run along the opposite borders of the ribs, and inosculate with the intercostal arteries from the aorta. They are at first situated between the pleura and the internal Intercostal muscles, and then between the two layers of these muscles. They supply the Intercostal and Pectoral muscles, and the mammary gland.

The *anterior* or *perforating arteries* correspond to the five or six upper intercostal spaces. They arise from the internal mammary, pass forwards through the intercostal spaces, and, curving outwards, supply the Pectoralis major, and the integument. Those which correspond to the first three spaces, are distributed to the mammary gland. In females, during lactation, these branches are of large size.

The *musculo-phrenic artery* is directed obliquely downwards and outwards, behind the cartilages of the false ribs, perforating the Diaphragm at the eighth or ninth rib, and terminating, considerably reduced in size, opposite the last intercostal space. It gives off anterior intercostal arteries to each of the intercostal spaces across which it passes; they diminish in size as the spaces decrease in length, and are distributed in a manner precisely similar to the anterior intercostals from the internal mammary. It also gives branches backwards to the Diaphragm, and downwards to the abdominal muscles.

The *superior epigastric* continues in the original direction of the internal mammary, descends behind the Rectus muscle, and, perforating its sheath, divides into branches which supply the Rectus, anastomosing with the epigastric artery from the external iliac. Some vessels perforate the sheath of the Rectus, and supply the muscles of the abdomen and the integument, and a small branch which passes inwards upon the side of the ensiform appendix, anastomoses in front of that cartilage with the artery of the opposite side.

The SUPERIOR INTERCOSTAL arises from the upper and back part of the subclavian artery, beneath the anterior scalenus on the right side, and to the inner side of the muscle on the left side. Passing backwards, it gives off the deep cervical branch, and then descends behind the pleura in front of the necks of the first two ribs, and inosculates with the first aortic intercostal. In the first intercostal space, it gives off a branch which is distributed in a similar manner with the aortic intercostals. The branch for the second intercostal space usually joins with one from the first aortic intercostal. Each intercostal gives off a branch to the posterior spinal muscles, and a small one, which passes through the corresponding intervertebral foramen to the spinal cord and its membranes.

The *deep cervical branch* (*profunda cervicis*) arises, in most cases, from the superior intercostal, and is analogous to the posterior branch of an aortic intercostal artery. Passing backwards, between the transverse process of the seventh cervical vertebra and the first rib, it ascends the back part of the neck, between the Complexus and Semi-spinalis colli muscles, as high as the axis, supplying these and adjacent muscles, and anastomosing with the arteria princeps cervicis of the occipital, and with branches which pass outwards from the vertebral.

## SURGICAL ANATOMY OF THE AXILLA.

The *Axilla* is a conical space, situated between the upper and lateral parts of the chest, and inner side of the arm.

*Boundaries.* Its *apex*, which is directed upwards towards the root of the neck, corresponds to the interval between the first rib internally, the superior border of the scapula externally, and the clavicle and Subclavius muscle in front. The *base*, directed downwards, is formed by the integument, and a thick layer of fascia, extending between the lower border of the Pectoralis major in front, and the lower border of the Latissimus dorsi behind; it is broad internally, at the chest, but narrow and pointed externally, at the arm. Its *anterior boundary* is formed by the Pectoralis major and minor muscles, the former covering the whole of the anterior wall of the axilla, the latter covering only its central part. Its *posterior boundary*, which extends somewhat lower than the anterior, is formed by the Subscapularis above, the Teres major and Latissimus dorsi below. On the *inner side* are the first four ribs and their corresponding Intercostal muscles, and part of the Serratus magnus. On the *outer side*, where the anterior and posterior boundaries converge, the space is narrow, and bounded by the humerus, the Coraco-brachialis and Biceps muscles.

*Contents.* This space contains the axillary vessels, and brachial plexus of nerves with their

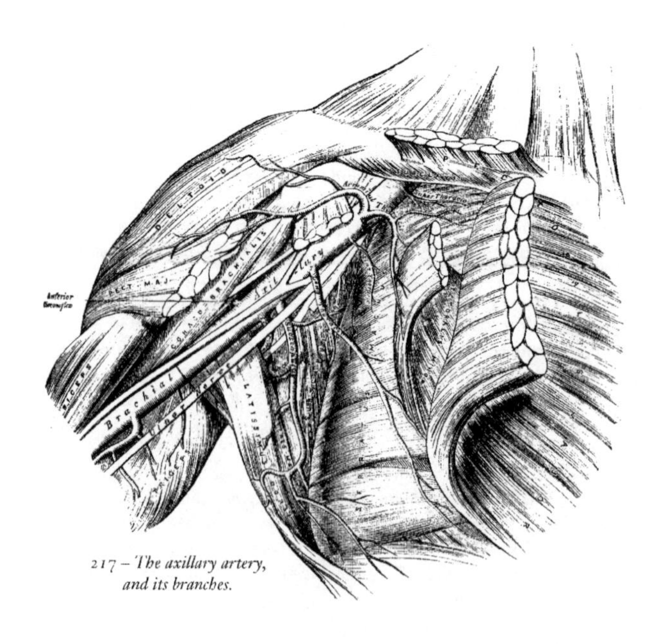

217 – *The axillary artery, and its branches.*

branches, some branches of the intercostal nerves, a large number of symphatic glands, all connected together by a quantity of fat and loose areolar tissue.

*Their Position.* The axillary artery and vein, with the brachial plexus of nerves, extend obliquely along the outer boundary of the axillary space, from its apex to its base, and are placed much nearer the anterior than the posterior wall, the vein lying to the inner or thoracic side of the artery, and altogether concealing it. At the fore part of the axillary space, in contact with the Pectoral muscles, are the thoracic branches of the axillary artery, and along the anterior margin of the axilla, the long thoracic artery extends to the side of the chest. At the back part, in contact with the lower margin of the Subscapularis muscle, are the subscapular vessels and nerves; winding around the lower border of this muscle, is the dorsalis scapulæ artery and veins; and towards the outer extremity of the muscle the posterior circumflex vessels and nerve are seen curving backwards to the shoulder.

Along the inner or thoracic side, no vessel of any importance exists, its upper part being crossed by a few small branches from the superior thoracic artery. There are some important nerves, however, in this situation; the posterior thoracic or external respiratory nerve, descending on the surface of the Serratus magnus, to which it is distributed; and perforating the upper and anterior part of this wall, are the intercosto-humeral nerves, which pass across the axilla to the inner side of the arm.

The cavity of the axilla is filled by a quantity of loose areolar tissue, a large number of small arteries and veins, all of which are, however, of inconsiderable size, and numerous lymphatic glands; these are from ten to twelve in number, and situated chiefly on the thoracic side, and lower and back part of this space.

The student should attentively consider the relation of the vessels and nerves in the several parts of the axilla; for it not unfrequently happens, that the surgeon is called upon to extirpate diseased glands, or to remove a tumour from this situation. In performing such an operation, it will be necessary to proceed with much caution in the direction of the outer wall and apex of the space, as here the axillary vessels will be in danger of being wounded. Towards the posterior wall it will be necessary to avoid the subscapular, dorsalis scapulæ, and posterior circumflex vessels, and, along the anterior wall, the thoracic branches. It is only along the inner or thoracic wall, and in the centre of the axillary cavity, that there are no vessels of any importance; a most fortunate circumstance, for it is in this situation more especially that tumours requiring removal, are most frequently situated.

## The Axillary Artery.

The axillary artery, the continuation of the subclavian, commences at the lower border of the first rib, and terminates at the lower border of the tendons of the Latissimus dorsi and Teres major muscles, when it becomes the brachial. Its direction varies with the position of the limb: when the arm lies by the side of the chest, the vessel forms a gentle curve, the convexity being upwards and outwards; when it is directed at right angles with the trunk, the vessel is nearly straight; and if elevated still higher, it describes a curve, the concavity of which is directed upwards. At its commencement the artery is very deeply situated but near its termination is superficial, being covered only by the skin and fascia. The description of the relations of this vessel may be facilitated by its division into three portions. The first portion being that above the Pectoralis minor; the second portion, beneath; and the third, below that muscle.

The *first portion* of the axillary artery is in relation, *in front*, with the clavicular portion of the Pectoralis major, the costo-coracoid membrane, and the cephalic vein; *behind*, with the first intercostal space, the corresponding Intercostal muscle, the first serration of the Serratus magnus, and the posterior thoracic nerve; on its *outer side* with the brachial plexus, from which it is separated by a little cellular interval; on its *inner*, or thoracic side, with the axillary vein.

### Relations of First Portion of the Axillary Artery.

*In front.*
Pectoralis major. Costo-coracoid membrane.
Cephalic vein.

*Outer side.*
Brachial plexus.

Axillary Artery. First portion

*Inner side.*
Axillary vein.

*Behind.*
First intercostal space, and Intercostal muscle.
First serration of Serratus magnus.
Posterior thoracic nerve.

The *second portion* of the axillary artery lies beneath the Pectoralis minor. It is covered, *in front*, by the Pectoralis major and minor muscles; *behind*, it is separated from the Subscapularis by a cellular interval; on the *inner side*, it is in contact with the axillary vein. The brachial plexus of nerves surrounds the artery, and separates it from direct contact with the vein and adjacent muscles.

### Relations of Second Portion of the Axillary Artery.

*In front.*
Pectoralis major and minor.

*Outer side.*
Brachial plexus.

Axillary Artery. Second portion.

*Inner side.*
Axillary vein.

*Behind.*
Subscapularis.

The *third portion* of the axillary artery lies below the Pectoralis minor. It is in relation, *in front*, with the lower border of the Pectoralis major above, being covered only by the integument and fascia below; *behind*, with the lower part of the Subscapularis, and the tendons of the Latissimus dorsi and Teres major; on its *outer side*, with the Coraco-brachialis; on its *inner*, or thoracic side, with the axillary vein. The brachial plexus of nerves bears the following relation to the artery in this part of its course: on the *outer side* is the median nerve, and the musculo-cutaneous for a short distance; on the *inner side*, the ulnar, the internal, and lesser internal cutaneous nerves; and *behind*, the musculo-spiral, and circumflex, the latter extending only to the lower border of the Subscapularis muscle.

### Relations of Third Portion of the Axillary Artery.

*In front.*
Integument and fascia. Pectoralis major.

*Outer side.*
Coraco-brachialis.
Median nerve.
Musculo-cutaneous nerve.

Axillary Artery. Third portion.

*Inner side.*
Ulnar nerve.
Internal cutaneous nerves.
Axillary vein.

*Behind.*
Subscapularis. Tendons of Latissimus dorsi, and Teres major.
Musculo-spiral, and circumflex nerves.

*Peculiarities.* The axillary artery, in about one case out of every ten, gives off a large branch, which forms either one of the arteries of the fore-arm, or a large muscular trunk. In the first set of cases, this artery is most frequently the radial (1 in 33), sometimes the ulnar (1 in 72), and, very rarely, the interosseous (1 in 506). In the second set of cases, the trunk gave origin to the subscapular, circumflex, and profunda arteries of the arm. Sometimes, only one of the circumflex, or one of the profunda arteries, arose from the trunk. In these cases, the brachial plexus surrounded the trunk of the branches, and not the main vessel.

*Surgical Anatomy.* The student having carefully examined the relations of the axillary artery in its various parts, should now consider in what situation compression of this vessel may be most easily effected, and the best position for the application of a ligature to it when necessary.

*Compression* of this vessel is required in the removal of tumours, or in amputation of the upper part of the arm; and the only situation in which this can be effectually made, is in the lower part of its course; on compressing it in this situation from within outwards upon the humerus, the circulation may be efficiently suspended.

The *application of a ligature to the axillary artery* may be required in cases of aneurism of the upper part of the brachial; and there are only two situations in which it may be secured, viz., in the upper, or in the lower part of its course; for the axillary artery at its central part is so deeply seated, and, at the same time, so closely surrounded with large nervous trunks, that the application of a ligature to it in this situation, would be almost impracticable.

In the *lower part* of its course, the operation is more simple, and may be performed in the following manner:—The patient being placed on a bed, and the arm separated from the side, with the hand supinated, the head of the humerus is felt for, and an incision made through the integument over it, about two inches in length, a little nearer to the anterior than the posterior fold of the axilla. After carefully dissecting through the areolar tissue and fascia, the median nerve and axillary vein are exposed; the former having been displaced to the outer, and the latter to the inner side of the arm, the elbow being at the same time bent, so as to relax these structures, and facilitate their separation; the ligature may be passed round the artery from the ulnar to the radial side. This portion of the artery is occasionally crossed by a muscular slip derived from the Latissimus dorsi, which may mislead the surgeon during an operation. It may easily be recognised by the transverse direction of its fibres (see p. 192).

The *upper portion* of the axillary artery may be tied, in cases of aneurism encroaching so far upwards that a ligature cannot be applied in the lower part of its course. Notwithstanding that this operation has been performed in some few cases, and with success, its performance is attended with much difficulty and danger. The student will remark, that in this situation, it would be necessary to divide a thick muscle, and, after separating the costo-coracoid membrane, the artery would be exposed at the bottom of a more or less deep space, with the cephalic and axillary veins in such relation with it as must render the application of a ligature to this part of the vessel particularly hazardous. Under such circumstances, it is an easier, and at the same time, more advisable operation, to tie the subclavian artery in the third part of its course.

In a case of wound of this vessel the general practice of cutting down upon, and tying it above and below the wounded point, should be adopted in all cases.

The branches of the axillary artery are, the

| | |
|---|---|
| *From 1st Part.* | Superior thoracic.<br>Acromial thoracic. |
| *From 2nd Part.* | Thoracica longa.<br>Thoracica alaris. |
| *From 3rd Part.* | Subscapular.<br>Anterior circumflex.<br>Posterior circumflex. |

The *superior thoracic* is a small artery, which arises from the axillary, or by a common trunk with the acromial thoracic. Running forwards and inwards along the upper border of the Pectoralis minor, it passes between it and the Pectoralis major to the side of the chest. It supplies these muscles, and the parietes of the thorax, anastomosing with the internal mammary and intercostal arteries.

The *acromial thoracic* is a short trunk, which arises from the fore part of the axillary artery. Projecting forwards to the upper border of the Pectoralis minor, it divides into three sets of branches, thoracic, acromial, and descending. The thoracic branches, two or three in number, are distributed to the Serratus magnus and Pectoral muscles, anastomosing with the intercostal branches of the internal mammary. The acromial branches are directed outwards towards the acromion, supplying the Deltoid muscle, and anastomosing, on the surface of the acromion, with the supra-scapular and posterior circumflex arteries. The descending branch passes in the interspace between the Pectoralis major and Deltoid, accompanying the cephalic vein, and supplying both muscles.

The *long thoracic* passes downwards and inwards along the lower border of the Pectoralis minor to the side of the chest, supplying the Serratus magnus, the Pectoral muscles, and mammary gland, and sending branches across the axilla to the axillary glands and Subscapularis, which anastomose with the internal mammary and intercostal arteries.

The *thoracica alaris* is a small branch, which supplies the glands and areolar tissue of the axilla. Its place is frequently supplied by branches from some of the other thoracic arteries.

The *subscapular*, the largest branch of the axillary artery, arises opposite the lower border of the Subscapularis muscle, and passes downwards and backwards along its lower margin to the inferior angle of the scapula, where it anastomoses with the posterior scapular, a branch of the subclavian. It distributes branches to the Subscapularis, Serratus magnus, Teres major, and Latissimus dorsi muscles, and gives off, about an inch and a-half from its origin, a large branch, the dorsalis scapulæ. This vessel curves round the inferior border of the scapula, leaving the axilla in the interspace between the Teres minor above, the Teres major below, and the long head of the Triceps in front; and divides into three branches, a subscapular, which enters the subscapular fossa beneath the Subscapularis which it supplies, anastomosing with the subscapular and supra-scapular arteries; an infra-spinous branch (dorsalis scapulæ), which turns round the axillary border of the scapula, between the Teres minor and the bone, enters the infra-spinous fossa, supplies the Infra-spinatus muscle, and anastomoses with the supra-scapular and posterior scapular arteries; and a median branch, which is continued along the axillary border of the scapula, between the Teres major and minor, and, at the dorsal surface of the inferior angle of the bone, anastomoses with the supra-scapular.

The *circumflex arteries* wind round the neck of the humerus.

The *posterior circumflex*, the larger of the two, arises from the back part of the axillary, opposite the lower border of the Subscapularis muscle, and, passing backwards with the circumflex veins and nerve, through the quadrangular space bounded by the Teres major and minor, the scapular head of the Triceps and the humerus, winds round the neck of that bone, is distributed to the Deltoid muscle and shoulder-joint, anastomosing with the anterior circumflex, supra-scapular, and acromial thoracic arteries.

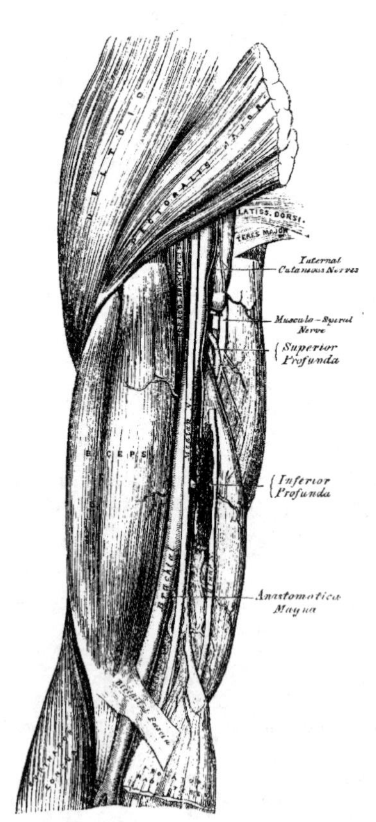

218 – *The surgical anatomy of the brachial artery.*

The *anterior circumflex*, considerably smaller than the preceding, arises just below that vessel, from the outer side of the axillary artery. It passes horizontally outwards, beneath the Coraco-brachialis and short head of the Biceps, lying upon the fore part of the neck of the humerus, and, on reaching the bicipital groove, gives off an ascending branch, which passes upwards along it, to supply the head of the bone and the shoulder-joint. The trunk of the vessel is then continued outwards beneath the Deltoid which it supplies, and anastomoses with the posterior circumflex, and acromial thoracic arteries.

## Brachial Artery (Fig. 218).

The brachial artery commences at the lower margin of the tendon of the Teres major, and, passing down the inner and anterior aspect of the arm, terminates about half an inch below the bend of the elbow, where it divides into the radial and ulnar arteries.

The direction of this vessel is marked by a line drawn from the outer side of the axillary space between the folds of the axilla, to a point midway between the condyles of the humerus, which corresponds to the depression along the inner border of the Coraco-brachialis and Biceps muscles. In the upper part of its course, this vessel lies internal to the humerus but below, it is in front of the bone.

*Relations.* This artery is superficial throughout its entire extent, being covered, *in front*, by the integument, the superficial and deep fasciæ; the bicipital fascia separates it opposite the elbow from the median basilic vein; the median nerve crosses it at its centre; and the basilic vein lies in the line of the artery but separated from it by the fascia in the lower half of its course. *Behind*, it is separated from the inner side of the humerus above, by the long and inner heads of the Triceps, the musculo-spiral nerve and superior profunda artery intervening; and from the front of the bone below, by the insertion of the Coraco-brachialis and the Brachialis anticus muscles. By its *outer side*, it is in relation with the commencement of the median nerve, and the Coraco-brachialis and Biceps muscles, which slightly overlap the artery. By its *inner side*, with the internal cutaneous and ulnar nerves, its upper half; the median nerve its lower half. It is accompanied by two veins, the venæ comites: they lie in close contact with the artery, being connected together at intervals by short transverse communicating branches.

## Plan of the Relations of the Brachial Artery.

*In front.*
Integument and fasciæ.
Bicipital fascia, median basilic vein.
Median nerve.

*Outer side.*
Median nerve.
Coraco-brachialis.
Biceps.

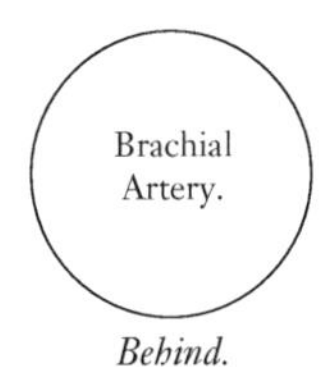


*Inner side.*
Internal cutaneous.
Ulnar and median nerves.

*Behind.*
Triceps.
Musculo-spiral nerve.
Superior profunda artery.
Coraco-brachialis.
Brachialis anticus.

## Bend of the Elbow.

At the bend of the elbow, the brachial artery sinks deeply into a triangular interval, the base of which is directed upwards towards the humerus, and the sides of which are bounded, externally, by the Supinator longus; internally, by the Pronator radii teres; its floor is formed by the Brachialis anticus, and Supinator brevis. This space contains the brachial artery, with its accompanying veins; the radial and ulnar arteries; the median and musculo-spiral nerves; and the tendon of the Biceps. The brachial artery occupies the middle line of this space, and divides opposite the coronoid process of the ulna into the radial and ulnar arteries; it is covered, *in front*, by the integument, the superficial fascia, and the median basilic vein, the vein being separated from direct contact with the artery by the bicipital fascia. *Behind*, it lies on the Brachialis anticus, which separates it from the elbow-joint. The median nerve lies on the inner side of the artery, but separated from it below by an interval of half an inch. The tendon of the Biceps lies to the outer side of the space, and the musculo-spiral nerve still more externally, lying upon the Supinator brevis, and partly concealed by the Supinator longus.

*Peculiarities of the Artery as regards its Course.* The brachial artery, accompanied by the median nerve, may leave the inner border of the Biceps, and descend towards the inner condyle of the humerus, where it usually curves round a prominence of bone, to which it is connected by a fibrous band; it then inclines outwards, beneath or through the substance of the Pronator teres muscle, to the bend of the elbow. This variation bears considerable analogy with the normal condition of the artery in some of the carnivora (see p. 79).

*As regards its Division.* Occasionally, the artery is divided for a short distance at its upper part into two trunks, which are united above and below. A similar peculiarity occurs in the main vessel of the lower limb.

The point of bifurcation may be above or below the usual point, the former condition being by far the most frequent. Out of 481 examinations recorded by Mr. Quain, some made on the right, and some on the left side of the body, in 386 the artery bifurcated in its normal position. In one case only was the place of division lower than usual, being two or three inches below the elbow-joint. In ninety cases out of 481, or about 1 in $5\frac{1}{9}$, there were two arteries instead of one in some part, or in the whole of the arm.'

There appears, however, to be no correspondence between the arteries of the two arms, with respect to their irregular division; for in sixty-one bodies it occurred on one side only in forty-three; on both sides, in different positions, in thirteen; on both sides, in the same position, in five.

The point of bifurcation takes place at different parts of the arm, being most frequent in the upper part, less so in the lower part, and least so in the middle, the most usual point for the application of a ligature; under any of these circumstances, two large arteries would be found in the arm instead of one. The most frequent (in three out of four) of these peculiarities is the high division of the radial. It often arises from the inner side of the brachial, and runs parallel with the main trunk to the elbow, where it crosses it, lying beneath the fascia; or it may perforate the fascia, and pass over the artery, immediately beneath the integument.

The ulnar sometimes arises from the brachial high up, and then occasionally leaves that vessel at the lower part of the arm, and descends towards the inner condyle. In the fore-arm, it generally lies beneath the deep fascia, superficial to the flexor muscles; occasionally between the integument and deep fascia, and very rarely beneath the flexor muscles.

The interosseous artery sometimes arises from the upper part of the brachial or axillary: as it descends the arm, it lies behind the main trunk, and, at the bend of the elbow, regains its usual position.

In some cases of high division of the radial, the remaining trunk (ulnar interosseous) occasionally passes, together with the median nerve, along the inner margin of the arm to the inner condyle, and then passing from within outwards, beneath or through the Pronator teres, regains its usual position at the end of the elbow.

Occasionally, the two arteries representing the brachial are connected at the bend of the elbow by a short transverse branch, and are even sometimes reunited.

Sometimes, long slender vessels, *vasa aberrantia*, connect the brachial or axillary arteries with one of the arteries of the fore-arm, or a branch from them. These vessels usually join the radial.

*Varieties in Muscular Relations.** The brachial artery is occasionally concealed, in some part of its course, by muscular or tendinous slips derived from various sources. In the upper third of the arm, the brachial vessels and median nerve have been seen concealed to the extent of three inches by a muscular layer of considerable thickness derived from the Coraco-brachialis, which passed round to the inner side of the vessel, and joined the internal head of the Triceps. In the lower half of the arm it is occasionally concealed by a broad thin head to the Biceps muscle (see p. 218). A narrow fleshy slip from the Biceps has been seen to cross the artery, concealing it for an inch and a half, its tendon ending in the aponeurosis covering the Pronator teres. A muscular and tendinous slip has been seen to arise from the external bicipital ridge by a long tendon, cross obliquely behind the long tendon of the Biceps, end in a fleshy belly, which appears on the inner side of the arm between the Biceps and Coraco-brachialis, passes down along the inner edge of the former, and crosses the artery very obliquely, so as to lie in front of it for three inches, and, finally, end in a narrow flattened tendon, which is inserted into the aponeurosis over the Pronator teres. A tendinous slip, arising from the deep part of the tendon of the Pectoralis major, has been seen to cross the artery obliquely at or below the Coraco-brachialis, and join the intermuscular septum above the inner condyle. The Brachialis anticus not infrequently projects at the outer side of the artery, occasionally overlaps it, sending inwards, across the artery, an aponeurosis which binds the vessel down upon the Brachialis anticus. Sometimes, a fleshy slip from the muscle covers the vessel, in one case to the extent of three inches. In some cases of high origin of the Pronator teres, an aponeurosis extends from it to join the Brachialis anticus external to the artery; a kind of arch being thus formed under which the principal artery and median nerve pass, so as to be concealed for half an inch above the transverse level of the condyle.

*Surgical Anatomy.* Compression of the brachial artery is required in cases of amputation of the arm or fore-arm, in resection of the elbow-joint, and the removal of tumours; and it will be observed, that it may be effected in almost any part of its course; if pressure is made in the upper part of the limb it should be directed from within outwards, and if in the lower part, from before backwards, as the artery lies on the inner side of the humerus above, and in front of it below. The most favourable situation is either above or below the insertion of the Coraco-brachialis.

The application of a ligature to the brachial artery may be required in cases of wounds of the vessel, or in wounds of the palmar arch, where compression of the radial and ulnar arteries fails to arrest the hæmorrhage. It is also necessary in cases of aneurism of the brachial, the radial, ulnar, or interosseous arteries; and may be secured in any part of its course. The chief guides in determining its position are the surface-markings produced by the inner margin of the Coraco-brachialis and Biceps, the known course of the vessel, and its pulsation, which should be carefully felt for before any operation is performed, as the vessel occasionally deviates from its usual position in the arm. In whatever situation the operation is performed, great care is necessary on account of the extreme thinness of the parts covering the artery, and the intimate connection which the vessel has throughout its whole course with important nerves and veins. Sometimes a thin layer of muscular fibre is met with concealing the artery; if such is the case, it must be divided across, in order to expose the vessel.

*In the upper third of the arm* the artery may be exposed in the following manner. The patient being placed horizontally upon a table, the affected limb should be raised from the side, and the hand supinated. An incision about two inches in length should be made on the ulnar side of the Coraco-brachialis muscle, and the subjacent fascia cautiously divided so as to avoid wounding the internal cutaneous nerve or basilic vein, which sometimes run on the surface of the artery as high as the axilla. The fascia having been divided, it should be remembered, that the ulnar and internal cutaneous nerves lie on the inner side of the artery, the median on the outer side, the latter nerve being occasionally superficial to the artery in this situation, and that the venæ comites are also in relation with the vessel, one on either side. These being carefully separated, the aneurism needle should be passed round the artery from the ulnar to the radial side.

If two arteries are present in the arm in consequence of a high division, they are usually placed side by side;

* Struther's *Anatomical and Physiological Observations.*

and if they are exposed in an operation, the surgeon should endeavour to ascertain, by alternately pressing on one or the other vessel, which of the two communicates with the wound or aneurism, when a ligature may be applied accordingly; or if pulsation or hæmorrhage ceases only when both vessels are compressed, both vessels may be tied, as it may be concluded that the two communicate above the seat of disease or are reunited.

It should also be remembered, that two arteries may be present in the arm in a case of high division, and that one of these may be found along the inner intermuscular septum, in a line towards the inner condyle of the humerus, or in its usual position, but deeply placed beneath the common trunk: a knowledge of these facts will at once suggest the precautions necessary in every case, and indicate the necessary measures to be adopted when met with.

*In the middle of the arm* the brachial artery may be exposed by making an incision along the inner margin of the Biceps muscle. The fore-arm being bent so as to relax the muscle, it should be drawn slightly aside, and the fascia being carefully divided, the median nerve will be exposed lying upon the artery (sometimes beneath); this being drawn inwards and the muscle outwards, the artery should be separated from its accompanying veins and secured. In this situation the inferior profunda may be mistaken for the main trunk, especially if enlarged, from the collateral circulation having become established; this may be avoided by directing the incision externally towards the Biceps rather than inwards or backwards towards the Triceps.

*The lower part of the brachial artery* is of extreme interest in a surgical point of view, on account of the relation which it bears to those veins most commonly opened in venesection. Of these vessels, the median basilic is the largest and most prominent, and, consequently, the one usually selected for the operation. It should be remembered, that this vein runs parallel with the brachial artery, from which it is separated by the bicipital fascia, and that in no case should this vessel be selected for venesection, except in a part which is not in contact with the artery.

*Collateral Circulation.* After the application of a ligature to the brachial artery in the upper third of the arm, the circulation is carried on by branches from the circumflex and subscapular arteries, anastomosing with ascending branches from the superior profunda. If the brachial is tied *below* the origin of the profunda arteries the circulation is maintained by the branches of the profundæ, anastomosing with the recurrent radial, ulnar, and interosseus arteries. In two cases described by Mr. South,* in which the brachial artery had been tied some time previously, in one 'a long portion of the artery had been obliterated, and sets of vessels are descending on either side from above the obliteration, to be received into others which ascend in a similar manner from below it. In the other, the obliteration is less extensive, and a single curved artery about as big as a crow-quill passes from the upper to the lower open part of the artery.'

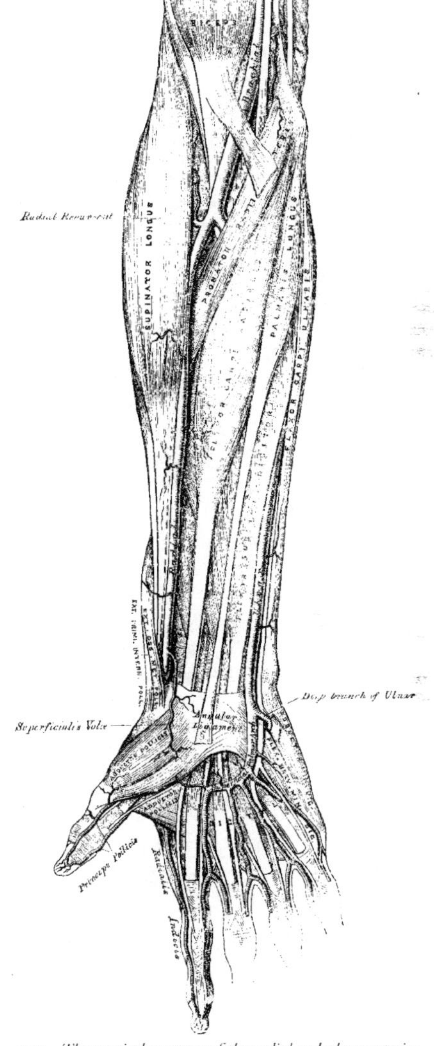


219 – *The surgical anatomy of the radial and ulnar arteries.*

The branches of the brachial artery are the

| Superior profunda. | Inferior profunda. |
|---|---|
| Nutrient artery. | Anastomotica magna. |
| Muscular. | |

The *superior profunda* arises from the inner and back part of the brachial, opposite the lower border

* Chelius' *Surgery*, p. 254.

of the Teres major, and passes backwards to the interval between the outer and inner heads of the Triceps muscle, accompanied by the musculo-spiral nerve; it winds round the back part of the shaft of the humerus in the spiral groove, between the Triceps and the bone, and descends on the outer side of the arm to the space between the Brachialis anticus, and Supinator longus, as far as the elbow, where it anastomoses with the recurrent branch of the radial artery. It supplies the Deltoid, Coraco-brachialis, and Triceps muscles, and whilst in the groove, between the Triceps and the bone, it gives off the posterior articular artery, which descends perpendicularly between the Triceps and the bone, to the back part of the elbow-joint, where it anastomoses with the interosseous recurrent branch, and, on the inner side of the arm, with the posterior ulnar recurrent, and with the anastomotica magna or inferior profunda (fig. 221).

The *nutrient artery* of the shaft of the humerus arises from the brachial, about the middle of the arm. Passing downwards, it enters the nutritious canal of that bone, near the insertion of the Coraco-brachialis muscle.

The *inferior profunda*, of small size, arises from the brachial, a little below the middle of the arm; piercing the internal intermuscular septum, it descends on the surface of the inner head of the Triceps muscle, to the space between the inner condyle and olecranon, accompanied by the ulnar nerve, and terminates by anastomosing with the posterior ulnar recurrent, and anastomotica magna.

The *anastomotica magna* arises from the brachial, about two inches above the elbow-joint. It passes transversely inwards upon the Brachialis anticus, and piercing the internal intermuscular septum, winds round the back part of the humerus, between the Triceps and the bone, forming an arch above the olecranon fossa, by its junction with the posterior articular branch of the superior profunda. As this vessel lies on the Brachialis anticus, an offset passes between the internal condyle and olecranon, which anastomoses with the inferior profunda and posterior ulnar recurrent arteries. Other branches ascend to join the inferior profunda; and some descend in front of the inner condyle, to anastomose with the anterior ulnar recurrent.

The *muscular* are three or four large branches, which are distributed to the muscles in the course of the artery. They supply the Coraco-brachialis, Biceps and Brachialis anticus muscles.

## Radial Artery.

The Radial artery appears, from its direction, to be the continuation of the brachial but, in size, it is smaller than the ulnar. It commences at the bifurcation of the brachial just below the bend of the elbow, and passes along the radial side of the fore-arm; the wrist; it then winds backwards, round the outer side of the carpus, beneath the extensor tendons of the thumb, and, running forwards, passes between the two heads of the first Dorsal interoseous muscle, into the palm of the hand. It then crosses the metacarpal bones to the ulnar border of the hand forming the deep palmar arch and, at its termination, inosculates with the deep branch of the ulnar artery. The relations of this vessel may thus be conveniently divided into three parts, viz., in front of the fore-arm, at the back of the wrist, and in the hand.

*Relations. In the fore-arm* this vessel extends from opposite the neck of the radius to the fore part of the styloid process, being placed to the inner side of the shaft of that bone above, and in front of it below. It is superficial throughout its entire extent being covered by the integument, the superficial and deep fasciæ, and slightly overlapped above by the Supinator longus. In its course downwards it lies upon the tendon of the Biceps, the Supinator brevis the Pronator radii teres, radial origin of the Flexor sublimis digitorum, the Flexor longus pollicis, Pronator quadratus and the lower extremity of the radius. In the upper third of its course, it lies between the Supinator longus and the Pronator radii teres; in its lower two thirds, between the tendons of the Supinator longus and the Flexor carpi radialis. The radial nerve lies along the outer side of the artery, in the middle third of its course; and some filaments of the musculo-cutaneous nerve, after piercing the deep fascia run along the lower part of the artery as it winds round the wrist. The vessel is accompanied by venæ comites throughout its whole course.

## Plan of the Relations of the Radial Artery in the Fore-arm.

*In front.*
Integument—superficial and deep fasciæ.
Supinator longus.

*Inner side.*
Pronator radii teres.
Flexor carpi radialis.

Radial
Artery in
Fore-arm.

*Outer side.*
Supinator longus.
Radial nerve (middle third.)

*Behind.*
Tendon of Biceps.
Supinator brevis.
Pronator radii teres.
Flexor sublimis digitorum.
Flexor longus pollicis.
Pronator quadratus.
Radius.

*At the wrist*, as it winds round the outer side of the carpus, from the styloid process to the first interosseous space, it lies upon the external lateral ligament being covered by the extensor tendons of the thumb, subcutaneous veins, some filaments of the radial nerve, and the integument. It is accompanied by two veins, and a filament of the musculo-cutaneous nerve.

*In the hand*, it passes from the upper end of the first interosseous space, between the heads of the Abductor indicis, transversely across the palm, to the base of the metacarpal bone of the little finger, where it inosculates with the communicating branch from the ulnar artery, forming the deep palmar arch. It lies upon the carpal extremities of the metacarpal bones and the Interossei muscles, being covered by the flexor tendons of the fingers, the Lumbricales, the muscles of the little finger, and the Flexor brevis pollicis, and is accompanied by the deep branch of the ulnar nerve.

*Peculiarities.* The origin of the radial artery varies in the proportion nearly of one in eight cases. In one case the origin was lower than usual. In the other cases, the upper part of the brachial was a more frequent source of origin than the axillary. The variations in the position of this vessel in the arm, and at the bend of the elbow, have been already mentioned. In the fore-arm it deviates less frequently from its position than the ulnar. It has been found lying over the fascia, instead of beneath it. It has also been observed on the surface of the Supinator longus, instead of along its inner border: and in turning round the wrist, it has been seen lying over, instead of beneath, the Extensor tendons.

*Surgical Anatomy.* The operation of tying the radial artery is required in cases of wounds either of its trunk, or of some of its branches, or for aneurism: and it will be observed, that the vessel may be easily exposed in any part of its course through the fore-arm. This operation in the middle or inferior third of this region is easily performed; but in the upper third, near the elbow, the operation is attended with some difficulty, from the greater depth of the vessel, and from its being overlapped by the Supinator longus and Pronator teres muscles.

To tie the artery in this situation, an incision three inches in length should be made through the integument, from the bend of the elbow obliquely downwards and outwards, on the radial side of the fore-arm, avoiding the branches of the median vein; the fascia of the arm being divided, and the Supinator longus drawn a little outwards, the artery will be exposed. The venæ comites should be carefully separated from the vessel, and the ligature passed from the radial to the ulnar side.

In the middle third of the fore-arm the artery may be exposed by making an incision of similar length on the inner margin of the Supinator longus. In this situation the radial nerve lies in close relation with the outer side of the artery, and should, as well as the veins, be carefully avoided.

In the lower third, the artery is easily secured by dividing the integument and fascia in the interval between the tendons of the Supinator longus and Flexor carpi radials muscles.

The branches of the radial artery may be divided into three groups, corresponding with the three regions in which this vessel is situated.

In the Fore-arm. { Radial recurrent. Muscular. Superficialis volæ. Anterior carpal.

Wrist. { Posterior carpal. Metacarpal. Dorsales pollicis. Dorsalis indicis.

Hand. { Princeps pollicis.
Radialis indicis.
Perforantes.
Interossei. }

The *radial recurrent* is given off immediately below the elbow. It ascends between the branches of the musculo-spiral nerve, lying on the Supinator brevis and then between the Supinator longus and Brachialis anticus, supplying these muscles, the elbow-joint, and anastomosing with the terminal branches of the superior profunda.

The *muscular branches* are distributed to the muscles on the radial side of the fore-arm.

The *superficialis volæ* arises from the radial artery, just where this vessel is about to wind round the wrist. Running forwards, it passes between the muscles of the thumb, which it supplies, and anastomoses with the termination of the ulnar artery, completing the superficial palmar arch. This vessel varies considerably in size, usually it is very small, and terminates in the muscles of the thumb; sometimes it is as large as the continuation of the radial.

The *carpal branches* supply the joints of the wrist. The *anterior carpal* is a small vessel which arises from the radial artery near the lower border of the Pronator quadratus, and running inwards in front of the radius, anastomoses with the anterior carpal branch of the ulnar artery. From the arch thus formed, branches descend to supply the articulations of the wrist.

The *posterior carpal* is a small vessel which arises from the radial artery beneath the extensor tendons of the thumb; crossing the carpus transversely to the inner border of the hand, it anastomoses with the posterior carpal branch of the ulnar. It sends branches upwards, which anastomose with the termination of the anterior interosseous artery; other branches descend to the metacarpal spaces; they are the dorsal interosseous arteries for the third and fourth interosseous spaces; they anastomose with the posterior perforating branches from the deep palmar arch.

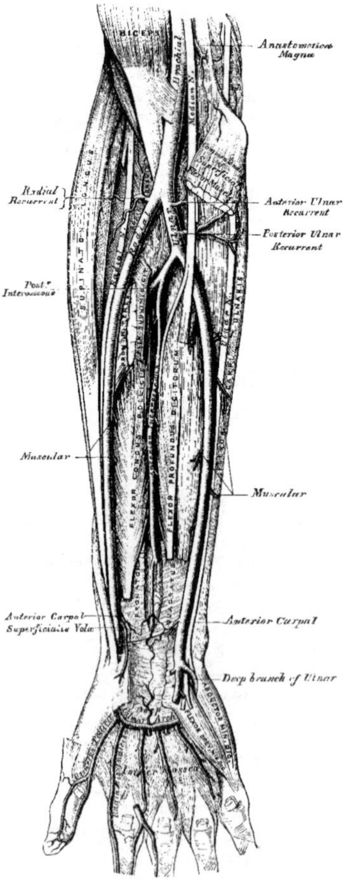

The *metacarpal* (*first dorsal interosseous branch*) arises beneath the extensor tendons of the thumb, sometimes with the posterior carpal artery; running forwards on the second dorsal interosseous muscle; it communicates, behind, with the corresponding perforating branch of the deep palmar arch; and, in front, inosculates with the digital branch of the superficial palmar arch, and supplies the adjoining sides of the index and middle fingers.

The *dorsales pollicis* are two small vessels which run along the sides of the dorsal aspect of the thumb. They sometimes arise separately, or occasionally by a common trunk, near the base of the first metacarpal bone.

The *dorsalis indicis*, also a small branch, runs along the radial side of the back of the index finger, sending a few branches to the Abductor indicis.

The *princeps pollicis* arises from the radial just as it turns inwards to the deep part of the hand; it descends between the Abductor indicis and Adductor pollicis, along the ulnar side of the metacarpal bone of the thumb, to the base of the first phalanx, where it divides into two branches, which run along the sides of the palmar aspect of the thumb, and form an arch on the under surface of the last phalanx, from which branches are distributed to the integument and cellular membrane of the thumb.

The *radialis indicis* arises close to the preceding, descends between the Abductor indicis and Adductor pollicis, and runs along the radial side of



220 – *Ulnar and radial arteries. Deep view.*

the index finger to its extremity, where it anastomoses with the collateral digital artery from the superficial palmar arch. At the lower border of the Adductor pollicis this vessel anastomoses with the princeps pollicis, and gives a communicating branch to the superficial palmar arch.

The *perforantes*, three in number, pass backwards between the heads of the last three Dorsal interossei muscles, to inosculate with the dorsal interosseous arteries.

The *palmar interossei*, three or four in number, are branches of the deep palmar arch; they run forwards upon the Interossei muscles, and anastomose at the clefts of the fingers with the digital branches of the superficial arch.

## Ulnar Artery.

The *Ulnar Artery*, the larger of the two sub-divisions of the brachial, commences a little below the bend of the elbow, and crosses the inner side of the forearm obliquely to the commencement of its lower half; it then runs along its ulnar border to the wrist, crosses the annular ligament on the radial side of the pisiform bone, and passes across the palm of the hand, forming the superficial palmar arch, which terminates by inosculating with the superficialis volæ.

*Relations in the Fore-arm.* In its *upper half*, it is deeply seated, being covered by all the superficial flexor muscles, excepting the Flexor carpi ulnaris; crossed by the median nerve, which, for about an inch lies to its inner side; and it lies upon the Brachialis anticus and Flexor profundus digitorum muscles. In the *lower half* of the fore-arm, it lies upon the Flexor profundus, being covered by the integument, the superficial and deep fasciæ, and is placed between the Flexor carpi ulnaris and Flexor sublimis digitorum muscles. It is accompanied by two veins, which lie one on each side of the vessel; the ulnar nerve lies on its inner side for the lower two-thirds of its extent, and a small branch from it descends on the lower part of the vessel to the palm of the hand.

### Plan of Relations of the Ulnar Artery in the Fore-arm.

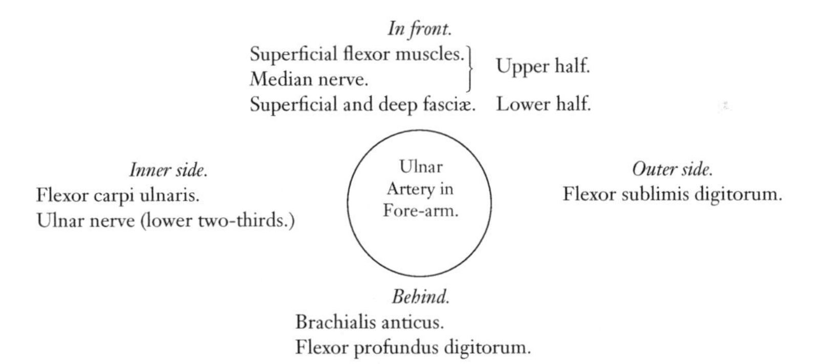


*At the wrist*, the ulnar artery is covered by the integument and fascia, and lies upon the anterior annular ligament. On its inner side is the pisiform bone. The ulnar nerve lies at the inner side, and somewhat behind the artery.

*In the palm of the hand*, the continuation of the ulnar artery is called the superficial palmar arch; it passes obliquely outwards to the interspace between the ball of the thumb and the index finger, where it anastomoses with the superficialis volæ, and a branch from the radialis indicis, thus completing the superficial palmar arch. The convexity of this arch is directed towards the fingers, its concavity towards the muscles of the thumb.

The superficial palmar arch is covered by the Palmaris brevis, the palmar fascia and integument; and lies upon the annular ligament, the muscles of the little finger, the tendons of the superficial flexor, and the divisions of the median and ulnar nerves, the latter accompanying the artery a short part of its course.

## Relations of the Superficial Palmar Arch.

| *In front.* | Ulnar Artery in Hand. | *Behind.* |
|---|---|---|
| Integument. | | Annular ligament. |
| Palmaris brevis. | | Origin of muscles of little finger. |
| Palmar fascia. | | Superficial flexor tendons. |
| | | Divisions of median and ulnar nerves. |

*Peculiarities.* The ulnar artery was found to vary in its origin nearly in the proportion of one in thirteen cases, in one case arising lower than usual, about two or three inches below the elbow, and in all the other cases much higher, the brachial being a more frequent source of origin than the axillary.

Variations in the position of this vessel are more frequent than in the radial. When its origin is normal, the course of the vessel is rarely changed. When it arises high up, its position in the fore-arm is almost invariably superficial to the flexor muscles, lying commonly beneath the fascia, more rarely between the fascia and integument. In a few cases, its position was subcutaneous in the upper part of the fore-arm, sub-aponeurotic in the lower part.

*Surgical Anatomy.* The application of a ligature to this vessel is required in cases of wound of the artery, or of its branches, or in consequence of aneurism. In the upper half of the fore-arm, the artery is deeply seated beneath the superficial flexor muscles, and their division would be requisite in a case of recent wound of the artery in this situation in order to secure it, but under no other circumstances. In the middle and lower third of the fore-arm, this vessel may be easily secured by making an incision on the radial side of the tendon of the Flexor carpi ulnaris; the deep fascia being divided, and the Flexor carpi ulnaris and its companion muscle, the Flexor sublimis, being separated from each other, the vessel will be exposed, accompanied by its venæ comites, the ulnar nerve lying on its inner side. The veins being separated from the artery, the ligature should be passed from the ulnar to the radial side, taking care to avoid the ulnar nerve.

The branches of the ulnar artery may be arranged into the following groups,

- *Fore-arm.*
  - Anterior ulnar recurrent.
  - Posterior ulnar recurrent.
  - Interosseous
    - Anterior interosseous.
    - Posterior interosseous.
  - Muscular.
- *Wrist.*
  - Anterior carpal.
  - Posterior carpal.
- *Hand.*
  - Deep or communicating branch.
  - Digital.

*The anterior ulnar recurrent* arises immediately below the elbow-joint, passes upwards and inwards between the Brachialis anticus and Pronator radii teres, supplies these muscles, and, in front of the inner condyle, anastomoses with the anastomotica magna and inferior profunda.

The *posterior ulnar recurrent* is much larger, and arises somewhat lower than the preceding. It passes backwards and inwards, beneath the Flexor sublimis, and ascends behind the inner condyle of the humerus. In the interval between this eminence and the olecranon, it lies beneath the Flexor carpi ulnaris, ascending between the heads of that muscle, beneath the ulnar nerve; it supplies the neighbouring muscles and joint, and anastomoses with the inferior profunda, anastomotica magna, and interosseous recurrent arteries.

The *interosseous artery* is a short trunk, about an inch in length, and of considerable size, which arises immediately below the tuberosity of the radius, and, passing backwards to the upper border of the interosseous membrane, divides into two branches, the anterior, and posterior interosseous.

The *anterior interosseous* passes down the fore-arm on the anterior surface of the interosseous membrane, to which it is connected by a thin aponeurotic arch. It is accompanied by the interosseous branch of the median nerve, and overlapped by the contiguous margins of the Flexor profundus digitorum and Flexor longus pollicis muscles, giving off in this situation musclar branches, and the nutrient arteries of the radius and ulna. At the upper border of the Pronator quadratus, a branch descends in front of that muscle, to anastomose in front of the carpus with branches from the anterior carpal and deep palmar arch. The continuation of the artery passes behind the Pronator quadratus, and, piercing the interosseous membrane, descends to the back of the wrist, where it anastomoses with the posterior carpal branches of the radial and ulnar arteries. The anterior interosseous gives off a long, slender branch, which accompanies the median nerve and gives off-sets to its substance. This, the median artery, is sometimes much enlarged.

The *posterior interosseous artery* passes backwards through the interval between the oblique ligament and the upper border of the interosseous membrane, and passes down the back part of the fore-arm, between the superficial and deep layer of muscles, to both of which it distributes branches. Descending to the back of the wrist, it anastomoses with the termination of the anterior interosseous, and with the posterior carpal branches of the radial and ulnar arteries. This artery gives off, near its origin, the posterior interosseous recurrent branch, a large vessel, which ascends to the interval between the external condyle and olecranon, beneath the Anconeus and Supinator brevis, anastomosing with a branch from the superior profunda, and with the posterior ulnar recurrent artery.

The *muscular branches* are distributed to the muscles along the ulnar side of the fore-arm.

The *carpal branches* are intended for the supply of the wrist-joint.

The *anterior carpal* is a small vessel, which crosses the front of the carpus beneath the tendons of the Flexor profundus, and inosculates with a corresponding branch of the radial artery.

The *posterior carpal* arises immediately above the pisiform bone, winding backwards beneath the tendon of the Flexor carpi ulnaris; it gives off a branch which passes across the dorsal surface of the carpus beneath the extensor tendons, anastomosing with a corresponding branch of the radial artery, and forming the posterior carpal arch; it is then continued along the metacarpal bone of the little finger, forming its dorsal branch.

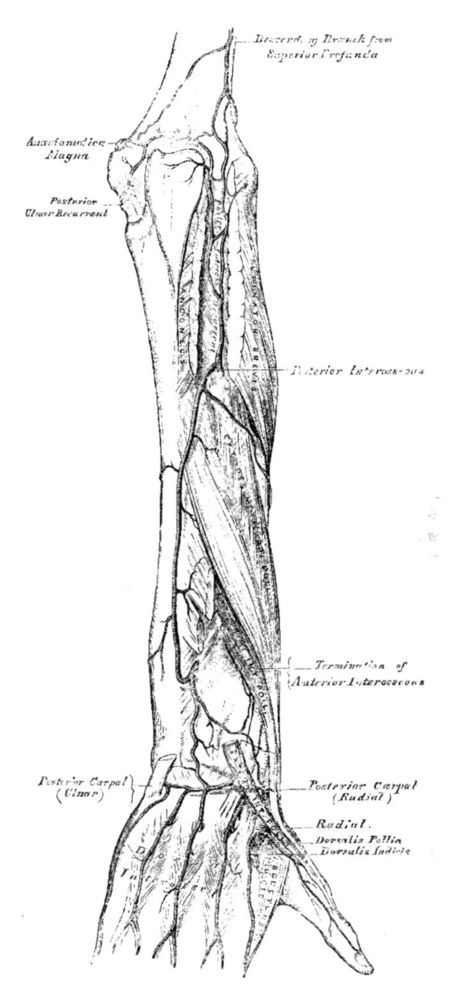


221 – *Arteries of the back of the fore-arm and hand.*

The *deep* or *communicating branch* arises at the commencement of the palmar arch, passing deeply inwards between the Abductor minimi digiti and Flexor brevis minimi digiti, near their origins; it anastomoses with the termination of the radial artery, completing the deep palmar arch.

The *digital branches*, four in number, are given off from the convexity of the superficial palmar arch. They supply the ulnar side of the little finger, and the adjoining sides of the ring, middle, and index-fingers; the radial side of the index-finger and thumb being supplied from the radial artery. The digital arteries at first lie superficial to the flexor tendons, but as they pass forwards with the digital nerves to the clefts between the fingers, they lie between them, and are there joined by the interosseous branches from the deep palmar arch. The digital arteries on the sides of the fingers lie beneath the digital nerves; and, about the middle of the last phalanx, the two branches for each finger form an arch, from the convexity of which branches pass to supply the matrix of the nail.

## The Descending Aorta.

The descending aorta is divided into two portions, the thoracic, and abdominal, in correspondence with the two great cavities of the trunk in which it is situated.

The *Thoracic Aorta* commences at the lower border of the third dorsal vertebra, on the left side, and terminates at the aortic opening in the Diaphragm, in front of the last dorsal vertebra. At its commencement, it is situated on the left side of the spine; it approaches the median line as it descends; and, at its termination, lies directly in front of the column. The direction of this vessel being influenced by the spine, upon which it rests, it is concave forwards in the dorsal region, and, as the branches given off from it are small, the diminution in the size of the vessel is inconsiderable. It is contained in the back part of the posterior mediastinum, being in relation, *in front*, from above downwards, with the left pulmonary artery, the left bronchus, the pericardium, and the œsophagus; *behind*, with the vertebral column, and the vena azygos minor; on the *right side*, with the vena azygos major, and thoracic duct; on the *left side*, with the left pleura, and lung. The œsophagus, with its accompanying nerves, lies on the right side of the aorta *above:* in front of this vessel, in the middle of its course; whilst, at its *lower part*, it is on the left side, on a plane anterior to it.

## Plan of the Relations of the Thoracic Aorta.

*In front.*
Left pulmonary artery.
Left bronchus.
Pericardium.
Œsophagus.

*Right side.*
Œsophagus (above).
Vena azygos major.
Thoracic duct.

Thoracic Aorta.

*Left side.*
Pleura.
Left lung.
Œsophagus (below).

*Behind.*
Vertebral column.
Vena azygos minor.

*Surgical Anatomy*. The student should now consider the effects likely to be produced by aneurism of the thoracic aorta, a disease of common occurrence. When we consider the great depth of the vessel from the surface, and the number of important structures which surround it on every side, it may be easily conceived what a variety of obscure symptoms may arise from disease of this part of the arterial system, and how they may be liable to be mistaken for those of other affections. Aneurism of the thoracic aorta most usually extends backwards, along the left side of the spine, producing absorption of the bodies of the vertebræ, causing extensive curvature of the spine; whilst the irritation or pressure on the cord, will give rise to pain, either in the chest, back, or loins, with radiating pain in the left upper intercostal spaces, from pressure on the intercostal nerves; at the same time, the tumour may project back on each side of the spine, beneath the integument, as a pulsating swelling, simulating abscess connected with diseased bone; or it may displace the œsophagus, and compress the lung on one or the other side. If the tumour extend forward, it may press upon and displace the heart, giving rise to palpitation, and other symptoms of disease of that organ; or it may displace, or even compress, the œsophagus, causing pain and difficulty of swallowing, as in stricture of that tube, and ultimately even open into it by ulceration, producing fatal hæmorrhage. If the disease make way to either side, it may press upon the thoracic duct; or it may burst into the pleural cavity, or into the trachea or lung; and lastly, it may open into the posterior mediastinum.

## Branches of the Thoracic Aorta.

Pericardiac. Œsophageal.
Bronchial. Posterior mediastinal.
Intercostal.

The *pericardiac* are a few small vessels, irregular in their origin, distributed to the pericardium.

The *bronchial* arteries are the nutrient vessels of the lungs, and vary in number, size, and origin. That of the right side arises from the first aortic intercostal, or by a common trunk with the left bronchial, from the front of the thoracic aorta. Those of the left side, usually two in number, arise from the thoracic aorta, one a little lower than the other. Each vessel is directed to the back part of the corresponding bronchus, along which they run, dividing and subdividing, upon the bronchial tubes, supplying them, the cellular tissue of the lungs, the bronchial glands, and the œsophagus.

The *œsophageal arteries*, usually four or five in number, arise from the front of the aorta, and pass obliquely downwards to the œsophagus, forming a chain of anastomoses along that tube, anastomosing with the œsophageal branches of the inferior thyroid arteries above, and with ascending branches from the phrenic and gastric arteries below.

The *posterior mediastinal arteries* are numerous small vessels which supply the glands and loose areolar tissue in the mediastinum.

The *Intercostal arteries* arise from the back part of the aorta. They are usually ten in number on each side, the superior intercostal space (and occasionally the second one) being supplied by the superior intercostal, a branch of the subclavian. The right intercostals are longer than the left, on account of the position of the aorta to the left side of the spine; they pass outwards, across the bodies of the vertebræ, to the intercostal spaces, being covered by the pleura, the œsophagus, thoracic duct, sympathetic nerve, and the vena azygos major; the left passing beneath the superior intercostal vein, the vena azygos minor, and sympathetic. In the intercostal spaces, each artery divides into two branches, an anterior, or proper intercostal branch; and a posterior, or dorsal branch.

The *anterior branch* passes outwards, at first lying upon the External intercostal muscle, covered in front by the pleura, and a thin fascia. It then passes between the two layers of Intercostal muscles, and, having ascended obliquely to the lower border of the rib above, divides, near the angle of that bone, into two branches; of these, the larger runs in the groove, on the lower border of the rib above; the smaller branch along the upper border of the rib below; passing forward, they supply the Intercostal muscles, and anastomose with the anterior intercostal branches of the internal mammary, and with the thoracic branches of the axillary artery. The first aortic intercostal anastomoses with the superior intercostal, and the last three pass between the abdominal muscles, inosculating with the epigastric in front, and with the phrenic, and lumbar arteries. Each intercostal artery is accompanied by a vein and nerve, the former being above, and the latter below, except in the upper intercostal spaces, where the nerve is at first above the artery. The arteries are protected from pressure during the action of the Intercostal muscles, by fibrous arches thrown across, and attached by each extremity to the bone.

The *posterior*, or *dorsal branch*, of each intercostal artery, passes backwards to the inner side of the anterior costo-transverse ligament, and divides into a spinal branch, which supplies the vertebræ, the spinal cord and its membranes and a muscular branch, which is distributed to the muscles and integument of the back.

## The Abdominal Aorta (Fig. 222).

The *Abdominal Aorta* commences at the aortic opening of the Diaphragm, in front of the body of the last dorsal vertebra, and, descending a little to the left side of the vertebral column, terminates on the left side of the body of the fourth lumbar vertebra, where it divides into the two common iliac arteries. As it lies upon the bodies of the vertebræ, it is convex forwards, the greatest convexity corresponding to the third lumbar vertebra, which is a little above and to the left side of the umbilicus.

*Relations.* It is covered, *in front*, by the lesser omentum and stomach, behind which are the branches of the cœliac axis, and the solar plexus; below these, by the splenic vein, the pancreas, the left renal vein, the transverse portion of the duodenum, the mesentery, and aortic plexus. *Behind*, it is separated from the lumbar vertebræ by the left lumbar veins, the receptaculum chyli, and thoracic duct. On the *right side*, with the inferior vena cava (the right crus of the Diaphragm being interposed above), the vena azygos, thoracic duct, and right semi-lunar ganglion. On the *left side*, with the sympathetic nerve, and left semilunar ganglion.

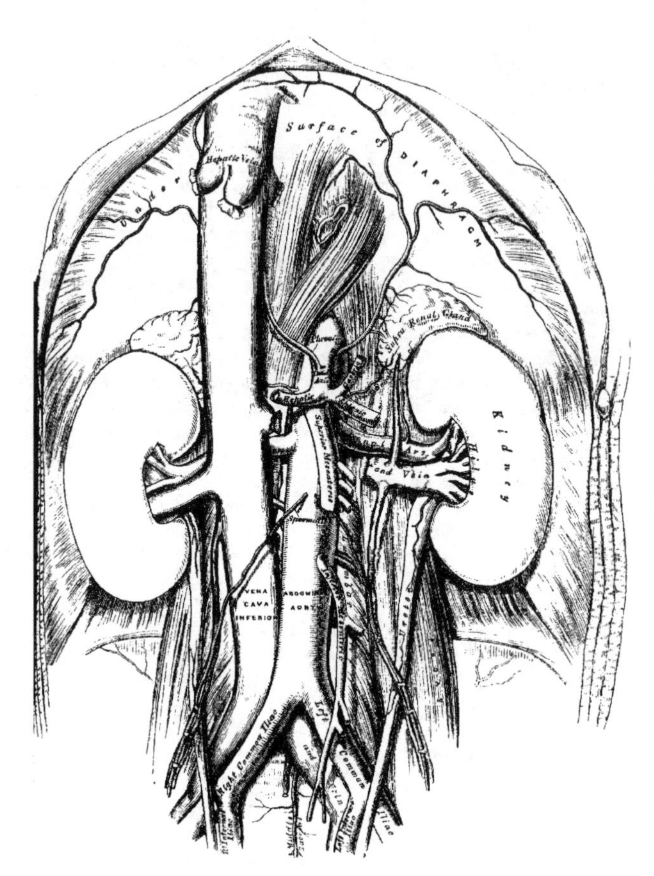


222 – *The abdominal aorta and its branches.*

## Plan of the Relations of the Abdominal Aorta.

*In front.*

Lesser omentum and stomach.
Branches of cœliac axis and solar plexus.
Splenic vein.
Pancreas.
Left renal vein.
Transverse duodenum.
Mesentery.
Aortic plexus.

*Right side.*

Right crus of Diaphragm.
Inferior vena cava.
Vena azygos.
Thoracic duct.
Right semi-lunar ganglion.

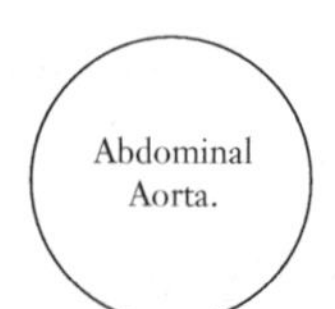


*Left side.*

Sympathetic nerve.
Left semi-lunar ganglion.

*Behind.*

Left lumbar veins.
Receptaculum chyli.
Thoracic duct.
Vertebral column.

*Surgical Anatomy*. Aneurisms of the abdominal aorta near the cœliac axis communicate in nearly equal proportion with the anterior and posterior parts of this vessel.

When an aneurismal sac is connected with the back part of the abdominal aorta, it usually produces absorption of the bodies of the vertebræ, and forms a pulsating tumour, that presents itself in the left hypochondriac or epigastric regions, accompanied by symptoms of disturbance of the alimentary canal. Pain is invariably present, and is usually of two kinds, a fixed and constant pain in the back, caused by the tumour pressing on or displacing the branches of the solar plexus and splanchnic nerves, and a sharp lancinating pain, radiating along those branches of the lumbar nerves pressed on by the tumour; hence the pain in the loins, the testes, the hypogastrium, and in the lower limb (usually of the left side). This form of aneurism usually bursts into the peritoneal cavity, or behind the peritoneum, in the left hypochondriac region; or it may form a large aneurismal sac, extending down as low as Poupart's ligament; hæmorrhage in these case being generally very extensive, but slowly produced, and never rapidly fatal.

When an aneurismal sac is connected with the front of the aorta near the cœliac axis, it forms a pulsating tumour in the left hypochondriac or epigastric regions, usually attends with symptoms of disturbance of the alimentary canal, as sickness, dyspepsia, or constipation and accompanied by pain, which is constant but nearly always fixed in the loins, epigastrium, or some part of the abdomen; the radiating pain being rare, as the lumbar nerves are seldom implicated. This form of aneurism may burst into the peritoneal cavity, or behind the peritoneum, between the layers of the mesentery, or, more rarely, into the duodenum; it rarely extends backwards so as to affect the spine.

## Branches of the Abdominal Aorta.

Phrenic.
Cœliac axis { Gastric. Hepatic. Splenic.
Superior mesenteric.
Supra-renal.

Renal.
Spermatic.
Inferior mesenteric.
Lumbar.
Sacra media.

The branches may be divided into two sets: 1. Those supplying the viscera. 2. Those distributed to the walls of the abdomen.

*Visceral Branches.*

Cœliac axis { Gastric. Hepatic. Splenic.
Superior mesenteric.
Inferior mesentric.
Supra-renal. Renal. Spermatic.

*Parietal Branches.*

Phrenic.
Lumbar.
Sacra media.

## Cœliac Axis.

To expose this artery, raise the liver, draw down the stomach, and then tear through the layers of the lesser omentum.

The Cœliac axis is a short thick trunk, about half an inch in length, arising from the aorta, opposite the margin of the Diaphragm, and passing nearly horizontally forwards (in the erect posture), divides into three large branches, the gastric, hepatic, and splenic, occasionally giving off one of the phrenic arteries.

*Relations.* It is covered, by the lesser omentum. On the *right side*, it is in relation with the right semilunar ganglion, and the lobus Spigelii. On the *left side*, with the right semilunar ganglion and cardiac end of the stomach. *Below*, it rests upon the upper border of the pancreas.

The Gastric Artery (*Coronaria ventriculi*), the smallest of the three branches of the cœliac axis, passes upwards and to the left side, to the cardiac orifice of the stomach, distributing branches to the œsophagus, which anastomose with the aortic œsophageal arteries; others supply the cardiac end of the stomach, inosculating with branches of the splenic artery: it then passes from left to right, along the lesser curvature of the stomach to the pylorus, lying in its course between the layers of the lesser omentum, and giving branches to both surfaces of the organ; at its termination it anastomoses with the pyloric branch of the hepatic.

The Hepatic Artery in the adult is intermediate in size between the gastric and splenic; in the fœtus, it is the largest of the three branches of the cœliac axis. It passes upwards to the right side,

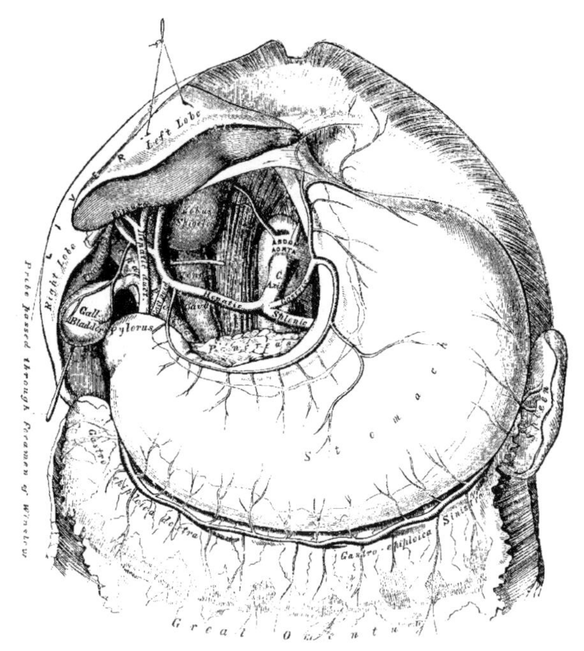

223 – *The cœliac axis and its branches, the liver having been raised, and the lesser omentum removed.*

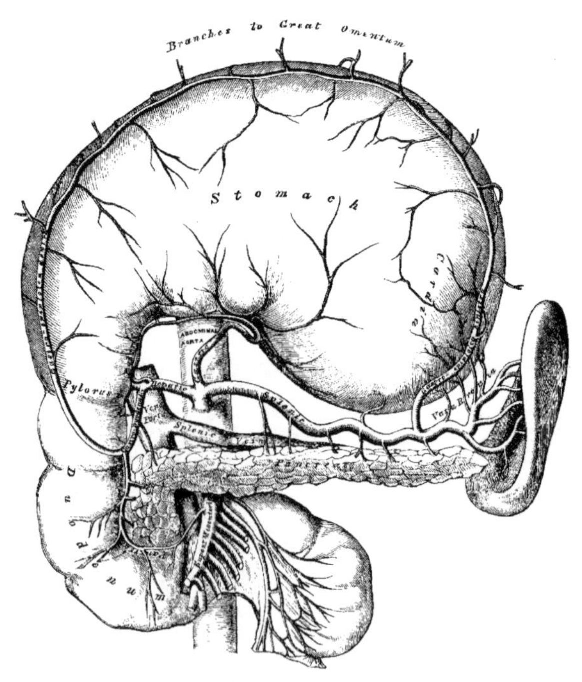

224 – *The cœliac axis and its branches, the stomach having been raised, and the transverse meso-colon removed.*

between the layers of the lesser omentum, and in front of the foramen of Winslow, to the transverse fissure of the liver, where it divides into two branches (right and left), which supply the corresponding lobes of that organ, accompanying the ramifications of the vena portæ and hepatic duct. The hepatic artery, in its course along the right border of the lesser omentum, is in relation with the ductus communis choledocus and portal vein, the former lying to the right of the artery, and the vena portæ behind.

Its branches are the

Pyloric.
Gastro-duodenalis { Gastro-epiploica dextra. / Pancreatico-duodenalis. }
Cystic.

The *pyloric branch* arises from the hepatic, above the pylorus, descends to the pyloric end of the stomach; and passes from right to left along its lesser curvature supplying it with branches, and inosculating with the gastric artery.

The *gastro-duodenalis* is a short but large branch, which descends behind the duodenum, near the pylorus, and divides at the lower border of the stomach into two branches, the gastro-epiploica dextra and the pancreatico-duodenalis. Previous to its division, it gives off two or three small inferior pyloric branches to the pyloric end of the stomach and pancreas.

The *gastro-epiploica dextra* runs from right to left along the greater curvature of the stomach, between the layers of the great omentum, anastomosing about the middle of the lower border of this organ with the gastro-epiploica sinistra from the splenic artery. This vessel gives off numerous branches, some of which ascend to supply both surfaces of the stomach, whilst others descend to supply the great omentum.

The *pancreatico-duodenalis* descends along the contiguous margins of the duodenum and pancreas. It supplies both these organs, and anastomoses with the inferior pancreatico-duodenal branch of the superior mesenteric artery.

In ulceration of the duodenum, which frequently occurs in connexion with severe burns, this artery is often involved, and death may occur from hæmorrhage into the intestinal canal.

The *cystic artery*, usually a branch of the right hepatic, passes upwards and forwards along the neck of the gall bladder, and divides into two branches, one of which ramifies on its free surface, the other, between it and the substance of the liver.

The Splenic Artery, in the adult, is the largest of the three branches of the cœliac axis, and is remarkable for the extreme tortuosity of its course. It passes horizontally to the left side behind the upper border of the pancreas, accompanied by the splenic vein, which lies below it; and on arriving near the spleen, divides into branches, some of which enter the hilus of that organ to be distributed to its structure, whilst others are distributed to the great end of the stomach.

The branches of this vessel are:

Pancreaticæ parvæ. — Gastric (Vasa brevia).
Pancreatica magna. — Gastro-epiploica sinistra.

The *pancreatic* are numerous small branches derived from the splenic as it runs behind the upper border of the pancreas, supplying its middle and left parts. One of these, larger than the rest, is given off from the splenic near the left extremity of the pancreas; it runs from left to right near the posterior surface of the gland following the course of the pancreatic duct, and is called the *pancreatica magna*. These vessels anastomose with the pancreatic branches of the pancreatico-duodenal arteries.

The *gastric* (*vasa brevia*) consist of from five to seven small branches, which arise either from the termination of the splenic artery, or from its terminal branches; and passing from left to right, between the layers of the gastro-splenic omentum, are distributed to the great curvature of the stomach; anastomosing with branches of the gastric and gastro-epiploica sinistra arteries.

The *gastro-epiploica sinistra*, the largest branch of the splenic, runs from left to right along the great curvature of the stomach, between the layers of the great omentum; and anastomoses with the gastro-epiploica dextra. In its course, it distributes several branches to the stomach, which ascend upon both surfaces; others descend to supply the omentum.

## Superior Mesenteric Artery.

In order to expose this vessel, raise the great omentum and transverse colon, draw down the small intestines, and if the peritoneum is divided where the transverse meso-colon and mesentery join, this artery will be exposed just as it issues beneath the lower border of the pancreas.

The Superior Mesenteric Artery (fig. 225) supplies the whole length of the small intestine, except the first part of the duodenum; it also supplies the cæcum, ascending and transverse colon; it is a vessel of large size, arising from the fore part of the aorta, about a quarter of an inch below the cœliac axis; being covered, at its origin, by the splenic vein and pancreas. It passes forwards, between the pancreas and transverse portion of the duodenum, crosses in front of this portion of the intestine, and descends between the layers of the mesentery to the right iliac fossa, where it terminates considerably diminished in size. In its course it forms an arch, the convexity being directed forwards and downwards to the left side, the concavity backwards and upwards to the right. It is accompanied by the superior mesenteric vein, and is surrounded by the superior mesenteric plexus of nerves. Its branches are the

| | |
|---|---|
| Inferior pancreatico-duodenal. | Ileo-colic. |
| Vasa intestini tenuis. | Colica dextra. |
| Colica media. | |

The *inferior pancreatico-duodenal* is given off from the superior mesenteric below the pancreas, and is distributed to the head of the pancreas, and the transverse and descending portions of the duodenum; anastomosing with the pancreatico-duodenal artery.

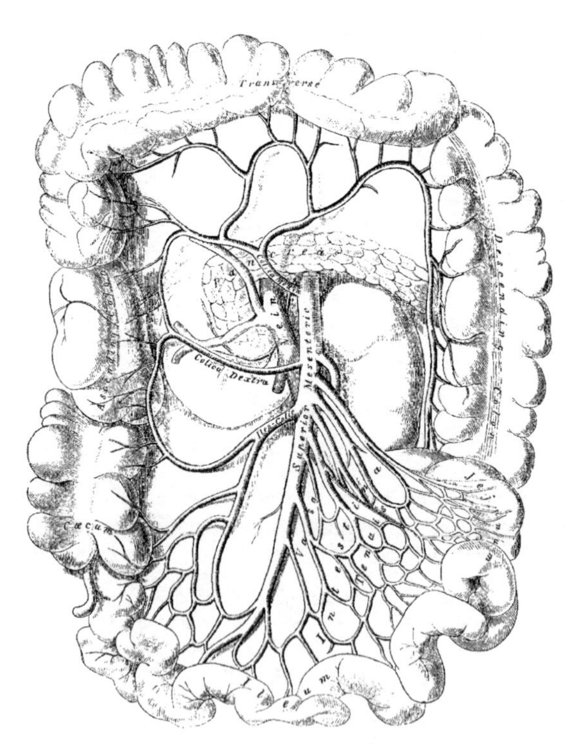

225 – *The superior mesenteric artery and its branches.*

The *vasa intestini tenuis* arise from the convex side of the superior mesenteric artery. They are usually from twelve to fifteen in number, and are distributed to the jejunum and ileum. They run parallel with one another between the layers of the mesentery; each vessel dividing into two branches, which unite with a similar branch on each side, forming a series of arches, the convexities of which are directed towards the intestine. From this first set of arches branches arise, which again unite with similar branches from either side, and thus a second series of arches is formed; and from these latter, a third, and even a fourth or fifth series of arches are constituted, diminishing in size the nearer they approach the intestine. From the terminal arches numerous small straight vessels arise which encircle the intestine, upon which they are minutely distributed, ramifying between its coats.

The *ileo-colic artery* is the lowest branch given off from the concavity of the superior mesenteric artery. It descends between the layers of the mesentery to the right iliac fossa, where it divides into two branches. Of these the inferior one inosculates with the lowest branches of the vasa intestini tenuis, from the convexity of which branches proceed to supply the termination of the ileum, the cæcum and appendix cœci, and the ileo-cœcal and ileo-colic valves. The superior division inosculates with the colica dextra, and supplies the commencement of the colon.

The *colica dextra* arises from about the middle of the concavity of the superior mesenteric artery, and passing beneath the peritoneum to the middle of the ascending colon, divides into two branches; a descending branch, which inosculates with the ileo-colic; and an ascending branch, which anastomoses with the colica media. These branches form arches, from the convexity of which vessels are distributed to the ascending colon. The branches of this vessel are covered with peritoneum only on their anterior aspect.

The *colica media* arises from the upper part of the concavity of the superior mesenteric, and, passing forwards between the layers of the transverse meso-colon, divides into two branches; the one on the right side inosculating with the colica dextra; that on the left side, with the colica sinistra, a branch of the inferior mesenteric. From the arches formed by their inosculation, branches are distributed to the transverse colon. The branches of this vessel lie between two layers of peritoneum.

## Inferior Mesenteric Artery.

In order to expose this vessel, draw the small intestines and mesentery over to the right side of the abdomen, raise the transverse colon towards the thorax, and divide the peritoneum covering the left side of the aorta.

The Inferior Mesenteric Artery (fig. 226) supplies the descending and sigmoid flexure of the colon, and greater part of the rectum. It is smaller than the superior mesenteric; and arises from the left side of the aorta, between one and two inches above its division into the common iliacs. It passes downwards to the left iliac fossa, and then descends, between the layers of the meso-rectum, into the pelvis, under the name of the *superior hæmorrhoidal artery*. It lies at first in close relation with the left side of the aorta, and then passes in front of the left common iliac artery. Its branches are the

Colica sinistra. Sigmoid.
Superior hæmorrhoidal.

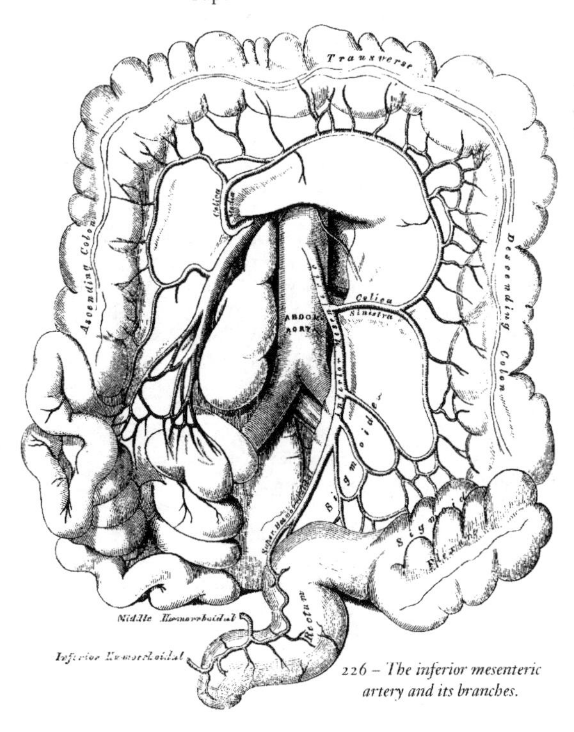


226 – *The inferior mesenteric artery and its branches.*

The *colica sinistra* passes behind the peritoneum, in front of the left kidney, to reach the descending colon, and divides into two branches; an ascending branch, which inosculates with the colica media; and a descending branch, which anastomoses with the sigmoid artery. From the arches formed by these inosculations, branches are distributed to the descending colon.

The *sigmoid artery* runs obliquely downwards across the Psoas muscle to the sigmoid flexure of the colon, and divides into branches which supply that part of the intestine; anastomosing above, with the colica sinistra; and below, with the superior hæmorrhoidal artery. This vessel is sometimes replaced by three or four small branches.

The *superior hæmorrhoidal artery*, the continuation of the inferior mesenteric, descends into the pelvis between the layers of the meso-rectum, crossing, in its course, the ureter, and left common iliac vessels. Opposite the middle of the sacrum it divides into two branches, which descend one on each side of the rectum, where they divide into several small branches, which are distributed between the mucous and muscular coats of that tube, to near its lower end; anastomosing with each other, with the middle hæmorrhoidal arteries, branches of the internal iliac, and with the inferior hæmorrhoidal, branches of the internal pudic.

The student should especially remark, that the trunk of the vessel descends along the back part of the rectum as far as the middle of the sacrum before it divides; this is about a finger's length or four inches from the anus. In disease of this tube, the rectum should never be divided beyond this point in that direction, for fear of involving this artery.

The Supra-Renal Arteries are two small vessels which arise, one on each side of the aorta, opposite the superior mesenteric artery. They pass obliquely upwards and outwards, to the under surface of the supra-renal capsules, to which they are distributed, anastomosing with capsular branches from the phrenic and renal arteries. In the adult these arteries are of small size; in the fœtus they are as large as the renal arteries.

The Renal Arteries are two large trunks, which arise from the sides of the aorta, immediately below the superior mesenteric artery. Each is directed outwards, so as to form nearly a right angle with the aorta. The right one longer than the left, on account of the position of the aorta, passes behind the inferior vena cava. The left is somewhat higher than the right. Previously to entering the kidney, each artery divides into four or five branches, which are distributed to its substance. At the hilus, these branches lie between the renal vein and ureter, the vein being usually in front, the ureter behind. Each vessel gives off some small branches to the supra-renal capsules, the ureter, and to the surrounding cellular membrane and muscles.

The Spermatic Arteries are distributed to the testes in the male, and to the ovaria in the female. They are two slender vessels, of considerable length, which arise from the front of the aorta, a little below the renal arteries. Each artery passes obliquely outwards and downwards, behind the peritoneum, crossing the ureter, and resting on the Psoas muscle, the right spermatic lying in front of the inferior vena cava, the left behind the sigmoid flexure of the colon. On reaching the margin of the pelvis, each vessel passes in front of the corresponding external iliac artery, and takes a different course in the two sexes.

*In the male*, it is directed outwards, to the internal abdominal ring, and accompanies the other constituents of the spermatic cord along the spermatic canal to the testis, where it becomes tortuous, and divides into several branches, two or three of which accompany the vas deferens, and supply the epididymis, anastomosing with the artery of the vas deferens; others pierce the back part of the tunica albuginea, and supply the substance of the testis.

*In the female*, the spermatic arteries (ovarian) are shorter than in the male, and do not pass out of the abdominal cavity. On arriving at the margins of the pelvis, each artery passes inwards, between the two laminæ of the broad ligament of the uterus, to be distributed to the ovary. One or two small branches supply the Fallopian tube; another passes on to the side of the uterus, and anastomoses with the uterine arteries. Other offsets are continued along the round ligament, through the inguinal canal, to the integument of the labium and groin.

At an early period of fœtal life, when the testes lie by the side of the spine, below the kidneys, the spermatic arteries are short; but as these organs descend from the abdomen into the scrotum, they become gradually lengthened.

The Phrenic Arteries are two small vessels, which present much variety in their origin. They may arise separately from the front of the aorta, immediately below the cœliac axis, or by a common trunk, which may spring either from the aorta, or from the cœliac axis. Sometimes one is derived from the aorta, and the other from one of the renal arteries. In only one out of thirty-six cases, did these arteries arise as two separate vessels from the aorta. They diverge from one another across the crura of the Diaphragm, and then pass obliquely upwards and outwards upon its under surface. The left phrenic passes behind the œsophagus, and runs forwards on the left side of the œsophageal opening. The right phrenic passes behind the liver and inferior vena cava, and ascends along the right side of the aperture for transmitting that vein. Near the back part of the central tendon, each vessel divides into two branches. The internal branch runs forwards to the front of the thorax, supplying the Diaphragm, and anastomosing with its fellow of the opposite side, and with the musculo-phrenic, a branch of the internal mammary. The external branch passes towards the side of the thorax, and inosculates with the intercostal arteries. The internal branch of the right phrenic gives off a few vessels to the inferior vena cava; and the left one some branches to the œsophagus. Each vessel also sends capsular branches to the supra-renal capsule of its own side. The spleen on the left side, and the liver on the right, also receive a few branches from these vessels.

The Lumbar Arteries are analogous to the intercostal. They are usually four in number on each side, and arise from the back part of the aorta, nearly at right angles with that vessel. They pass outwards and backwards, around the sides of the body of the corresponding lumbar vertebra, behind the sympathetic nerve and the Psoas muscle; those on the right side being covered by the inferior vena cava, and the two upper ones on each side by the crura of the Diaphragm. In the interval between the transverse processes of the vertebræ, each artery divides into a dorsal and an abdominal branch.

The *dorsal branch* gives off, immediately after its origin, a spinal branch, which enters the spinal canal; it then continues its course backwards, between the transverse processes, and is distributed to the muscles and integument of the back, anastomosing with each other, and with the posterior branches of the intercostal arteries.

The *spinal branch*, besides supplying offsets which run along the nerves to the dura mater and cauda equina, anastomosing with the other spinal arteries, divides into two branches, one of which ascends on the posterior surface of the body of the vertebra above, and the other descends on the posterior surface of the body of the vertebra below, both vessels anastomosing with similar branches from neighbouring spinal arteries. The inosculations of these vessels on each side, throughout the whole length of the spine, form a series of arterial arches behind the bodies of the vertebræ, which are connected with each other, and with a median longitudinal vessel, extending along the middle of the posterior surface of the bodies of the vertebræ, by transverse branches. From these vessels offsets are distributed to the periosteum and bones.

The *abdominal branches* pass outwards, behind the Quadratus lumborum, the lowest branch occasionally in front of that muscle, and, being continued between the abdominal muscles, anastomose with branches of the epigastric and internal mammary *in front*, the intercostals *above*, and those of the ilio-lumbar, and circumflex iliac, *below*.

The Middle Sacral Artery is a small vessel, about the size of a crow-quill, which arises from the back part of the aorta, just at its bifurcation. It descends upon the last lumbar vertebra, and along the middle line of the front of the sacrum, to the upper part of the coccyx, where it terminates by anastomosing with the lateral sacral arteries. From it, branches arise which run through the meso-rectum, to supply the posterior surface of the rectum. Other branches are given off on each side, which anastomose with the lateral sacral arteries, and send off small offsets which enter the anterior sacral foramina.

## Common Iliac Arteries.

The abdominal aorta divides into the two common iliac arteries. The bifurcation usually takes place on the left side of the body of the fourth lumbar vertebra. This point corresponds to the left side of the umbilicus, and is on a level with a line drawn from the highest point of one iliac crest to the other. The common iliac arteries are about two inches in length; diverging from the termination of the aorta, they pass downwards and outwards to the margin of the pelvis, and divide opposite the intervertebral substance, between the last lumbar vertebra and the sacrum, into two branches, the external and internal iliac arteries; the former supplying the lower extremity; the latter, the viscera, and parietes of the pelvis.

The *right common iliac* is somewhat larger than the left, and passes more obliquely across the body of the last lumbar vertebra. It is covered by the peritoneum, the ileum, the branches of the sympathetic nerve; and crossed, at its point of division, by the ureter. *Behind*, it is separated from the last lumbar vertebra, by the two common iliac veins. On its *outer side*, it is in relation with the inferior vena cava, and right common iliac vein, above; and the Psoas magnus muscle, below.

The *Left common iliac* is in relation, in front, with the peritoneum, branches of the sympathetic

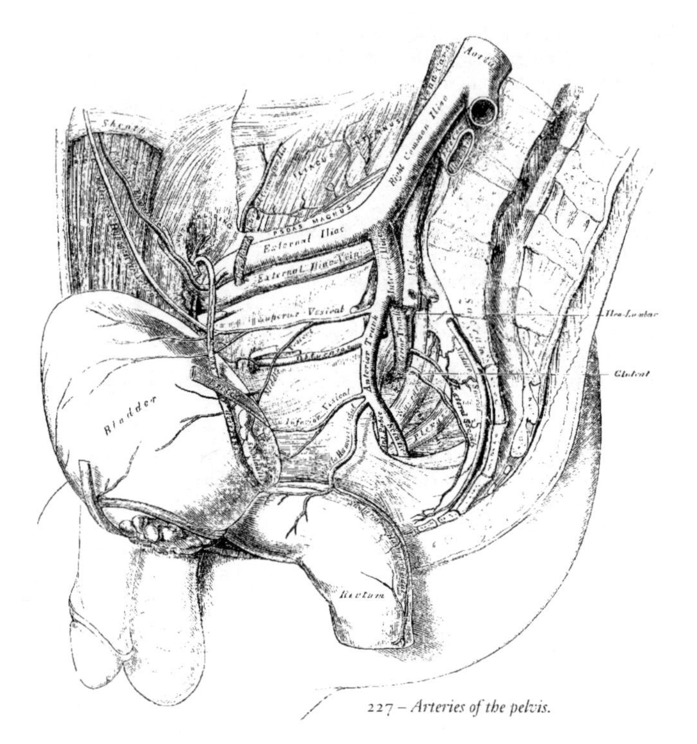


227 – *Arteries of the pelvis.*

nerve, the rectum and superior hæmorrhoidal artery; and crossed, at its point of bifurcation, by the ureter. The left common iliac vein lies partly on the inner side, and part beneath the artery; on its outer side, it is in relation with the Psoas magnus.

*Branches.* The common iliac arteries give off small branches to the peritoneum, Psoæ muscles, ureters, and to the surrounding cellular membrane, and occasionally give origin to the ilio-lumbar, or renal arteries.

*Peculiarities.* Its *point of origin* varies according to the bifurcation of the aorta. In three-fourths of a large number of cases, the aorta bifurcated either upon the fourth lumbar vertebra, or upon the intervertebral disc between it and the fifth; one case in nine being below, and one in eleven above this point. In ten out of every thirteen cases, the vessel bifurcated within half an inch above or below the level of the crest of the ilium; more frequently below than above.

*The point of division* is subject to great variety. In two-thirds of a large number of cases, it was between the last lumbar vertebra and the upper border of the sacrum; in one case in eight being above, and in one in six below that point. The left common iliac artery divides lower down more frequently than the right.

The *relative length*, also, of the two common iliac arteries varies. The right common iliac was longest in sixty-three cases; the left, in fifty-two; whilst they were both equal in fifty-three. The length of the arteries varied in five-sevenths of the cases examined, from an inch and a-half to three inches; in about half of the remaining cases, the artery was longer; and in the other half, shorter; the minimum length being less than half an inch, the maximum, four and a-half inches. In one instance, the right common iliac was found wanting, the external and internal iliacs arising directly from the aorta.

*Surgical Anatomy.* The application of a ligature to the common iliac artery may be required on account of aneurism or hæmorrhage, implicating the external or internal iliacs, or on account of secondary hæmorrhage after amputation of the thigh high up. It has been seen, that the origin of this vessel corresponds to the left side of the umbilicus on a level with a line drawn from the highest point of one iliac crest to the opposite one, and its course to a line extending from the left side of the umbilicus downwards towards the middle of Poupart's ligament. The line of incision required in the first steps of an operation for securing this vessel, would materially depend upon the nature of the disease. If the surgeon select the iliac region, a curved incision, about five inches in length, may be made, commencing on the left side of the umbilicus, carried outwards towards the anterior superior iliac spine, and then along the upper border of Poupart's ligament, as far as its middle. But if the aneurismal tumour should extend high up in the abdomen, along the external iliac, it is better to select the side of the abdomen, approaching the artery from above, by making an incision from four to five inches in length, from about two inches above and to the left of the umbilicus, carried outwards in a curved direction towards the lumbar region, and terminating a little below the anterior superior iliac spine. The abdominal muscles (in either case) having been cautiously divided in succession, the transversalis fascia must be carefully cut through, and the peritoneum, together with the ureter, separated from the artery, and pushed aside: the sacro-iliac articulation must then be felt for, and upon it the vessel will be felt pulsating, and may be fully exposed in close connection with its accompanying vein. On the right side, both common iliac veins, as well as the inferior vena cava, are in close connection with the artery, and must be carefully avoided. On the left side, the vein usually lies on the inner side, and behind the artery; but it occasionally happens, that the two common iliac veins are joined on the left instead of the right side, which would add much to the difficulty of an operation in such a case. If the common iliac artery is so short that danger is to be apprehended from secondary hæmorrhage if a ligature is applied to it, it would be preferable, in such a case, to tie both the external and internal iliacs near their origin. This operation has been performed in 17 cases, 9 of which were cured, and 8 died.

*Collateral Circulation.* The principal agents in carrying on the collateral circulation after the application of a ligature to the common iliac, are, the anastomoses of the hæmorrhoidal branches of the internal iliac, with the superior hæmorrhoidal from the inferior mesenteric; and by the anastomoses of the uterine and ovarian arteries, and of the vesical arteries of opposite sides; of the lateral sacral, with the middle sacral artery; of the epigastric, with the internal mammary, inferior intercostal and lumbar arteries; of the ilio-lumbar, with the last lumbar artery; of the obturator artery, by means of its pubic branch, with the vessel of the opposite side, and with the internal epigastric; and of the gluteal, with the posterior branches of the sacral arteries.

## Internal Iliac Artery.

The internal iliac artery supplies the walls and viscera of the pelvis, the generative organs, and inner side of the thigh. It is a short, thick vessel, smaller than the external iliac, and about an inch and a-half in length, which arises at the point of bifurcation of the common iliac; and, passing downwards to the upper margin of the great sacro-sciatic foramen, divides into two large trunks, an anterior, and posterior; a partially obliterated cord, the hypogastric artery, extending from the extremity of the vessel forwards to the bladder.

*Relations. In front*, with the ureter, which separates it from the peritoneum. *Behind*, with the internal iliac vein, the lumbo-sacral nerve, and Pyriformis muscle. By its *outer side*, near its origin, with the Psoas muscle.

### Plan of the Relations of the Internal Iliac Artery.

*In front.*
Peritoneum.
Ureter.

*Outer side.*
Psoas magnus.

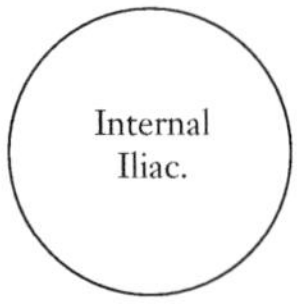

Internal
Iliac.

*Behind.*
Internal iliac vein.
Lumbo-sacral nerve.
Pyriformis muscle.

*In the fœtus*, the internal iliac artery (hypogastric), is twice as large as the external iliac, and appears the continuation of the common iliac. Passing forwards to the bladder, it ascends along the side of that viscus to its summit, to which it gives branches; it then passes upwards along the back part of the anterior wall of the abdomen to the umbilicus, converging towards its fellow of the opposite side. Having passed through the umbilical opening, the two arteries twine round the umbilical vein, forming with it the umbilical cord; and, ultimately, ramify in the placenta. That portion of the vessel placed within the abdomen, is called the hypogastric artery; and that external to that cavity, the umbilical artery.

*At birth*, when the placental circulation ceases, that portion of the hypogastric artery which extends from the umbilicus to the summit of the bladder, contracts, and ultimately dwindles to a solid fibrous cord; the portion of the same vessel extending from the summit of the bladder to within an inch and a-half of its origin, is not totally impervious, though it becomes considerably reduced in size; and serves to convey blood to the bladder, under the name of the superior vesical artery.

*Peculiarities, as regards its length.* In two-thirds of a large number of cases, the length of the internal iliac varied between an inch and an inch and a-half; in the remaining third, it was more frequently longer than shorter, the maximum length being three inches, the minimum, half an inch.

The lengths of the common and internal iliac arteries bear an inverse proportion to each other, the internal iliac artery being long when the common iliac is short, and *vice versâ.*

*As regards its place of division.* The place of division of the internal iliac varies between the upper margin of the sacrum, and the upper border of the sacro-sciatic foramen.

The arteries of the two sides in a series of cases often differed in length, but neither seemed constantly to exceed the other.

*Surgical Anatomy.* The application of a ligature to the internal iliac artery may be required in cases of aneurism or hæmorrhage affecting one of its branches. This vessel may be secured by making an incision through the abdominal parietes in the iliac region, in a direction and to an extent similar to that for securing the common iliac; the transversalis fascia having been cautiously divided, and the peritoneum pushed inwards from the iliac fossa towards the pelvis, the finger may feel the pulsation of the external iliac at the bottom of the wound; and, by tracing this vessel upwards, the internal iliac is arrived at, opposite the sacro-iliac articulation. It should be remembered that the vein lies behind, and on the right side, a little external to the artery, and in close contact with it; the ureter and peritoneum, which lie in front, must also be avoided. The degree of facility in applying a ligature to this vessel, will mainly depend upon its length. It has been seen that, in the great majority of the cases examined, the artery was short, varying from an inch to an inch and a-half; in these cases, the artery is deeply seated in the pelvis; when, on the contrary, the vessel is longer, it is found partly above that cavity. If the artery is very short, which occasionally happens, it would be preferable to apply a ligature to the common iliac, or upon the external and internal iliacs at their origin. This operation has been performed in 7 cases, 4 of which recovered, and 3 died.

*Collateral Circulation.* In Mr. Owen's dissection of a case in which the internal iliac artery had been tied by Stevens ten years before death, for aneurism of the sciatic artery, the internal iliac was found impervious for about an inch above the point where the ligature had been applied; but the obliteration did not extend to the origin of the external iliac as the ilio-lumbar artery arose just above this point. Below the point of obliteration, the artery resumed its natural diameter, and continued so for half an inch; the obturator, lateral sacral, and gluteal, arising in succession from the latter portion. The obturator artery was entirely obliterated. The lateral sacral artery was as large as a crow's quill, and had a very free anastomosis with the artery of the opposite side, and with the middle sacral artery. The sciatic artery was entirely obliterated as far as its point of connection with the aneurismal tumour; but, on the distal side of the sac, it was continued down along the back of the thigh nearly as large in size as the femoral, being pervious about an inch below the sac by receiving an anastomosing vessel from the superior profunda.* In addition to the above, the circulation in the parts supplied by the internal iliac would be carried on by the anastomoses of the uterine and ovarian arteries; of the opposite vesical arteries; of the hæmorrhoidal branches of the internal iliac, with those from the inferior mesenteric; of the obturator artery, by means of its pubic branch, with the vessel of the opposite side, and with the epigastric and internal circumflex; by the anastomoses of the circumflex, and perforating branches of the femoral, with the sciatic; of the gluteal, with the posterior branches of the sacral arteries; of the ilio-lumbar, with the last lumbar; of the lateral sacral, with the middle sacral; and by the anastomoses of the circumflex iliac, with the iliac-lumbar and gluteal.

## Branches of the Internal Iliac.

| *From the Anterior Trunk.* | | *From the Posterior Trunk.* |
|---|---|---|
| | Superior vesical. | Gluteal. |
| | Middle vesical. | Ilio-lumbar. |
| | Inferior vesical. | Lateral sacral. |
| | Middle hæmorrhoidal. | |
| | Obturator. | |
| | Internal pudic. | |
| | Sciatic. | |
| *In female.* | Uterine. | |
| | Vaginal. | |

The *superior vesical* is that part of the fœtal hypogastric artery, which remains pervious after birth. It extends to the side of the bladder, distributing numerous branches to the body and fundus of this

* *Medico-Chirurgical Trans.*, vol. xvi.

organ. From one of these, a slender vessel is derived, which accompanies the vas deferens in its course to the testis, where it anastomoses with the spermatic artery. This is the artery of the vas deferens. Other branches supply the ureter.

The *middle vesical*, usually a branch of the superior, is distributed to the base of the bladder, and under surface of the vesiculæ seminales.

The *inferior vesical* arises from the anterior division of the internal iliac, in common with the middle hæmorrhoidal, and is distributed to the base of the bladder, the prostate gland, and vesiculæ seminales. Those branches distributed to the prostate, communicate with the corresponding vessel of the opposite side.

The *middle hæmorrhoidal artery* usually arises together with the preceding vessel. It supplies the rectum, anastomosing with the other hæmorrhoidal arteries.

The *uterine artery* passes downwards from the anterior trunk of the internal iliac to the neck of the uterus. Ascending, in a tortuous course, on the side of this viscus, between the layers of the broad ligament, it distributes branches to its substance, anastomosing, near its termination, with a branch from the ovarian artery. Branches from this vessel are also distributed to the bladder and ureter.

The *vaginal artery* is analogous to the inferior vesical in the male; it descends upon the vagina, supplying its mucous membrane, and sending branches to the neck of the bladder, and contiguous part of the rectum.

The Obturator Artery usually arises from the anterior trunk of the internal iliac, frequently from the posterior. It passes forwards below the brim of the pelvis, to the canal in the upper border of the obturator foramen, and, escaping from the pelvic cavity through this aperture, divides into an internal and an external branch. In the pelvic cavity, this vessel lies upon the pelvic fascia, beneath the peritoneum, and a little below the obturator nerve; and, whilst passing through the obturator foramen, is contained in an oblique canal, formed by the horizontal branch of the pubes, above; and the arched border of the obturator membrane, below.

*Branches. Within the pelvis*, the obturator artery gives off an *iliac branch* to the iliac fossa, which supplies the bone and the Iliacus muscle, and anastomoses with the ilio-lumbar artery; a *vesical branch*, which runs backwards to supply the bladder; and a *pubic branch*, which is given off from the vessel just before it leaves the pelvic cavity. This branch ascends upon the back of the pubes, communicating with offsets from the epigastric artery, and with the corresponding vessel of the opposite side. This branch is placed on the inner side of the femoral ring. *External to the pelvis*, the obturator artery divides into an external and an internal branch, which are deeply situated beneath the Obturator externus muscle; skirting the circumference of the obturator foramen, they anastomose at the lower part of this aperture with each other, and with branches of the internal circumflex artery.

The *internal branch* curves inwards along the inner margin of the obturator foramen, distributing branches to the Obturator muscles, Pectineus, Adductors, and Gracilis, and anastomoses with the external branch, and with the internal circumflex artery.

The *external branch* curves round the outer margin of the foramen, to the space between the Gemellus inferior and Quadratus femoris, where it anastomoses with the sciatic artery. It supplies the Obturator muscles, anastomoses, as it passes backwards, with the internal circumflex, and sends a branch to the hip-joint through the cotyloid notch, which ramifies on the round ligament as far as the head of the femur.

*Peculiarities.* In two out of every three cases the obturator arises from the internal iliac. In one case in $3\frac{1}{2}$ from the epigastric; and in about one in seventy-two cases by two roots from both vessels. It arises in about the same proportion from the external iliac artery. The origin of the obturator from the epigastric is not commonly found on both sides of the same body.

When the obturator artery arises at the front of the pelvis from the epigastric, it descends almost vertically downwards to the upper part of the obturator foramen. The artery in this course usually descends in contact with the external iliac vein, and lies on the outer side of the femoral ring (fig. 228); in such cases it would not be endangered in the operation for femoral hernia. Occasionally, however, it curves inwards along the free margin of Gimbernat's ligament (fig. 229), and under such circumstances would almost completely encircle the neck of a hernial sac (supposing a hernia to exist in such a case), and would be in great danger of being wounded if an operation was performed.

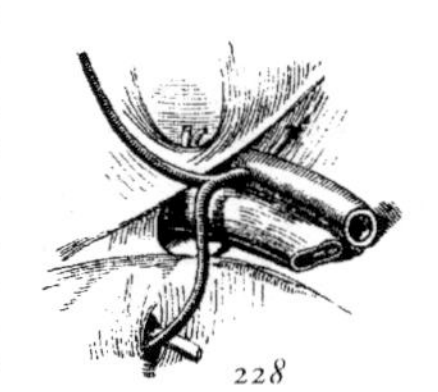
228

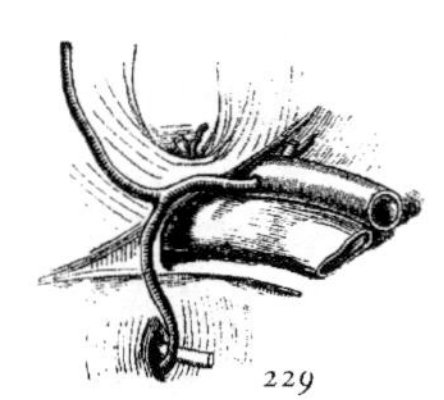
229

*Variations in origin and course of obturator artery.*

The *Internal Pudic* is the smaller of the two terminal branches of the anterior trank of the internal iliac, and supplies the external organs of generation. It passes downwards and outwards to the lower border of the great sacro-sciatic foramen, and emerges from the pelvis between the Pyriformis and Coccygeus muscles; it then crosses the spine of the ischium, and re-enters the pelvis through the lesser sacro-sciatic foramen. The artery now crosses the Obturator internus muscle, to the ramus of the ischium, being covered by the obturator fascia, and situated about an inch and a half from the margin of the tuberosity; it then ascends forwards and upwards along the ramus of the ischium, pierces the posterior layer of the deep perinæal fascia, and runs forwards along the inner margin of the ramus of the pubes; finally, it perforates the anterior layer of the deep perinæal fascia, and divides into its two terminal branches, the dorsal artery of the penis, and the artery of the corpus cavernosum.

*Relations.* In the first part of its course, within the pelvis, it lies in front of the Pyriformis muscle and sacral plexus of nerves, and on the outer side of the rectum (on the left side). As it crosses the spine of the ischium, it is covered by the Gluteus maximus, and great sacro-sciatic ligament. And when it enters the pelvis, it lies on the outer side of the ischio-rectal fossa, upon the surface of the Obturator internus muscle, contained in a fibrous canal formed by the obturator fascia and the falciform process of the great sacro-sciatic ligament. It is accompanied by the pudic veins, and the internal pudic nerve.

*Peculiarities.* The internal pudic is sometimes smaller than usual, or fails to give off one or two of its usual branches; in such cases, the deficiency is supplied by branches derived from an additional vessel, the *accessory pudic,* which generally arises from the pudic artery before its exit from the great sacro-sciatic foramen, and passes forwards near the base of the bladder, on the upper part of the prostate gland, to the perinæum, where it gives off those branches usually derived from the pudic artery. The deficiency most frequently met with, is that in which the internal pudic ends as the artery of the bulb; the artery of the corpus cavernosum and arteria dorsalis penis being derived from the accessory pudic. Or the pudic may terminate as the superficial perinæal, the artery of the bulb being derived, with the other two branches, from the accessory vessel.

The relation of the accessory pudic to the prostate gland and urethra, is of the greater interest in a surgical point of view, as this vessel is in danger of being wounded in the lateral operation of lithotomy.

*Branches. Within the pelvis,* the internal pudic gives off several small branches which supply the muscles, sacral nerves, and viscera in this cavity. *In the perinæum* the following branches are given off.

| | |
|---|---|
| Inferior or external hæmorrhoidal. | Artery of the bulb. |
| Superficial perinæal. | Artery of the corpus cavernosum. |
| Transverse perinæal. | Dorsal artery of the penis. |

The *external hæmorrhoidal* are two or three small arteries, which arise from the internal pudic as it passes above the tuberosity of the ischium. Crossing the ischio-rectal fossa, they are distributed to the muscles and integument of the anal region.

The *superficial perinæal artery* supplies the scrotum, and muscles and integament of the perinæum. It arises from the internal pudic, in front of the preceding branches, and piercing the lower border of the deep perinæal fascia, runs across the Transversus perinæi, and through the triangular space between the Accelerator urinæ and Erector penis, both of which it supplies, and is finally distributed to the skin of the scrotum and dartos. In its passage through the perinæum it lies beneath the superficial perinæal fascia.

The *transverse perinæal* is a small branch which arises either from the internal pudic, or from the superficial perinæal artery as it crosses the Transversus perina muscle. Piercing the lower border of the deep perinæal fascia, it runs transversely inwards along the cutaneous surface of the Transversus perinæi muscle which it supplies, as well as the structures between the anus and bulb of the urethra.

The *artery of the bulb* is a large but very short vessel, arising from the internal pudic between the two layers of the deep perinæal fascia, and passing near transversely inwards, pierces the bulb of the urethra, in which it ramifies. It gives off a small branch which descends to supply Cowper's gland. This artery is of considerable importance in a surgical point of view, as it is in danger of being wounded in the lateral operation of lithotomy, an accident usually attended with severe and alarming hæmorrhage. This vessel is sometimes very small, occasionally wanting, or even double. It sometimes arises from the internal pudic earlier than usual, and crosses the perinæum to reach the back part of the bulb. In such a case the vessel could hardly fail to be wounded in the performance of the lateral operation for lithotomy. If, on the contrary, it should arise from an accessory pudic, it lies more forward than usual, and is out of danger in the operation.

The *artery of the corpus cavernosum*, one of the terminal branches of the internal pudic, arises from that vessel while it is situated between the crus penis and the ramus of the pubes; piercing the crus penis obliquely, it runs forwards in the corpus cavernosum by the side of the septum pectiniforme, to which its branches are distributed.

The *dorsal artery of the penis* ascends between the crus and pubic symphysis, and piercing the suspensory ligament, runs forward on the dorsum of the penis to the glans, where it divides into two branches, which supply the glans and prepuce. On the dorsum of the penis, it lies immediately beneath the integument, parallel with the dorsal vein and corresponding artery of the opposite side. It supplies the integument and fibrous sheath of the corpus cavernosum.

The *internal pudic artery in the female* is smaller than in the male. Its origin and course are similar, and there is considerable analogy in the distribution of its branches. The superficial artery supplies the labia pudenda; the artery of the bulb supplies the erectile tissue of the bulb of the vagina, whilst the two terminal branches supply the clitoris; the artery of the corpus cavernosum, the cavernous body of the clitoris; and the arteria dorsalis clitoridis, the dorsum of that organ.

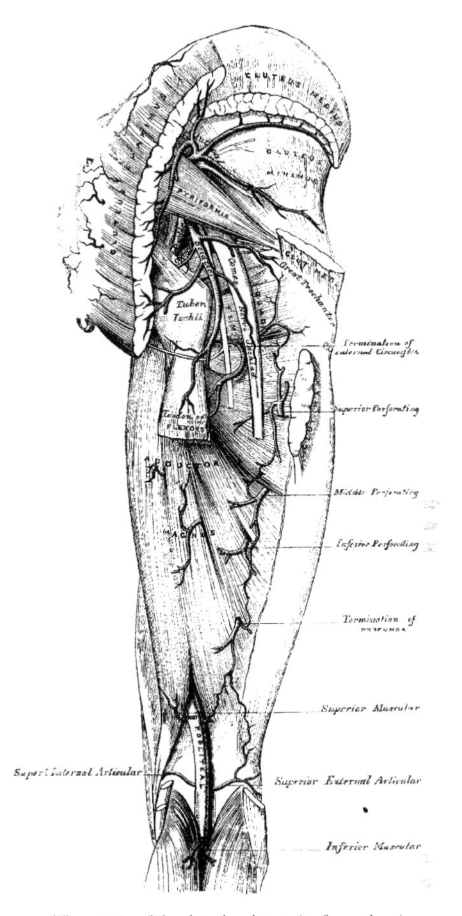


230 – *The arteries of the gluteal and posterior femoral regions.*

The *Sciatic Artery* (fig. 230), the larger of the two terminal branches of the anterior trunk of the internal iliac, is distributed to the muscles on the back of the pelvis. It passes down to the lower part of the great sacro-sciatic foramen, behind the internal pudic, resting on the sacral plexus of nerves and Pyriformis muscle, and escapes from the pelvis between the Pyriformis and Coccygeus. It then descends in the interval between the trochanter major and tuberosity of the ischium, accompanied by the sciatic nerves, and covered by the Gluteus maximus, and divides into branches, which supply the deep muscles at the back of the hip.

*Within the pelvis* it distributes branches to the Pyriformis, Coccygeus, and Levator ani muscles; some hæmorrhoidal branches, which supply the rectum, and occasionally take the place of the middle hæmorrhoidal artery; and vesical branches to the base and neck of the bladder, vesiculæ seminales, and prostate gland. *External to the pelvis* it gives off the coccygeal, inferior gluteal, comes nervi ischiadici, muscular, and articular branches.

The *coccygeal branch* runs inwards, pierces the great sacro-sciatic ligament, and supplies the Gluteus maximus, the integument, and other structures on the back of the coccyx.

The *inferior gluteal branches*, three or four in number, supply the Gluteus maximus muscle.

The *comes nervi ischiadici* is a long slender vessel, which accompanies the great sciatic nerve for a short distance; it then penetrates it, and runs in its substance to the lower part of the thigh.

The *muscular branches* supply the muscles on the back part of the hip, anastomosing with the gluteal, internal and external circumflex, and superior perforating arteries.

Some *articular branches* are distributed to the capsule of the hip-joint.

The *Gluteal Artery* is the largest branch of the internal iliac, and appears to be the continuation of the posterior division of that vessel. It is a short thick trunk, which passes out of the pelvis above the upper border of the Pyriformis muscle, and immediately divides into a superficial and deep branch. Within the pelvis, it gives off a few muscular branches to the Iliacus, Pyriformis, and Obterator internus, and just previous to quitting that cavity a nutritious artery, which enters the ilium.

The *superficial branch* passes beneath the Gluteus maximus, and divides into numerous branches, some of which supply this muscle, whilst others perforate its tendinous origin, and supply the integument covering the posterior surface of the sacrum, anastomosing with the posterior branches of the sacral arteries.

The *deep branch* runs between the Gluteus medius and minimus, and subdivides into two. Of these, the *superior division*, continuing the original course of the vessel, passes along the upper border of the Gluteus minimus to the anterior superior spine of the ilium, anastomosing with the circumflex iliac and ascending branches of the external circumflex artery. The *inferior division* crosses the Gluteus minimus obliquely to the trochanter major, distributing branches to the Glutei muscles, and inosculates with the external circumflex artery. Some branches pierce the Gluteus minimus to supply the hip-joint.

The *Ilio-Lumbar Artery* ascends beneath the Psoas muscle and external iliac vessels, to the upper part of the iliac fossa, where it divides into a lumbar and an iliac branch.

The *lumbar branch* supplies the Psoas and Quadratus lumborum muscles, anastomosing with the last lumbar artery, and sends a small spinal branch through the intervertebral foramen, between the last lumbar vertebra and the sacrum, into the spinal canal, to supply the spinal cord and its membranes.

The *iliac branch* descends to supply the Iliacus internus, some offsets running between the muscle and the bone, one of which enters an oblique canal to supply the diploë, whilst others run along the crest of the ilium, distributing branches to the Gluteal and Abdominal muscles, and anastomosing in their course with the gluteal, circumflex iliac, external circumflex, and epigastric arteries.

The *Lateral Sacral Arteries* are usually two in number on each side, superior and inferior.

The *superior*, which is of large-size, passes inwards, and after anastomosing with branches from the middle sacral, enters the first or second sacral foramen, is distributed to the contents of the sacral canal, and escaping by the corresponding posterior sacral foramen, supplies the skin and muscles on the dorsum of the sacrum.

The *inferior branch* passes obliquely across the front of the Pyriformis muscle and sacral nerves to the inner side of the anterior sacral foramina, descends on the front of the sacrum, and anastomoses over the coccyx with the sacra-media and opposite lateral sacral arteries. In its course, it gives off branches, which enter the anterior sacral foramina, these, after supplying the bones and membranes of the interior of the spinal canal, escape by the posterior sacral foramina, and are distributed to the muscles and skin on the dorsal surface of the sacrum.

## External Iliac Artery.

The external iliac artery is the chief vessel which supplies the lower limb. It is larger in the adult than the internal iliac, and passes obliquely downwards and outwards along the inner border of the Psoas muscle, from the bifurcation of the common iliac to the femoral arch, where it enters the thigh, and becomes the femoral artery. The course of this vessel would be indicated by a line drawn from the left side of the umbilicus to a point midway between the anterior superior spinous process of the ilium and the symphysis pubis.

*Relations. In front*, with the peritoneum, sub-peritoneal areolar tissue, the intestines, and a thin layer of fascia, derived from the iliac fascia, which surrounds the artery and vein. At its origin it is occasionally crossed by the ureter. The spermatic vessels descend for some distance upon it near its termination, and it is crossed in this situation by a branch of the genito-crural nerve and the circumflex iliac vein; the vas deferens curves down along its inner side. *Behind*, it is in relation with the external iliac vein, which, at the femoral arch, lies at its inner side; on the left side the vein is altogether internal to the artery. *Externally*, it rests against the Psoas muscle, from which it is separated by the iliac fascia. The artery rests upon this muscle near Poupart's ligament. Numerous lymphatic vessels and glands are found lying on the front and inner side of the vessel.

## Plan of the Relations of the External Iliac Artery.

*In front.*

Near Poupart's Ligament.
- Peritoneum, intestines, and iliac fascia.
- Spermatic vessels.
- Genito-crural nerve.
- Circumflex iliac vein.
- Lymphatic vessels and glands.

*Outer side.*
Psoas magnus.
Iliac fascia.

External Iliac.

*Inner side.*
External iliac vein and vas deferens at femoral arch.

*Behind.*
External iliac vein.

*Surgical Anatomy.* The application of a ligature to the external iliac may be required in cases of aneurism of the femoral artery, or in cases of secondary hæmorrhage, after the latter vessel has been tied for popliteal aneurism. This vessel may be secured in any part of its course, excepting near its upper end, on account of the circulation through the internal iliac, and near its lower end, on account of the origin of the epigastric and circumflex iliac vessels. One of the chief points in the performance of the operation is to secure the vessel without injury to the peritoneum. The patient having been placed in the recumbent position, an incision should be made, commencing about an inch above and to the inner side of the anterior superior spinous process of the ilium, and running downwards and outwards to the outer end of Poupart's ligament, and parallel with its outer half, to a little above its middle. The abdominal muscles and transversalis fascia having been cautiously divided, the peritoneum should be separated from the iliac fossa and pushed towards the pelvis; and on introducing the finger to the bottom of the wound the artery may be felt pulsating along the inner border of the Psoas muscle. The external iliac vein is situated along the inner side of the artery, and must be cautiously separated from it by the finger-nail, or handle of the knife, and the aneurism needle should be introduced on the inner side, between the artery and vein.

*Collateral Circulation.* The principal anastomoses in carrying on the collateral circulation, after the application of a ligature to the external iliac, are—the ilio-lumbar with the circumflex iliac; the gluteal with the external circumflex; the obturator with the internal circumflex; the sciatic with the profunda artery; the internal pudic with the external pudic, and with the internal circumflex. When the obturator arises from the epigastric, it is supplied with blood by branches, either from the internal iliac, the lateral sacral, or from the internal pudic. The epigastric receives its supply from the internal mammary and inferior intercostal arteries, and from the internal iliac, by the anastomoses of its branches with the obturator.

*Branches.* Besides several small branches to the Psoas muscle and the neighbouring lymphatic glands, the external iliac gives off two branches of considerable size, the

Epigastric. Circumflex iliac.

The *epigastric artery* arises from the external iliac, a few lines above Poupart's ligament. It at first descends to reach this ligament, and then ascends obliquely upwards and inwards between the peritoneum and transversalis fascia, to the margin of the sheath of the Rectus muscle. Having perforated the sheath near its lower third, it ascends vertically upwards behind the Rectus, to which it is distributed, dividing into numerous branches, which anastomose above the umbilicus with the terminal branches of the internal mammary and inferior intercostal arteries. It is accompanied by two veins, which usually unite into a single trunk before their termination in the external iliac vein. As this artery ascends from Poupart's ligament to the Rectus, it lies behind the inguinal canal, to the inner side of the internal abdominal ring, and immediately above the femoral ring, the vas deferens in the male, and the round ligament in the female, crossing behind the artery in descending into the pelvis.

*Branches.* The branches of this vessel are the *cremasteric*, which accompanies the spermatic cord, and supplies the Cremaster muscle, anastomosing with the spermatic artery. A *pubic branch*, which runs across Poupart's ligament, and then descends behind the pubes to the inner side of the femoral ring, and anastomoses with offsets from the obturator artery. *Muscular branches*, some of which are distributed to the abdominal muscles and peritoneum, anastomosing with the lumbar and circumflex iliac arteries; others perforate the tendon of the External oblique and supply the integument, anastomosing with branches of the external epigastric.

*Peculiarities*. The origin of the epigastric may take place from any part of the external iliac between Poupart's ligament and two inches and a half above it; or it may arise below this ligament, from the femoral, or from the deep femoral.

*Union with Branches*. It frequently arises from the external iliac by a common trunk with the obturator. Sometimes the epigastric arises from the obturator, the latter vessel being furnished by the internal iliac, or the epigastric may be formed of two branches, one derived from the external iliac, the other from the internal iliac.

The *circumflex iliac artery* arises from the outer side of the external iliac, nearly opposite the epigastric artery. It ascends obliquely outwards behind Poupart's ligament, and runs along the inner surface of the crest of the ilium to about its middle, where it pierces the Transversalis, and runs backwards between this muscle and the Internal oblique, to anastomose with the ilio-lumbar and gluteal arteries. Opposite the anterior superior spine of the ilium, it gives off a large branch, which ascends between the Internal oblique and Transversalis muscles, supplying them and anastomosing with the lumbar and epigastric arteries. The circumflex iliac artery is accompanied by two veins, which, uniting into a single trunk, crosses the external iliac artery just above Poupart's ligament, and enters the external iliac vein.

## Femoral Artery.

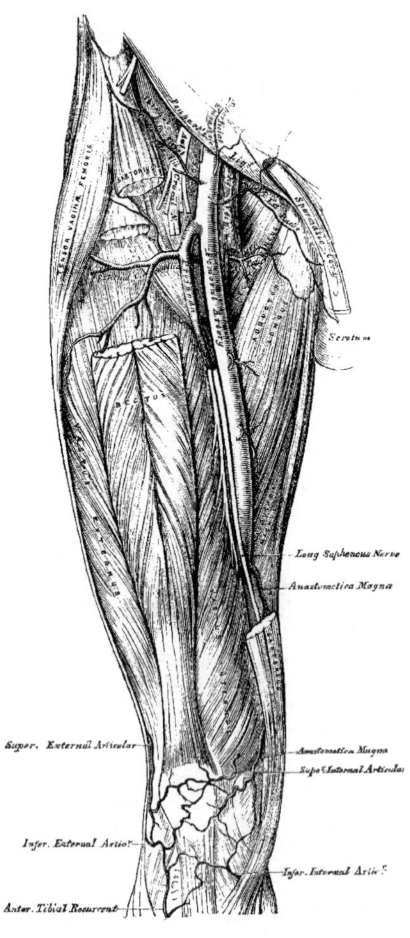


231 – *Surgical anatomy of the femoral artery.*

The femoral artery is the continuation of the external iliac. It commences immediately beneath Poupart's ligament, midway between the anterior superior spine of the ilium and the symphysis pubis, and passing down the fore part and inner side of the thigh, terminates at the opening in the Adductor magnus, at the junction of the middle with the lower third of the thigh, where it becomes the popliteal artery. A line drawn from a point midway between the anterior superior spine of the ilium and the symphysis pubis to the inner side of the inner condyle of the femur, will be nearly parallel with the course of the artery. This vessel, at the upper part of the thigh, lies a little internal to the head of the femur; in the lower part of its course, on the inner side of the shaft of this bone, and between these two points, the vessel is separated from the bone by a considerable interval.

*In the upper third of the thigh* the femoral artery is very superficial, being covered by the integument, inguinal glands, and by the superficial and deep fasciæ, and is contained in a triangular space, called 'Scarpa's triangle.'

*Scarpa's Triangle.* Scarpa's triangle corresponds to the depression seen immediately below the fold of the groin. It is a triangular space, the apex of which is directed downwards, and the sides of which are formed externally by the Sartorius, internally by the Adductor longus, and the base, by Poupart's ligament. The floor of this space is formed from without inwards by the Iliacus, Psoas, Pectineus, Adductor longus, and a small part of the Adductor brevis muscles; and it is divided into two nearly equal parts by the femoral vessels, which extend from the middle of its base to its apex: the artery giving off in this situation its cutaneous and profunda branches, the vein receiving the deep femoral and internal saphenous veins. In this space, the femoral artery rests on the inner margin of the Psoas muscle, which separates it from the capsular ligament of the hip-joint. The artery in this situation is crossed in front by the crural branch of the genito-crural serve, and behind by the branch to the Pectineus from the anterior crural. The femoral vein lies at its inner side, between the margins of the Pectineus and Psoas muscles. The anterior crural nerve lies about half an inch to the outer side of the femoral artery, deeply imbedded between the Iliacus and Psoas muscles; and on the Iliacus muscle, internal to the anterior superior spinous process of the ilium, is the external cutaneous nerve. The femoral artery and vein are enclosed in a strong fibrous sheath, formed by fibrous and cellular tissue, and by a process of fascia sent inwards from the fascia lata; the vessels are separated, however, from one another by thin fibrous partitions.

*In the middle third of the thigh*, the femoral artery is more deeply seated, being covered by the integument, the superficial and deep fasciæ, and the Sartorius, and is contained in an aponeurotic canal, formed by a dense fibrous band, which extends transversely from the Vastus internus to the tendons of the Adductor longus and magnus muscles. In this part of its course it lies in a depression, bounded externally by the Vastus internus, internally by the Adductor longus and Adductor magnus. The femoral vein lies on the outer side of the artery, in close apposition with it, and, still more externally, is the internal (long) saphenous nerve.

*Relations. From above downwards*, the femoral artery rests upon the Psoas muscle, which separates it from the margin of the pelvis and capsular ligament of the hip; it is next separated from the Pectineus by the profunda vessels and femoral vein; it then lies upon the Adductor longus; and lastly, upon the tendon of the Adductor magnus, the femoral vein being interposed. To its *inner side*, it is in relation, above, with the femoral vein, and lower down, with the Adductor longus, and Sartorius. To its *outer side*, the Vastus internus separates it from the femur, in the lower part of its course.

The *femoral vein*, at Poupart's ligament, lies close to the inner side of the artery, separated from it by a thin fibrous partition, but, as it descends, gets behind it, and then to its outer side.

The *internal saphenous nerve* is situated on the outer side of the artery, in the middle third of the thigh, beneath the aponeurotic covering, but not within the sheath of the vessels. Small cutaneous nerves cross the front of the sheath.

*Peculiarities. Double femoral re-united.* Four cases are at present recorded, in which the femoral artery divided into two trunks below the origin of the profunda, and became re-united near the opening in the Adductor magnus, so as to form a single popliteal artery. One of them occurred in a patient operated upon for popliteal aneurism.

*Change of Position.* A similar number of cases have been recorded, in which the femoral artery was situated at the back of the thigh, the vessel being continuous above with the internal iliac, escaping from the pelvis through the great sacro-sciatic foramen, and accompanying the great sciatic nerve to the popliteal space, where its division occurred in the usual manner.

*Position of the Vein.* The femoral vein is occasionally placed along the inner side of the artery, throughout the entire extent of Scarpa's triangle; or it may be slit, so that a large vein is placed on each side of the artery for a greater or less extent.

*Origin of the Profunda.* This vessel occasionally arises from the inner side, and more rarely, from the back of the common trunk; but the more important peculiarity, in a surgical point of view, is that which relates to the height at which the vessel arises from the femoral. In three-fourths of a large number of cases, it arose between one and two inches below Poupart's ligament; in a few cases, the distance was less than an inch; more rarely, opposite the ligament; and in one case, above Poupart's ligament, from the external iliac. Occasionally, the distance between the origin of the vessel and Poupart's ligament exceeds two inches, and in one case, it was found to be as much as four inches.

*Surgical Anatomy. Compression* of the femoral artery, which is constantly requisite in amputations, or other operations on the lower limb, is most effectually made immediately below Poupart's ligament. In this situation, the artery is very superficial, and is merely separated from the margin of the acetabulum and front of the head of the femur, by the Psoas muscle; so that the surgeon, by means of his thumb, or any other resisting body may effectually control the circulation through it. This vessel may also be compressed in the middle third of the thigh, by placing a compress over the artery, beneath the tourniquet, and directing the pressure from within outwards, so as to compress the vessel on the inner side of the shaft of the femur.

The *application of a ligature* to the femoral artery may be required in cases of wound or aneurism of the arteries of the leg, of the popliteal or femoral; and the vessel may be exposed and tied in any part of its course. The great depth of this vessel in the lower part of its course, its close connection with important structures, and the density of its sheath, render the operation in this situation one of much greater difficulty than the application of a ligature at its upper part, where it is more superficial.

Ligature of the femoral artery, within two inches of its origin, is usually considered very unsafe, on account of the connection of large branches with it, the epigastric and circumflex iliac arising just above its origin; the profunda, from one to two inches below, occasionally, also, one of the circumflex arteries, arises from the vessel in the interspace between these. The profunda sometimes arises higher than the point above-mentioned, and rarely between two or three inches (in one case four), below Poupart's ligament. It would appear, then, that the most favourable situation for the application of a ligature to this vessel, is between four and five inches from its point of origin. In order to expose the artery in this situation, an incision, between two and three inches long should be made in the course of the vessel, the patient lying in the recumbent position with the limb slightly flexed and abducted. A large vein is frequently met with, passing in the course of the artery to join the saphena; this must be avoided, and the fascia lata having been cautiously divided, and the Sartorius exposed, this muscle must be drawn outwards, in order to fully expose the sheath of the vessels. The finger being introduced into the wound, and the pulsation of the artery felt, the sheath should be divided over it to a sufficient extent to allow of the introduction of the ligature, but no further; otherwise the nutrition of the coats of the vessel may be interfered with, or muscular branches which arise from the vessel at irregular intervals may be divided. In this part of the operation, a small nerve which crosses the sheath should be avoided. The aneurism needle must be carefully introduced and kept close to the artery, to avoid the femoral vein, which lies behind the vessel in this part of its course.

To expose the artery in the middle of the thigh, an incision should be made through the integument, between three and four inches in length, over the inner margin of the Sartorius, taking care to avoid the internal saphenous vein, the situation of which may be previously known by compressing it higher up in the thigh. The fascia lata having been divided, and the Sartorius muscle exposed, it should be drawn outwards, when the strong fascia which is stretched across from the Adductors to the Vastus internus, will be exposed, and must be freely divided; the sheath of the vessels is now seen, and must be opened, and the artery secured by passing the aneurism needle between the vein and artery, in the direction from within outwards. The femoral vein in this situation lies on the outer side of the artery, the long saphenous nerve on its anterior and outer side.

It has been seen that the femoral artery occasionally divides into two trunks, below the origin of the profunda. If, in the operation for tying the femoral two vessels are met with, the surgeon should alternately compress each, in order to ascertain which vessel is connected with the aneurismal tumour, or with the bleeding from the wound, and that one only tied which controls it. If, however, it is necessary to compress both vessels before the circulation in the tumour is controlled, both should be tied, as it would be probable that they became re-united, as is mentioned above.

*Collateral Circulation*. The principal agents in carrying on the collateral circulation after ligature of the femoral artery are according to Sir A. Cooper, as follows.*

'The arteria profunda formed the new channel for the blood.' 'The first artery sent off passed down close to the back of the thigh bone, and entered the two superior articular branches of the popliteal artery.'

'The second new large vessel arising from the profunda at the same part with the former, passed down by the inner side of the Biceps muscle, to an artery of the popliteal which was distributed to the Gastrocnemius muscle; whilst a third artery dividing into several branches passed down with the sciatic nerve behind the knee-joint, and some of its branches united themselves with the inferior articular arteries of the popliteal, with some recurrent branches of those arteries, with arteries passing to the Gastrocnemii, and lastly, with the origin of the anterior and posterior tibial arteries.'

'It appears then that it is those branches of the profunda which accompany the sciatic nerve, that are the principal supporters of the new circulation.'

*Branches.* The branches of the femoral artery are the:

Superficial epigastric.
Superficial circumflex iliac.
Superficial external pudic.
Deep external pudic.
Profunda. { External circumflex. Internal circumflex. Three perforating.
Muscular.
Anastomotica magna.

The *superficial epigastric* arises from the femoral, about half an inch below Poupart's ligament, and, passing through the saphenous opening in the fascia lata, ascends on to the abdomen, in the superficial fascia covering the External oblique muscle, nearly as high as the umbilicus. It distributes branches to the inguinal glands, the superficial fascia and integument, anastomosing with branches of the deep epigastric, and internal mammary arteries.

The *superficial circumflex iliac*, the smallest of the cutaneous branches, arises close to the preceding, and, piercing the fascia lata, runs outwards, parallel with Poupart's ligament, as far as the crest of the ilium, dividing into branches which supply the integument of the groin, the superficial fascia, and inguinal glands, anastomosing with the circumflex iliac, and with the gluteal and external circumflex arteries.

The *superficial external pudic* (superior), arises from the inner side of the femoral artery, close to

* *Med. Chir. Trans.*, vol. ii. 1811.

the preceding vessels, and, after piercing the fascia lata at the saphenous opening, passes inwards, across the spermatic cord, to be distributed to the integument on the lower part of the abdomen, and of the penis and scrotum in the male, and to the labia in the female, anastomosing with branches of the internal pudic.

The *deep external pudic* (inferior), more deeply seated than the preceding passes inwards on the Pectineus muscle, covered by the fascia lata, which it pierces opposite the ramus of the pubes, its branches being distributed, in the male, to the integument of the scrotum and perinæum, and in the female, to the labium, anastomosing with branches of the superficial perinæal artery.

The *Profunda Femoris* (deep femoral artery), nearly equals the size of the superficial femoral. It arises from the outer and back part of the femoral artery, from one to two inches below Poupart's ligament. It at first lies on the outer side of the superficial femoral, and then passes beneath it and the femoral vein to the inner side of the femur, and terminates at the lower third of the thigh in a small branch, which pierces the Adductor magnus, to be distributed to the flexor muscles, on the back of the thigh, anastomosing with branches of the popliteal and inferior perforating arteries.

*Relations. Behind*, it lies first upon the Iliacus, and then on the Adductor brevis and Adductor magnus muscles. *In front*, it is separated from the femoral artery, above by the femoral and profunda veins, and below by the Adductor longus. On its *outer side*, the insertion of the Vastus internus separates it from the femur.

## Plan of the Relations of the Profunda Artery.

*In front.*
Femoral and profunda veins.
Adductor longus.

*Outer side.*
Vastus internus.

*Behind.*
Iliacus.
Adductor brevis.
Adductor magnus.

The *External Circumflex Artery* supplies the muscles on the front of the thigh. It arises from the outer side of the profunda, passes horizontally outwards, between the divisions of the anterior crural nerve, and beneath the Sartorius and Rectus muscles, and divides into three sets of branches, ascending, transverse, and descending.

The *ascending branches* pass upwards, beneath the Tensor vaginæ femoris muscle, to the outer side of the hip, anastomosing with the terminal branches of the gluteal, and circumflex iliac arteries.

The *descending branches*, three or four in number, pass downwards, beneath the Rectus, upon the Vasti muscles, to which they are distributed, one or two passing beneath the Vastus externus as far as the knee, anastomosing with the superior articular branches of the popliteal artery.

The *transverse branches*, the smallest and least numerous, pass outwards over the Cruræus, pierce the Vastus externus, and wind round the femur to its back part, just below the great trochanter, anastomosing at the back of the thigh with the internal circumflex, sciatic, and superior perforating arteries.

The *Internal Circumflex Artery*, smaller than the external, arises from the inner and back part of the profunda, and winds round the inner side of the femur, between the Pectineus and Psoas muscles. On reaching the tendon of the obturator externus, it divides into two branches; one, ascending, is distributed to the Adductor muscles, the Gracilis, and Obturator externus, anastomosing with the obturator artery, a descending branch which passes beneath the Adductor brevis, to supply it and the great Adductor; the continuation of the vessel passing backwards, between the Quadratus femoris and upper border of the Adductor magnus, anastomosing with the sciatic, external circumflex, and superior perforating arteries. Opposite the hip-joint, this branch gives off an articular vessel, which enters the joint beneath the transverse ligament; and, after supplying the adipose tissue, passes along the round ligament to the head of the bone.

The *Perforating Arteries* (fig. 230), usually three in number, are so called from their perforating the tendons of the Adductor brevis and magnus muscles to reach the back of the thigh. The first is given off above the Adductor brevis, the second in front of that muscle, and the third immediately below it.

The *first* or *superior perforating artery* passes backwards between the Pectineus and Adductor brevis (sometimes perforates the latter); it then pierces the Adductor magnus close to the linea aspera, and divides into branches which supply both Adductors, the Biceps, and Gluteus maximus muscle; anastomosing with the sciatic, internal circumflex, and middle perforating arteries.

The *second* or *middle perforating artery*, larger than the first, pierces the tendons of the Adductor brevis and Adductor magnus muscles, divides into asending and descending branches, which supply the flexor muscles of the thigh; anastomosing with the superior and inferior perforantes. The nutrient artery of the femur is usually given off from this branch.

The *third* or *inferior perforating artery* is given off below the Adductor brevis; it pierces the Adductor magnus, and divides into branches which supply the flexor muscles of the thigh; anastomosing with the perforating arteries, above, and with the terminal branches of the profunda, below.

*Muscular Branches* are given off from the superficial femoral throughout its entire course. They vary from two to seven in number, and supply chiefly the Sartorius and Vastus internus.

The *Anastomotica Magna* arises from the femoral artery just before it passes through the tendinous opening in the Adductor magnus muscle, and divides into a superficial and deep branch.

The *superficial branch* accompanies the long saphenous nerve, beneath the Sartorius, and piercing the fascia lata, is distributed to the integument.

The *deep branch* descends in the substance of the Vastus internus, lying in front of the tendon of the Adductor magnus, to the inner side of the knee, where it anastomoses with the superior internal articular artery and recurrent branch of the anterior tibial. A branch from this vessel crosses outwards above the articular surface of the femur forming an anastomotic arch with the superior external articular artery, and supplies branches to the knee-joint.

## Popliteal Artery.

The popliteal artery commences at the termination of the femoral, at the opening in the Adductor magnus, and, passing obliquely downwards and outwards behind the knee-joint to the lower border of the Popliteus muscle, divides into the anterior and posterior tibial arteries. Through this extent the artery lies in the popliteal space.

## The Popliteal Space.

*Dissection.* A vertical incision about eight inches in length should be made along the back part of the knee-joint, connected above and below by a transverse incision pasing from the inner to the outer side of the limb. The flaps of integument included between these incisions should be reflected in the direction shown in fig. 189.

On removing the integument, the superficial fascia is exposed, and ramifying in it along the middle line are found some filaments of the small sciatic nerve, and towards the inner part, some offsets from the internal cutaneous nerve.

The superficial fascia having been removed, the fascia lata is brought into view. In this region it is strong and dense, being strengthened by transverse fibres, and firmly attached to the tendons on the inner and outer sides of the space. It is perforated below by the external saphenous vein. This fascia having been reflected back in the same direction as the integument, the small sciatic nerve and external saphenous vein are seen immediately beneath it, in the middle line. If the loose adipose tissue is now removed, the boundaries and contents of the space may be examined.

*Boundaries.* The popliteal space, or the ham, occupies the lower third of the thigh and the upper fifth of the leg; extending from the aperture in the Adductor magnus, to the lower border of the Popliteus muscle. It is a lozenge-shaped space, being widest at the back part of the knee-joint, and deepest above the articular end of the femur. It is bounded, externally, above the joint, by the Biceps and below the articulation, by the Plantaris and external head of the Gastrocnemius. Internally, above the joint, by the Semi-membranosus, Semi-tendinosus, Gracilis, and Sartorius; below the joint, by the inner head of the Gastrocnemius.

Above, it is limited by the apposition of the inner and outer hamstring muscles; below by the junction of the two heads of the Gastrocnemius. The floor is formed by the lower part of the posterior surface of the shaft of the femur, the posterior ligament of the knee-joint, the upper end of the tibia, and the fascia covering the Popliteus muscle, and the space is covered in by the fascia lata.

*Contents.* It contains the popliteal vessels and their branches, together with the termination of the external saphenous vein, the internal and external popliteal nerves and their branches, the small sciatic nerve, the articular branch from the obturator nerve, a few small lymphatic glands, and a considerable quantity of loose adipose tissue.

*Position of contained parts.* The internal popliteal nerve descends in the middle line of the space, lying superficial, and a little external to the vein and artery. The external popliteal nerve descends on the outer side of the space, lying close to the tendon of the Biceps muscle. More deeply at the bottom of the space are the popliteal vessels, the vein lying superficial and a little external to the artery to which it is closely united by dense areolar tissue; sometimes the vein is placed on the inner instead of the outer side of the artery; or the vein may be double, the artery then lies between them, the two veins being usually connected by short transverse branches. More deeply, and close to the surface of the bone, is the popliteal artery, and passing off from it at right angles are its articular branches. The articular branch from the obturator nerve descends upon the popliteal artery to supply the knee; and occasionally there is found deep in the space an articular filament from the great sciatic nerve. The popliteal lymphatic glands, four or five in number, are found surrounding the artery; one usually lies superficial to the vessel, another is situated between it and the bone, and the rest are placed on either side of it. In health, these glands are small; but when enlarged and indurated from inflammation, the pulsation communicated to them from the popliteal artery makes them resemble so closely an aneurismal tumour, that it requires a very careful examination to discriminate between them.

The POPLITEAL ARTERY (fig. 232), in its course downwards from the aperture in the Adductor magnus, to the lower border of the Popliteus muscle, rests first on the inner, and then on the posterior surface of the femur; in the middle of its course on the posterior ligament of the knee-joint; and below, on the fascia covering the Popliteus muscle. *Superficially*, it is covered, above, by the Semi-membranosus; in the middle of its course, by a quantity of fat, which separates it from the deep fascia and integument; and below, it is overlapped by the Gastrocnemis Plantaris and Soleus muscles, the popliteal vein, and the internal popliteal nerve The popliteal vein, which is intimately attached to the artery, lies superficial and external to it, until near its termination, when it crosses it and lies to its inner side. The popliteal nerve is still more superficial and external, crossing, however, the artery below the joint, and lying on its inner side. *Laterally*, it is bounded by the muscles which form the boundaries of the popliteal space.

*Peculiarities in point of division.* Occasionally the popliteal artery divides prematurely into its terminal branches; this division occurs most frequently opposite the knee-joint.

*Unusual branches.* This artery sometimes divides into the anterior tibial and peroneal, the posterior tibial being wanting, or very small. In a single case, this artery divided into three branches, the anterior and posterior tibial, and peroneal.

*Surgical Anatomy.* Ligature of the popliteal artery is required in cases of wound of that vessel, but for aneurism of the posterior tibial, it is preferable to tie the superficial femoral. The popliteal may be tied in the upper or lower part of its course; but in the middle of the space the operation is attended with considerable difficulty, from the great depth of the artery, and from the extreme degree of tension of its lateral boundaries.

In order to expose the vessel in the upper part of its course, the patient should be placed in the prone position, with the limb extended. An incision about three inches in length should then be made through the integument, along the posterior margin of the Semi-membranosus, and the fascia lata having been divided, this muscle must be drawn inwards, when the pulsation of the vessel will be detected with the finger; the nerve lies on the outer or fibular side of the artery, the vein, superficial and also to its outer side; having cautiously separated it from the artery, the aneurism needle should be passed around the latter vessel from without inwards.

To expose the vessel in the lower part of its course, where the artery lies between the two heads of the Gastrocnemius, the patient should be placed in the same position as in the preceeding operation. An incision should then be made through the integument in the middle line, commencing opposite the bend of the knee-joint, care being taken to avoid the external saphenous vein and nerve. After dividing the deep fascia and separating some dense cellular membrane, the artery, vein, and nerve will be exposed, descending between the two heads of the Gastrocnemius. Some muscular branches of the popliteal should be avoided if possible, or if divided, tied immediately. The leg being now flexed, in order the more effectually to separate the two heads of the Gastrocnemius, the nerve should be drawn inwards and the vein outwards, and the aneurism needle passed between the artery and vein from without inwards.

The branches of the popliteal artery are, the

Muscular { Superior.
Inferior or Sural.
Cutaneous.
Superior external articular.
Superior internal articular.
Azygos articular.
Inferior external articular.
Inferior internal articular.

The *superior muscular branches*, two or three in number, arise from the upper part of the popliteal artery, and are distributed to the Vastus externus and flexor muscles of the thigh; anastomosing with the inferior perforating, and terminal branches of the profunda.

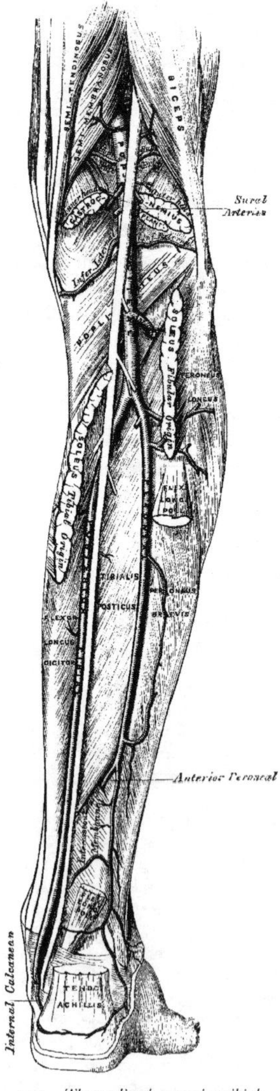

232 – *The popliteal, posterior tibial, and peroneal arteries.*

The *inferior muscular* (*Sural*) are two large branches, which are distributed to the two heads of the Gastrocnemius and Plantaris muscles. They arise from the popliteal artery opposite the knee-joint.

*Cutaneous branches* descend on each side and in the middle of the limb, between the Gastrocnemius and integument; they arise separately from the popliteal artery, or from some of its branches, and supply the integument of the calf.

The *superior articular arteries*, two in number, arise one on either side of the popliteal, and wind round the femur immediately above its condyles to the front of the knee-joint.

The *internal branch* passes beneath the tendon of the Adductor magnus, and divides into two, one of which supplies the Vastus internus, inosculating with the anastomotica magna and inferior internal articular; the other ramifies close to the surface of the femur, supplying it and the knee-joint, and anastomosing with the superior external articular artery.

The *external branch* passes above the outer condyle, beneath the tendon of the Biceps, and divides into a superficial and deep branch: the superficial branch supplies the Vastus externus, and anastomoses with the descending branch of the external circumflex artery; the deep branch supplies the lower part of the femur and knee-joint, and forms an anastomotic arch across the bone with the anastomotica magna artery.

The *azygos articular* is a small branch arising from the popliteal artery opposite the bend of the knee-joint. It pierces the posterior ligament, and supplies the ligaments and synovial membrane in the interior of the articulation.

The *inferior articular arteries*, two in number, arise from the popliteal, beneath the Gastrocnemius, and wind round the head of the tibia, below the joint.

The *internal* one passes below the inner tuberosity, beneath the internal lateral ligament, at the anterior border of which it ascends to the front and inner side of the joint, to supply the head of the tibia and the articulation of the knee.

The *external* one passes outwards above the head of the fibula, to the front of the knee-joint, lying in its course beneath the outer head of the Gastrocnemius, the external lateral ligament, and the tendon of the Biceps muscle, and divides into branches, which anastomose with the artery of the opposite side, the superior articular, and the recurrent branch of the anterior tibial.

## Anterior Tibial Artery.

The anterior tibial artery commences at the bifurcation of the popliteal, at the lower border of the Popliteus muscle, passes forwards between the two heads of the Tibialis posticus, and through the aperture left between the bones at the upper part of the interosseous membrane, to the deep part of the front of the leg; it then descends on the anterior surface of the interosseous ligament, and of the tibia, to the front of the ankle-joint, where it lies more superficially, and becomes the dorsalis pedis. A line drawn from the inner side of the head of the fibula to midway between the two malleoli, will be parallel with the course of the artery.

*Relations.* In the upper two-thirds of its extent, it rests upon the interosseous ligament, to which it is connected by delicate fibrous arches thrown across it. In the lower third, upon the front of the tibia, and the anterior ligament of the ankle-joint. In the upper third of its course, it lies between the Tibialis anticus and Extensor longus digitorum; in the middle third, between the Tibialis anticus and Extensor proprius pollicis. In the lower third, it is crossed by the tendon of the Extensor proprius pollicis, and lies between it and the innermost tendon of the Extensor longus digitorum. It is covered, in the upper two-thirds of its course, by the muscles which lie on either side of it, and by the deep fascia; in the lower third, by the integument, annular ligament, and fascia.

The anterior tibial artery is accompanied by two veins (venæ comites), which lie one on either side of the artery; the anterior tibial nerve lies at first to its outer side, and about the middle of the leg is placed superficial to it; at the lower part of the artery, the nerve is on the outer side.

### Plan of the Relations of the Anterior Tibial Artery.

*In front.*
Integument, superficial and deep fasciæ.
Tibialis anticus.
Extensor longus digitorum.
Extensor proprius pollicis.
Anterior tibial nerve.

*Inner side.*
Tibialis anticus.
Extensor proprius pollicis.

Anterior Tibial.

*Outer side.*
Anterior tibial nerve.
Extensor longus digitorum.
Extensor proprius pollicis.

*Behind*
Interosseous membrane.
Tibia.
Anterior ligament of ankle-joint.

*Peculiarities in Size.* This vessel may be diminished in size, or it may be deficient to a greater or less extent, or it may be entirely wanting, its place being supplied by perforating branches from the posterior tibial, or by the anterior division of the peroneal artery.

*Course.* This artery occasionally deviates in its course towards the fibular side of the leg, regaining its usual position beneath the annular ligament at the front of the ankle. In two instances, this vessel has approached the surface in the middle of the leg, from this point onwards being covered merely by the integument and fascia.

*Surgical Anatomy.* The anterior tibial artery may be tied in the upper or lower part of the leg. In the upper part, the operation is attended with great difficulty, on account of the depth of the vessel from the surface. An incision, about four inches in length, should be made through the integument, midway between the spine of the tibia and the outer margin of the fibula, the fascia and intermuscular septum between the Tibialis anticus and Extensor communis digitorum being divided to the same extent. The foot must be flexed to relax these muscles, and they must be separated from each other by the finger. The artery is then exposed, deeply seated, lying upon the interosseous membrane, the nerve lying externally, and one of the venæ comites on either side; these must be separated from the artery before the aneurism needle is passed round it.

To tie this vessel in the lower third of the leg above the ankle-joint, an incision about three inches in length should be made through the integument between the tendons of the Tibialis anticus and Extensor proprius pollicis muscles, the deep fascia being divided to the same extent; the tendon on either side should be held aside, when the vessel will be seen lying upon the tibia, with the nerve superficial to it, and one of the venæ comites on either side.

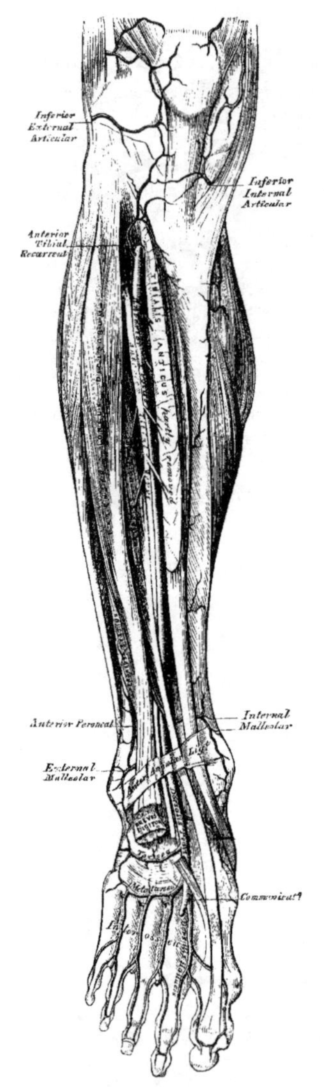


233 – *Surgical anatomy of the anterior tibial and dorsalis pedis arteries.*

In order to secure this vessel over the instep, an incision should be made on the fibular side of the tendon of the Extensor proprius pollicis, between it and the innermost tendon of the long Extensor; the deep fascia having been divided, the artery will be exposed, the nerve lying either superficial to it, or to its outer side.

The branches of the anterior tibial artery are, the

Recurrent tibial.
Muscular.
Internal malleolar.
External malleolar.

The *recurrent branch* arises from the anterior tibial, as soon as that vessel has passed through the interosseous space; it ascends in the Tibialis anticus muscle, and ramifies on the front and sides of the knee-joint, anastomosing with the articular branches of the popliteal.

The *muscular branches* are numerous; they are distributed to the muscles which lie on either side of the vessel, some piercing the deep fascia to supply the integument, others passing through the interosseous membrane, and anastomosing with branches of the posterior tibial and peroneal arteries.

The *malleolar arteries* supply the ankle-joint.

The *internal* arises about two inches above the articulation, passes beneath the tendon of the Tibialis anticus to the inner ankle, upon which it ramifies, anastomosing with branches of the posterior tibial and internal plantar arteries.

The *external* passes beneath the tendons of the Extensor longus digitorum and Extensor proprius pollicis, and supplies the outer ankle, anastomosing with the anterior peroneal artery, and with ascending branches from the tarsea branch of the dorsalis pedis.

## Dorsalis Pedis Artery.

The dorsalis pedis, the continuation of the anterior tibial, passes forwards from the bend of the ankle along the tibial side of the foot to the back part of the first interosseous space, where it divides into two branches, the dorsalis hallucis and communicating.

*Relations.* This vessel in its course forwards, rests upon the astragalus, scaphoid, and internal cuneiform bones, and the ligaments connecting them, being covered by the integument and fascia, and crossed near its termination by the innermost tendon of the Extensor brevis digitorum. On its *tibial side* is the tendon of the Extensor proprius pollicis; on its *fibular side*, the innermost tendon of the Extensor longus digitorum. It is accompanied by two veins, and by the anterior tibial nerve, which lies on its outer side.

## Plan of the Relations of the Dorsalis Pedis Artery.

*In front.*
Integument and fascia.
Innermost tendon of Extensor brevis digitorum.

*Tibial side.*
Extensor proprius pollicis.

*Fibular side.*
Extensor longus digitorum.
Anterior tibial nerve.

*Behind.*
Astragalus.
Scaphoid.
Internal cuneiform, and their ligaments.

*Peculiarities in Size.* The dorsal artery of the foot may be larger than usual, to compensate for a deficient plantar artery; or it may be deficient in its terminal branches to the toes, which are then derived from the internal plantar; or its place may be supplied altogether by a large anterior peroneal artery.

*Position.* This artery frequently curves outwards, lying external to the line between the middle of the ankle and the back part of the first interosseous space.

*Surgical Anatomy.* This artery may be tied, by making an incision through the integument, between two and three inches in length, on the fibular side of the tendon of the Extensor proprius pollicis, in the interval between it and the inner border of the short Extensor muscle. The incision should not extend further forwards than the back part of the first interosseous space, as the artery divides in this situation. The deep fascia being divided to the same extent, the artery will be exposed, the nerve lying upon its outer side.

*Branches.* The branches of the dorsalis pedis are, the

Tarsea. Interosseæ.
Metatarsea. Dorsalis pollicis.
Communicating.

The *tarsea artery* arises from the dorsalis pedis, as that vessel crosses the scaphoid bone; it passes in an arched direction outwards, lying upon the tarsal bones, and covered by the Extensor brevis digitorum; it supplies that muscle and the articulations of the tarsus, and anastomoses with branches from the metatarsea, external malleolar, peroneal, and external plantar arteries.

The *metatarsea* arises a little anterior to the preceding; it passes outwards to the outer part of the foot, over the bases of the metatarsal bones, beneath the tendons of the short Extensor, its direction being influenced by its point of origin; and it anastomoses with the tarsea and external plantar arteries. This vessel gives off three branches, the interosseæ, which pass forwards upon the three outer Dorsal interossei muscles, and, in the clefts between the toes, divide into two dorsal collateral branches for the adjoining toes. At the back part of each interosseous space these vessels receive the posterior perforating branches from the plantar arch; and at the fore part of each interosseous space, they are joined by the anterior perforating branches, from the digital arteries. The outer-most interosseous artery gives off a branch which supplies the outer side of the little toe.

The *dorsalis hallucis* runs forwards along the outer border of the first metatarsal bone, and, at the cleft between the first and second toes, divides into two branches, one of which passes inwards, beneath the tendon of the Flexor longus pollicis, and is distributed to the inner border of the great toe; the other branch bifurcating to supply the adjoining sides of the great and second toes.

The *communicating artery* dips down into the sole of the foot, between the two heads of the first Dorsal interosseous muscle, and inosculates with the termination of the external plantar artery, to complete the plantar arch. It here gives of two digital branches; one runs along the inner side of the great toe, on its plantar surface; the other passes forwards along the first metatarsal space, and bifurcates for the supply of the adjacent sides of the great and second toes.

## Posterior Tibial Artery.

The posterior tibial is an artery of large size, which extends obliquely downwards from the lower border of the Popliteus muscle, along the tibial side of the leg, to the fossa between the inner ankle and the heel, where it divides beneath the origin of the Abductor pollicis, into the internal and external plantar arteries. At its origin it lies opposite the interval, between the tibia and fibula; as it descends, it approaches the inner side of the leg, lying behind the tibia, and, in the lower part of its course, is situated midway between the inner malleolus and, in the tuberosity of the os calcis.

*Relations.* It lies successively upon the Tibialis posticus, the Flexor longus digitorum, and below, upon the tibia and back part of the ankle-joint. It is *covered* by the intermuscular fascia, which separates it above from the Gastrocnemius and Soleus muscles. In the lower third, where it is more superficial, it is covered only by the integument and fascia, and runs parallel with the inner border of the tendo Achillis. It is accompanied by two veins, and by the posterior tibial nerve, which lies at first to the inner side of the artery, but soon crosses it, and is, in the greater part of its course, on its outer side.

## Plan of the Relations of the Posterior Tibial Artery.

*In front.*
Tibialis posticus.
Flexor longus digitorum.
Tibia.
Ankle-joint.

*Inner side.*
Posterior tibial nerve, upper third.

Posterior Tibial.

*Outer side.*
Posterior tibial nerve, lower two-thirds.

*Behind.*
Gastrocnemius.
Soleus.
Deep fascia and integument.

*Behind the Inner Ankle*, the tendons and blood-vessels are arranged in the following order, from within outwards: First, the tendons of the Tibialis posticus and Flexor longus digitorum, lying in the same groove, behind the inner malleolus, the former being the most internal. External to these is the posterior tibial artery, having a vein on either side; and, still more externally, the posterior tibial nerve. About half an inch nearer the heel is the tendon of the Flexor longus pollicis.

*Peculiarities in Size.* The posterior tibial is not unfrequently smaller than usual, or absent, its place being compensated for by a large peroneal artery, which passes inwards at the lower end of the tibia, and either joins the small tibial artery, or continues alone to the sole of the foot.

*Surgical Anatomy.* The *application of a ligature* to the posterior tibial may be required in cases of wound of the sole of the foot, attended with great hæmorrhage, when the vessel should be tied at the inner ankle. In cases of wound of the posterior tibial, it will be necessary to enlarge the wound so as to expose the vessel at the wounded point (excepting where the vessel is injured by a punctured wound from the front of the leg). In cases of aneurism from wound of the artery low down, the vessel should be tied in the middle of the leg. But in aneurism of the posterior tibial high up, it would be better to tie the femoral artery.

To tie the posterior tibial artery at the ankle, a semi-lunar incision should be made through the integument, about two inches and a half in length, midway between the heel and inner ankle, but a little nearer the latter. The subcutaneous cellular membrane having been divided, a strong and dense fascia, the internal annular ligament, is exposed. This ligament is continuous above with the deep fascia of the leg, covers the vessels and nerves, and is intimately adherent to the sheaths of the tendons. This having been cautiously divided upon a director, the sheath of the vessels is exposed, and being opened, the artery is seen with one of the venæ comites on each side. The aneurism needle should be passed round the vessel from the heel towards the ankle, in order to avoid the posterior tibial nerve, care being at the same time taken not to include the venæ comites.

The vessel may also be tied in the lower third of the leg, by making an incision about three inches in length, parallel with the inner margin of the tendo Achillis. The internal saphenous vein being carefully avoided, the two layers of fascia must be divided upon a director, when the artery is exposed along the inner margin of the Flexor longus digitorum, with one of its venæ comites on either side, and the nerve lying external to it.

To tie the posterior tibial in the middle of the leg, is a very difficult operation, on account of the great depth of the vessel from the surface, and from its being covered in by the Gastrocnemius and Soleus muscles. The patient being placed in the recumbent position, the injured limb should rest on its outer side, the knee being partially bent, and the foot extended, so as to relax the muscles of the calf. An incision about four inches in length should then be made through the integument, along the inner margin of the tibia, taking care to avoid the internal saphenous vein. The deep fascia having been divided, the margin of the Gastrocnemius is exposed, and must be drawn aside, and the tibial attachment of the Soleus divided, a director being previously passed beneath it. The artery may now be felt pulsating beneath the deep fascia, about an inch from the margin of the tibia. The fascia having been divided, and the limb placed in such a position as to relax the muscles of the calf as much as possible, the veins should be separated from the artery, and the aneurism needle passed round the vessel from without inwards, so as to avoid wounding the posterior tibial nerve.

The branches of the posterior tibial artery are, the

Peroneal. Nutritious.
Muscular. Communicating.
Internal calcanean.

The *Peroneal Artery* lies, deeply seated, along the back part of the fibular side of the leg. It arises from the posterior tibial, about an inch below the lower border of the Popliteus muscle, passes obliquely outwards to the fibula, and then descends along the inner border of this bone to the lower third of the leg, where it gives off the anterior peroneal. It then passes across the articulation, between the tibia and fibula, to the outer side of the os calcis, supplying the neighbouring muscles and back of the ankle, and anastomosing with the external malleolar, tarsal, and external plantar arteries.

*Relations.* This vessel rests at first upon the Tibialis posticus, and, in the greater part of its course, in the fibres of the Flexor longus pollicis, in a groove between the interosseous ligament and the bone. It is *covered*, in the upper part of its course, by the Soleus and deep fascia; *below*, by the Flexor longus pollicis.

## Plan of the Relations of the Peroneal Artery.

*In front.*
Tibialis posticus.
Flexor longus pollicis.

*Outer side.*
Fibula.

Peroncal
Artery.

*Behind.*
Soleus.
Deep fascia.
Flexor longus pollicis.

*Peculiarities in Origin.* The peroneal artery may arise three inches below the Popliteal, or from the posterior tibial high up, or even from the popliteal.

*Its Size* is more frequently increased than diminished, either reinforcing the posterior tibial by its junction with it, or by altogether taking the place of the posterior tibial, in the lower part of the leg and foot, the latter vessel only existing as a short muscular branch. In those rare cases, where the peroneal artery is smaller than usual, a branch from the posterior tibial supplies its place, and a branch from the anterior tibial compensates for the diminished anterior peroneal artery. In one case, the peroneal artery has been found entirely wanting.

The anterior peroneal is sometimes enlarged, and takes the place of the dorsal artery of the foot.

The peroneal artery, in its course, gives off branches to the Soleus, Tibialis posticus, Flexor longus pollicis, and Peronei muscles, and a nutrient branch to the fibula.

The *Anterior Peroneal* pierces the interosseous membrane, about two inches above the outer malleolus, to reach the fore part of the leg, and passing down beneath the Peroneus tertius to the outer ankle, ramifies on the front and outer side of the tarsus, anastomosing with the external malleolar and tarsal arteries.

The *nutritious artery* of the tibia arises from the posterior tibial near its origin and after supplying a few muscular branches, enters the nutritious canal of the bone, which it traverses obliquely from above downwards. This is the largest nutrient artery of bone in the body.

The *muscular branches* are distributed to the Soleus and deep muscles along the back of the leg.

The *communicating branch* to the peroneal runs transversely across the back of the tibia, about two inches above its lower end, passing beneath the Flexor longus pollicis.

The *internal calcanean* consists of several large branches, which arise from the posterior tibial just before its division; they are distributed to the fat and integument behind the tendo Achillis and about the heel, and to the muscles on the inner side of the sole, anastomosing with the peroneal and internal malleolar arteries.

The *Internal Plantar Artery*, much smaller than the external, passes forwards along the inner side of the foot. It is at first situated above the Abductor pollicis and then between it and the Flexor brevis digitorum, both of which it supplies. At the base of the first metatarsal bone, where it has become much diminished in size, it passes along the inner border of the great toe, inosculating with its digital branches.

The *External Plantar Artery*, much larger than the internal, passes obliquely outwards and forwards to the base of the fifth metatarsal bone. It then turns obliquely inwards to the interval between the bases of the first and second metatarsal bones, where it inosculates with the communicating branch from the dorsalis pedis artery, thus completing the plantar arch. As this artery passes outwards it is at first placed between the os calcis and Abductor pollicis, and then between the Flexor brevis digitorum and Flexor accessorius; and as it passes forwards to the base of the little toe, it lies more superficially between the Flexor brevis digitorum and Abductor minimi digiti, covered by the deep fascia and integument. The remaining portion of the vessel is deeply situated: it extends from the base of the metatarsal bone of the little toe to the back part of the first interosseous space, and forms the plantar arch; it is convex forwards, lies upon the Interossei muscles, opposite the tarsal ends of the metatarsal bones, and is covered by the Adductor pollicis, the flexor tendons of the toes, and the Lumbricales.

*Branches.* The plantar arch, besides distributing numerous branches to the muscles, integument, and fasciæ in the sole, gives off the following branches:

Posterior perforating. Digital—Anterior perforating.

The *Posterior Perforating* are three small branches, which ascend through the back part of the three outer interosseous spaces, between the heads of the Dorsal interossei muscles, and anastomose with the interosseous branches from the metatarsal artery.

The *Digital Branches* are four in number, and supply the three outer toes and half the second toe. The *first* passes outwards from the outer side of the plantar arch, and is distributed to the outer side of the little toe, passing in its course beneath the Abductor and short Flexor muscles. The *second*, *third*, and *fourth* run forwards along the metatarsal spaces, and on arriving at the clefts between the toes, divide into collateral branches, which supply the adjacent sides of the three outer toes and the outer side of the second. At the bifurcation of the toes, each digital artery sends upwards, through the fore part of the corresponding metatarsal space, a small branch, which inosculates with the interosseous branches of the metatarsal artery. These are the anterior perforating branches.

From the arrangement already described of the distribution of the vessels to the toes, it will be seen that both sides of the three outer toes, and the outer side of the second toe, are supplied by branches from the plantar arch; both sides of the great toe, and the inner side of the second, being supplied by the dorsal artery of the foot.

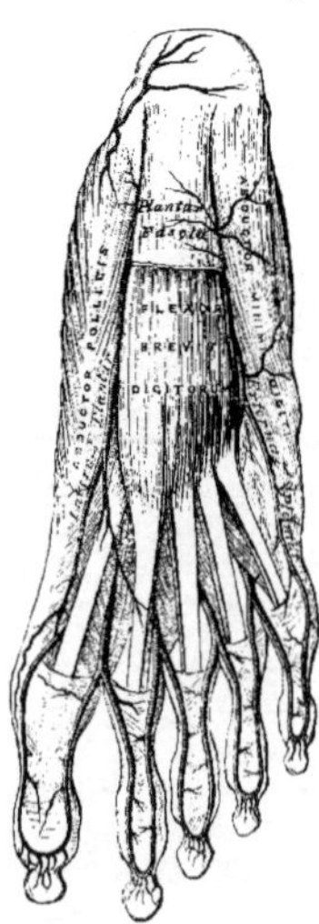
234 – *The plantar arteries. Superficial view.*

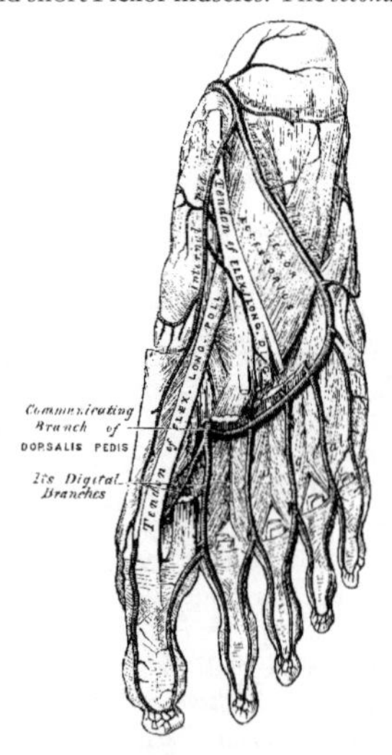

235 – *The plantar arteries. Deep view.*

## Pulmonary Artery.

The pulmonary artery conveys the venous blood from the right side of the heart to the lungs. It is a short wide vessel, about two inches in length, arising from the left side of the base of the right ventricle, in front of the ascending aorta. It ascends obliquely upwards, backwards, and to the left side, as far as the under surface of the arch of the aorta, where it divides into two branches of nearly equal size, the right and left pulmonary arteries.

*Relations.* The greater part of this vessel is contained, together with the ascending part of the arch of the aorta, in the pericardium, being enclosed with it in a tube of serous membrane, continued upwards from the base of the heart, and has attached to it, above, the fibrous layer of this membrane. Behind, it rests at first upon the ascending aorta, and higher up in front of the left auricle. On either side of its origin is the appendix of the corresponding auricle, and a coronary artery; and higher up it passes to the left side of the ascending aorta. A little to the left of its point of bifurcation, it is connected to the under surfaces of the arch of the aorta by a short fibrous cord, the remains of a vessel peculiar to fœtal life, the ductus arteriosus.

The *right pulmonary artery*, longer and larger than the left, runs horizontally outwards, behind the ascending aorta and superior vena cava, to the root of the right lung, where it divides into two branches, of which the lower, the larger, supplies the lower lobe; the upper giving a branch to the middle lobe.

The *left pulmonary artery*, shorter but somewhat smaller than the right, passes horizontally in front of the descending aorta and left bronchus to the root of the left lung, where it divides into two branches for the two lobes.

---

The author has to acknowledge valuable aid derived from the following works:—Harrison's 'Surgical Anatomy of the Arteries of the Human Body.' Dublin, 1824—Richard Quain's 'Anatomy of the Arteries of the Human Body.' London, 1844—Sibson's 'Medical Anatomy'; and the other works on General and Microscopic Anatomy before referred to.

# Of the Veins.

The Veins are the vessels which serve to return the blood from the capillaries of the different parts of the body to the heart. They consist of two distinct sets of vessels, the pulmonary and systemic.

The *Pulmonary Veins*, unlike other vessels of this kind, contain arterial blood, which they return from the lungs to the left auricle of the heart.

The *Systemic Veins* return the venous blood from the body generally to the right auricle of the heart.

The *Portal Vein*, an appendage to the systemic venous system, is confined to the abdominal cavity, returning the venous blood from the viscera of digestion, and carrying it to the liver by a single trunk of large size, the vena portæ. From this organ, the same blood is conveyed to the inferior vena cava by means of the hepatic veins.

The veins, like the arteries, are found in nearly every tissue of the body; they commence by minute plexuses, which communicate with the capillaries, the branches from which, uniting together, constitute trunks, which increase in size as they pass towards the heart, from the termination of larger branches in them. The veins are larger and altogether more numerous than the arteries; hence, the entire capacity of the venous system is much greater than the arterial; the pulmonary veins excepted, which do not exceed in capacity the pulmonary arteries. From the combined area of the smaller venous branches being greater than the main trunks, it results, that the venous system represents a cone, the summit of which corresponds to the heart; its base, to the circumference of the body. In form, the veins are not perfectly cylindrical, like the arteries, their walls being collapsed when empty, and the uniformity of their surface being interrupted at intervals by slight contractions, which indicate the existence of valves in their interior. They usually retain, however, the same calibre as long as they receive no neighbouring branches.

The veins communicate very freely with one another, especially in certain regions of the body; and this communication exists between the larger trunks as well as between the smaller branches. Thus, in the cavity of the cranium, and between the veins of the neck, where obstruction of the cerebral venous system would be attended with imminent danger, we find that the sinuses and larger veins have large and very frequent anastomoses. The same free communication exists between the veins throughout the whole extent of the spinal canal, and between the veins composing the various venous plexuses in the abdomen and pelvis, as the spermatic, uterine, vesical, prostatic, etc.

The veins are subdivided into three sets: superficial, deep, and sinuses.

The *Superficial* or *Cutaneous Veins* are found between the layers of superficial fascia, immediately beneath the integument; they return the blood from these structures, and communicate with the deep veins by perforating the deep fascia.

The *Deep Veins* accompany the arteries, and are usually enclosed in the same sheath with those vessels. In the smaller arteries, as the radial, ulnar, brachial, tibial, peroneal, they exist generally in pairs, one lying on each side of the vessel, and are called *venæ comites*. The larger arteries, as the axillary, subclavian, popliteal and femoral, have usually only one accompanying vein. In certain organs of the body, however, the deep veins do not accompany the arteries; for instance, the veins in the skull and spinal canal, the hepatic veins in the liver, and the larger veins returning blood from the osseous tissue.

*Sinuses* are venous channels, which, in their structure and mode of distribution, differ altogether from the veins. They are found only in the interior of the skull, and are formed by a subdivision of the layers of the dura mater; their outer coat consisting of fibrous tissue, their inner of a serous membrane continuous with the serous membrane of the veins.

Veins have thinner walls than the arteries, which is due to the small amount of elastic and muscular tissues which they contain. The superficial veins usually have thicker coats than the deep veins, and the veins of the lower limb are thicker than those of the upper.

Veins are composed of three coats: internal, middle, and external.

The *internal coat* is similar in structure to that of the arteries. In the smallers veins, it consists of epithelium and nucleated connective tissue, arranged so as to form an outer and an inner layer; the latter, which is the thinnest, representing the middle coat. As these vessels approach the capillaries, the epithelium and outer layer of connective tissue become gradually lost. On the contrary, in those of rather larger size, there is superadded a layer of muscular fibre-cells, a circular fibrous coat, with areolar elastic tissue beneath the epithelium, and in the muscular and external coats. In middle-sized veins, the internal coat consists of epithelium supported on one or more striped nucleated lamellæ, external to which is a layer of elastic fibrous tissue. In the veins of the gravid uterus, and in the long saphenous and popliteal veins, muscular tissue is one of the component parts of the inner coat. In the largest veins, as the inferior vena cava, the trunks of the hepatic, and in the innominate veins, the internal coat has a similar structure to that already mentioned; but is somewhat thicker, owing to the increase in the number of the striped lamellæ, and the greater thickness of the elastic fibrous coat.

The *middle coat* is thin, and differs in structure from the middle coat of arteries in containing a smaller amount of elastic and muscular tissues, and more connective tissue. In the smallest veins, as already mentioned, it consists merely of a thin layer of nucleated connective tissue, the fibres of which run in a longitudinal direction; to which is added, in those of rather larger size, a layer of muscular tissue, the cells of which are disposed transversely. In middle-sized veins, such as the cutaneous and deep veins of the limbs, as far as the brachial and popliteal, and the visceral veins, the middle coat is of a reddish-yellow colour, remarkable for its great thickness, being more developed than the same coat in the large veins. It consists of a thick inner layer of connective tissue with elastic fibres, having intermixed in some veins a transverse layer of muscular fibres; and an outer layer consisting of longitudinal elastic lamellæ, varying from five to ten in number, alternating with layers of transverse muscular fibres and connective tissue, which resembles somewhat in structure the middle coat of large arteries. In the large veins, as in the commencement of the vena portæ, in the upper part of the abdominal portion of the inferior vena cava, and in the large hepatic trunks within the liver, the middle coat is thick, and its structure similar to that of the middle coat in medium-sized veins; but its muscular tissue is scanty, and the longitudinal elastic networks less distinctly lamellated. The muscular tissue of this coat is best marked in the splenic and portal veins, it is absent in certain parts of the vena cava below the liver, and wanting in the subclavian vein and terminal parts of the two cavæ.

The *external coat* is usually the thickest, increasing in thickness with the size of the vessel; it is similar in structure to the external coat of arteries; but its chief peculiarity is, that in some veins it contains a longitudinal network of muscular fibres. In the smallest veins, it consists of a thick layer of nucleated connective tissue. In middle-sized veins, it is much thicker than the middle coat, and consists of elastic and connective tissues, the fibres of which are longitudinally arranged. In the largest veins, this coat is from two to five times thicker than the middle coat, and contains a large number of longitudinal muscular fibres. This is most distinct in the hepatic part of the inferior vena cava, and at the termination of this vein in the heart; in the trunks of the hepatic veins; in all the large trunks of the vena portæ; in the splenic, superior mesenteric, external iliac, renal, and azygos veins. Where the middle coat is absent, this muscular layer extends as far as the inner coat. In the renal and portal veins, it extends through the whole thickness of the outer coat; but in the other veins mentioned, a layer of connective and elastic tissues is found external to the muscular fibres. All the large veins which open into the heart, are covered for a short distance by a layer of muscular tissue continued on to them from the heart.

Muscular tissue is wanting in the veins: 1. Of the maternal part of the placenta. 2. In most of the cerebral veins and sinuses of the dura mater. 3. In the veins of the retina. 4. In the veins of the cancellous tissue of bones. 5. In the venous spaces of the corpora cavernosa. The veins of the above-mentioned parts consist of an internal epithelial lining, supported on one or more layers of areolar tissue.

Most veins are provided with valves, which serve to prevent the reflux of the blood. They are formed by a reduplication of the middle and inner coats, and consist of connective tissue and elastic fibres, covered on both surfaces by epithelium; their form is semilunar. They are attached, by their convex edge, to the wall of the vein; the concave margin is free, directed in the course of the venous current, and lies in close apposition with the wall of the vein as long as the current of blood takes its natural course; if, however, any regurgitation takes place, the valves become distended, their opposed edges are brought into contact, and the current is intercepted. Most commonly two such valves are found, placed opposite one another, more especially in the smaller veins, or in the larger trunks at the point where they are joined by small branches; occasionally there are three, and sometimes only one. The wall of the vein immediately above the point of attachment of each segment of the valve, is expanded into a pouch or sinus, which gives to the vessel, when injected or distended with blood, a knotted appearance. The valves are very numerous in the veins of the extremities, especially the lower ones, these vessels having to conduct the blood against the force of gravity. They are absent in the very small veins, also in the venæ cavæ, the hepatic vein, portal vein and its branches, the renal, uterine, and ovarian veins. A few valves are found in the spermatic veins, and one also at their point of junction with the renal vein and inferior cava in both sexes. The cerebral and spinal veins, the veins of the cancellated tissue of bone, the pulmonary veins, and the umbilical vein and its branches, are also destitute of valves. They are occasionally found, few in number, in the venæ azygos and intercostal veins.

The veins are supplied with nutrient vessels, *vasa vasorum*, like the arteries; but nerves are not generally found distributed upon them. The only vessels upon which they have at present been traced, are the sinuses of the dura mater; on the spinal veins; on the venæ cavæ; on the common jugular, iliac, and crural veins; and on the hepatic veins. (Kölliker).

The veins may be arranged into three groups: 1. Those of the head and neck, upper extremity, and thorax, which terminate in the superior vena cava. 2. Those of the lower limb, pelvis, and abdomen, which terminate in the inferior vena cava. 3. The cardiac veins, which open directly into the right auricle of the heart.

## Veins of the Head and Neck.

The veins of the head and neck may be subdivided into three groups. 1. The veins of the exterior of the head. 2. The veins of the neck. 3. The veins of the diploë and interior of the cranium.

The veins of the exterior of the head are, the

| | |
|---|---|
| Facial. | Temporo-maxillary. |
| Temporal. | Posterior auricular. |
| Internal Maxillary. | Occipital. |

The Facial Vein passes obliquely across the side of the face, extending from the inner angle of the orbit, downwards and outwards, to the anterior margin of the Masseter muscle. It lies to the outer side of the facial artery, and is not so tortuous as that vessel. It commences in the frontal region, where it is called the *frontal vein;* at the inner angle of the eye it has received the name of the *angular vein*; and from this point to its termination, the *facial vein*.

The *frontal vein* commences on the anterior part of the skull, by a venous plexus, which communicates with the anterior branches of the temporal vein; the veins converge to form a single trunk, which descends along the middle line of the forehead parallel with the vein of the opposite side, and unites with it at the root of the nose by a transverse trunk, called the *nasal arch*. Occasionally the frontal veins join to form a single trunk which bifurcates at the root of the nose into the two angular veins. At the nasal arch the branches diverge, and run along the side of the root of the nose. The frontal vein as it descends upon the forehead, receives the supra-orbital vein; the dorsal veins of the nose terminate in the nasal arch; and the angular vein receives, on its inner side, the veins of the ala nasi; on its outer side, the superior palpebral veins; it moreover communicates with the ophthalmic vein, which establishes an important anastomosis between this vessel and the cavernous sinus.

The *facial vein* commences at the inner angle of the orbit, being a continuation of the angular vein. It passes obliquely downwards and outwards, beneath the great Zygomatic muscle, descends along the anterior border of the Masseter, crosses over the body of the lower jaw, with the facial artery, and, passing obliquely outwards and backwards, beneath the Platysma and cervical fascia, unites with a branch of communication from the temporo-maxillary vein, to form a trunk of large size which enters the internal jugular.

*Branches.* The facial vein receives, near the angle of the mouth, communicating branches from the pterygoid-plexus. It is also joined by the inferior palpebral, the superior and inferior labial veins, the buccal veins from the cheek, and the masseteric veins. Below the jaw, it receives the submental, the inferior palatine, which returns the blood from the plexus around the tonsil and soft palate; the submaxillary vein, which commences in the submaxillary gland; and lastly, the ranine vein.

The *Temporal Vein* commences by a minute plexus on the side and vertex of the skull, which communicates with the frontal vein in front, the corresponding vein of the opposite side, and the posterior auricular and occipital veins behind. From this network, anterior and posterior branches are formed which unite above the zygoma, forming the trunk of the vein. This trunk is joined in this

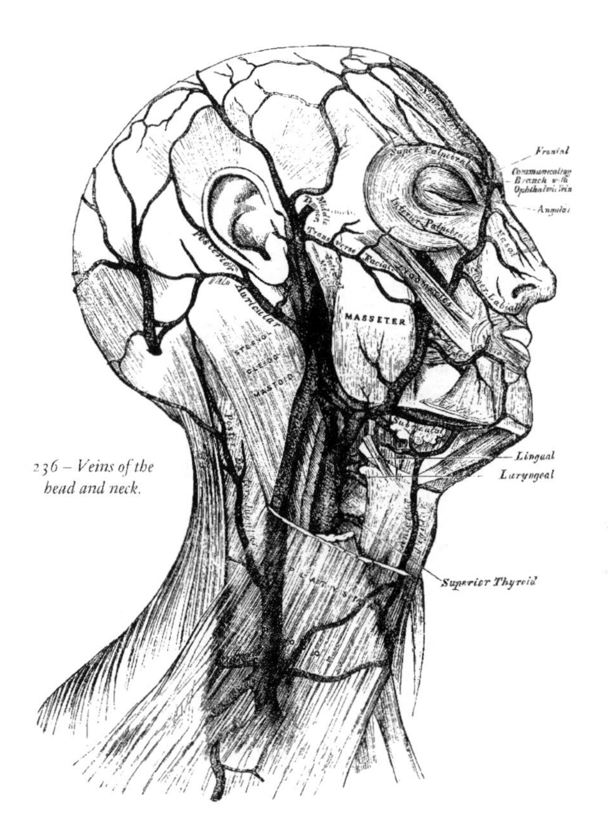


236 – *Veins of the head and neck.*

situation by a large vein, the middle temporal, which receives the blood from the substance of the Temporal muscle and pierces the fascia at the upper border of the zygoma. The temporal vein then descends between the external auditory meatus and the condyle of the jaw, enters the substance of the parotid gland, and unites with the internal maxillary vein, to form the temporo-maxillary.

*Branches.* The temporal vein receives in its course some parotid veins, an articular branch from the articulation of the jaw, anterior auricular veins from the external ear, and a vein of large size, the transverse facial, from the side of the face.

The *Internal Maxillary Vein* is a vessel of considerable size, receiving branches which correspond with those derived from the internal maxillary artery. Thus it receives the middle meningeal veins, the deep temporal, the pterygoid, masseteric, and buccal, some palatine veins, and the inferior dental. These branches form a large plexus, the pterygoid, which is placed between the Temporal and External pterygoid, and partly between the Pterygoid muscles. This plexus communicates very freely with the facial vein, and with the cavernous sinus, by branches through the base of the skull. The trunk of the vein then passes backwards, behind the neck of the lower jaw, and unites with the temporal vein, forming the temporo-maxillary.

The *Temporo-Maxillary Vein*, formed by the union of the temporal and internal maxillary vein, descends in the substance of the parotid gland, between the ramus of the jaw and the Sterno-mastoid muscle, and divides into two branches, one of which passes inwards to join the facial vein, the other is continuous with the external jugular. It receives near its termination the posterior auricular vein.

The *Posterior Auricular Vein* commences upon the side of the head, by a plexus which communicates with the branches of the temporal and occipital veins; descending behind the external ear it joins the temporo-maxillary, just before that vessel terminates in the external jugular. This vessel receives the stylo-mastoid vein, and some branches from the back part of the external ear.

The *Occipital Vein* commences at the back part of the vertex of the skull, by a plexus in a similar manner with the other veins. It follows the course of the occipital artery, passing deeply beneath the muscles of the back part of the neck, and terminates in the internal jugular, occasionally in the external jugular. As this vein passes opposite the mastoid process, it receives the mastoid vein, which establishes a communication with the lateral sinus.

## Veins of the Neck.

The veins of the neck, which return the blood from the head and face, are the

External jugular. Anterior jugular.
Posterior external jugular. Internal jugular.
Vertebral.

The *External Jugular Vein* receives the greater part of the blood from the exterior of the cranium and deep parts of the face, being a continuation of the temporo-maxillary and posterior auricular veins. It commences in the substance of the parotid gland, on a level with the angle of the lower jaw, and runs perpendicularly down the neck, in the direction of a line drawn from the angle of the jaw to the middle of the clavicle. In its course, it crosses the Sterno-mastoid muscle, and runs parallel with its posterior border as far as its attachment to the clavicle, where it perforates the deep fascia, and terminates in the subclavian vein, on the outer side of the internal jugular. As it descends the neck, it is separated from the Sterno-mastoid by the anterior layer of the deep cervical fascia, and is covered by the Platysma, the superficial fascia, and the integument. This vein is crossed about its centre by the superficial cervical nerve, and its upper half is accompanied by the auricularis magnus nerve. The external jugular vein varies in size, bearing an inverse proportion to that of the other veins of the neck: it is occasionally double. It is provided with two pairs of valves, the lower pair being placed at its entrance into the subclavian vein, the upper pair in most cases about an inch a half above the clavicle. These valves do not prevent the regurgitation of the blood, or the passage of injection from below upwards.*

*Branches.* This vein receives the occipital, the posterior external jugular, and, near its termination, the supra-scapular and transverse cervical veins. It communicates with the anterior jugular, and, in the substance of the parotid, receives a large branch of communication from the internal jugular.

The *Posterior External Jugular Vein* returns the blood from the integument and superficial muscles in the upper and back part of the neck, lying between the Splenius and Trapezius muscles. It descends the back part of the neck, and opens into the external jugular just below the middle of its course.

The *Anterior Jugular Vein* collects the blood from the integument and muscles in the middle of the anterior region of the neck. It passes down between the median line and the anterior border of the Sterno-mastoid, and, at the lower part of the neck, passes beneath that muscle to open into the subclavian vein, near the termination of the external jugular. This vein varies considerably in size, bearing almost always an inverse proportion to the external jugular. Most frequently there are two anterior jugulars, a right and left; but occasionally only one. This vein receives some laryngeal branches, and occasionally an inferior thyroid vein. Just above the sternum, the two anterior jugular veins communicate by a transverse trunk, which receives branches from the inferior thyroid veins. It also communicates with the external and with the internal jugular. There are no valves in this vein.

The *Internal Jugular Vein* collects the blood from the interior of the cranium, from the superficial parts of the face, and from the neck. It commences at the jugular foramen, in the base of the skull, being formed by the coalescence of the lateral and inferior petrosal sinuses. At its origin it is somewhat dilated, and this dilatation is called the sinus, or gulf of the internal jugular vein. It runs down the side of the neck in a vertical direction, lying at first on the outer side of the internal carotid, and then on the outer side of the common carotid, and at the root of the neck unites with the subclavian vein, to form the vena innominata. The internal jugular vein, at its commencement, lies upon the Rectus lateralis, behind, and at the outer side of the internal carotid, and the eighth and ninth pairs of nerves; lower down, the vein and artery lie upon the same plane, the glosso-pharyngeal and hypoglossal nerves passing forwards between them: the pneumogastric descends between and behind them, in the same sheath; and the spinal accessory passes obliquely outwards, behind the vein. At the root of the neck the vein of the right side is placed at a little distance from the

* The student may refer to an interesting paper by Dr. Struthers, 'On Jugular Venesection in Asphyxia, Anatomically and Experimentally Considered, including the Demonstration of Valves in the Veins of the Neck,' in the *Edinburgh Medical Journal*, for November, 1856.

artery; on the left side, it usually crosses it at its lower part. This vein is of considerable size, but varies in different individuals, the left one being usually the smallest. It is provided with a pair of valves, which are placed at its point of termination, or from half to three quarters of an inch above it.

*Branches.* This vein receives in its course, the facial, lingual, pharyngeal, superior and middle thyroid veins, and the occipital. At its point of junction with the branch common to the temporal and facial veins, it becomes greatly increased in size.

The *lingual veins* commence on the dorsum, sides, and under surface of the tongue, and passing backwards, following the course of the lingual artery and its branches, terminate in the internal jugular.

The *pharyngeal vein* commences in a minute plexus, the pharyngeal, at the back part and sides of the pharynx, and after receiving meningeal branches, and the Vidian and spheno-palatine veins, terminates in the internal jugular. It occasionally opens into the facial, lingual, or superior thyroid vein.

The *superior thyroid vein* commences in the substance and on the surface of the thyroid gland, by branches corresponding with those of the superior thyroid artery, and terminates in the upper part of the internal jugular vein.

The *middle thyroid vein* collects the blood from the lower part of the lateral lobe of the thyroid gland, and, being joined by some branches from the larynx and trachea, terminates in the lower part of the internal jugular vein.

The *Vertebral Vein* commences by numerous small branches in the occipital region, from the deep muscles at the upper and back part of the neck, passes outwards, and enters the foramen in the transverse process of the atlas, and descends by the side of the vertebral artery, in the canal formed by the transverse processes of the cervical vertebræ. Emerging from the foramen in the transverse process of the sixth cervical, it terminates at the root of the neck in the back part of the innominate vein near its origin, its mouth being guarded by a pair of valves. This vein, in the lower part of its course, occasionally divides into two branches, one emerges with the artery at the sixth cervical vertebra; the other escapes through the foramen in the seventh cervical.

*Branches.* This vein receives in its course the posterior condyloid vein, muscular branches from the muscles in the prevertebral region; dorsi-spinal veins, from the back part of the cervical portion of the spine; meningo-rachidian veins, from the interior of the spinal canal; and lastly, the ascending and deep cervical veins.

## Veins of the Diploë.

The diploë of the cranial bones is channelled, in the adult, with a number of tortuous canals, which are lined by a more or less complete layer of compact tissue. The veins they contain are large and capacious, their walls being thin, and formed only of epithelium, resting upon a layer of elastic tissue, and they present, at irregular intervals, pouch-like dilatations, or *culs de sac*, which serve as reservoirs for the blood. These are the veins of the diploë, they can only be displayed by removing the outer table of the skull.

In adult life, as long as the cranial bones are distinct and separable, these veins are confined to the particular bones; but in old age, when the sutures are united, they communicate with each other, and increase in size. These vessels communicate, in the interior of the cranium, with the meningeal veins, and with the sinuses of the dura mater; and on the exterior of the skull, with the veins of the pericranium. They are divided into the *frontal*, which opens into the supra-orbital vein, by an aperture at the supra-orbital notch; the *anterior temporal*, which is confined chiefly to the frontal bone, and opens into

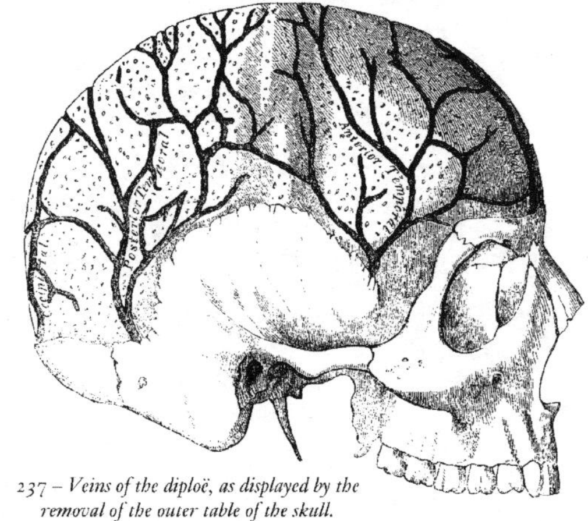


237 – *Veins of the diploë, as displayed by the removal of the outer table of the skull.*

one of the deep temporal veins, after escaping by an aperture in the great wing of the sphenoid; the *posterior temporal*, which is confined to the parietal bone, terminates in the lateral sinus by an aperture at the posterior inferior angle of the parietal bone; and the *occipital* which is confined to the occipital bone, and opens either into the occipital vein, or the occipital sinus.

## Cerebral Veins.

The *Cerebral Veins* are remarkable for the extreme thinness of their coats, from the muscular tissue in them being wanting, and for the absence of valves. They may be divided into two sets, the superficial, which are placed on the surface, and the deep veins, which occupy the interior of the organ.

The *Superficial Cerebral Veins* ramify upon the surface of the brain, being lodged in the sulci, between the convolutions, a few running across the convolutions. They receive branches from the substance of the brain, and terminate in the sinuses. They are named from the position they occupy, superior, inferior, internal, or external.

The *Superior Cerebral Veins*, seven or eight in number on each side, pass forwards and inwards towards the great longitudinal fissure, where they receive the internal cerebral veins, which return the blood from the convolutions of the flat surface of the corresponding hemisphere; passing obliquely forwards, they become invested with a tubular sheath of the arachnoid membrane, and open into the superior longitudinal sinus, in the opposite direction to the course of the blood.

The *Inferior Anterior Cerebral Veins* commence on the under surface of the anterior lobes of the brain, and terminate in the cavernous sinuses.

The *Inferior Lateral Cerebral Veins* commence on the lateral parts of the hemispheres and at the base of the brain: they unite to form from three to five veins, which open into the lateral sinus from before backwards.

The *Inferior Median Cerebral Veins*, which are very large, commence at the fore part of the under surface of the cerebrum, and from the convolutions of the posterior lobe, and terminate in the straight sinus behind the venæ Galeni.

The *Deep Cerebral*, or *Ventricular Veins* (venæ Galeni), are two in number, one from the right, the other from the left, ventricle. They are each formed by two veins, the vena corporis striati, and the choroid vein. They pass backwards, parallel with one another, enclosed within the velum interpositum, and pass out of the brain at the great transverse fissure, between the under surface of the corpus callosum and the tubercula quadrigemina, and enter the straight sinus.

The *vena corporis striati* commences in the groove between the corpus striatum and thalamus opticus, receives numerous veins from both of these parts, and unites behind the anterior pillar of the fornix with the choroid vein, to form one of the venæ Galeni.

The *choroid vein* runs along the whole length of the outer border of the choroid plexus, receiving veins from the hippocampus major, the fornix and corpus callosum, and unites, at the anterior extremity of the choroid plexus, with the vein of the corpus striatum.

The *Cerebellar Veins* occupy the surface of the cerebellum, and are disposed in three sets, superior, inferior, and lateral. The superior pass forwards and inwards, across the superior vermiform process, and terminate in the straight sinus: some open into the venæ Galeni. The inferior cerebellar veins, of large size, run transversely outwards, and terminate by two or three trunks in the lateral sinuses. The lateral anterior cerebellar veins, terminate in the superior petrosal sinuses.

## Sinuses of the Dura Mater.

The sinuses of the dura mater are venous channels, analogous to the veins, their outer coat being formed by the dura mater; their inner, by a continuation of the serous membrane of the veins. They are twelve in number, and are divided into two sets. 1. Those situated at the upper and back part of the skull. 2. The sinuses at the base of the skull. The former are the

| | |
|---|---|
| Superior longitudinal. | Straight sinus. |
| Inferior longitudinal. | Lateral sinuses. |

Occipital sinuses.

The *Superior Longitudinal Sinus* occupies the attached margin of the falx cerebri. Commencing at the crista Galli, it runs from before backwards, grooving the inner surface of the frontal, the adjacent margins of the two parietal, and the superior division of the crucial ridge of the occipital bone, and terminates by dividing into the two lateral sinuses. This sinus is triangular in form, narrow in front, and gradually increasing in size as it passes backwards. On examining its inner surface, it presents the internal openings of the cerebral veins, these vessels are, for the most part, directed from behind forwards, and chiefly open at the back part of the sinus, their orifices being concealed by fibrous areolæ; numerous fibrous bands (*chordæ Willisi*) are also seen, which extend transversely across its inferior angle; and lastly, some small, white, projecting bodies, the glandulæ Pacchioni. This sinus receives the superior cerebral veins, numerous veins from the diploë and dura mater, and, at the posterior extremity of the sagittal suture, the parietal veins from the pericranium.

The point where the superior longitudinal and lateral sinuses are continuous is called the *confluence of the sinuses*, or the *torcular Herophili*. It presents considerable dilatation, of very irregular form, and is the point of meeting of six sinuses, the superior longitudinal, the two lateral, the two occipital, and the straight.

The *Inferior Longitudinal Sinus*, more correctly described as the *inferior longitudinal vein*, is contained in the posterior part of the free margin of the falx cerebri. It is of a circular form, increases in size as it passes backwards, and terminates in the straight sinus. It receives several veins from the falx cerebri and occasionally a few from the flat surface of the hemispheres.

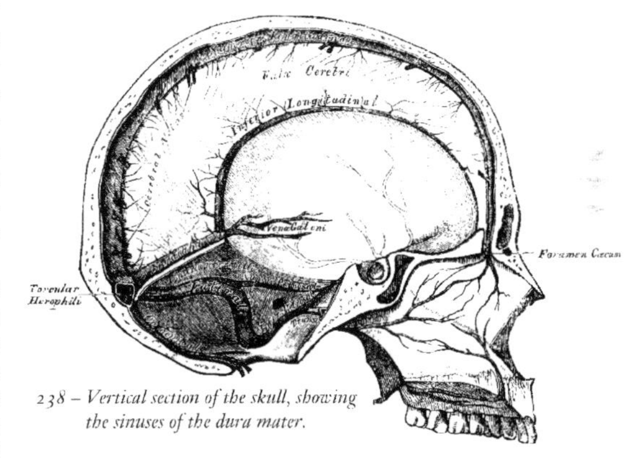

*238 – Vertical section of the skull, showing the sinuses of the dura mater.*

The *Straight Sinus* is situated at the line of junction of the falx cerebri with the tentorium. It is triangular in form, increases in size as it proceeds backwards, and runs obliquely downwards and backwards from the termination of the inferior longitudinal sinus to the torcular Herophili. Besides the inferior longitudinal sinus, it receives the venæ Galeni, the inferior median cerebral veins, and the superior cerebellar. A few transverse bands cross its interior.

The *Lateral Sinuses* are of large size, and situated in the attached margin of the tentorium cerebelli. They commence at the torcular Herophili, and passing horizontally outwards to the base of the petrous portion of the temporal bone, curve downwards and inwards on each side to reach the jugular foramen, where they terminate in the internal jugular vein. Each sinus rests, in its course, upon the inner surface of the occipital, the posterior inferior angle of the parietal, the mastoid portion of the temporal, and on the occipital again just before its termination. These sinuses are of unequal size, the right being the larger, and they increase in size as they proceed from behind forwards. The horizontal portion is of a triangular form, the curved portion semi-cylindrical; their inner surface is smooth, and not crossed by the fibrous bands found in the other sinuses. These sinuses receive blood from the superior longitudinal, the straight, and the occipital sinuses; and in front they communicate with the superior and inferior petrosal. They communicate with the veins of the pericranium by means of the mastoid and posterior condyloid veins, and they receive the inferior cerebral and inferior cerebellar veins, and some from the diploë.

The *Occipital* are the smallest of the cranial sinuses. They are usually two in number, and situated in the attached margin of the falx cerebelli. They commence by several small veins around the posterior margin of the foramen magnum, which communicate with the posterior spinal veins, and terminate by separate openings (sometimes by a single aperture) in the torcular Herophili.

The sinuses at the base of the skull are the

| | |
|---|---|
| Cavernous. | Inferior petrosal. |
| Circular. | Superior petrosal. |
| Transverse. | |

The *Cavernous Sinuses* are named from their presenting a reticulated structure. They are two in number, of large size, and placed one on each side of the sella Turcica, extending from the

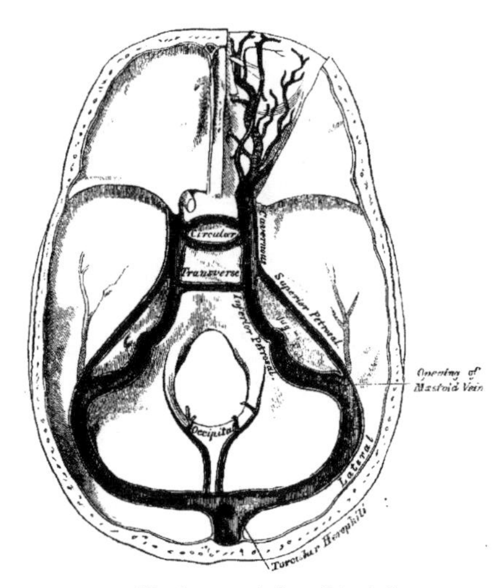


*239 – The sinuses at the base of the skull.*

sphenoidal fissure to the apex of the petrous portion of the temporal bone: they receive anteriorly the ophthalmic vein through the sphenoidal fissure, communicate behind with the petrosal sinuses, and with each other by the circular and transverse sinuses. On the inner wall of each sinus is found the internal carotid artery, accompanied by filaments of the carotid plexus and by the sixth nerve; and on its outer wall, the third, fourth, and ophthalmic nerves. These parts are separated from the blood flowing along the sinus by the lining membrane, which is continuous with the inner coat of the veins. The cavity of the sinus, which is larger behind than in front, is intersected by filaments of fibrous tissue and small vessels. The cavernous sinuses receive the inferior anterior cerebral veins; they communicate with the lateral sinuses by means of the superior and inferior petrosal, and with the facial vein through the ophthalmic.

The *ophthalmic* is a large vein, which connects the frontal vein at the inner angle of the orbit with the cavernous sinus; it pursues the same course as the ophthalmic artery, and receives branches corresponding to those derived from that vessel. Forming a short single trunk, it passes through the inner extremity of the sphenoidal fissure, and terminates in the cavernous sinus.

The *Circular Sinus* completely surrounds the pituitary body, and communicates on each side with the cavernous sinuses. Its posterior half is larger than the anterior; and in old age it is more capacious than at an early period of life. It receives veins from the pituitary body, and from the adjacent bone and dura mater.

The *Inferior Petrosal Sinus* is situated in the groove formed by the junction of the inferior border of the petrous portion of the temporal with the basilar process of the occipital. It commences in front at the termination of the cavernous sinus, and opens behind, into the jugular foramen, forming with the lateral sinus the commencement of the internal jugular vein. These sinuses are semi-cylindrical in form.

The *Transverse Sinus* is placed transversely across the fore part of the basilar process of the occipital bone serving to connect the two inferior petrosal and cavernous sinuses. A second is occasionally found opposite the foramen magnum.

The *Superior Petrosal Sinus* is situated along the upper border of the petrous portion of the temporal bone, in the front part of the attached margin of the tentorium. It is small and narrow, and connects together the cavernous and lateral sinuses at each side. It receives a cerebral vein (inferior lateral cerebral) from the under part of the middle lobe, and a cerebellar vein (anterior lateral cerebellar) from the anterior border of the cerebellum.

## VEINS OF THE UPPER EXTREMITY.

The veins of the upper extremity are divided into two sets: 1. The superficial veins. 2. The deep veins.

The *Superficial Veins* are placed immediately beneath the integument between the two layers of superficial fascia; they commence in the hand chiefly on its dorsal aspect, where they form a more or less complete arch.

The *Deep Veins* accompany the arteries, and constitute the venæ comites of those vessels.

Both sets of vessels are provided with valves, which are more numerous in the deep than in the superficial.

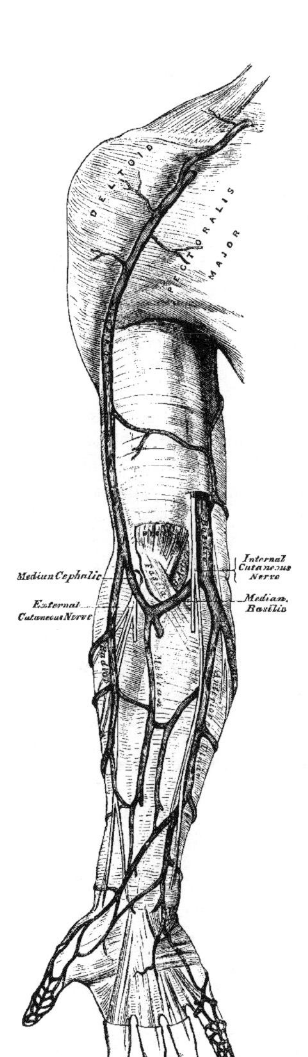


240 – *The superficial veins of the upper extremity.*

The superficial veins of the upper extremity are the

| | |
|---|---|
| Anterior ulnar. | Cephalic. |
| Posterior ulnar. | Median. |
| Basilic. | Median basilic. |
| Radial. | Median cephalic. |

The *Anterior Ulnar Vein* commences on the anterior surface of the wrist and ulnar side of the hand, and ascends along the inner side of the fore-arm to the bend of the elbow, where it joins with the posterior ulnar vein to form the basilic. It communicates with branches of the median vein in front, and with the posterior ulnar behind.

The *Posterior Ulnar Vein* commences on the posterior surface of the ulnar side of the hand, and from the vein of the little finger (vena salvatella), situated over the fourth metacarpal space. It ascends on the posterior surface of the ulnar side of the fore-arm, and just below the elbow unites with the anterior ulnar vein to form the basilic.

The *Basilic* is a vein of considerable size, formed by the coalescence of the interior and posterior ulnar veins; ascending along the inner side of the elbow, it receives the median basilic vein, and passing upwards along the inner side of the arm, pierces the deep fascia, and ascends in the course of the brachial artery, terminating either in one of the venæ comites of that vessel, or in the axillary vein.

The *Radial Vein* commences from the dorsal surface of the thumb, index finger, and radial side of the hand, by branches which communicate with the vena salvatella. They form by their union a large vessel, which ascends along the radial side of the fore-arm, receiving numerous branches from both its surfaces. At the bend of the elbow it receives the median cephalic, when it becomes the cephalic vein.

The *Cephalic Vein* ascends along the outer border of the Biceps muscle, to the upper third of the arm; it then passes in the interval between the Pectoralis major and Deltoid muscles, accompanied by the descending branch of the thoracica acromialis artery, and terminates in the axillary vein just below the clavicle. This vein is occasionally connected with the external jugular or subclavian, by a branch which passes from it upwards in front of the clavicle.

The *Median Vein* collects the blood from the superficial structures in the palmar surface of the hand and middle line of the fore-arm, communicating with the anterior ulnar and radial veins. At the bend of the elbow, it receives a branch of communication from the deep veins, accompanying the brachial artery, and divides into two branches, the median cephalic and median basilic, which diverge from each other as they ascend.

The *Median Cephalic*, the smaller of the two, passes outwards in the groove between the Supinator longus and Biceps muscles, and joins with the cephalic vein. The branches of the external cutaneous nerve pass behind this vessel.

The *Median Basilic* vein passes obliquely inwards, in the groove between the Biceps and Pronator radii teres, and joins with the basilic. This vein passes in front of the brachial artery, from which it is separated by a fibrous expansion, given off from the tendon of the Biceps to the fascia covering the Flexor muscles of the fore-arm. Filaments of the internal cutaneous nerve pass in front as well as behind this vessel.

The *Deep Veins of the Upper Extremity* follow the course of the arteries, forming their venæ comites. They are generally two in number, one lying on each side of the corresponding artery, and they are connected at intervals by short transverse branches.

There are two digital veins, accompanying each artery along the sides of the fingers; these, uniting at their base, pass along the interosseous spaces in the palm, and terminate in the two superficial palmar veins. Branches from these vessels on the radial side of the hand accompany the superficialis volæ, and on the ulnar side, terminate in the deep ulnar veins. The deep ulnar veins, as they pass in front of the wrist, communicate with the interosseous and superficial veins, and unite at the elbow, with the deep radial veins, to form the venæ comites of the brachial artery.

The *Interosseous Veins* accompany the anterior and posterior interosseous arteries. The anterior interosseous veins commence in front of the wrist, where they communicate with the deep radial and ulnar veins; at the upper part of the fore-arm they receive the posterior interosseous veins, and terminate in the venæ comites of the ulnar artery.

The *Deep Palmar Veins* accompany the deep palmar arch, being formed by branches which accompany the ramifications of this vessel. They communicate with the superficial palmar veins at the inner side of the hand; and on the outer side, terminate in the venæ comites of the radial artery. At the wrist, they receive a dorsal and a palmar branch from the thumb, and unite with the deep radial veins. Accompanying the radial artery, these vessels terminate in the venæ comites of the brachial artery.

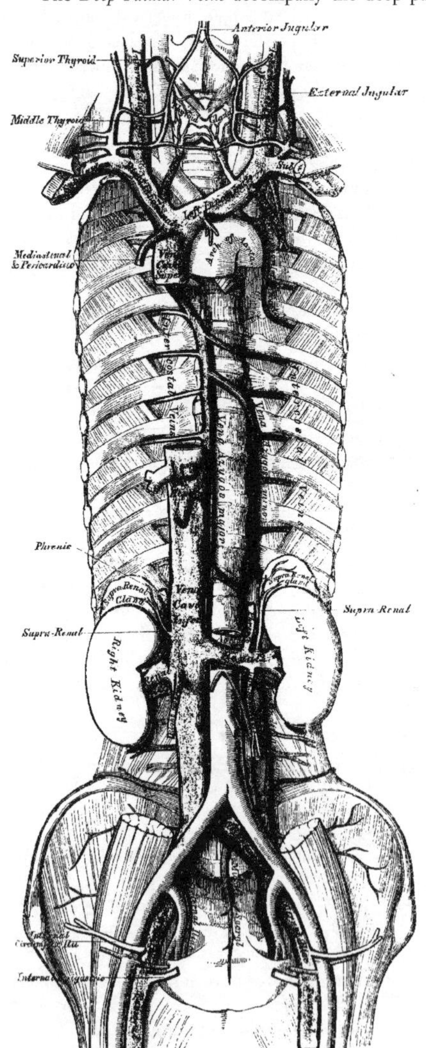


241 – *The venæ cavæ and azygos veins, with their formative branches.*

The *Brachial Veins* are placed one on each side of the brachial artery, receiving branches corresponding with those given off from this vessel; at the lower margin of the axilla they unite with the basilic to form the axillary vein.

The deep veins have numerous anastomoses, not only with each other, but also with the superficial veins.

The Axillary Vein is of large size and formed by the continuation upwards of the basilic vein. It commences at the lower part of the axillary space, and increasing in size as it ascends, by receiving branches corresponding with those of the axillary artery, terminates immediately beneath the clavicle at the outer margin of the first rib, and becomes the subclavian vein. This vessel is covered in front by the Pectoral muscles and costo-coracoid membrane, and lies on the thoracic side of the axillary artery. Opposite the Subscapularis, it is joined by a large vein, formed by the junction of the venæ comites of the brachial; and near its termination it receives the cephalic vein. This vein is provided with a pair of valves, opposite the lower border of the Subscapularis muscle; valves are also found at the termination of the cephalic and subscapular veins.

The Subclavian Vein, the continuation of the axillary, extends from the outer margin of the first rib to the inner end of the sterno-clavicular articulation, where it unites with the internal jugular, to form the vena innominata. It is in relation, in front, with the clavicle and Subclavius muscle; behind, with the subclavian artery, from which it is separated internally by the Scalenus anticus and phrenic nerve. Below, it rests in a depression on the first rib and upon the pleura. Above, it is covered by the cervical fascia and integument.

The subclavian vein occasionally rises in the neck to a level with the third part of the subclavian artery, and in two instances, has been seen passing with this vessel behind the Scalenus anticus. This vessel is provided with valves about an inch from its termination in the innominate, just external to the entrance of the external jugular vein.

*Branches.* It receives the external and anterior jugular veins and a small branch from the cephalic, outside the Scalenus; and on the inner side of this muscle, the internal jugular veins.

The VENÆ INNOMINATE (fig. 241) are two large trunks, placed one on each side of the root of the neck, and formed by the union of the internal jugular and subclavian veins of the corresponding side.

The *Right Vena Innominata* is a short vessel, about an inch and a half in length, which commences at the inner end of the clavicle, and, passing almost vertically downwards, joins with the left vena innominata just below the cartilage of the first rib, to form the superior vena cava. It lies superficial and external to the arteria innominata; on its right side the pleura is interposed between it and the apex of the lung. This vein, at its angle of junction with the subclavian, receives the right vertebral vein, and right lymphatic duct; and, lower down, the right internal mammary, right inferior thyroid, and right superior intercostal veins.

The *Left Vena Innominata*, about three inches in length, and larger than the right, passes obliquely from right to left across the upper and front part of the chest, to unite with its fellow of the opposite side, forming the superior vena cava. It is in relation, in front, with the sternal end of the left clavicle, the left sterno-clavicular articulation, and with the first piece of the sternum, from which it is separated by the Sterno-hyoid and Sterno-thyroid muscles, the thymus gland or its remains, and some loose areolar tissue. Behind, it lies across the roots of the three large arteries arising from the arch of the aorta. This vessel is joined by the left vertebral, left inferior thyroid, left internal mammary, and the left superior intercostal veins, and occasionally some thymic and pericardiac veins. There are no valves in the venæ innominatæ.

*Peculiarities.* Sometimes the innominate veins open separately into the right auricle; in such cases the right vein takes the ordinary course of the superior vena cava, but the left vein, after communicating by a small branch with the right one, passes in front of the root of the left lung, and turning to the back of the heart, receives the cardiac veins, and terminates in the back of the right auricle. This occasional condition of the veins in the adult, is a regular one in the fœtus at an early period, and the two vessels are persistent in birds and some mammalia. The subsequent changes which take place in these vessels are the following. The communicating branch between the two trunks enlarges and forms the future left innominate vein; the remaining part of the left trunk is obliterated as far as the heart, where it remains pervious, and forms the coronary sinus; a remnant of the obliterated vessel is seen in adult life as a fibrous band passing along the back of the left auricle and in front of the root of the left lung, called by Mr. Marshall, the vestigial fold of the pericardium.

The *internal mammary veins*, two in number to each artery, follow the course of that vessel, and receive branches corresponding with those derived from it. The two veins unite into a single trunk, which terminates in the innominate vein.

The *inferior thyroid veins*, two, frequently three or four in number, arise in the venous plexus, on the thyroid body, communicating with the middle and superior thyroid veins. The left one, descends in front of the trachea, behind the Sterno-thyroid muscles, communicating with its fellow by transverse branches, and terminates in the left vena innominata. The right one, which is placed a little to the right of the median line, opens into the right vena innominata, just at its junction with the superior cava. These veins receive tracheal and inferior laryngeal branches, and are provided with valves at their termination in the innominate veins.

The Superior Intercostal Veins return the blood from the upper intercostal spaces.

The *right superior intercostal*, much smaller than the left, closely corresponds with the superior intercostal artery, receiving the blood from the first, or first and second intercostal spaces, and terminates in the right vena innominata. Sometimes it passes down, and opens into the vena azygos major.

The *left superior intercostal* is always larger than the right, but varies in size in different subjects, being small when the left upper azygos vein is large, and *vice versâ*. It is usually formed by branches from the two or three upper intercostal spaces, and, passing across the arch of the aorta, terminates in the left vena innominata. The left bronchial vein opens into it.

The SUPERIOR VENA CAVA receives the blood which is conveyed to the heart from the whole of the upper half of the body. It is a short trunk, varying from two inches and a half to three inches in length, formed by the junction of the two venæ innominatæ. It commences immediately below the cartilage of the first rib on the right side, and, descending vertically downwards, enters the pericardium about an inch and a half above the heart, and terminates in the upper part of the right auricle. In its course, it describes a slight curve, the convexity of which is turned to the right side.

*Relations. In front*, with the thoracic fascia, which separates it from the thymus gland, and from the sternum; *behind*, with the root of the right lung. On its *right side*, with the phrenic nerve and the pleura of the right side; on its *left side*, with the ascending part of the aorta. The portion contained within the pericardium, is covered by the serous layer of that membrane, in its anterior three-fourths. It receives the vena azygos major, just before it enters the pericardium, and several small veins from the pericardium and parts in the mediastinum. The superior vena cava has no valves.

The AZYGOS VEINS connect together the superior and inferior venæ cavæ, supplying the place of these vessels in that part of the trunk in which they are deficient, on account of their connection with the heart.

The larger, or *right azygos vein*, commences opposite the first or second lumbar vertebra, by receiving a branch from the right lumbar veins; sometimes by a branch from the renal vein, or from the inferior vena cava. It enters the thorax through the aortic opening in the Diaphragm, and passes along the right side of the vertebral column to the third dorsal vertebra, where it arches forward, over the root of the right lung, and terminates in the superior vena cava, just before that vessel enters the pericardium. Whilst passing through the aortic opening of the Diaphragm, it lies with the thoracic duct on the right side of the aorta; and in the thorax, it lies upon the intercostal arteries, on the right side of the aorta and thoracic duct, covered by the pleura.

*Branches.* It receives nine or ten lower intercostal veins of the right side, the vena azygos minor, several œsophageal, mediastinal, and vertebral veins; near its termination, the right bronchial vein; and it is occasionally connected with the right superior intercostal vein. A few imperfect valves are found in this vein; but its branches are provided with complete valves.

The intercostal veins on the left side, below the two or three upper intercostal spaces, usually form two trunks, named the left lower, and the left upper, azygos veins.

The *left lower*, or *smaller azygos vein*, commences in the lumbar region, by a branch from one of the lumbar veins, or from the left renal. It passes into the thorax, through the left crus of the Diaphragm, and, ascending on the left side of the spine, as high as the sixth or seventh vertebra, passes across the column, behind the aorta and thoracic duct, to terminate in the right azygos vein. It receives the four or five lower intercostal veins of the left side, and some œsophageal and mediastinal veins.

The *left upper azygos vein* varies according to the size of the left superior intercostal. It receives veins from the intercostal spaces between the left superior intercostal vein, and highest branch of the left lower azygos. They are usually two or three in number, and join to form a trunk which ends in the right azygos vein, or in the left lower azygos. When this vein is small, or altogether wanting, the left superior intercostal vein will extend as low as the fifth or sixth intercostal space.

The *bronchial veins* return the blood from the substance of the lungs; that of the right side opens into the vena azygos major, near its termination; that of the left side, in the left superior intercostal vein.

## THE SPINAL VEINS.

The numerous venous plexuses placed upon and within the spine, may be arranged into four sets.

1. Those placed on the exterior of the spinal column (the dorsi-spinal veins).
2. Those situated in the interior of the spinal canal, between the vertebræ and the theca vertebralis (meningo-rachidian veins).
3. The veins of the bodies of the vertebræ.
4. The veins of the spinal cord (medulli spinal).

1. The *Dorsi-Spinal Veins* commence by small branches, which receive their blood from the integument of the back of the spine, and from the muscles in the vertebral grooves. They form a complicated network, which surrounds the spinous processes, laminæ, and the transverse and articular processes of all the vertebræ. At the bases of the transverse processes, they communicate,

by means of ascending and descending branches, with the veins surrounding the contiguous vertebræ, and they join with the veins in the spinal canal by branches which perforate the ligamenta subflava; in the intervals between the arches of the vertebræ, they terminate in the vertebral veins in the neck, in the intercostal veins in the thorax, in the lumbar and sacral veins in the loins and pelvis.

2. The veins contained in the spinal canal, are situated between the theca vertebralis and the vertebræ. They consist of two longitudinal plexuses, one of which runs along the posterior surface of the bodies of the vertebræ, throughout the entire length of the spinal canal (anterior longitudinal spinal veins), receiving the veins belonging to the bodies of the vertebræ (venæ basis vertebrarum). The other plexus (posterior longitudinal spinal veins) is placed on the inner, or anterior surface of the laminæ of the vertebræ, and extends also along the entire length of the spinal canal.

The *Anterior Longitudinal Spinal Veins* consist of two large, tortuous, venous canals, which extend along the whole length of the vertebral column, from the foramen magnum to the base of the coccyx, being placed one on each side of the posterior surface of the bodies of the vertebræ, external to the posterior common ligament. These veins communicate together opposite each vertebra, by transverse trunks, which pass beneath the ligament, and receive the large venæ basis vertebrarum, from the interior of the body of each vertebra. The anterior longitudinal spinal veins are least developed in the cervical and sacral regions. They are not of uniform size throughout, being alternately enlarged and constricted. At the intervertebral foramina, they communicate with the dorsi-spinal veins, and with the vertebral veins in the neck, with the intercostal veins in the dorsal region, and with the lumbar and sacral veins in the corresponding regions.

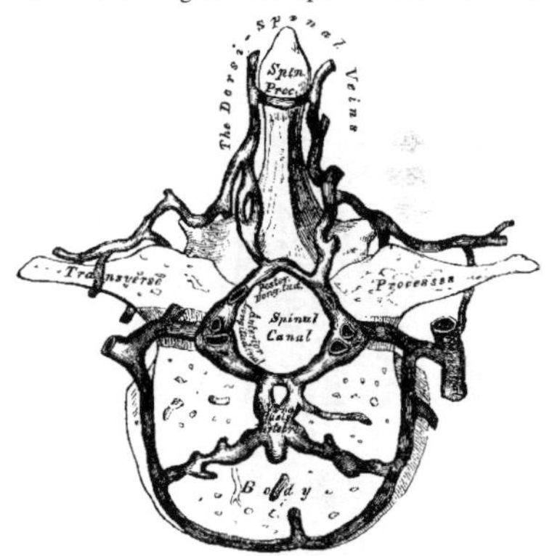

242 – *Transverse section of a dorsal vertebra, showing the spinal veins.*

The *Posterior Longitudinal Spinal Veins*, smaller than the anterior, are situated one on either side, between the inner surface of the laminæ and the theca vertebralis. They communicate (like the anterior), opposite each vertebra, by transverse trunks; and with the anterior longitudinal veins, by lateral transverse branches, which pass from behind forwards. These veins, at the intervertebral foramina, join with the dorsi-spinal veins.

3. The *Veins of the Bodies of the Vertebræ* (venæ basis vertebrarum), emerge from the foramina on their posterior surface, and join the transverse trunk connecting the anterior longitudinal spinal veins. They are contained in large, tortuous channels, in the substance of the bones, similar in every respect to those found in the diploë of the cranial bones. These canals lie parallel to the upper and lower surface of the bones, arise from the entire circumference of the vertebra, communicate with veins which enter through the foramina, on the anterior surface of the bodies, and converge to the principal canal, which is sometimes double towards its posterior part. They become greatly developed in advanced age.

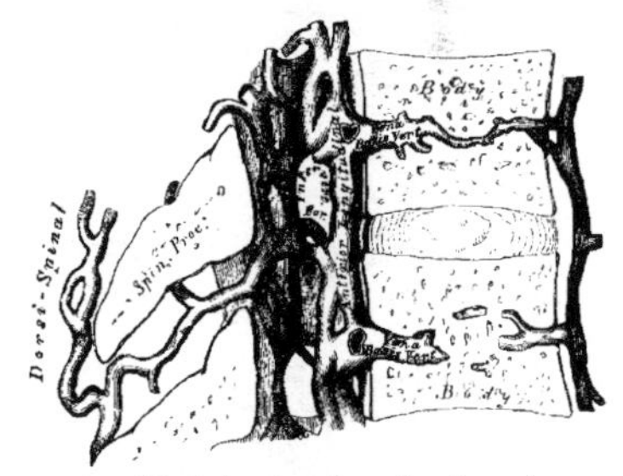

243 – *Vertical section of two dorsal vertebræ, showing the spinal veins.*

4. The *Veins of the Spinal Cord* (medulli spinal), consist of a minute tortuous venous plexus, which covers the entire surface of the cord, being situated between the pia mater and arachnoid. These vessels emerge chiefly from the posterior median furrow, and are largest in the lumbar region. Near the base of the skull they unite, and form two or three small trunks, which communicate with the vertebral veins, and then terminate in the inferior cerebellar veins, or in the petrosal sinuses. Each of the spinal nerves is accompanied by a branch as far as the intervertebral foramina, where they join the other veins from the spinal canal. There are no valves in the spinal veins.

# VEINS OF THE LOWER EXTREMITY.

The veins of the lower extremity are subdivided, like those of the upper, into two sets, superficial and deep; the superficial veins being placed beneath the integument, between the two layers of superficial fascia; the deep veins accompanying the arteries, and forming the venæ comites of those vessels. Both sets of veins are provided with valves, which are more numerous in the deep than in the supericial set. These valves are also more numerous in the lower than in the upper limbs.

The *Superficial Veins* of the lower extremity are the internal or long saphenous, and the external or short saphenous.

The *internal saphenous vein* (fig. 244) commences from a minute plexus, which covers the dorsum and inner side of the foot; it ascends in front of the inner ankle, and along the inner side of the leg, behind the inner margin of the tibia, accompanied by the internal saphenous nerve. At the knee, it passes backwards behind the inner condyle of the femur, ascends along the inside of the thigh, and, passing through the saphenous opening in the fascia lata, terminates in the femoral vein, an inch and a half below Poupart's ligament. This vein receives in its course cutaneous branches from the leg and thigh, and at the saphenous opening, the superficial epigastric, superficial circumflex iliac, and external pudic veins. The veins from the inner and back part of the thigh frequently unite to form a large vessel, which enters the main trunk near the saphenous opening; and sometimes those on the outer side of the thigh join to form a large branch; so that occasionally three large veins are seen converging from different parts of the thigh towards the saphenous opening. The internal saphenous vein communicates in the foot with the internal plantar vein; in the leg, with the posterior tibial veins, by branches which perforate the tibial origin of the Soleus muscle, and also with the anterior tibial veins; at the knee, with the articular veins; in the thigh, with the femoral vein by one or more branches. The valves in this vein vary from two to six in number; they are more numerous in the thigh than in the leg.

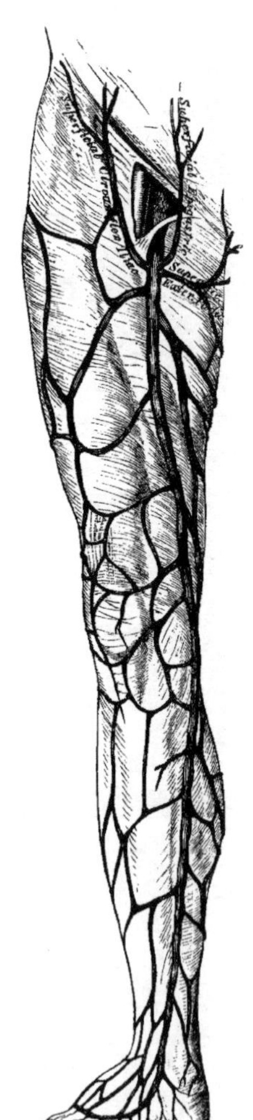


*244 – The internal or long saphenous vein and its branches.*

The *external* or *short saphenous vein* is formed by branches which collect the blood from the dorsum and outer side of the foot; it ascends behind the outer ankle, and along the outer border of the tendo Achillis, across which it passes at an acute angle to reach the middle line of the posterior aspect of the leg. Ascending directly upwards, it perforates the deep fascia in the lower part of the popliteal space, and terminates in the popliteal vein, between the heads of the Gastrocnemius muscle. It is accompanied by the external saphenous nerve. It receives numerous large branches from the back part of the leg, and communicates with the deep veins on the dorsum of the foot, and behind the outer malleolus. This vein has only two valves, one of which is always found near its termination in the popliteal vein.

The *Deep Veins* of the lower extermity accompany the arteries and their branches, and are called the *venæ comites* of those vessels.

The external and internal plantar veins unite to form the posterior tibial. They accompany the posterior tibial artery, and are joined by the peroneal veins.

The *anterior tibial veins* are formed by a continuation upwards of the venæ dorsales pedis. They perforate the interosseous membrane at the upper part of the leg, and form, by their junction with the posterior tibial, the popliteal vein.

The valves in the deep veins are very numerous.

The Popliteal Vein is formed by the junction of the venæ comites of the anterior and posterior tibial vessels; it ascends through the popliteal space to the tendinous aperture in the Adductor magnus, where it becomes the femoral vein. In the lower part of its course, it is placed internal to the artery; between the heads of the Gastrocnemius, it is superficial to that vessel; but above the knee-joint, it is close to its outer side. It receives the sural veins from the Gastrocnemius muscle, the articular veins, and the external saphenous. The valves in this vein are usually four in number.

The Femoral Vein accompanies the femoral artery through the upper two-thirds of the thigh. In the lower part of its course, it lies external to the artery; higher up, it is behind it; and beneath Poupart's ligament, it lies to its inner side, and on the same plane as that vessel. It receives numerous muscular branches; the profunda femoris joins it about an inch and a-half below Poupart's ligament, and near its termination the internal saphenous vein. The valves in this vein are four or five in number.

The External Iliac Vein commences at the termination of the femoral, beneath the crural arch, and passing upwards along the brim of the pelvis, terminates opposite the sacro-iliac symphysis, by uniting with the internal iliac to form the common iliac vein. On the right side, it lies at first along the inner side of the external iliac artery; but as it passes upwards, gradually inclines behind it. On the left side, it lies altogether on the inner side of the artery. It receives, immediately above Poupart's ligament, the epigastric and circumflex iliac veins. It has no valves.

The Internal Iliac Vein is formed by the venæ comites of the branches of the internal iliac artery, the umbilical arteries excepted. It receives the blood from the exterior of the pelvis by the gluteal, sciatic, internal pudic, and obturator veins; and from the organs in the cavity of the pelvis by the hæmorrhoidal and vesico-prostatic plexuses in the male, and the uterine and vaginal plexuses in the female. The vessels forming these plexuses are remarkable for their large size, their frequent anastomoses, and the number of valves which they contain. The internal iliac vein lies at first on the inner side and then behind the internal iliac artery, and terminates opposite the sacro-iliac articulation, by uniting with the external iliac, to form the common iliac vein. This vessel has no valves.

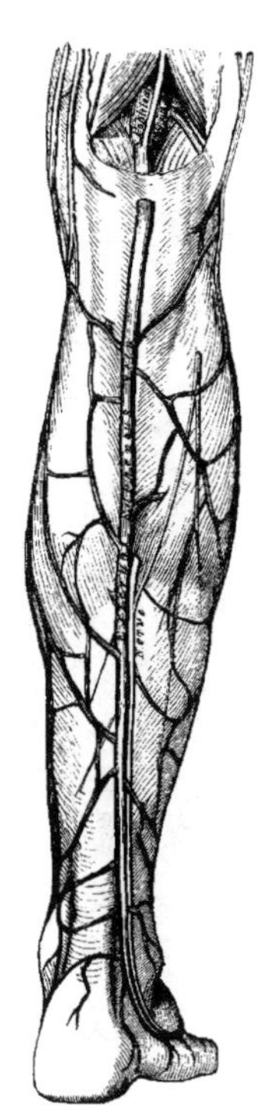

245 – *External, or short saphenous vein.*

The *hæmorrhoidal plexus* surrounds the lower end of the rectum, being formed by the superior hæmorrhoidal veins, branches of the inferior mesenteric, and the middle and inferior hæmorrhoidal, which terminate in the internal iliac. The portal and general venous systems have a free communication by means of the branches composing this plexus.

The *vesico-prostatic plexus* surrounds the neck and base of the bladder and prostate gland. It communicates with the hæmorrhoidal plexus behind, and receives the dorsal vein of the penis, which enters the pelvis beneath the sub-pubic ligament. This plexus is supported upon the sides of the bladder by a reflection of the pelvic fascia. The veins composing it, are very liable to become varicose, and often contain hard earthy concretions, called *phlebolites*.

The *dorsal vein of the penis* is a vessel of large size, which returns the blood from the body of this organ. At first it consists of two branches, which are contained in the groove on the dorsum of the penis, and receives veins from the glans, the corpus spongiosum, and numerous superficial veins; these unite near the root of the penis into a single trunk, which pierces the triangular ligament beneath the pubic arch, and divides into two branches, which enter the prostatic plexus.

The *vaginal plexus* surrounds the mucous membrane of the vagina, being especially developed at the orifice of this canal; it communicates with the vesical plexus in front, and with the hæmorrhoidal plexus behind.

The *uterine plexus* is situated along the sides and superior angles of the uterus, receiving large venous canals (the uterine sinuses) from its substance. The veins composing this plexus anastomose frequently with each other and with the ovarian veins. They are not tortuous like the arteries.

Each Common Iliac Vein is formed by the union of the external and internal iliac veins in front of the sacro-vertebral articulation; passing obliquely upwards towards the right side, they terminate upon the intervertebral substance between the fourth and fifth lumbar vertebræ, where they unite at an acute angle to form the inferior vena cava. The *right common iliac* is shorter than the left, nearly vertical in its direction, and ascends behind and then to the outer side of its corresponding artery. The *left common iliac*, longer and more oblique in its course, is at first situated at the inner side of the corresponding artery, and then behind the right common iliac. Each common iliac receives the ilio-lumbar, and sometimes the lateral sacral veins. The left one receives, in addition, the middle sacral vein. No valves are found in these veins.

The *middle sacral vein* accompanies its corresponding artery along the front of the sacrum, and terminates in the left common iliac vein; occasionally in the commencement of the inferior vena cava.

*Peculiarities.* The left common iliac vein, instead of joining with the right one in its usual position, occasionally ascends on the left side of the aorta as high as the kidney, where, after receiving the left renal vein, it crosses over the aorta, and then joins with the right vein to form the vena cava. In these cases, the two common iliacs are connected by a small communicating branch at the spot where they are usually united.

The Inferior Vena Cava returns to the heart the blood from all the parts below the Diaphragm. It is formed by the junction of the two common iliac veins on the right side of the intervertebral substance, between the fourth and fifth lumbar vertebræ. It passes upwards along the front of the spine, on the right side of the aorta, and having reached the under surface of the liver, is contained in a groove in its posterior border. It then perforates the tendinous centre of the Diaphragm, enters the pericardium, where it is covered by its serous layer, and terminates in the lower and back part of the right auricle. At its termination in the auricle, it is provided with a valve, the Eustachian, which is of large size during fœtal life.

*Relations. In front*, from below upwards, with the mesentery, transverse portion of the duodenum, the pancreas, portal vein, and the posterior border of the liver, which partly and occasionally completely surrounds it; *behind*, it rests upon the vertebral column, the right crus of the Diaphragm, the right renal and lumbar arteries; on the *left side*, it is in relation with the aorta. It receives in its course the following branches:

| | |
|---|---|
| Lumbar. | Supra-renal. |
| Right spermatic. | Phrenic. |
| Renal. | Hepatic. |

*Peculiarities. In Position.* This vessel is sometimes placed on the left side of the aorta as high as the left renal vein, after receiving which, it crosses over to its usual position on the right side; or it may be placed altogether on the left side of the aorta, as far upwards as its termination in the heart: in such cases, the abdominal and thoracic viscera, together with the great vessels, are all transposed.

*Point of Termination.* Occasionally the inferior vena cava joins the right azygos vein which is then of large size. In such cases, the superior cava receives the whole of the blood from the body before transmitting it to the right auricle, the blood from the hepatic veins excepted, these vessels terminating directly in the right auricle.

The *lumbar veins*, three or four in number on each side, collect the blood by dorsal branches from the muscles and integument of the loins, and by abdominal branches from the walls of the abdomen, where they communicate with the epigastric veins. At the spine, they receive branches from the spinal plexuses, and then pass forwards round the sides of the bodies of the vertebræ beneath the Psoas magnus, and terminate at the back part of the inferior cava. The left lumbar veins are longer than the right, and pass behind the aorta. The lumbar veins communicate with each other by branches which pass in front of the transverse process. Occasionally two or more of these veins unite to form a single trunk, the ascending lumbar, which serves to connect the common iliac, ilio-lumbar, lumbar, and azygo veins of the corresponding side of the body.

The *spermatic veins* emerge from the back of the testis, and receive branches from the epididymis; they form a branched and convoluted plexus, called the *spermatic plexus* (plexus pampiniformis), below the abdominal ring: the vessels composing this plexus are very numerous, and ascend along the cord in front of the vas deferens; having entered the abdomen, they coalesce to form two branches which ascend on the Psoas muscle, behind the peritoneum, lying one on each side of the spermatic artery, and unite to form a single vessel, which opens on the right side in the inferior vena cava, piercing this vessel obliquely; on the left side in the left renal vein, terminating at right angles with this vein. The spermatic veins are provided with valves. The left spermatic vein passes behind the sigmoid flexure of the colon; this circumstance, as well as the indirect communication of the vessel with the inferior vena cava, may serve to explain the more frequent occurrence of varicocele on the left side.

The *ovarian veins* are analogous to the spermatic in the male; they form a plexus near the ovary, and in the broad ligament and Fallopian tube, communicating with the uterine plexus. They terminate as in the male. Valves are occasionally found in these veins. These vessels, like the uterine veins, become much enlarged during pregnancy.

The *renal veins* are of large size, and placed in front of the divisions of the renal arteries. The left is longer than the right, and passes in front of the aorta just below the origin of the superior mesenteric artery. It receives the left spermatic and left inferior phrenic veins. It usually opens into the vena cava, a little higher than the right.

The *supra-renal vein* terminates, on the right side, in the vena cava; on the left side, in the left renal or phrenic vein.

The *phrenic veins* follow the course of the phrenic arteries. The *two superior* of small size, accompany the corresponding nerve and artery; the right terminating opposite the junction of the two venæ innominatæ, the left in the left superior intercostal or left internal mammary. The *two inferior phrenic veins* follow the verse of the inferior phrenic arteries, and terminate, the right in the inferior vena cava, the left in the left renal vein.

The *hepatic veins* commence in the substance of the liver, in the capillary terminations of the vena portæ: these branches, gradually uniting, form three large veins, which converge towards the posterior border of the liver, and open into the inferior vena cava, whilst that vessel is situated in the groove at the back part of this organ. Of these three veins, one from the right, and another from the left lobes, open obliquely into the vena cava; that from the middle of the organ and Spigelii having a straight course. The hepatic veins run singly, and are in direct contact with the hepatic tissue. They are destitute of valves.

## Portal System of Veins.

The portal venous system is composed of four large veins, which collect the venous blood from the viscera of digestion. The trunk formed by their union (vena portæ) enters the liver, ramifies throughout its substance, and its branches again emerging from that organ as the hepatic veins, terminate in the inferior vena cava. The branches of this vein are in all cases single, and destitute of valves.

The veins forming the portal system are the

| | |
|---|---|
| Inferior mesenteric. | Splenic. |
| Superior mesenteric. | Gastric. |

The *inferior mesenteric vein* returns the blood from the rectum, sigmoid flexure, and descending colon, corresponding with the ramifications of the branches of the inferior mesenteric artery. Ascending beneath the peritoneum in the lumbar region, it passes behind the transverse portion of the duodenum and pancreas, and terminates in the splenic vein. Its hæmorrhoidal branches inosculate with those of the internal iliac, and thus establish a communication between the portal and the general venous system.

The *superior mesenteric vein* returns the blood from the small intestines, and from the cæcum and ascending and transverse portions of the colon, corresponding with the distribution of the branches of the superior mesenteric artery. The large trunk formed by the union of these branches ascends along the right side and in front of the corresponding artery, passes in front of the transverse portion of the

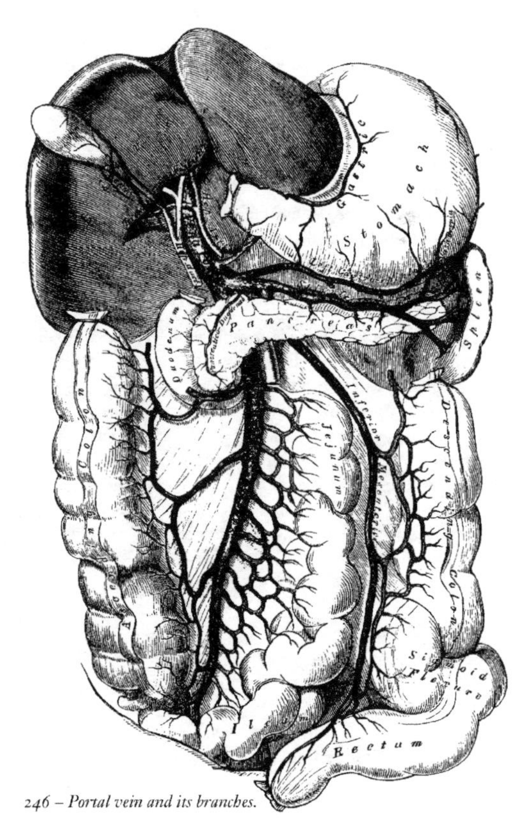

246 – *Portal vein and its branches.*

duodenum, and unites behind the upper border of the pancreas with the splenic vein, to form the vena portæ.

The *splenic vein* commences by five or six large branches, which return the blood from the substance of the spleen. These uniting form a single vessel, which passes from left to right behind the upper border of the pancreas, and terminates at its greater end by uniting at a right angle with the superior mesenteric to form the vena portæ. The splenic vein is of large size, and not tortuous like the artery. It receives the vasa brevia from the left extremity of the stomach, the left gastro-epiploic vein, pancreatic branches from the pancreas, the pancreatico-duodenal vein, and the inferior mesenteric vein.

The *gastric* is a vein of small size, which accompanies the gastric artery from left to right along the lesser curvature of the stomach, and terminates in the vena portæ.

The *Portal Vein* is formed by the junction of the superior mesenteric and splenic veins, their union taking place in front of the vena cava, and behind the upper border of the great end of the pancreas. Passing upwards through the right border of the lesser omentum to the under surface of the liver, it enters the transverse fissure, where it is somewhat enlarged, forming the sinus of the portal vein, and divides into two branches, which accompany the ramifications of the hepatic artery and hepatic duct throughout its substance. Of these two branches the right is the larger but the shorter of the two. The portal vein is about four inches in length, and, whilst contained in the lesser omentum, lies behind and between the hepatic duct and artery, the former being to the right, the latter to the left. These structures are accompanied by filaments of the hepatic plexus and numerous lymphatics, surrounded by a quantity of loose areolar tissue, the capsule of Glisson, and placed between the layers of the lesser omentor. The vena portæ receives the gastric and cystic veins; the latter vein sometimes terminates in the right branch of the vena portæ. Within the liver, the portal vein receives the blood from the branches of the hepatic artery.

## Cardiac Veins.

The veins which return the blood from the substance of the heart are, the

| | |
|---|---|
| Great cardiac vein. | Anterior cardiac veins. |
| Posterior cardiac vein. | Venæ Thebesii. |

The *Great Cardiac Vein* is a vessel of considerable size, which commences at the apex of the heart, and ascends along the anterior inter-ventricular groove to the base of the ventricles. It then curves to the left side, around the auriculo-ventricular groove, between the left auricle and ventricle, to the back part of the heart, and opens into the coronary sinus, its aperture being guarded by two valves. It receives the posterior cardiac vein, and the left cardiac veins from the left auricle and ventricle, one of which, ascending along the left margin of the ventricle, is of large size. The branches joining it are provided with valves.

The *Posterior Cardiac Vein* commences, by small branches, at the apex of the heart, communicating with those of the preceding. It ascends along the posterior inter-ventricular groove to the base of the heart, and terminates in the coronary sinus, its orifice being guarded by a valve. It receives the veins from the posterior surface of both ventricles.

The *Anterior Cardiac Veins* are three or four small branches, which collect the blood from the anterior surface of the right ventricle. One of these, larger than the rest, runs along the right border of the heart, the vein of Galen. They open separately into the lower part of the right auricle.

The *Venæ Thebesii* are numerous minute veins, which return the blood directly from the muscular substance, without entering the venous current. They open, by minute orifices, (*foramina Thebesii*), on the inner surface of the right auricle.

The Coronary Sinus is that portion of the great cardiac vein which is situated in the posterior part of the left auriculo-ventricular groove. It is about an inch in length, presents a considerable dilatation, and is covered by the muscular fibres of the left auricle. It receives the great cardiac vein, the posterior cardiac vein, and an oblique vein from the back part of the left auricle, the remnant of the obliterated left innominate trunk of the fœtus, described by Mr. Marshall. The coronary sinus terminates in the right auricle, between the inferior vena cava and the auriculo-ventricular aperture, its orifice being guarded by a semilunar fold of the lining membrane of the heart, the coronary valve. All the branches joining this vessel, excepting the oblique vein, above-mentioned, are provided with valves.

## The Pulmonary Vein.

The *Pulmonary Veins* return the arterial blood from the lungs to the left auricle of the heart. They are four in number, two for each lung. The pulmonary differ from other veins in several respects. 1. They carry arterial, instead of venous blood. 2. They are destitute of valves. 3. They are only slightly larger than the arteries they accompany. 4. And they accompany those vessels singly. They commence in a capillary network, upon the parietes of the bronchial cells, where they are continuous with the ramifications of the pulmonary artery, and, uniting together, form a single trunk for each lobule. These branches, successively uniting, form a single trunk for each lobe, three for the right, and two for the left, lung. The vein of the middle lobe of the right lung unites with that from the upper lobe, in most cases, forming two trunks on each side, which open separately into the left auricle. Occasionally they remain separate; there are then three veins on the right side. Not unfrequently, the two left pulmonary veins terminate by a common opening.

*Within the lung*, the branches of the pulmonary artery are *in front*, the veins *behind*, and the bronchi between the two.

*At the root of the lung*, the veins are *in front*, the artery *in the middle*, and the bronchus *behind*.

*Within the pericardium*, their anterior surface is invested by the serous layer of this membrane, the right pulmonary veins pass behind the right auricle and ascending aorta; the left pass in front of the thoracic aorta, with the left pulmonary artery.

# Of the Lymphatics.

The Lymphatics have derived their name from the appearance of the fluid contained in their interior (*lympha*, water). They are also called *absorbents*, from the property they possess of absorbing certain materials for the replenishing of the blood, and conveying them into the circulation.

The lymphatic system includes not only the lymphatic vessels and the glands through which they pass, but also the *lacteal*, or *chyliferous* vessels. The lacteals are the lymphatic vessels of the small intestine, and differ in no respect from the lymphatics generally, excepting that they contain a milk-white fluid, the chyle, during the process of digestion, and convey it into the blood through the thoracic duct.

The lymphatics are exceedingly delicate vessels, the coats of which are so transparent, that the fluid they contain is readily seen through them. They retain a nearly uniform size, being interrupted at intervals by constrictions, which give to them a knotted or beaded appearance. These constrictions correspond to the presence of valves in their interior. Lymphatics are found in nearly every texture and organ of the body, with the exception of the substance of the brain and spinal cord, the eyeball, cartilage, tendon, membranes of the ovum, the placenta, and umbilical cord, the nails, cuticle, and hair. Their existence in the substance of bone is doubtful.

The lymphatics are arranged into a superficial and deep set. The superficial lymphatics, on the surface of the body, are placed immediately beneath the integument, accompanying the superficial veins; they join the deep lymphatics in certain situations by perforating the deep fascia. In the interior of the body, they lie in the sub-mucous areolar tissue, throughout the whole length of the gastro-pulmonary and genito-urinary tracts; or in the sub-serous areolar tissue, beneath the serous membrane covering the various organs in the cranial, thoracic, and abdominal cavities. These vessels probably arise in the form of a dense plexiform net-work interspersed among the proper elements and blood-vessels of the several tissues; the vessels composing which, as well as the meshes between them, are much larger than those of the capillary plexus. From these net-works small vessels emerge which pass, either to a neighbouring gland, or to join some larger lymphatic trunk. The deep lymphatics, fewer in number, and larger than the superficial, accompany the deep blood-vessels. Their mode of origin is not known; it is, however, probable, similar to that of the superficial vessel. The lymphatics of any part or organ exceed, in number, the veins; but in size, they are much smaller. Their anastomoses also, especially of the large trunks, are more frequent, and are effected by vessels equal in diameter to those which they connect, the continuous trunks retaining the same diameter.

The lymphatic vessels, like arteries and veins, are composed of three coats.

The *internal* is an epithelial and elastic coat. It is thin, transparent, slightly elastic, and ruptures sooner than the other coats. It is composed of a layer of elongated epithelial cells, supported on a simple net-work of elastic fibres.

The *middle* coat is composed of smooth muscular and fine elastic fibres disposed in a transverse direction.

The *external*, or areolar-fibrous coat, consists of filaments of areolar tissue, intermixed with smooth muscular fibres, longitudinally or obliquely disposed. It forms a protective covering to the other coats, and serves to connect the vessel with the neighbouring structures.

The lymphatics are supplied by nutrient vessels, which are distributed to their outer and middle coats; but no nerves have at present been traced into them.

The lymphatics are very generally provided with valves, which assist materially in effecting the circulation of the fluid they contain. They are formed of a thin layer of fibrous tissue, lined on both surfaces with scaly epithelium. Their form is semilunar; they are attached by their convex edge to the sides of the vessel, the concave edge being free, and directed in the course of the contained current. Usually, two such valves, of equal size, are found placed opposite one another; but occasionally exceptions occur, especially at or near the anastomoses of lymphatic vessels. Thus one valve may be of very rudimentary size, the other increased in proportion. In other cases, the semilunar flaps have been found directed transversely across the vessel, instead of obliquely, so as to impede the circulation in both directions, but not to completely arrest it in either; or the semilunar flaps, taking the same direction, have been united on one side, so that they formed, by their union, a transverse septum, having a partial transverse slit; and sometimes the flap was constituted of a circular fold, attached to the entire circumference of the vessel, and having in its centre a circular or elliptical aperture, the arrangements of the flaps being similar to those composing the ilio-cæcal valve.

The valves in the lymphatic vessels are placed at much shorter intervals than in the veins. They are most numerous near the lymphatic glands, and they are found more frequently in the lymphatics of the neck and upper extremity, than in the lower. The wall of the lymphatics, immediately above the point of attachment of each segment of a valve, is expanded into a pouch or sinus, which gives to these vessels, when distended, the knotted or beaded appearance which they present. Valves are wanting in the vessels composing the plexiform net-work in which the lymphatics originate.

There is no satisfactory evidence to prove that any natural communication exists between the lymphatics of glandular organs and their ducts, or between the lymphatics and the capillary vessels.

The lymphatic or absorbent glands, named also *conglobate glands*, are small solid glandular bodies, situated in the course of the lymphatic and lacteal vessels. They are found in the neck and on the external parts of the head; in the upper extremity, in the axilla and front of the elbow; in the lower extremity, in the groin and popliteal space. In the abdomen, they are found in large numbers in the mesentery, and along the side of the aorta, vena cava, and iliac vessels; and in the thorax, in the anterior and posterior mediastina. They are somewhat flattened and of a round or oval form. In size, they vary from a hemp-seed to an almond and their colour, on section, is of a pinkish grey tint, excepting the bronchial glands which in the adult are mottled with black. Each gland has a layer of cellular tissue investing it, forming a capsule, from which prolongations dip into its substance forming partitions. The lymphatic and lacteal vessels pass through these bodies in their passage to the thoracic and lymphatic ducts. A lymphatic or lacteal, previous to entering a gland, divides into several small branches, which are named *afferent vessels*. As they enter, their external coat becomes continuous with the capsule of the gland, and the vessels, much thinned, and consisting only of their internal coat and epithelium, pass into the gland, where subdividing they pursue a tortuous course; and, finally anastomosing form a plexus. The vessels composing this plexus, unite to form two or more *efferent* vessels, which on emerging from the gland are again invested with their external coat. Within the lymphatic vessels, as supposed by Kölliker, Goodsir, and others, or lying between them, grouped in cells, like the acini of secreting glands, are a large number of minute dotted corpuscles. They are spheroidal, or disk-shaped pellucid particles, about $\frac{1}{3000}$ of an inch in diameter, having two or three minute dark particles in their interior. It is probable that they play an important part in the more complete elaboration of the lymph or chyle traversing the glands. Capillary vessels are abundantly distributed on the walls of the lymphatics in the glands.

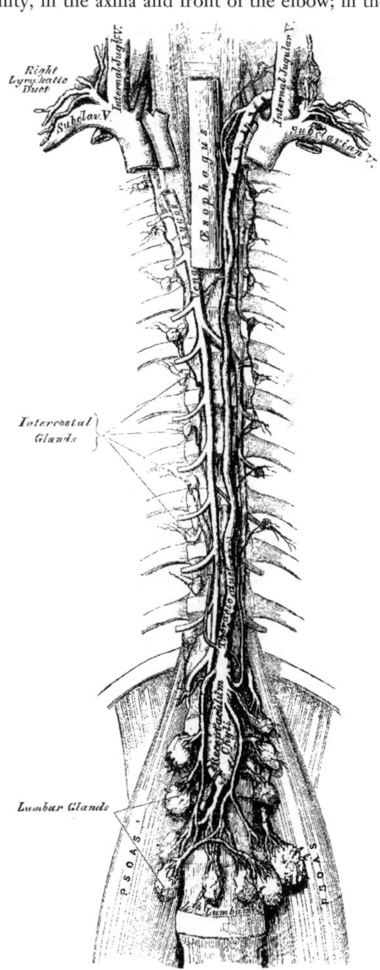


247 – The thoracic and right lymphatic ducts.

## Thoracic Duct.

The thoracic duct (fig. 247) conveys the great mass of the lymph and chyle into the blood. It is the common trunk of all the lymphatic vessels of the body, excepting those of the right side of the head, neck, and thorax, and right upper extremity, the right lung, right side of the heart, and the convex surface of the liver. It varies from eighteen to twenty inches in length in the the adult, and extends from the second lumbar vertebra to the root of the neck. It commences in the abdomen by a triangular dilatation, the receptaculum chyli (reservoir or cistern of Pecquet), which is situated upon the front of the body of the second lumbar vertebra, to the right side and behind the aorta, by the side of the right crus of the Diaphragm. It ascends into the thorax through the aortic opening in the Diaphragm, and is placed in the posterior mediastinum in front of the vertebral column, lying between the aorta and vena azygos. Opposite the fourth dorsal vertebra it inclines towards the left side and ascends behind the arch of the aorta, on the left side of the œsophagus, and behind the first portion of the left subclavian artery, to the upper orifice of the thorax. Opposite the upper border of the seventh cervical vertebra it curves downwards above the subclavian artery, and in front of the Scalenus muscle, so as to form an arch; and terminates near the angle of junction of the left internal jugular and subclavian veins. The thoracic duct, at its commencement, is about equal in size to the diameter of a goose-quill, diminishes considerably in its calibre in the middle of the thorax, and is again dilated just before its termination. It is generally flexuous in its course, and constricted at intervals so as to present a varicose appearance. The thoracic duct not unfrequently divides in the middle of its course into two branches of unequal size which soon re-unite, or into several branches which form a plexiform interlacement. It occasionally bifurcates, at its upper part, into two branches, the left one terminating in the usual manner, the right one opening into the right subclavian vein, in connection with the right lymphatic duct. The thoracic duct has numerous valves throughout its whole course, but they are more numerous in the upper than in the lower part; at its termination it is provided with a pair of valves, the free borders of which are turned towards the vein, so as to prevent the regurgitation of venous blood into the duct.

*Branches.* The thoracic duct at its commencement receives four or five large trunks from the abdominal lymphatic glands, and also the trunk of the lacteal vessels. Within the thorax, it is joined by the lymphatic vessels from the left half of the wall of the thoracic cavity, the lymphatics from the sternal and intercostal glands, those of the left lung, left side of the heart, trachea, and œsophagus; and just before its termination, receives the lymphatics of the left side of the head and neck, and left upper extremity.

*Structure.* The thoracic duct is composed of three coats, which differ in some respects from those of the lymphatic vessels. The *internal coat* consists of a layer of epithelium, resting upon some striped lamellæ, and an elastic fibrous coat, the fibres of which run in a longitudinal direction. The *middle coat* consists of a layer of connective tissue, beneath which are several laminæ of muscular tissue, the fibres of which are

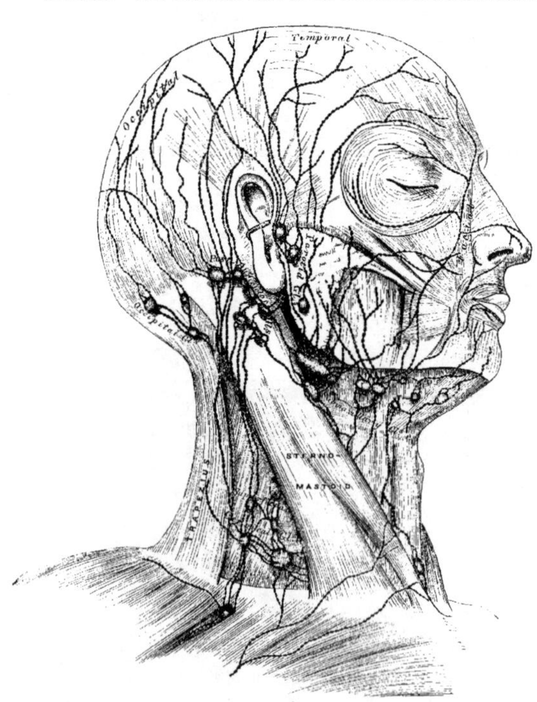


*248 – The superficial lymphatics and glands of the head, face, and neck.*

disposed transversely, and intermixed with fine elastic fibres. The *external coat* is composed of areolar tissue, with elastic fibres and isolated fasciculi of muscular fibres.

The *Right Lymphatic Duct* is a short trunk, about an inch in length, and a line or a line and a half in diameter, which receives the lymph from the right side of the head and neck, the right upper extremity, the right side of the thorax, the right lung and right side of the heart, and from the convex surface of the liver, and terminates at the angle of union of the right subclavian and right internal jugular veins. Its orifice is guarded by two semilunar valves, which prevent the entrance of blood from the veins.

## Lymphatics of the Head, Face, and Neck.

The *Superficial Lymphatic Glands of the Head* (fig. 248) are of small size, few in number, and confined to its posterior region. They are the *occipital*, placed at the back of the head along the attachment of the Occipito-frontalis; and the *posterior auricular*, near the upper end of the Sterno-mastoid. These glands become considerably enlarged in cutaneous affections and other diseases of the scalp. In the face, the superficial lymphatic glands are more numerous: they are the *parotid*, some of which are superficial and others deeply placed in its substance; the *zygomatic*, situated under the zygoma; the *buccal*, on the surface of the Buccinator muscle; and the *submaxillary*, the largest, beneath the body of the lower jaw.

The *superficial lymphatics of the head* are divided into an anterior and a posterior set, which follow the course of the temporal and occipital vessels. The temporal set accompany the temporal artery in front of the ear, to the parotid lymphatic glands, from which they proceed to the lymphatic glands of the neck. The occipital set follow the course of the occipital artery, descend to the occipital and posterior auricular lymphatic glands, and from thence join the cervical glands.

The *superficial lymphatics of the face* are more numerous than those of the head. They commence over its entire surface, those from the frontal region accompanying the frontal vessels; they then pass obliquely across the face, accompanying the facial vein, pass through the buccal glands on the surface of the Buccinator muscle, and join the submaxillary lymphatic glands. The latter receive the lymphatic vessels from the lips, and are often found enlarged in cases of malignant disease of these parts.

The *deep lymphatics of the face* are derived from the pituitary membrane of the nose, the mucous membrane of the mouth and pharynx, and the contents of the temporal and orbital fossæ; they accompany the branches of the internal maxillary artery, and terminate in the deep parotid and cervical lymphatic glands.

The *deep lymphatics of the cranium* consist of two sets, the meningeal and cerebral. The meningeal lymphatics accompany the meningeal vessels, escape through foramina at the base of the skull, and join the deep cervical lymphatic glands. The cerebral lymphatics are described by Fohmann as being situated between the arachnoid and pia mater, as

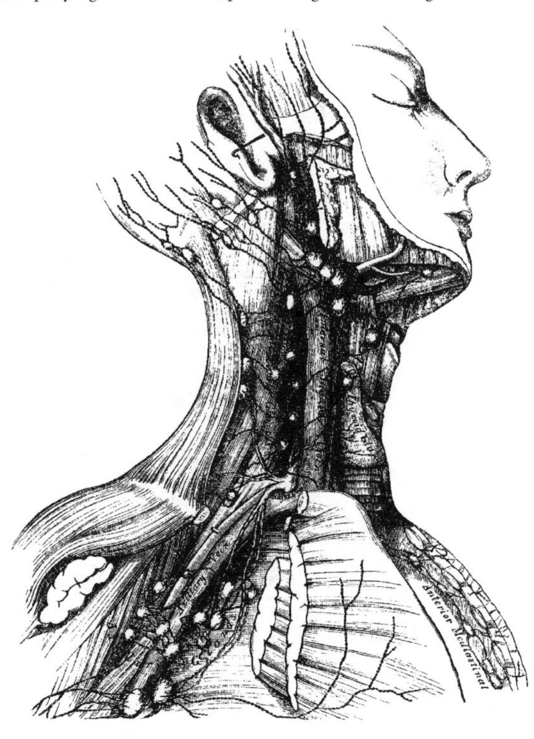


249 – *The deep lymphatics and glands of the neck and thorax.*

well as in the choroid plexuses of the lateral ventricles; they accompany the trunks of the carotid and vertebral arteries, and probably pass through foramina at the base of the skull, to terminate in the deep cervical glands. They have not at present been demonstrated in the dura mater, or in the substance of the brain.

The *Lymphatic Glands of the Neck* are divided into two sets, superficial and deep.

The *superficial cervical glands* are placed in the course of the external jugular vein, between the Platysma and Sterno-mastoid. They are most numerous at the root of the neck, in the triangular interval between the clavicle, the Sterno-mastoid, and the Trapezius, where they are continuous with the axillary glands. A few small glands are also found on the front and sides of the larynx.

The *deep cervical glands* (fig. 249) are numerous and of large size; they form an uninterrupted chain along the sheath of the carotid artery and internal jugular vein, lying by the side of the pharynx, œsophagus, and trachea, and extending from the base of the skull to the thorax, where they communicate with the lymphatic glands in this cavity.

The *superficial and deep cervical lymphatics* are a continuation of those already described on the cranium and face. After traversing the glands in those regions, they pass through the chain of glands which lie along the sheath of the carotid vessels, being joined by the lymphatics from the pharynx, œsophagus, larynx, trachea, and thyroid gland. At the lower part of the neck, after receiving some lymphatics from the thorax, they unite into a single trunk, which terminates on the left side, in the thoracic duct; on the right side, in the right lymphatic duct.

## Lymphatics of the Upper Extremity.

The *Lymphatic Glands* of the upper extremity (fig. 250) may be subdivided into two sets, superficial and deep.

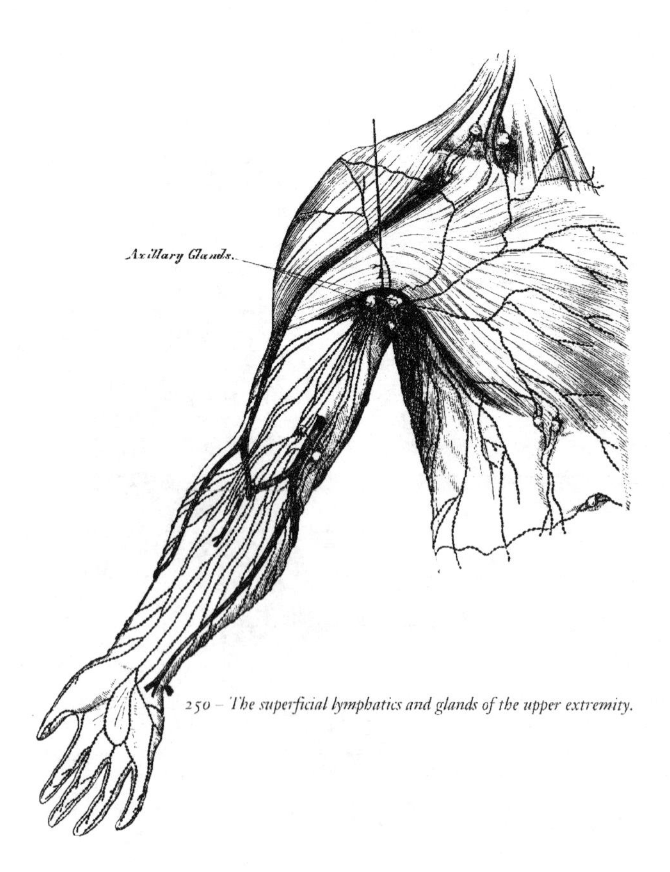


250 – *The superficial lymphatics and glands of the upper extremity.*

The *superficial lymphatic glands* are few, and of small size. There are occasionally two or three in front of the elbow, and one or two above the internal condyle of the humerus, near the basilic vein.

The *deep lymphatic glands* are also few in number. In the fore-arm a few small ones are occasionally found in the course of the radial and ulnar vessels; and in the arm, there is a chain of small glands along the inner side of the brachial artery.

The *axillary glands* are of large size, and usually ten or twelve in number. A chain of these glands surrounds the axillary vessels imbedded in a quantity of loose areolar tissue; they receive the lymphatic vessels from the arm: others are dispersed in the areolar tissue of the axilla: the remainder are arranged in two series, a small chain running along the lower border of the Pectoralis major, as far as the mammary gland, receiving the lymphatics from the front of the chest and mamma; and others are placed along the lower margin of the posterior wall of the axilla, which receive the lymphatics from the integument of the back. Two or three subclavian lymphatic glands are placed immediately beneath the clavicle; it is through these that the axillary and deep cervical glands communicate with each other. One is figured by Mascagni near the umbilicus. In malignant diseases, tumours, or other affections implicating the upper part of the back and shoulder, the front of the chest and mamma, the upper part of the front and side of the abdomen, or the hand, fore-arm, and arm, the axillary glands are usually found enlarged.

The *superficial lymphatics* of the upper extremity arise from the skin of the hand, and run along the sides of the fingers chiefly on the dorsal surface of the hand; they then pass up the fore-arm, and subdivide into two sets, which take the course of the subcutaneous veins. Those from the inner border of the hand accompany the ulnar veins along the inner side of the fore-arm to the bend of the elbow, where they join with some lymphatics from the outer side of the fore-arm; they then follow the course of the basilic vein, communicate with the glands immediately above the elbow, and terminate in the axillary glands, joining with the deep lymphatics. The superficial lymphatics from the outer and back part of the hand accompany the radial veins to the bend of the elbow, being less numerous than the preceding. Here the greater number join the basilic group; the rest ascend with the cephalic vein on the outer side of the arm, some crossing obliquely the upper part of the Biceps to terminate in the axillary glands, whilst one or two accompany the cephalic vein in the cellular interval between the Pectoralis major and Deltoid, and enter the subclavian lymphatic glands.

The *deep lymphatics* of the upper extremity accompany the deep blood-vessels. In the fore-arm, they consist of three sets, corresponding with the radial, ulnar, and interosseous arteries; they pass through the glands occasionally found in the course of these vessels, and communicate at intervals with the superficial lymphatics. In their ascent upwards, some of them pass through the glands which lie upon the brachial artery; they then enter the axillary and subclavian glands, and at the root of the neck terminate, on the left side in the thoracic duct, and on the right side in the right lymphatic duct.

## Lymphatics of the Lower Extremity.

The *Lymphatic Glands* of the lower extremity may be subdivided into two sets, superficial and deep.

The *superficial lymphatic glands* are confined to the inguinal region.

The *superficial inguinal glands*, placed immediately beneath the integument, are of large size, and vary from eight to ten in number. They are divisible into two groups; an upper, disposed irregularly along Poupart's ligament, receiving the lymphatic vessels from the integument of the scrotum, penis, parietes of the abdomen, perinæum, and gluteal regions; and an inferior group, which surround the saphenous opening in the fascia lata, a few being sometimes continued along the saphenous vein to a variable extent. The latter receive the superficial lymphatic vessels from the lower extremity. These glands frequently become enlarged in diseases implicating the parts from which their efferent lymphatics originate. Thus, in malignant or syphilitic affections of the prepuce and penis, the labis majora in the female, in cancer scroti, in abscess in the perinæum, or in any other disease affecting the integument and superficial structures in these parts, or the sub-umbilical part of the abdomen or gluteal region, the upper chain of glands is almost invariably enlarged, the lower chain being implicated in diseases affecting the lower limb.

The *deep lymphatic glands* are, the anterior tibial, popliteal, deep inguinal, gluteal, and ischiatic.

The *anterior tibial gland* is not constant in its existence. It is generally found by the side of the anterior tibial artery, upon the interosseous membrane at the upper part of the leg. Occasionally, two glands are found in this situation.

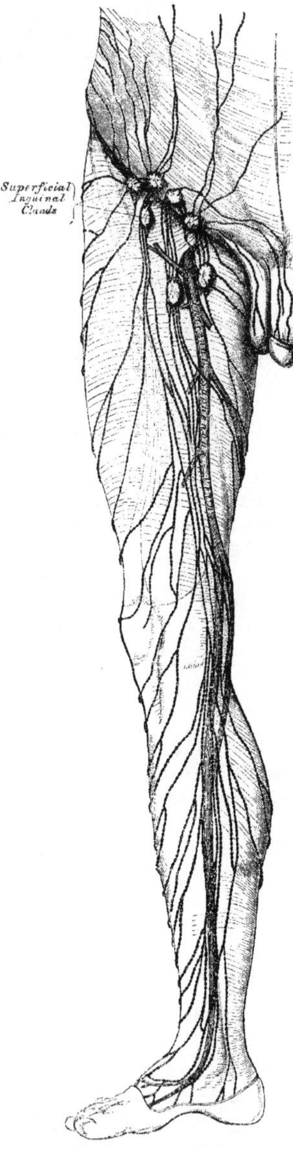


251 – *The superficial lymphatics and glands of the lower extremity.*

The *deep popliteal glands*, four or five in number, are of small size; they surround the popliteal vessels, imbedded in the cellular tissue and fat of the popliteal space.

The *deep inguinal glands* are placed beneath the deep fascia around the femoral artery and vein. They are of small size, and communicate with the superficial inguinal glands through the saphenous opening.

The *gluteal* and *ischiatic glands* are placed the former above, the latter below the Pyriformis mucle, resting on their corresponding vessels as they pass through the great sacro-sciatic foramen.

The *Lymphatics* of the lower extremity, like the veins, may be divided into two sets, superficial and deep.

The *superficial lymphatics* are placed between the integument and superficial fascia, and are divisible into two groups, an internal group, which follow the course of the internal saphenous vein; and an external group, which accompany the external saphenous.

The *internal group*, the largest, commence on the inner side and dorsum of the foot; they pass, some in front, and some behind the inner ankle, ascend the leg with the internal saphenous vein, pass with it behind the inner condyle of the femur, and accompany it to the groin, where they terminate in the group of inguinal glands which surround the saphenous opening. Some of the efferent vessels from these glands pierce the cribriform fascia and sheath of the femoral vessels, and terminate in a lymphatic gland contained in the femoral canal, thus establishing a communication between the lymphatics of the lower extremity and those of the trunk; others pierce the fascia lata, and join the deep inguinal glands.

The *external group* arise from the outer side of the foot, ascend in front of the leg, and, just below the knee, cross the tibia from without inwards, to join the lymphatics on the inner side of the thigh. Others commence on the outer side of the foot, pass behind the outer malleolus, and accompany the external saphenous vein along the back of the leg, where they enter the popliteal glands.

The *deep lymphatics* of the lower extremity are few in number, and accompany the deep blood-vessels. In the leg, they consist of three sets, the anterior tibial, peroneal, and posterior tibial, which accompany the corresponding vessels, being two or three in number to each; they ascend with the blood-vessels, and enter the lymphatic glands in the popliteal space; the efferent vessels from these glands accompany the femoral vein, and join the deep inguinal glands; from these, the vessels pass beneath Poupart's ligament, and communicate with the chain of glands surrounding the external iliac vessels.

The deep lymphatics of the gluteal and ischiatic regions follow the course of the blood-vessels, and join the gluteal and ischiatic glands at the great sacro-sciatic foramen.

## Lymphatics of the Pelvis and Abdomen.

The *deep lymphatic glands in the pelvis* are, the external iliac, the internal iliac, and the sacral. Those of the abdomen are the lumbar glands.

The *external iliac glands* form an uninterrupted chain round the external iliac vessels, three being placed round the commencement of the vessel just behind the crural arch. They communicate below with the femoral lymphatics, and above with the lumbar glands.

The *internal iliac glands* surround the internal iliac vessels; they receive the lymphatics corresponding to the branches of the internal iliac artery, and communicate with the lumbar glands.

The *sacral glands* occupy the sides of the anterior surface of the sacrum, some being situated in the meso-rectal fold. These and the internal iliac glands become greatly enlarged in malignant disease of the bladder, rectum, or uterus.

The *lumbar glands* are very numerous; they are situated on the front of the lumbar vertebræ, surrounding the common iliac vessels, the aorta, and vena cava: they receive the lymphatic vessels from the lower extremities and pelvis, as well as from the testes and some of the abdominal viscera; the efferent vessels from these glands unite into a few large trunks, which, with the lacteals, form the commencement of the thoracic duct. In some cases of malignant disease, these glands become enormously enlarged, completely surrounding the aorta and vena cava, and occasionally greatly contracting the calibre of these vessels. In all cases of malignant disease of the testis, and in malignant disease of the lower limb, before any operation is attempted, careful examination of the abdomen should be made, in order to ascertain if any enlargement exists; and if any should be detected, all operative measures are fruitless.

The *lymphatics of the pelvis and abdomen* may be divided into two sets, superficial and deep.

The *superficial lymphatics of the walls of the abdomen and pelvis* follow the course of the superficial blood-vessels. Those derived from the integument of the lower part of the abdomen below the umbilicus, follow the course of the superficial epigastric vessels, and converge to the superior group

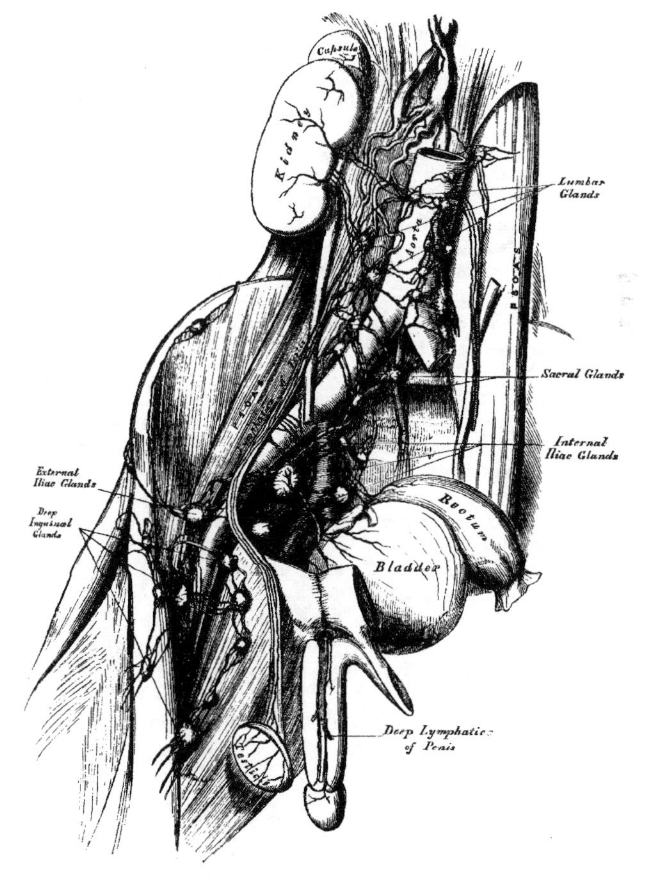


252 – *The deep lymphatic vessels and glands of the abdomen and pelvis.*

of the superficial inguinal glands; the deep set accompany the deep epigastric vessels, and communicate with the external iliac glands. The superficial lymphatics from the sides and lumbar part of the abdominal wall wind round the crest of the ilium, accompanying the superficial circumflex iliac vessels, to join the superior group of the superficial inguinal glands; the greater number, however, accompany the ilio-lumbar and lumbar vessels backwards, to join the lumbar glands.

The *superficial lymphatics of the gluteal region* turn horizontally round the outer side of the nates, and join the superficial inguinal glands.

The *superficial lymphatics of the scrotum and perinæum* follow the course of the external pudic vessels, and terminate in the superficial inguinal glands.

The *superficial lymphatics of the penis* occupy the sides and dorsum of the organ, the latter receiving the lymphatics from the skin covering the glans penis: they all converge to the upper chain of the superficial inguinal glands. The deep lymphatic vessels of the penis follow the course of the internal pudic vessels, and join the internal iliac glands.

In the female, the lymphatic vessels of the mucous membrane of the labia, symphæ, and clitoris, terminate in the upper chain of the inguinal lymphatic glands.

The *deep lymphatics of the pelvis and abdomen* take the course of the principal blood-vessels. Those of the parietes of the pelvis, which accompany the gluteal, ischiatic, and obturator vessels, follow the course of the internal iliac artery, and ultimately join the lumbar lymphatics.

The efferent vessels from the inguinal glands enter the pelvis beneath Poupart's ligament, where they lie in close relation with the femoral vein; they then pass through the chain of glands surrounding the external iliac vessels, and finally terminate in the lumbar glands. They receive the deep epigastric, circumflex iliac, and ilio-lumbar lymphatics.

The *lymphatics of the bladder* arise from the entire surface of the organ; the greater number run beneath the peritoneum on its posterior surface, and, after passing through the lymphatic glands in this situation, join with the lymphatics from the prostate and vesiculæ seminales, and enter the internal iliac glands.

The *lymphatics of the rectum* are of large size; after passing through some small glands that lie upon its outer wall and in the meso-rectum, they pass to the sacral or lumbar glands.

The *lymphatics of the uterus* consist of two sets, superficial and deep; the former being placed beneath the peritoneum, the latter in the substance of the organ. The lymphatics of the cervix uteri, together with those from the vagina enter the internal iliac and sacral glands; those from the body and fundus of the uterus pass outwards in the broad ligaments, and, being joined by the lymphatics from the ovaries, broad ligaments, and Fallopian tubes, ascend with the ovarian vessels to open into the lumbar glands. In the unimpregnated uterus, they are small; but during gestation, they become very greatly enlarged.

The *lymphatics of the testicle* consist of two sets, superficial and deep; the former commence on the surface of the tunica vaginalis, the latter in the epididymis and body of the testis. They form several large trunks, which ascend with the spermatic cord, and accompanying the spermatic vessels into the abdomen, open into the lumbar glands; hence the enlargement of these glands in malignant disease of the testis.

The *lymphatics of the kidney* arise on the surface, and also in the interior of the organ; they join at the hilus, and after receiving the lymphatic vessels from the ureter and supra-renal capsule, open into the lumbar glands.

The *lymphatics of the liver* are divisible into two sets, superficial and deep. The former arise in the sub-peritoneal areolar tissue over the entire surface of the organ. Those on the convex surface may be divided into four groups: 1. Those which pass from behind forwards, consisting of three or four branches, which ascend in the longitudinal ligament, and unite to form a single trunk, which passes up between the fibres of the Diaphragm, behind the ensiform cartilage, to enter the anterior mediastinal glands, and finally ascends to the root of the neck, to terminate in the right lymphatic duct. 2. Another group, which also incline from behind forwards, are reflected over the anterior margin of the liver to its under surface, and from thence pass along the longitudinal fissure to the glands in the gastro-hepatic omentum. 3. A third group incline outwards to the right lateral ligament, and uniting into one or two large trunks, pierce the Diaphragm, and run along its upper surface to enter the anterior mediastinal glands; or, instead of entering the thorax, turn inwards across the crus of the Diaphragm, and open into the commencement of the thoracic duct. 4. The fourth group incline outwards from the surface of the left lobe of the liver to the left lateral ligament, pierce the Diaphragm, and passing forwards, terminate in the glands in the anterior mediastinum.

The *superficial lymphatics on the under surface of the liver* are divided into three sets: 1. Those on the right side of the gall-bladder enter the lumbar glands. 2. Those surrounding the gall-bladder form a remarkable plexus, which accompanies the hepatic vessels, and open into the glands in the gastro-hepatie omentum. 3. Those on the left of the gall-bladder pass to the œsophageal glands and to those placed along the lesser curvature of the stomach.

The *deep lymphatics* accompany the branches of the portal vein and the hepatic artery and duct through the substance of the liver; passing out at the transverse fissure, they enter the lymphatic glands along the lesser curvature of the stomach and behind the pancreas, or join with one of the lacteal vessels previous to its termination in the thoracic duct.

The *lymphatic glands of the stomach* are of small size; they are placed along the lesser and greater curvatures, some within the gastro-splenic omentum, whilst others surround its cardiac and pyloric orifices.

The *lymphatics of the stomach* consist of two sets, superficial and deep; the former originating in the subserous, and the latter in the submucous coat. They follow the course of the blood-vessels, and may consequently be arranged into three groups. The *first group* accompany the coronary vessels along the lesser curvature, receiving branches from both surfaces of the organ, and pass to the glands around the pylorus. The *second group* pass from the great end of the stomach, accompany the vasa brevia, and enter the splenic lymphatic glands. The *third group* run along the greater curvature with the right gastro-epiploic vessels, and terminate at the root of the mesentery in one of the principal lacteal vessels.

The *lymphatic glands of the spleen* occupy the hilus. Its *lymphatic vessels* consist of two sets, superficial and deep; the former are placed beneath its peritoneal covering, the latter in the substance of the organ: they accompany the blood-vessels, passing through a series of small glands, and after receiving the lymphatics from the pancreas, ultimately pass into the thoracic duct.

## The Lymphatic System of the Intestines.

The *lymphatic glands of the small intestine* are placed between the layers of the mesentery, occupying the meshes formed by the superior mesenteric vessels, and hence called *mesenteric glands*. They vary in number from a hundred to a hundred and fifty; and in size, from that of a pea to that of a small almond. These glands are most numerous, and largest, superiorly near the duodenum, and inferiorly opposite the termination of the ileum in the colon. This latter group becomes greatly enlarged and infiltrated with deposit in cases of fever accompanied with ulceration of the intestines.

The *lymphatic glands of the large intestine* are much less numerous than the mesenteric glands; they are situated along the vascular arches formed by the arteries previous to their distribution, and even sometimes upon the intestine itself. They are fewest in number along the transverse colon, where they form an uninterrupted chain with the mesenteric glands.

The *lymphatics of the small intestine* are called *lacteals*, from the milk-white fluid they usually contain; they consist of two sets, superficial and deep; the former lie beneath the peritoneal coat, taking a longitudinal course along the outer side of the intestine; the latter occupy the submucous tissue, and course transversely round the intestine, accompanied by the branches of the mesenteric vessels: they pass between the layers of the mesentery, enter the mesenteric glands, and finally unite to form two or three large trunks, which terminate in the thoracic duct.

The *lymphatics of the large intestine* consist of two sets: those of the cæcum, ascending and transverse colon, which, after passing through their proper glands, enter the mesenteric glands; and those of the descending colon and rectum, which pass to the lumbar glands.

## The Lymphatics of the Thorax.

The *deep lymphatic glands of the thorax* are the intercostal, internal mammary, anterior mediastinal, and posterior mediastinal.

The *intercostal glands* are small, irregular in number, and situated on each side of the spine, near the costo-vertebral articulations, some being placed between the two planes of intercostal muscles.

The *internal mammary glands* are placed at the anterior extremity of each intercostal space, by the side of the internal mammary vessels.

The *anterior mediastinal glands* are placed in the loose areolar tissue of the anterior mediastinum, some lying upon the Diaphragm in front of the pericardium, and others round the great vessels at the base of the heart.

The *posterior mediastinal glands* are situated in the areolar tissue in the posterior mediastinum, forming a continuous chain by the side of the aorta and œsophagus; they communicate on each side with the intercostal, below with the lumbar glands, and above with the deep cervical.

The *superficial lymphatics of the front of the thorax* run across the great Pectoral muscle, and those on the back part of this cavity lie upon the Trapezius and Latissimus dorsi; they all converge to the axillary glands. The lymphatics from the mamma run along the lower border of the Pectoralis major, through a chain of small lymphatic glands, and communicate with the axillary glands.

The *deep lymphatics of the thorax* are the intercostal, internal mammary, and diaphragmatic.

The *intercostal lymphatics* follow the course of the intercostal vessels, receiving lymphatics from the Intercostal muscles and pleura; they pass backwards to the spine, and unite with lymphatics from the back part of the thorax and spinal canal. After traversing the intercostal glands, they incline down the spine, and terminate in the thoracic duct.

The *internal mammary lymphatics* follow the course of the internal mammary vessels; they commence in the muscles of the abdomen above the umbilicus, communicating with the epigastric lymphatics, ascend between the fibres of the Diaphragm at its attachment to the ensiform appendix, and in their course behind the costal cartilages are joined by the intercostal lymphatics, terminating on the right side in the right lymphatic duct, on the left side in the thoracic duct.

The *lymphatics of the Diaphragm* follow the course of their corresponding vessels, and terminate, some in front in the anterior mediastinal and internal mammary glands, some behind in the intercostal and hepatic lymphatics.

The *bronchial glands* are situated round the bifurcation of the trachea and roots of the lungs. They are ten or twelve in number, the largest being placed opposite the bifurcation of the trachea, the smallest round the bronchi and their primary divisions for some little distance within the substance of the lungs. In infancy, they present the same appearance as lymphatic glands in other situations, in the adult they assume a brownish tinge, and in old age a deep black colour. Occasionally they become sufficiently enlarged to compress and narrow the canal of the bronchi; and they are often the seat of tubercle or deposits of phosphate of lime.

The *lymphatics of the lung* consist of two sets, superficial and deep: the former are placed beneath the pleura, forming a minute plexus, which covers the outer surface of the lung; the latter accompany the blood-vessels, and run along the bronchi: they both terminate at the root of the lungs in the bronchial glands. The efferent vessels from these glands, two or three in number, ascend upon the trachea to the root of the neck, traverse the tracheal and œsophageal glands, and terminate on the left side in the thoracic duct, and on the right side in the right lymphatic duct.

The *cardiac lymphatics* consist of two sets, superficial and deep; the former arise in the subserous areolar tissue of the surface, and the latter beneath the internal lining membrane of the heart. They follow the course of the coronary vessels; those of the right side unite into a trunk at the root of the aorta, which ascending across the arch of that vessel, passes backwards to the trachea, upon which it ascends, to terminate at the root of the neck in the right lymphatic duct. Those of the left side unite into a single vessel at the base of the heart, which passing along the pulmonary artery, and traversing some glands at the root of the aorta, ascends on the trachea to terminate in the thoracic duct.

The *thymic lymphatics* arise from the spinal surface of the thymus gland, and terminate on each side in the internal jugular veins.

The *thyroid lymphatics* arise from either lateral lobe of this organ; they converge to form a short trunk, which terminates on the right side in the right lymphatic duct, on the left side in the thoracic duct.

The *lymphatics of the œsophagus* form a plexus round that tube, traverse the glands in the posterior mediastinum, and, after communicating with the pulmonary lymphatic vessels near the roots of the lungs, terminate in the thoracic duct.

Printed in the United States
by Baker & Taylor Publisher Services